# 21st Joint Workshop on Electron Cyclotron Emission and Electron Cyclotron Resonance Heating (EC21)



EPJ Web of Conferences Volume 277 (2023)

ITER Organization Headquarters, France
20 – 24 June 2022

**Editors:**

**Emanuele Poli**
**Yong Liu**
**Victor Udintsev**

ISBN: 978-1-7138-7051-7

**Printed from e-media with permission by:**

Curran Associates, Inc.
57 Morehouse Lane
Red Hook, NY 12571

**Some format issues inherent in the e-media version may also appear in this print version.**



For additional information, please contact EDP Sciences – Web of Conferences
at the address below.

EDP Sciences – Web of Conferences
17, Avenue du Hoggar
Parc d'Activité de Courtabœuf
BP 112
F-91944 Les Ulis Cedex A
France

Phone: +33 (0) 1 69 18 75 75
Fax: +33 (0) 1 69 28 84 91

contact-edps@webofconferences.org

**Additional copies of this publication are available from:**

Curran Associates, Inc.
57 Morehouse Lane
Red Hook, NY 12571 USA
Phone: 845-758-0400
Fax: 845-758-2633
Email: curran@proceedings.com
Web: www.proceedings.com

# TABLE OF CONTENTS

# Preface

*Emanuele* Poli[1], *Yong* Liu[2,3], and *Victor* Udintsev[2]

[1]Max-Planck-Institut für Plasmaphysik, 85748 Garching bei München, Germany
[2]ITER Organization, Route de Vinon-sur-Verdon, CS 90046, 13067 St. Paul Lez Durance Cedex, France
[3]Institute of Plasma Physics, Chinese Academy of Sciences, Hefei 230031, China

This issue of EPJ Web of Conferences collects the proceedings of the 21st Joint Workshop on Electron Cyclotron Emission and Electron Cyclotron Resonance Heating, which was held on the premises of the ITER Organization, Cadarache, France, on June 20-24, 2022.

Due to the pandemic, the meeting took place four years after the previous one, hosted by IPP Greifswald in 2018, while the usual cycle for this workshop series is two years. Given the specific nature of the EC workshop, intended as a forum for a direct exchange of information, as well as for discussion and planning of future activities, the organizers discarded the option of a purely virtual meeting in 2020 and 2021. The EC21 was held as a hybrid event, with a small number of presentations delivered remotely. In order to give a good coverage of the work of the past four years, the EC21 programme was extended to five instead of the traditional four days. The workshop was attended by 117 participants (28 remote) from 15 countries and organizations. There were 15 invited, 39 oral and 36 poster presentations. The present proceedings collect a total of 42 papers.

The programme was organized around the traditional four main topics represented by EC theory, EC experiments, EC diagnostics and EC-related technology. Considerable progress was achieved not only in preparation of ITER (e.g. in the physics and control software, and in the development of gyrotrons and loads), but also in the design and modelling of two machines to be built in Europe in the next years, namely DTT and STEP, which will heavily rely on EC waves. Impressive reports were delivered about the refinement of the diagnostic capabilities related to EC emission, e.g. for turbulence detection and characterization. In general, a remarkable development that can be noted over the past years is the fact that theoretical and experimental activities become increasingly intertwined. We hope that these proceedings provide a representative snapshot of the status of the progress of the EC community in the last four years. We acknowledge the help of many colleagues (both among the participants and beyond) who accepted to review the papers published here.

We would like to thank the local organizing committee for the flawless management of both the meeting and the supporting programme. Taking a look into the ITER tokamak pit from the port that will house the EC equatorial launcher was a haunting and highly motivating experience for many colleagues. We look forward to the next meeting (EC22), which will take place in Korea in spring 2024.

The members of the Local Organizing Committee were
Yong Liu (scientific coordinator)

Victor Udintsev (scientific coordinator)
Julie Juif (conference secretary, registration)

The members of the International Programme Committee were
Max Austin (IFS University of Texas at Austin, USA)
Young-Soon Bae (NFRI, Korea)
Gregory Denisov (IAP Nizhny Novgorod, Russia)
Daniela Farina (CNR Milano, Italy)
Franco Gandini (ITER Organization)
Mei Huang (SWIP, China)
Shin Kubo (NIFS, Japan)
John Lohr (GA San Diego, USA)
Burkhard Plaum (IGVP University of Stuttgart, Germany)
Emanuele Poli (IPP Germany, chair)
Braj K. Shukla (IPR, India)
Victor Udintsev (ITER Organization)

Emanuele Poli, Yong Liu, Victor Udintsev (editors)
Garching/Cadarache, January 2023

EDP Sciences

Web of Conferences

# Statement of Peer review

In submitting conference proceedings to *Web of Conferences,* the editors of the proceedings certify to the Publisher that

1. They adhere to its **Policy on Publishing Integrity** in order to safeguard good scientific practice in publishing.
2. Ali articles have been subjected to peer review administered by the proceedings editors.
3. Reviews have been conducted by expert referees, who have been requested to provide unbiased and constructive comments aimed, whenever possible, at improving the work.
4. Proceedings editors have taken all reasonable steps to ensure the quality of the materials they publish and their decision to accept or reject a paper for publication has been based only on the merits of the work and the relevance to the journal.

Title, date and place of the conference

21st Joint Workshop on Electron Cyclotron Emission (ECE) and Electron Cyclotron Resonance Heating (ECRH)
20-24 June 2022, ITER Organization, Cadarache (France)

**Proceedings editor(s):**

Emanuele Poli (Max-Planck-Institut für Plasmaphysik, Garching, Germany)
Yong Liu (ITER Organization, Cadarache, France),
Victor Udintsev (ITER Organization, Cadarache, France)

**Date and editor's signature**

https://doi.org/10.1051/epjconf/202327701001

# Radiation pressure by electron cyclotron waves on density fluctuations

*Abhay* Ram[1,*] and *Kyriakos* Hizanidis[2,**]

[1]Plasma Science and Fusion Center, Massachusetts Institute of Technology, Cambridge, MA 02139, USA
[2]National Technical University of Athens, 157 73 Zographou, Greece

**Abstract.** The presence of turbulence in the form of large density fluctuations and coherent filamentary structures in the edge region of fusion plasmas has been well documented. Radio frequency waves, launched from structures near the wall of a tokamak, have to propagate through this turbulent plasma before reaching the core. These density fluctuations can reflect, refract, and diffract the electromagnetic waves, thereby modifying the flow of energy and momentum to the core plasma. Conversely, the radiation pressure of the radio frequency waves can modify the turbulence, whether it is in the edge region or in the core. This article examines some consequences of the radiation force induced by electron cyclotron waves in plasmas. The effect of waves on two different representations of density fluctuations are studied. In the first representation, suitable for both edge and core plasmas, it is assumed that a planar interface separates two different density regimes. The physics basis for the radiation force on an interface separating two different scalar dielectric media was first elucidated by Poynting in 1905 [J. H. Poynting, The London, Edinburgh, and Dublin Philosophical Magazine and Journal of Science **9**, 393-406 (1905)]. Poynting's results are explained within the context of Snell's law and Fresnel equations, and, subsequently, extended to magnetized plasmas. The analysis shows that electron cyclotron waves lead to peaking of the density profile – the interface is pushed towards the region of higher density. The planar interface approximation is the basis of Kirchhoff theory [P. Beckmann and A. Spizzichino, *The Scattering of Electromagnetic Waves from Rough Surfaces* (Artech, Massachusetts, 1987) Chapter 3] used to study wave scattering by turbulent media. In the second representation, appropriate for coherent structures in edge plasmas, the radiation force on a cylindrical filament embedded in a background plasma is determined using the Maxwell stress tensor. A detailed study reveals that the radiation force has a different effect on filaments – those with densities higher than the background density are pulled in towards the source launching the waves, while the lower density filaments are pushed away. The reaction on a filament is large enough to be observed experimentally.

## 1 Introduction

The edge plasma of toroidal fusion devices is replete with large density fluctuations in the form of turbulence and filamentary structures. The effect of such fluctuations on the propagation of radio frequency (RF) waves has been extensively studied theoretically and computationally [1–7]. Fluctuations tend to scatter RF waves through reflections, refractions, and diffractions. On the other hand, the radiation pressure of electromagnetic RF waves can have an impact on density fluctuations [8]. In this paper we discuss some of our initial thoughts and ideas on RF induced radiation pressure; specifically, by waves in the electron cyclotron (EC) range of frequencies. In section 2 we set up the physics basis of radiation pressure at an interface separating two scalar dielectric media. The transition to interfaces separating two different electron densities in a magnetized plasma is carried out in section 3. The permittivity of a plasma is a tensor but the physical consequences of the radiation force are similar to those in a scalar dielectric. Planar interfaces are not a good representation of turbulent structures that manifest themselves as filament or blobs – the geometry of the scatterer plays a role in the scattering process. In section 4, we model the scattering of EC waves by a cylindrical filament. The radiation force is evaluated using the Maxwell stress tensor and found to exhibit properties which are in contrast to those for planar interfaces. Besides reflection and transmission, common to planar interfaces, a filament leads to side-scattering, diffraction, and shadowing, as well as non-planar wave fields within the filament. Consequently, there is angular variation of the stress force along the interface which, in turn, changes the physics of the forces acting on the surface of a filament. Even though the theoretical analysis is for a step-function transition between two different plasma densities, numerical simulations have shown that there is no significant difference if the transition is smooth [4].

## 2 Radiation pressure at an interface separating two dielectrics

Consider a planar interface in the $y$-$z$ plane at $x = 0$ that separates two different scalar dielectric media with refrac-

*e-mail: abhay@mit.edu
**e-mail: kyriakos@central.ntua.gr

 https://doi.org/10.1051/epjconf/202327701001

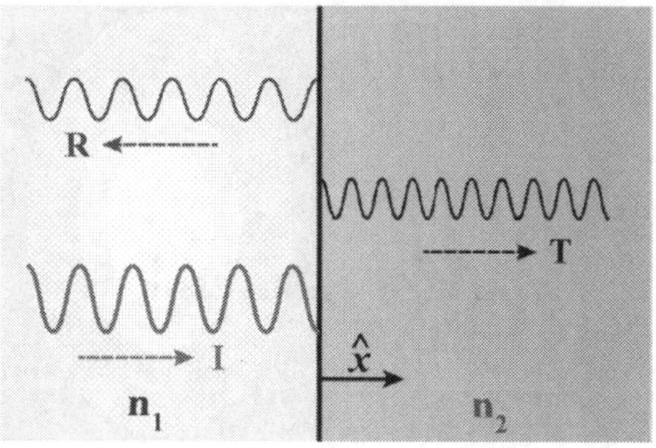


**Figure 1.** The scattering of an incident $I$ wave to a reflected $R$ and transmitted (refracted) $T$ wave at an interface separating two different scalar dielectrics with refractive indices $n_1$ and $n_2$. The interface is in the $y$-$z$ plane; $\hat{x}$ is the outward pointing normal to the surface.

tive indices $n_1$ and $n_2$ (figure 1). The normal to the interface is along $\hat{x}$ while $\hat{z}$ is normal to the plane of the figure. A plane electromagnetic wave (I) incident on the interface will get scattered into a reflected (R) and transmitted (T) plane waves. The electric field of an incident wave propagating along $\hat{x}$ is $\vec{E}_I = E_{I0}\, e^{ik_1 x - i\omega t}\, \hat{y}$, where $E_{I0}$ is the amplitude of the wave, $k_1$ is the magnitude of the propagation vector along $\hat{x}$, $\omega$ is the angular frequency of the wave, and $(\hat{x}, \hat{y}, \hat{z})$ are the unit vectors along the $(x, y, z)$ directions, respectively. From Snell's law, the reflected and the transmitted waves will also propagate along $\hat{x}$, and their electric fields are $\vec{E}_R = E_{R0}\, e^{-ik_1 x - i\omega t}\, \hat{y}$ and $\vec{E}_T = E_{T0}\, e^{ik_2 x - i\omega t}\, \hat{y}$, respectively. Here $E_{R0}$ and $E_{T0}$ are the electric field amplitudes of the reflected and transmitted waves, respectively, and $k_2$ is the magnitude of the propagation vector of the transmitted wave along $\hat{x}$. From Maxwell's equations, $\omega = v_1 k_1$ and $\omega = k_2 v_2$ are the dispersion relations for the waves in the two dielectric media; $v_1 = c/n_1$ and $v_2 = c/n_2$ with $c$ being the speed of light in vacuum. The continuity of the tangential components of the electric and magnetic fields at the interface yields the Fresnel equations [9],

$$\frac{E_R}{E_I} = \frac{n_1 - n_2}{n_1 + n_2}, \qquad \frac{E_T}{E_I} = \frac{2n_1}{n_1 + n_2}. \tag{1}$$

The energy, or Poynting, flux averaged over one time period $2\pi/\omega$ is,

$$\left\langle \vec{S} \right\rangle = \frac{1}{2\mu_0} \mathrm{Re}\left(\vec{E} \times \vec{B}^*\right), \tag{2}$$

where $\mu_0$ is the vacuum permeability. For waves in scalar dielectrics $\left\langle \vec{S} \right\rangle \parallel \vec{k} \parallel \hat{x}$. The reflection and transmission coefficients are, respectively,

$$R = \frac{|\langle S_x \rangle_R|}{|\langle S_x \rangle_I|} = \left(\frac{n_1 - n_2}{n_1 + n_2}\right)^2, \tag{3}$$

$$T = \frac{|\langle S_x \rangle_T|}{|\langle S_x \rangle_I|} = \frac{n_2}{n_1} \left(\frac{2n_1}{n_1 + n_2}\right)^2, \tag{4}$$

where the subscript $x$ indicates that the Poynting vector is along $\hat{x}$. From equations (3) and (4) it is straightforward to show that $R + T = 1$ – a statement on energy conservation.

The time-averaged momentum flux is defined as,

$$\left\langle \vec{\mathcal{P}} \right\rangle = \frac{\left\langle \vec{S} \right\rangle}{v}, \tag{5}$$

where $v$ is the speed of light in a dielectric medium. The normalized momenta for the reflected and transmitted waves are, respectively,

$$P_R = \frac{\langle \mathcal{P}_x \rangle_R}{\langle \mathcal{P}_x \rangle_I} = -\left(\frac{n_1 - n_2}{n_1 + n_2}\right)^2, \tag{6}$$

$$P_T = \frac{\langle \mathcal{P}_x \rangle_T}{\langle \mathcal{P}_x \rangle_I} = \frac{n_2^2}{n_1^2} \left(\frac{2n_1}{n_1 + n_2}\right)^2. \tag{7}$$

The negative sign in equation (6) is an indication that the reflected wave is propagating in a direction opposite to the incident wave. From equations (6) and (7), we note that $P_R + P_T \neq 1$; i.e., momentum is not conserved. Since neither dielectric allows for momentum dissipation, it is evident that the difference is the momentum $P_B$ imparted to the interface. From momentum conservation,

$$P_B = \frac{\langle \mathcal{P}_x \rangle_B}{\langle \mathcal{P}_x \rangle_I} = 1 - P_R - P_T = 2\left(\frac{n_1^2 - n_2^2}{n_1^2 + n_2^2}\right). \tag{8}$$

$P_B$ is a measure of the radiation force on the interface. From equation (8) we note that if $n_2 > n_1$, the force is in the $-\hat{x}$ direction, while if $n_1 > n_2$ it is in the $+\hat{x}$ direction. In other words, the force on the interface is always directed away from the medium with the higher refractive index. This interesting result was first noted by John Henry Poynting in 1905 [10]. In the paper Poynting states the following: "In any real refraction with ordinary light, there will be reflexion as well as refraction. The reflexion always produces a normal pressure, and the refraction a normal pull. But with unpolarized light, a calculation shows that the refraction pull, for glass at any rate, is always greater than the reflexion push, even at grazing incidence." The "push" and "pull" are with respect to the incoming wave which, with respect to figure 1, correspond to $+\hat{x}$ and $-\hat{x}$, respectively. For vacuum waves incident on a glass surface $n_2 > n_1$, and (8) confirms Poynting's observation. We would like to note that Poynting's argument was based on Planck's relation between energy and frequency, and not on the analysis given above. His reasoning can be interpreted as follows. The transmitted and reflected waves originate from the interface. From Newton's third law, the acquired momentum for each wave exerts a recoil force on the interface. The shorter wavelength wave has larger momentum associated with it; hence, the recoil force due to this wave dominates. For electromagnetic waves incident on a glass interface from vacuum, the transmitted wave has a shorter wavelength. Hence, the interface moves towards the vacuum region. The same argument holds for waves incident on the interface from inside the glass.

## 3 Radiation pressure at an interface separating two different electron densities in a magnetized plasma

In this section we extend the analysis on radiation pressure for simple dielectrics to a cold, magnetized plasmas. Referring to figure 1, the regions $n_1$ and $n_2$ correspond to two different electron densities. We assume that the magnetic field is uniform and pointing along $\hat{z}$, $\vec{B}_0 = B_0\hat{z}$.

### 3.1 Cold plasma dispersion relation

The dispersion tensor for plane electromagnetic waves propagating in a cold, magnetized plasma is [11],

$$\overleftrightarrow{\mathbf{D}} = \frac{c^2}{\omega^2}\left(\vec{k}\vec{k} - k^2\overleftrightarrow{\mathbf{I}}\right) + \overleftrightarrow{\mathbf{K}}, \tag{9}$$

where $\overleftrightarrow{\mathbf{I}}$ is the identity tensor, and $\vec{k}\vec{k}$ is a dyadic. The permittivity tensor $\overleftrightarrow{\mathbf{K}}$ is,

$$\overleftrightarrow{\mathbf{K}} = \begin{pmatrix} K_{xx} & -iK_{xy} & 0 \\ iK_{xy} & K_{xx} & 0 \\ 0 & 0 & K_{zz} \end{pmatrix}, \tag{10}$$

where,

$$\begin{aligned} K_{xx} &= 1 - \frac{\omega_{pe}^2}{\omega^2 - \omega_{ce}^2} - \sum_i \frac{\omega_{pi}^2}{\omega^2 - \omega_{ci}^2}, \\ K_{xy} &= -\frac{\omega_{ce}}{\omega}\frac{\omega_{pe}^2}{\omega^2 - \omega_{ce}^2} + \sum_i \frac{\omega_{ci}}{\omega}\frac{\omega_{pi}^2}{\omega^2 - \omega_{ci}^2}, \\ K_{zz} &= 1 - \frac{\omega_{pe}^2}{\omega^2} - \sum_i \frac{\omega_{pi}^2}{\omega^2}, \end{aligned} \tag{11}$$

$\omega_{pe}$ ($\omega_{pi}$) and $\omega_{ce}$ ($\omega_{ci}$) are the electron (ion) angular plasma and angular cyclotron frequencies, respectively, and the summation is over all the ion species in the plasma. The dispersion relation for waves in the plasma is given by $\mathcal{D}(\vec{k}, \omega) \equiv \det\left(\overleftrightarrow{\mathbf{D}}\right) = 0$, where det denotes the determinant of the second rank tensor. Without loss of generality, we choose $\vec{k} = k_\perp\hat{x} + k_\parallel\hat{z}$ – for a homogeneous medium, a simple rotation about the $z$-axis will ensure this particular form. Assuming that $k_\parallel$, $\omega$, $\omega_{pe}$, $\omega_{pi}$, $\omega_{ce}$, and $\omega_{ci}$ are all prescribed, the dispersion relation $\mathcal{D}(k_\perp, k_\parallel, \omega) = 0$ is biquadratic in $k_\perp$. A representative solution of the dispersion relation for EC waves in a cold plasma is plotted in figure 2. There are two primary differences between scalar and anisotropic tensor permittivities. First, the former has a dispersion relation that is quadratic in $k$ while the latter is biquadratic in $k_\perp$. Thus, there is an additional wave that is a normal mode in a plasma. Second, for waves in scalar dielectrics, $\vec{k}$, $\vec{E}$, and $\vec{B}$ are normal to each other. The continuity of tangential $\vec{E}$ and $\vec{B}$ at the interface leads to two independent boundary conditions. For plasma waves, it is $\vec{k}$, $\vec{D} = \epsilon_0\overleftrightarrow{\mathbf{K}}.\vec{E}$, and $\vec{B}$ that are normal to each other; $\vec{D}$ is the electric displacement field and $\epsilon_0$ is the vacuum permittivity. Combined with the anisotropy of the permittivity tensor, the continuity of tangential $\vec{E}$ and $\vec{B}$ at an interface leads to four independent boundary conditions [2]. This is consistent with the presence of two distinct waves in a plasma. As depicted in figure 3, the scattering of an incident wave by an interface separating two different densities in a magnetized plasma leads to two reflected and two transmitted waves. For electron cyclotron waves, regardless of the incident wave, the scattered waves are the extraordinary X and the ordinary O waves.

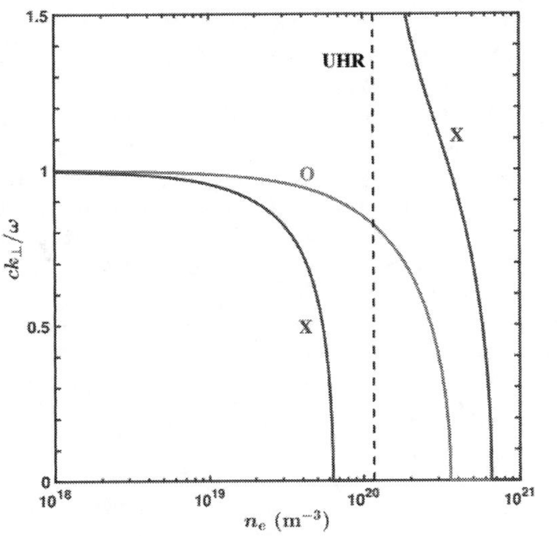


**Figure 2.** Dispersion properties of EC waves obtained from $\mathcal{D}(k_\perp, k_\parallel, \omega) = 0$: $ck_\perp/\omega$ is plotted as a function of the electron density for $f = \omega/2\pi = 170$ GHz, $B_0 = 5$ T, and $k_\parallel = 0$. The labels indicate the extraordinary X mode, the ordinary O mode, and the upper hybrid resonance UHR. The X mode is cutoff at $n_e \approx 6.34 \times 10^{19}$ m$^{-3}$, the UHR is located at $n_e \approx 1.15 \times 10^{20}$ m$^{-3}$, the O mode is cutoff at $n_e \approx 3.58 \times 10^{20}$ m$^{-3}$, and the second X mode is cutoff at $n_e \approx 6.53 \times 10^{20}$ m$^{-3}$.

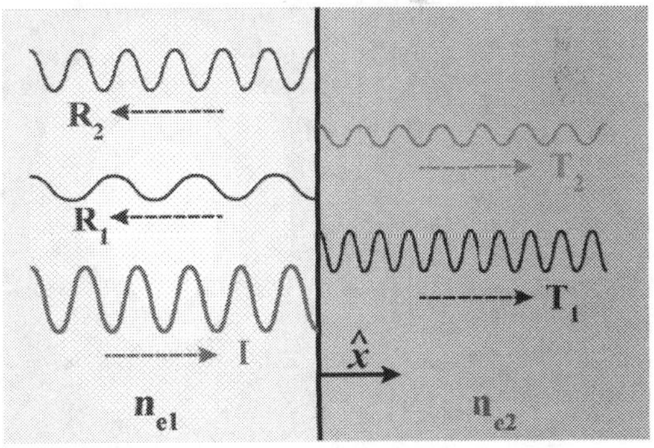


**Figure 3.** In contrast to figure 1, the scattering of an incident $I$ wave at an interface separating two different electron densities, $n_{e1}$ and $n_{e2}$, in a uniformly magnetized plasma results in two reflected, $R_1$ and $R_2$, and two transmitted, $T_1$ and $T_2$, waves.

### 3.2 Scattering and radiation pressure

We follow the same procedure as in section 2 to determine the radiation pressure at an interface separating two different electron densities. For the parameters indicated in figure 2, the direction of propagation of the incident wave – whether it is the O mode or the X mode – is normal to the interface. For this particular case, the scattered waves are the same as the incident wave; e.g., for an incident X mode the scattered waves are also X modes. The time-averaged Poynting flux is given by (5) in which $v$ is replaced by the phase speed of the wave $\omega/k_\perp$ [12]. We assume that the incident wave, either the O mode or the X mode, is approaching the interface from a region with density $n_{e1}$. Below we present results for three different combinations of $n_{e1}$ and $n_{e2}$ (figure 3).

#### 3.2.1 $n_{e1} < n_{e2}$ with O and X modes propagating in both regions

For $n_{e1} = 8 \times 10^{18}$ m$^{-3}$ and $n_{e2} = 10^{19}$ m$^{-3}$, figure 2 shows that the O and X modes are propagating on either side of the interface. The momentum transfer to the interface is obtained by matching the wave phases and the transverse componets of the wave electric and magnetic fields at the interface. We find that the normalized momentum transfer is,

$$P_{BO} = 2.86 \times 10^{-3}, \qquad P_{BX} = 1.06 \times 10^{-2}, \quad (12)$$

for an incident O mode and an incident X mode, respectively. The absolute values do not have an experimental significance at this point; we will put these numbers into proper context when we consider the radiation pressure on a well-defined cylindrical filament. However, two observations stand out. First, from the positive sign in front of the numerical values, we realize that both the X and O modes push the higher density region away from the antenna that launches these waves. This is in complete agreement with the direction of the momentum imparted to an interface separating two different scalar dielectrics when the electromagnetic wave is incident from a region of higher refractive index. For simple dielectrics, the index of refraction increases with density; see, for example, [13]. For plasmas, as shown in figure 2, the index of refraction decreases with density. Second, the push due to the incident X mode is stronger than that for the O mode.

#### 3.2.2 $n_{e1} > n_{e2}$ with O and X modes propagating in both regions

Here we consider the case when the initial wave is incident at the interface from a region of lower density. Let $n_{e1} = 10^{19}$ m$^{-3}$ and $n_{e2} = 8 \times 10^{18}$ m$^{-3}$– this is a reversal of the parameters in section 3.2.1. The normalized momentum transfer to the interface is,

$$P_{BO} = -2.86 \times 10^{-3}, \qquad P_{BX} = -1.06 \times 10^{-2}, \quad (13)$$

for an incident O mode and an incident X mode, respectively. The negative sign indicates that the interface is pulled towards the antenna. Again, this is consistent with the intuition developed from scalar dielectrics. As in section 3.2.1, the magnitude of the momentum due to the X mode is larger than that due to the O mode.

#### 3.2.3 $n_{e1} < n_{e2}$ – total internal reflection of the X mode

From figure 2, we note that the X mode is cutoff at a density below the upper hybrid resonance. We examine the affect of this cutoff on wave scattering by choosing $n_{e1} = 5 \times 10^{19}$ m$^{-3}$ and $n_{e2} = 7 \times 10^{19}$ m$^{-3}$; in the first region both modes are propagating while in the second region the X mode is cutoff. For an incident O mode or X mode, the normalized momentum transfer to the interface is, respectively,

$$P_{BO} = 0.314, \qquad P_{BX} = 2.. \quad (14)$$

The direction of the momentum for the O mode is in line with our previous reasoning. For the X mode, the incident wave is completely reflected and, because of normal incidence, there is no transfer of momentum to the O mode. The physics for total reflection of the X mode is similar to an elastic collision of a ball with a wall – the momentum imparted to the wall is twice the incoming momentum of the ball.

## 4 Radiation pressure on a cylindrical filament

The scattering of waves by an infinite plane simplifies the associated algebra while providing an insight into the physics of radiation pressure on the plane. When the scatterer is not an infinite plane, geometrical effects become important because boundary conditions on the electromagnetic fields have to be satisfied on the surface of scatterer. The edge region of fusion plasmas is populated with blobs and filaments [14] which can alter the propagation of RF waves and, in turn, be influenced by the radiation pressure of these waves.

### 4.1 Basic description of wave scattering by a cylindrical filament

We assume a cylindrical plasma filament of radius $a$, embedded in a uniform background, with its axis aligned along the magnetic field line. The density inside the filament is constant and different from the background density. The whole system is in a homogeneous magnetic field. The wave propagation is given by the Faraday-Ampere equation,

$$\nabla \times \left(\nabla \times \vec{E}(\vec{r})\right) - \frac{\omega^2}{c^2} \overleftrightarrow{\mathbf{K}}(\vec{r}) \cdot \vec{E}(\vec{r}) = 0, \quad (15)$$

where the time dependence of all fields is of the form $\exp(-i\omega t)$. The propagation vector of the incident RF wave is in the $x$-$y$ plane and of the form $\vec{k} = k_\perp \hat{x} + k_\parallel \hat{z}$ with $\hat{x}$ and $\hat{z}$ being the unit vectors along the $x$ and $z$ directions, respectively. The Faraday-Ampere equation is solved in cylindrical coordinates subject to the boundary conditions [2],

$$\hat{\rho} \times \left(\vec{E}_I + \vec{E}_S\right)\Big|_{\rho=a} = \hat{\rho} \times \vec{E}_F\Big|_{\rho=a}, \quad (16)$$

$$\hat{\rho} \times \left(\vec{B}_I + \vec{B}_S\right)\Big|_{\rho=a} = \hat{\rho} \times \vec{B}_F\Big|_{\rho=a}, \quad (17)$$

 https://doi.org/10.1051/epjconf/202327701001

where $\hat{\rho}$ is the outward pointing normal vector at the interface of the filament, $\rho$ is the radial coordinate, and all fields are evaluated at the surface of the filament $\rho = a$. The subscripts $I$, $S$, and $F$ denote the incoming RF field, the scattered field (outside the filament), and the field inside the filament, respectively. A sample result for the scattering is shown in figure 4 where an incident O mode is scattered by a filament of radius 1 cm. The parameters are as outlined in the caption. The filament leads to some reflection, refraction, diffraction, side-scattering, and shadowing. Even though the incoming wave is a plane wave, it is evident that the scattered fields are not plane waves. This is exemplified in figure 5 showing details of the field structure inside the filament.

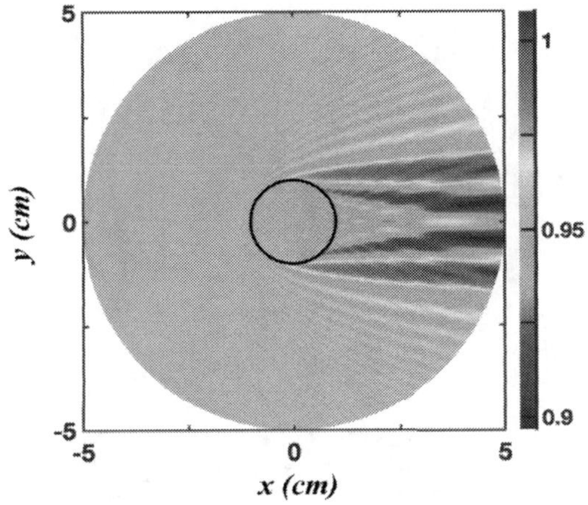


**Figure 4.** Scattering of an O mode, incident from the left, by a filament of radius 1 cm centered at the origin. The normalized, time-averaged Poynting flux in the $x$-direction $< S_x > /| < \vec{S}_I > |$ is plotted in the $x$-$y$ plane; $\vec{S}_I$ is the total Poynting vector of the incident wave. The electron density of the background plasma is $n_{eb} = 5 \times 10^{18}$ m$^{-3}$ while inside the filament it is $n_{ef} = 6.5 \times 10^{18}$ m$^{-3}$; the wave frequency is 170 GHz, the magnetic field is 5 T and pointing in the $z$ direction, and the incoming wave has $ck_{\|}/\omega = 0.3$.

## 4.2 Radiation force – Maxwell stress tensor

The electromagnetic fields obtained from the scattering equations (15)-(17) can be used to evaluate the Maxwell stress tensor and, subsequently, the radiation force on the filament [12, 15]. For the complex fields, the stress tensor is,

$$\overleftrightarrow{\mathbf{T}} = \vec{E}_R \vec{D}_R + \vec{H}_R \vec{B}_R - \frac{1}{2} \left( \vec{E}_R \,.\, \vec{D}_R + \vec{H}_R \,.\, \vec{B}_R \right) \overleftrightarrow{\mathbf{I}}, \quad (18)$$

where the subscript $R$ indicates the real part of the associated field; the first two terms on the right hand side are dyadics, $\overleftrightarrow{\mathbf{I}}$ is the identity tensor, $\vec{H}$ is the magnetic intensity, and $\vec{B} = \mu_0 \vec{H}$. Using the relationship between $\vec{D}$ and $\vec{E}$, the time average over one period of the wave cycle yields,

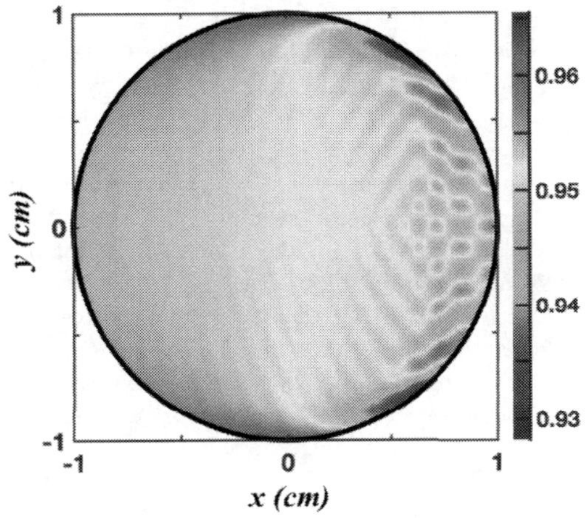


**Figure 5.** A magnified view of the normalized, time-averaged Poynting flux inside the filament.

$$\left\langle \overleftrightarrow{\mathbf{T}} \right\rangle = \frac{1}{2} \operatorname{Re} \left[ \epsilon_0 \vec{E} \overleftrightarrow{\mathbf{K}}^* \,.\, \vec{E}^* + \mu_0 \vec{H} \vec{H}^* - \frac{1}{2} \left( \epsilon_0 \vec{E} \,.\, \overleftrightarrow{\mathbf{K}}^* \,.\, \vec{E}^* + \mu_0 \left| \vec{H} \right|^2 \right) \overleftrightarrow{\mathbf{I}} \right], \quad (19)$$

where Re indicates the real part and $^*$ the complex conjugate.

In the cylindrical coordinate system, the time-averaged force on a filament of axial length $L$ is [12, 16],

$$\vec{\mathcal{F}} = \int \left\langle \overleftrightarrow{\mathbf{T}} \right\rangle . d\mathbf{A} = a \int_0^L \int_0^{2\pi} \left\langle \overleftrightarrow{\mathbf{T}} (\rho = a) \right\rangle . \hat{\rho}, \quad (20)$$

where the elemental area on the surface of the filament is $d\mathbf{A} = a \, d\phi \, dz \, \hat{\rho}$, with $\phi$ being the azimuthal coordinate around the axis of the filament. On the surface of the filament,

$$\left\langle \overleftrightarrow{\mathbf{T}} (\rho = a) \right\rangle . \hat{\rho} = \left\langle \overleftrightarrow{\mathbf{T}} (\rho = a) \right\rangle_b . \hat{\rho} - \left\langle \overleftrightarrow{\mathbf{T}} (\rho = a) \right\rangle_f . \hat{\rho}, \quad (21)$$

where the subscript $b$ indicates that the tensor is evaluated for the fields in the background plasma – the incident and the scattered fields, and the subscript $f$ is for fields inside the filament. The negative sign on the right hand side is due to the inward pointing normal for the filamentary fields. Equation (21) is a vector equation whose cylindrical components are determined using the boundary conditions (16) and (17). After some algebra we find that,

$$\hat{\rho} \,.\, \left\langle \overleftrightarrow{\mathbf{T}} (\rho = a) \right\rangle . \hat{\rho} = \frac{\epsilon}{4} \left[ K_{xx}^B \left| E_{I\rho} + E_{S\rho} \right|^2 - K_{xx}^F \left| E_{F\rho} \right|^2 + \left( K_{xx}^F - K_{xx}^B \right) \left| E_{F\phi} \right|^2 + \left( K_{zz}^F - K_{zz}^B \right) \left| E_{Fz} \right|^2 \right]_{\rho=a}, \quad (22)$$

$$\hat{\phi} \,.\, \left\langle \overleftrightarrow{\mathbf{T}} (\rho = a) \right\rangle . \hat{\rho} = 0, \quad (23)$$

$$\hat{z} \,.\, \left\langle \overleftrightarrow{\mathbf{T}} (\rho = a) \right\rangle . \hat{\rho} = 0. \quad (24)$$

The superscripts $B$ and $F$ indicate that these elements of the permittivity tensor are evaluated for plasma parameters corresponding to the background and filamentary plasmas, respectively. The subscripts in the electric fields include the components corresponding to the cylindrical coordinate system. The right hand side is evaluated at the surface of the filament. From equations (23)-(24) we note that the RF fields do not induce any shear forces on the filament; the only force is in the radial direction. The radial component (22) is a function of the azimuthal angle $\phi$ and is independent of $z$; the $z$ variation of the fields is $\exp(ik_z z)$ and is accounted for in the matching of wave phases at the surface of the filament. The radial force can lead to two possible effects – spatial displacement of the filament and its azimuthal distortion.

The only component of $\vec{F}$ in (20) is in the $\hat{\rho}$ direction. Since (22) has no variation along $z$, the radial force per unit axial length has two Cartesian components in the $x$ and $y$ directions,

$$\begin{pmatrix} \mathcal{F}_x \\ \mathcal{F}_y \end{pmatrix} = a \int_0^{2\pi} d\phi \begin{pmatrix} \cos\phi \\ \sin\phi \end{pmatrix} \hat{\rho} \cdot \left\langle \overleftrightarrow{\mathbf{T}}(\rho = a) \right\rangle \cdot \hat{\rho}. \quad (25)$$

We have redefined the force vector in (20) as force per unit axial length.

The power flux of the incident wave is

$$S_I = \left| \left\langle \vec{S}_I \right\rangle \right| = \frac{1}{2} \sqrt{\frac{\epsilon_0}{\mu_0}} \, |\mathcal{E}_0|^2 \quad (26)$$

where $\left\langle \vec{S}_I \right\rangle$ is the time-averaged Poynting vector and $\mathcal{E}_0$ is the electric field amplitude – both for the incident wave. The force on a filament per unit incident power flux is,

$$\begin{pmatrix} \widetilde{\mathcal{F}}_x \\ \widetilde{\mathcal{F}}_y \end{pmatrix} = \frac{1}{S_I} \begin{pmatrix} \mathcal{F}_x \\ \mathcal{F}_y \end{pmatrix} \frac{\mathrm{N\,m^{-1}}}{\mathrm{W\,m^{-2}}}. \quad (27)$$

The units expressed on the right hand side are Newton per meter (of axial length) for the force and Watt per square meter for the power flux. The stress force on the filament induces an acceleration,

$$\begin{pmatrix} \mathfrak{a}_x \\ \mathfrak{a}_y \end{pmatrix} = \frac{1}{\mathfrak{m}} \begin{pmatrix} \widetilde{\mathcal{F}}_x \\ \widetilde{\mathcal{F}}_y \end{pmatrix}, \quad (28)$$

where $\mathfrak{m}$ is the mass of a cylindrical filament of length 1 m and radius $a$ m.

### 4.3 Radiation force – numerical results

The analytical results in section 4.2 are used to illustrate various properties of the radiation force induced by EC waves on a filament of plasma. The primary term in calculating the force is the integrand in (25), $\left\langle T_{\rho\rho} \right\rangle \equiv \hat{\rho} \cdot \left\langle \overleftrightarrow{\mathbf{T}} \right\rangle \cdot \hat{\rho}$, given explicitly in (22). Figure 6 shows the variation of $c \left\langle T_{\rho\rho} \right\rangle / S_I$ as a function of $\phi$ along the surface of the filament. The variation along $\phi$ shows the radiation force is not uniform; the negative sign is an indication that the force is pulling the filament towards the source that launched the O mode.

In obtaining the results that follow, we assume a plasma containing deuterons as the single ion species. Then the mass of a filament 1 m long is $\mathfrak{m} = 1.05 \times 10^{-26} a^2 n_{ef}$ kg, where $n_{ef}$ is the electron density inside the filament. Furthermore, we set $a = 0.01$ m, $B_0 = 5$ T, and wave frequency at 170 GHz. There is an important point that needs to be made here – from all numerical simulations that we have carried out for different density regimes and, notably, for different waves (ion cyclotron, helicon, and lower hybrid waves), we find $\mathfrak{a}_y = 0$ [8]. Consequently, we will display results for $\mathfrak{a}_x$ only.

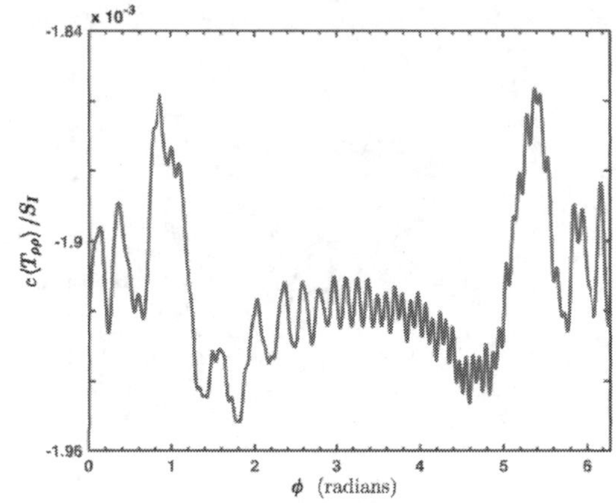


**Figure 6.** Radial force on the filament surface normalized to the power flux of the incident wave versus the azimuthal angle. The incident wave is an O mode and the parameters are the same as for figure 4.

#### 4.3.1 *Density inside the filament greater than the background density*

For an EC wave incident from a low density region and interacting with a higher density filament, we set the electron density in the background plasma and inside the filament to be $n_{eb} = 5 \times 10^{18}$ m$^{-3}$ and $n_{ef} = 6.5 \times 10^{18}$ m$^{-3}$, respectively. The stress force induced acceleration of the filament is plotted as a function of the parallel index of refraction $ck_\parallel/\omega$ for the O (blue) and X (red) modes in figure 7. The evaluated acceleration is for an incident power $S_I = 1$ kW m$^{-2}$. There are several points of note. The acceleration induced by the X mode is larger than that by the O mode. In both cases, the acceleration is negative which implies that the filament, which has a higher density relative to the background density, is being pulled in towards the source. This is a significant departure from the results obtained for planar scattering, where a wave incident from the low density region pushed the interface towards the higher density region. The difference can be attributed to the finite geometry of the scatterer – the bounded waves inside the filament are akin to cavity modes built from waves propagating in all directions. At the boundary, the internal modes couple to non-planar scattered fields. Consequently, the entire field structure does not possess the

Cartesian symmetry of the plane incoming wave. The azimuthal variation of the fields lacks uniformity and modifies the direction of the radiation force – figure 6 was an early sign of this feature.

For $ck_{\|}/\omega = 0.9$, the X mode is a propagating mode inside the filament, while for $ck_{\|}/\omega = 0.95$ it becomes evanescent. This change in the propagation of X mode results in a reversal of the slope. In contrast, for a planar interface, section 3.2.3, the X mode underwent total internal reflection. For $ck_{\|}/\omega \lessapprox 0.9$ the magnitude of the force decreases with $k_{\|}$ for the O mode, while it increases for the X mode.

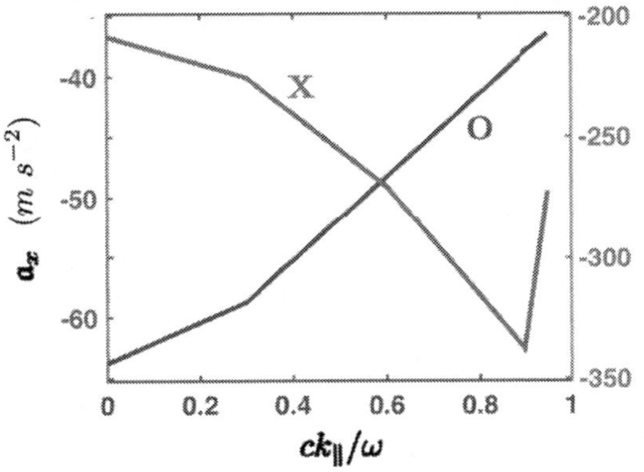


**Figure 7.** The acceleration of a filament induced by the radiation force of the O mode (left scale) and the X mode (right scale) as function of the index of refraction along the direction of the magnetic field. The density inside the filament is greater than the background density. The negative acceleration means that the filament is being pulled in towards the RF source. For $ck_{\|}/\omega = 0.95$ the X mode is evanescent inside the filament.

#### *4.3.2 Density inside the filament less than the background density*

In a reversal of the scenario in section 4.3.1, we choose $n_{eb} = 6.5 \times 10^{18}$ m$^{-3}$ and $n_{ef} = 5.0 \times 10^{18}$ m$^{-3}$; the incident wave is interacting with a filament having a lower density than the background plasma. Figure 8 plots $a_x$ versus $ck_{\|}/\omega$ for incident O (blue) and X (red) modes. The positive acceleration means that the radiation force pushes the lower density filament away from source of the incident wave, further exemplifying the difference in results from planar scattering. The force due to the X mode is larger than due to the O mode; the magnitudes of the two forces having opposite gradients with respect to $ck_{\|}/\omega$.

## 5 Conclusions

For normal incidence, the radiation force on a planar interface separating two different scalar dielectrics always pushes the interface towards the medium having the higher refractive index. This was first noted by Poynting and his argument was based on the back reaction of the reflected and transmitted waves on the interface. The direction of the radiation force had to be opposite to the direction of propagation of the wave with the shorter wavelength – from Planck's relation, the shorter wavelength wave had higher energy and momentum. An analysis based on Snell's law and Fresnel equations leads to the same conclusion.

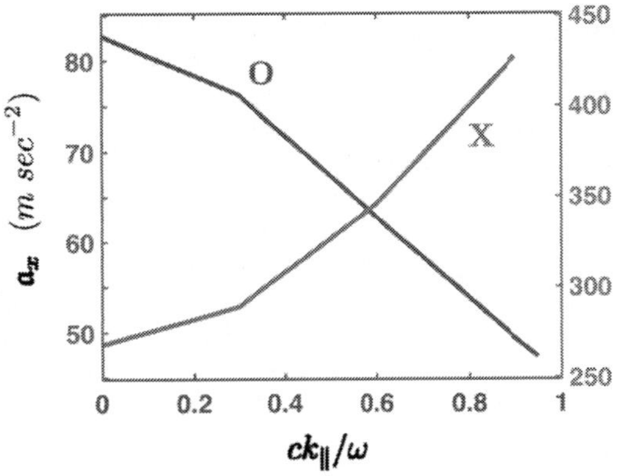


**Figure 8.** The acceleration of a filament induced by the radiation force of the O mode (left scale) and the X mode (right scale) as function of $ck_{\|}/\omega$. The density inside the filament is less than the background density. The positive acceleration means that the filament is being pushed away from the RF source.

In a cold magnetized plasma, whose permittivity is a tensor rather than a scalar, the situation is slightly different. First, there are two independent waves that can propagate in a plasma. The number of boundary conditions at the interface are doubled compared to scalar dielectrics. The scattering of any incoming wave at a planar interface will, in general, lead to two reflected and two transmitted waves. Second, unlike scalar dielectrics, over density ranges of interest, the refractive index of electron cyclotron waves decreases with plasma density. However, the physics of the radiation force on a planar interface separating two different plasma densities remains the same as for scalar dielectrics. For normal incidence, the direction of the force due to either the O mode or the X mode is such as to push the interface towards the higher density regions. Since the planar scattering model is applicable anywhere in a fusion plasma, this result leads to an interesting conclusion: the radiation force of the EC waves will lead to peaked density profiles. For the same power flux, the force due to X modes is greater than due to O modes.

A distinct advantage of the planar interface model is that the essential physics can be ascertained using plane waves. The edge region of tokamak plasmas is dominated by blobs and filaments. Any model based on planar scattering lacks validity as geometrical effects such as diffraction, side-scattering, and shadowing come into play. A full wave analysis in which a filament is modeled as a cylindrical plasma forms the basis for studying scattering of EC waves. The radiation pressure on the filament is evaluated using the Maxwell stress tensor. The force of EC waves on the surface of the filament is along the radial direction; there are no shear forces. The radial force is not uniformly distributed azimuthally, and can deform a filament . However, the integrated force is always such as to pull the filament towards the source of the incoming EC waves or push it away. The former occurs if the density inside the filament is greater than the background density; vice-versa for the latter case. This effect is in direct contrast to that for planar scattering. However, for scattering by planar interfaces or cylindrical filaments, the force due to the X mode is always stronger than that due to the O mode. For filaments, the variation in the force with the parallel index of refraction has opposite gradients for the O and X modes. It is quite possible to experimentally observe the influence of EC waves on filamentary structures, since the induced force leads to large accelerations even for modest input RF powers. The acceleration induced by the EC waves is instantaneous; however, the resulting dynamics of a filament through the background plasma merits a separate study.

The Kirchhoff theory is extensively used in studying the scattering of electromagnetic waves from rough surfaces [17, 18]. In this theory, also referred to as tangent plane or physical optics theory, the electromagnetic field at any point on the surface of a scatterer is assumed to be that on the tangent plane at that point. While not delving into the validity of this approximation, the Kirchhoff theory provides useful insight into the scattering process. In regions where the density fluctuations are not dominated by filaments and blobs, wave scattering by planar interfaces follows Kirchhoff theory. It is a constructive approach to understanding the effect of electron cyclotron waves on density fluctuations. The formalism leads to the conclusion that the radiation force of electron cyclotron waves on core fluctuations will lead to a peaking of the density profile. In experiments on ASDEX-U, peaking of the impurity density profile by EC waves has been observed [19].

For scattering off filamentary structures, the curvature effects are consequential, and the Kirchhoff theory is not applicable. The theoretical modeling has to resort to a full wave description. Electron cyclotron waves affect filaments differently compared to planar interfaces. However, the radiation force still leads to a redistribution of the density in the turbulent plasma.

## 6 Acknowledgement

AKR was supported by the US Department of Energy Grant numbers DE-FG02-91ER-54109 and DE-SC0018090.

## References

[1] A. K. Ram, K. Hizanidis, and Y. Kominis, Physics of Plasmas **20**, 056110-1–056110-10 (2013).

[2] A. K. Ram and K. Hizanidis, Physics of Plasmas **23**, 022504-1–022504-17 (2016).

[3] A. Köhn, E. Holzhauer, J. Leddy, M. B. Thomas, and R. G. L. Vann, Plasma Physics and Controlled Fusion **58**, 105008-1–105008-9 (2016).

[4] Z. C. Ioannidis, A. K. Ram, K. Hizanidis, and I. G. Tigelis, Physics of Plasmas **24**, 102115-1–102115-13 (2017).

[5] A. Snicker, E. Poli, O. Maj, L. Guidi, A. Köhn, H. Weber, G. Conway, M. Henderson, and G. Saibene, Nuclear Fusion **58**, 016002-1–016002-13 (2018).

[6] A. D. Papadopoulos, E. N. Glytsis, A. K. Ram, S. I. Valvis, P. Papagiannis, K. Hizanidis, and A. Zisis, Journal of Plasma Physics **85**, 905850309-1–905850309-20 (2019).

[7] S. I. Valvis, A. K. Ram, K. Hizanidis, P. Papagiannis, A. Papadopoulos, A. Zisis, I. G. Tigelis, and E. Glytsis, Journal of Plasma Physics **84**, 745840604-1–745840604-28 (2018).

[8] S. I. Valvis, A. K. Ram, and K. Hizanidis, Journal of Plasma Physics **87**, 905870612-1-905870612-37 (2021).

[9] D. J. Griffiths, *Introduction to Electrodynamics* (Pearson, Boston, 2013) Chapter 9.3.

[10] J. H. Poynting, The London, Edinburgh, and Dublin Philosophical Magazine and Journal of Science **9**, 393-406 (1905).

[11] A. Bers, *Plasma Physics and Fusion Plasma Electrodynamics, Volume 1* (Oxford, New York, 2016) Chapter 12.

[12] A. Bers, *Plasma Physics and Fusion Plasma Electrodynamics, Volume 1* (Oxford, New York, 2016) Chapter 19.

[13] A. Burnett, Mathematical Proceedings of the Cambridge Philosophical Society **23**, 907-911 (1927).

[14] O. Grulke, J. L. Terry, B. LaBombard, and S. J. Zweben, Physics of Plasmas **13**, 012306-1-012306-7 (2006).

[15] W. K. H. Panofsky and M. Phillips, *Classical Electricity and Magnetism* (Dover, New York, 2005) Chapter 10.

[16] J. A. Stratton, *Electromagnetic Theory* (McGraw-Hill, New York, 1941) Chapter II.

[17] P. Beckmann and A. Spizzichino, *The Scattering of Electromagnetic Waves from Rough Surfaces* (Artech, Massachusetts, 1987) Chapter 3.

[18] J. A. Ogilvy, *Theory of Wave Scattering from Random Rough Surfaces* (CRC Press, Philadelphia, 1991), Chapter 4.

[19] R. M. McDermott *et al.*, Nuclear Fusion **62**, 026006 (2022).

# Parametric Decay Instabilities during Electron Cyclotron Resonance Heating of Fusion Plasmas, Problems and Possibilities

*S.K.* Hansen[1,*], *S.K.* Nielsen[2], *J.* Stober[3], *J.* Rasmussen[2], *M.* Salewski[2], *M.* Willensdorfer[3], *M.* Hoelzl[3], *M.* Stejner[2], and the ASDEX Upgrade Team[4]

[1]Plasma Science and Fusion Center, Massachusetts Institute of Technology, Cambridge, MA 02139, USA
[2]Department of Physics, Technical University of Denmark, DK-2800 Kgs. Lyngby, Denmark
[3]Max-Planck-Institut für Plasmaphysik, D-85748 Garching b. München, Germany
[4]See Author List in H. Meyer for the ASDEX Upgrade Team, Nucl. Fusion **59**, 112014 (2019)

**Abstract.** We review parametric decay instabilities (PDIs) expected in connection with electron cyclotron resonance heating (ECRH) of magnetically confined fusion plasmas, with a specific focus on conditions relevant for the ITER tokamak. PDIs involving upper hybrid (UH) waves are likely to occur in O-mode ECRH scenarios at ITER if electron density profiles allowing trapping of UH waves near the ECRH frequency are present. Such PDIs may occur near the plasma center in ITER full-field scenarios heated by 170 GHz O-mode ECRH and on the high-field side of half-field ITER plasmas heated by 110 GHz or 104 GHz O-mode ECRH. Additionally, 110 GHz O-mode ECRH of half-field ITER scenarios may have low ECRH absorption, due to the electron cyclotron resonance being located on the high-field side of the main plasma. This potentially allows PDIs driven by a significant amount of ECRH radiation reaching the UH resonance in X-mode to occur, as X-mode radiation can be generated by reflection of unabsorbed O-mode radiation from the high-field side wall. The occurrence of PDIs during ECRH may damage microwave diagnostics, such as the electron cyclotron emission and low-field side reflectometer systems at ITER, as well as complicate the calculation of heating and current drive characteristics. However, if PDIs are induced in a controlled manner, they may provide novel diagnostic tools and allow the generation of a moderate fast ion population in plasmas heated only by ECRH.

## 1 Introduction

Parametric decay instabilities (PDIs) of millimeter wave radiation from gyrotron sources used for electron cyclotron resonance heating (ECRH) of magnetically confined fusion plasmas have been a topic of significant interest in recent years. The main reason for this is that PDIs can generate strong microwave signals capable of damaging microwave diagnostics [1], as well as transfering a significant fraction of the ECRH power from electromagnetic waves to quasi-electrostatic plasma waves, altering the heating and current drive characteristics from those expected based on linear ECRH theory [2, 3]. However, the microwave signals caused by PDIs can also serve as plasma diagnostics, e.g., allowing the occurrence of O-X-B heating to be confirmed [4, 5] and the properties of edge-localized modes to be explored [6]. Additionally, the low-frequency waves produced by some PDIs also allow generation of a moderate number of fast ions in pure ECRH plasmas [7–11]. In the present Paper, we provide an overview of the ECRH PDIs studied in recent years and assess their potential impact on the O-mode ECRH scenarios planned for ITER [12].

In a PDI, a strong quasi-coherent pump wave, e.g., the millimeter waves used for ECRH in magnetic confinement fusion research, decays to two daughter waves once the pump wave amplitude exceeds a threshold determined by the nonlinear interaction strength between the three waves. Energy conservation in the three-wave process requires the frequency, $f$, and wave vector, $\mathbf{k}$, of the pump wave to equal the sum of those of the daughter waves, i.e.,

$$f_0 = f_1 + f_2, \quad \mathbf{k}_0 = \mathbf{k}_1 + \mathbf{k}_2, \tag{1}$$

where subscript 0 refers to the pump wave, while subscripts 1 and 2 refer to the daughter waves. Equation (1) requires the daughter waves excited by PDIs to be shifted in frequency relative to the pump wave, meaning that their frequencies will often be outside the protective diagnostic filters around the ECRH frequency in the case of ECRH PDIs. This is the root cause of the microwave diagnostics damage associated with such PDIs [1]. Traditionally, magnetic confinement fusion ECRH PDIs were only expected in connection with extremely high power ECRH provided by pulsed free-electron maser sources [13] or in scenarios with lower ECRH power where a significant fraction of the ECRH power would reach the upper hybrid resonance (UHR) with X-mode polarization [4, 5, 9–11, 14–30]. PDIs are expected in the above cases due to the occurrence of electric fields with large amplitudes. These large amplitude electric fields allow the convective losses normally suppressing PDIs in inhomogeneous plas-

*e-mail: soerenkh@psfc.mit.edu

mas [31–33] to be overcome. In the free-electron maser case, the electric field amplitude is large enough to overcome the losses in the bulk plasma [13], while the UHR case relies on enhancement of the electric field amplitude associated with X-mode waves near the UHR, where their group velocity becomes very small [34]. Although PDIs involving X-mode waves reaching the UHR can occur at relatively low ECRH power levels without strict requirements on the plasma conditions [25, 26], they have generally only been considered to play a role in certain ECRH scenarios. This is due to the fact that X-mode waves cannot reach the UHR directly from the usual low-field side ECRH launcher position in tokamaks, owing to the R-cutoff always occurring on the low-field side of the UHR. The analysis of PDIs at the UHR is distinct for strongly overdense plasmas [18, 19, 22, 29], commonly found in spherical tokamaks, and underdense/weakly overdense plasmas [17, 24–28, 30], commonly found in conventional tokamaks and during plasma start-up, owing to the different dispersion properties of the electron Bernstein waves (EBWs) involved in the PDI in these cases. Specific scenarios in which PDIs at the UHR are of interest in connection with magnetic confinement fusion include fundamental X-mode [9–11, 17] and O-mode [9, 11, 21, 24–28] ECRH with optically thin/gray resonances, O-X-B heating [4, 5, 18–20, 29], and EBW start-up [35, 36].

In spite of previous expectations, strong PDI-like microwave signals were observed in connection with ECRH of plasma scenarios with tearing modes at the TEXTOR tokamak that did not fulfill any of the above criteria [37, 38]. This was explained in terms of ECRH PDIs involving trapped daughter waves [39], for which the convective losses normally limiting the occurrence of PDIs in inhomogeneous plasmas [31–33] are essentially removed. Specifically, the TEXTOR results at higher densities were explained in terms of decay of X-mode ECRH waves into two trapped UH waves near the O-point of the magnetic island [40–47]. PDIs of this kind have also been found at the ASDEX Upgrade tokamak [1, 6, 34], in low-temperature plasma filament experiments [2], and in particle-in-cell simulations [48–50]. The PDIs leading to microwave diagnostics damage at ASDEX Upgrade [1] and significant anomalous X-mode ECRH absorption in low-temperature plasma filaments [2] were both of the above type. More recently, an explanation of strong microwave signals observed at the Wendelstein 7-X stellarator in terms of decay of an X-mode ECRH wave into a trapped UH wave and a propagating X-mode wave around the O-point of a divertor island [51, 52], has also provided a potential explanation of the strong microwave signals observed at lower densities in TEXTOR [37, 38]. Apart from the PDIs driven by X-mode ECRH, PDIs driven by O-mode ECRH, relevant for the full-field ITER scenarios [12], have also been investigated. Particularly, decay of an O-mode ECRH wave to a trapped UH wave and a low-frequency lower hybrid wave has been investigated theoretically [53–56], in a low-temperature plasma filament [3], and at the FTU tokamak [57–59]. Additionally, it has recently been suggested that trapped low-frequency waves existing as a consequence of the strong gradients in the ITER and ASDEX Upgrade pedestals could lead to PDIs near the plasma edge during O-mode ECRH [60]. In this Paper, we review the basic conditions under which the above PDIs may occur and assess their relevance for the O-mode ECRH scenarios proposed for ITER [12], in a manner similar to the investigations in Section 4 of [1], but covering different scenarios.

The remainder of the Paper is organized as follows. Section 2 discusses the theoretical background of the employed criteria. In Section 3, the criteria are applied to the O-mode ECRH scenarios planned at ITER. Finally, Section 4 presents our conclusions.

## 2 Theoretical Background

ITER is a large tokamak with minor radius $a = 2\,\mathrm{m}$ and major radius $R_0 = 6.2\,\mathrm{m}$ [12]. The magnetic field strength, $B$, is approximately proportional to $1/R$, with $R$ being the distance from the symmetry axis. Thus,

$$B \approx B_0 \frac{R_0}{R}, \tag{2}$$

where $B_0$ is the field at the magnetic axis; in full-field scenarios, $B_0 = 5.3\,\mathrm{T}$, while $B_0 = 2.65\,\mathrm{T}$ in half-field scenarios [12]. For quasi-perpendicular ECRH injection in the non-relativistic limit, absorption occurs when $f_0$ is equal to multiples of the electron cyclotron frequency [61],

$$|f_{ce}| = \frac{eB}{2\pi m_e} \approx \frac{eB_0}{2\pi m_e}\frac{R_0}{R}, \tag{3}$$

where $e$ is the elementary charge and $m_e$ is the electron mass. Additionally in the cold plasma limit, O-mode radiation has a cutoff at the electron plasma frequency,

$$f_{pe} = \sqrt{\frac{e^2 n_e}{4\pi^2 \epsilon_0 m_e}}, \tag{4}$$

where $n_e$ is the electron density and $\epsilon_0$ is the permittivity of vacuum. The UHR occurs for X-mode radiation at the UH frequency,

$$f_{UH} = \sqrt{f_{pe}^2 + f_{ce}^2}, \tag{5}$$

in cold plasmas, although we note that its properties are substantially influenced by warm plasma effects at electron temperatures $T_e \gtrsim 3\,\mathrm{keV}$ [1, 62]. ECRH radiation with a frequency of $f_0$ will encounter the cold UHR ($f_0 = f_{UH}$) at an $n_e$ value of

$$n_e^{UH} \approx \frac{\epsilon_0 m_e}{e^2}\left(4\pi^2 f_0^2 - \frac{e^2 B_0^2}{m_e^2}\frac{R_0^2}{R^2}\right). \tag{6}$$

We investigate $n_e^{UH}$ for O-mode ECRH radiation at ITER, as such radiation may drive a PDI involving a trapped UH wave and a low-frequency lower hybrid wave in the presence of an inhomogeneous $n_e$ profile with a local maximum slightly above $n_e^{UH}$ [53–56]. Additionally, in connection with O-mode ECRH at ITER, we investigate whether scenarios are likely to occur in which a significant amount of X-mode radiation reaches the UHR upon reflection off the high-field side wall, as PDIs have been observed in such scenarios at ASDEX Upgrade, even without UH wave trapping [23–26]. We note that the trapping-$n_e$ of the UH wave is slightly different from the value in Eq. (6), since it is down-shifted relative to $f_0$ by the frequency of the low-frequency lower hybrid wave in accordance with Eq. (1). However, this shift can be ignored, as Eq. (6) already ignores finite-$T_e$ effects, which will modify the UHR condition in the core of ITER plasmas [1, 12, 62]. Equation (6) should thus be considered a rough estimate of the $n_e$-value at which UH wave trapping and field enhancement of X-mode ECRH waves may occur.

In addition to $n_e^{UH}$, we consider the $n_e$-limits and typical $n_e$-values at ITER. To avoid disruptions, $n_e$ should not exceed the Greenwald density,

$$n_e^G\,[10^{20}\,\mathrm{m}^{-3}] = \frac{I_p\,[\mathrm{MA}]}{\pi(a\,[\mathrm{m}])^2}, \tag{7}$$

at the plasma edge [63, 64]; $I_p$ is the plasma current, which is 15 MA in standard full-field ITER scenarios and 7.5 MA in half-field ITER scenarios [12]. We note that the core-$n_e$ can exceed $n_e^G$ in the case of peaked plasma profiles obtained by pellet fueling [64] and that the empirical $n_e^G$ given by Eq. (7) may underestimate the $n_e$-limit in ITER [65], meaning that $n_e^G$ does not represent a hard limit on $n_e$ in ITER, but still serves as a useful upper estimate of the edge-$n_e$. Further, the line-averaged $n_e$ values leading to the minimum L–H threshold, $\bar{n}_e^{\min}$, which are representative of the lower bound of $n_e$ in the core of typical H-mode plasmas, are given by $5\times10^{19}\,\mathrm{m}^{-3}$ in full-field ITER scenarios and $2.5\times10^{19}\,\mathrm{m}^{-3}$ in half-field ITER scenarios according to [12, 66].

## 3 Potential O-Mode ECRH PDI Scenarios at ITER

We plot $n_e^{UH}$, $n_e^G$, $\bar{n}_e^{\min}$, along with the (cold) fundamental ($f_0 = |f_{ce}|$) and second-harmonic ($f_0 = 2|f_{ce}|$) electron cyclotron resonances (ECRs) for the O-mode ECRH scenarios planned at ITER [12] in Fig. 1. The plots cover $R \in [4.11\,\mathrm{m}, 8.48\,\mathrm{m}]$, corresponding to the $R$-range of the ITER vacuum vessel in the midplane, while the $R$-ticks at $R_0 - a = 4.2\,\mathrm{m}$, $R_0 = 6.2\,\mathrm{m}$, and $R_0 + a = 8.2\,\mathrm{m}$ represent the extent of the main plasma in the midplane [12].

The top panel of Fig. 1 shows the situation for fundamental O-mode ECRH of the full-field ITER scenario ($B_0 = 5.3\,\mathrm{T}$, $I_p = 15\,\mathrm{MA}$) using the main ITER ECRH system operating at $f_0 = 170\,\mathrm{GHz}$ [12]. In this case, $n_e^{UH}$ attains values from 0 to $n_e^G$ in the central part of the plasma. Since the ECRH waves are injected from the low-field side (large $R$) [12] and thus encounter the region with $n_e^{UH} \in ]0, n_e^G[$ before they are absorbed around the fundamental ECR, PDIs involving decay of an O-mode ECRH wave to a trapped UH wave and a low-frequency lower hybrid wave are likely occur if a region allowing UH wave trapping is present near the plasma center in full-field ITER scenarios. It is also possible to operate plasmas with core $n_e$-values approaching $2n_e^G$ when the plasma is fueled by pellets [64], in which case trapped UH wave PDIs may be possible over a wider range of the full-field ITER plasmas; the requirement that $n_e < n_e^G$ at the plasma edge means that there will always be some point with $R < R_0 + a$ at which $f_0 = f_{UH}$ in stable plasmas. Trapped UH wave PDIs in the ITER full-field scenarios may create strong microwave signals with frequency shifts $\sim 10\,\mathrm{GHz}$ from $f_0 = 170\,\mathrm{GHz}$ [1]. This makes them a potential risk to the ITER electron cyclotron emission (ECE) and low-field side reflectometer systems, operating in the frequency ranges 70–1000 GHz [67, 68] and 30–165 GHz [69], respectively. PDIs involving a significant amount of X-mode ECRH power reflected from the high-field side wall reaching the UHR are unlikely to occur in the full-field ITER scenario, as the fundamental ECR is located close to the plasma center and therefore expected to be optically thick for the O-mode ECRH waves [61].

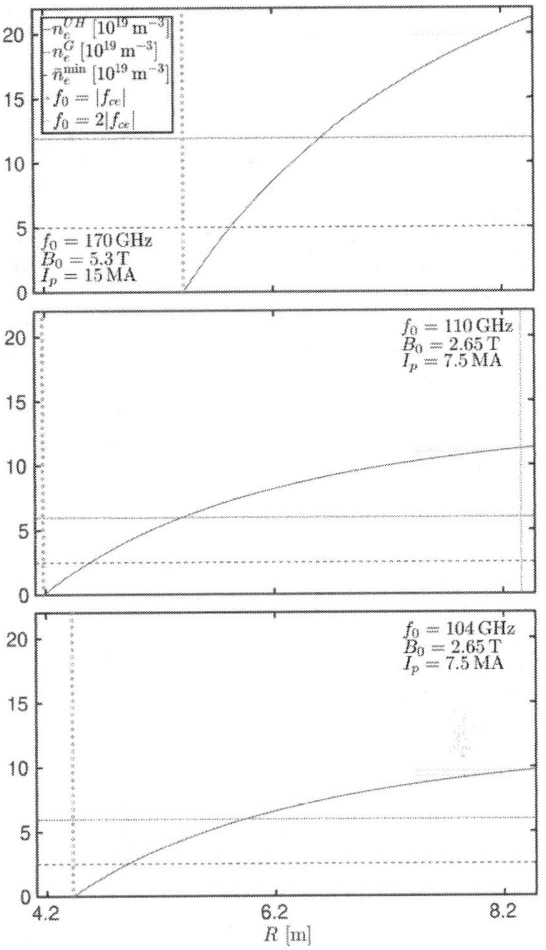


**Figure 1.** $n_e^{UH}$, $n_e^G$, $\bar{n}_e^{\min}$, $f_0 = |f_{ce}|$, and $f_0 = 2|f_{ce}|$ of the O-mode ECRH scenarios planned for ITER [12]. The top panel shows the situation for the full-field scenario ($B_0 = 5.3\,\mathrm{T}$, $I_p = 15\,\mathrm{MA}$) heated by the main ECRH system, $f_0 = 170\,\mathrm{GHz}$. The middle and bottom panels show the situations for O-mode ECRH of the half-field scenario ($B_0 = 2.65\,\mathrm{T}$, $I_p = 7.5\,\mathrm{MA}$) with $f_0 = 110\,\mathrm{GHz}$ and $f_0 = 104\,\mathrm{GHz}$, respectively.

The middle and bottom panels of Fig. 1 show the situation for fundamental O-mode ECRH of the half-field

ITER scenario ($B_0 = 2.65\,\mathrm{T}$, $I_p = 7.5\,\mathrm{MA}$) using the two lower-frequency ECRH options considered for ITER to enable ECRH start-up in the one-third-field scenarios [12], $f_0 = 110\,\mathrm{GHz}$ (middle panel) and $f_0 = 104\,\mathrm{GHz}$ (bottom panel). In these cases, $n_e^{UH}$ covers the range from 0 to $n_e^G$ on the high-field side of the plasma. Since the ECRH waves are still injected from the low-field side, they will encounter the region with $n_e^{UH} \in ]0, n_e^G[$ before potentially being absorbed at the fundamental ECR [61]; the second-harmonic ECR for $f_0 = 110\,\mathrm{GHz}$ which occurs far on the low-field side, outside the main plasma, is optically thin due to the low $n_e$ and $T_e$ in this region [61]. This means that 110 GHz and 104 GHz ECRH waves are also likely to drive PDIs involving decay of the O-mode ECRH waves into trapped UH waves and low-frequency lower hybrid waves in regions allowing UH wave trapping on the high-field side in ITER half-field scenarios. Such trapped UH wave PDIs would again pose a risk to the ECE [67, 68] and low-field side reflectometer [69] systems at ITER. Operation with pellet fueling above $n_e^G$ would allow the trapped UH wave PDIs to occur near the plasma center and on the low-field side in addition to the high-field side; once again, the requirement of $n_e < n_e^G$ at the plasma edge means that there will always be some point with $R < R_0 + a$ at which $f_0 = f_{UH}$ in stable plasmas.

While the $f_0 = 110\,\mathrm{GHz}$ and $f_0 = 104\,\mathrm{GHz}$ cases are similar in terms of their behavior with respect to PDIs involving a trapped UH wave, the conclusions regarding the possibility of having a significant amount of X-mode radiation reaching the UHR due to reflections from the high-field side wall and driving PDIs differ. For $f_0 = 110\,\mathrm{GHz}$, the middle panel of Fig. 1 shows that the fundamental ECR is located on the far high-field side, slightly outside the main plasma, meaning that its absorption is highly dependent on the precise value of $B_0$ and the angle at which the O-mode ECRH waves are launched [24–26, 61]. If virtually no particles inside the last closed flux surface are able to fulfill the ECR resonance condition [61], a significant amount of the injected O-mode radiation may reflect off the high-field side wall and (for a specific range of angles, which can be computed using [24]) re-enter the plasma as X-mode radiation that can drive PDIs at the UHR [17, 24–26]. This mechanism has been demonstrated experimentally in ASDEX Upgrade [24–26] and can be assessed for given ITER plasma conditions and ECRH beam geometries using the theory of [24, 30]. We note that by slightly increasing $B_0$, such that the fundamental ECR occurs inside the last closed flux surface at the midplane, the absorption of 110 GHz ECRH radiation from the midplane launchers can be made more certain, which should also limit the occurrence of PDIs relying on X-mode radiation reaching the UHR. The minimum value of $B_0$ at which ECR absorption around the midplane is guaranteed, $B_0^{\min}$, is given by the condition that $f_0 = |f_{ce}|$ at $R = R_0 - a$, from which Eq. (3) yields

$$B_0^{\min} \approx 2\pi f_0 \frac{m_e}{e} \frac{R_0 - a}{R_0}. \tag{8}$$

For $f_0 = 110\,\mathrm{GHz}$, we find $B_0^{\min} \approx 2.66\,\mathrm{T}$, so operating at $B_0$ slightly above 2.65 T should prevent the occurrence of PDIs relying on a significant amount of X-mode ECRH power reaching the UHR. In the case of $f_0 = 104\,\mathrm{GHz}$, shown in the bottom panel of Fig. 1, the fundamental ECR occurs well inside main plasma around the midplane for $B_0 = 2.65\,\mathrm{T}$, so no PDIs relying on X-mode ECRH waves reaching the UHR are expected in this case. Additionally, $B_0^{\min} \approx 2.52\,\mathrm{T}$ for $f_0 = 104\,\mathrm{GHz}$, meaning that operation at $B_0$ values somewhat below 2.65 T is also possible without the occurrence of PDIs relying on X-mode ECRH power around the UHR. Finally, we note that 104 GHz radiation might also be preferable for avoiding strong trapped UH wave PDIs involving X-mode ECRH pump waves [62].

## 4 Conclusion

We have investigated the potential occurrence of PDIs in O-mode ECRH scenarios at ITER [12], using an approach similar to that employed for the ITER X-mode ECRH scenarios in Section 4 of [1]. We found that PDIs are likely to occur near the UHR in connection with phenomena leading to non-monotonic $n_e$-profiles allowing UH wave trapping in a wide variety of O-mode ECRH scenarios at ITER, as well as in situations with X-mode radiation reflected off the high-field side wall reaching the UHR without restrictions on the $n_e$-profiles in scenarios with limited ECRH absorption. Specifically, trapped UH wave PDIs may occur near the plasma center during 170 GHz ECRH in full-field scenarios ($B_0 = 5.3\,\mathrm{T}$, $I_p = 15\,\mathrm{MA}$), while PDIs requiring limited ECRH absorption are unlikely in these scenarios. Trapped UH wave PDIs may also occur on the high-field side during 110 GHz or 104 GHz ECRH of half-field scenarios ($B_0 = 2.65\,\mathrm{T}$, $I_p = 7.5\,\mathrm{MA}$). Such scenarios may further allow PDIs requiring limited ECRH absorption at sufficiently low $B_0$. For 110 GHz ECRH, a $B_0 > 2.66\,\mathrm{T}$ was found to make such PDIs unlikely for ECRH beams launched from the outboard midplane, while a less strict requirement of $B_0 > 2.52\,\mathrm{T}$ was found to apply for 104 GHz ECRH. It is possible to mitigate the potential damage to microwave diagnostics due to ECRH PDIs by installing additional filters around the frequencies expected to be affected and by minimizing the abilitiy of the radiometers to pick up strong signals generated inside the ECRH beams [1].

While trapped UH wave PDIs have been found to lead to diagnostics damage [1] and significant anomalous ECRH absorption [2, 3], which would generally be detrimental to the operation of the ITER ECRH system, we note that PDIs may also allow novel diagnostics and heating schemes to be implemented. For instance, PDIs involving X-mode radiation reaching the UHR are being considered for generating fast ions with ECRH in spherical tokamaks [7, 8], although we note that the efficacy of this scheme has yet to be demonstrated, and trapped UH wave PDIs occurring near the plasma edge have already been used to investigate edge-localized mode and L-mode blob characteristics at ASDEX Upgrade [6]. Since PDIs involving low-frequency trapped waves in the density pedestal are expected in connection with O-mode ECRH at ITER [60], it would be of particular interest to assess their value as an edge diagnostic.

## Acknowledgements

S.K. Hansen acknowledges support by an Internationalisation Fellowship (CF19-0738) from the Carlsberg Foundation. This work was supported by a research grant (15483) from VILLUM FONDEN. This work has been carried out within the framework of the EUROfusion Consortium, funded by the European Union via the Euratom Research and Training Programme (Grant Agreement No 101052200 — EUROfusion). Views and opinions expressed are however those of the author(s) only and do not necessarily reflect those of the European Union or the European Commission. Neither the European Union nor the European Commission can be held responsible for them.

## References

[1] S.K. Hansen, A.S. Jacobsen, M. Willensdorfer, S.K. Nielsen, J. Stober, K. Höfler, M. Maraschek, R. Fischer, M. Dunne, the EUROfusion MST1 team et al., Plasma Phys. Control. Fusion **63**, 095002 (2021)

[2] A.B. Altukhov, V.I. Arkhipenko, A.D. Gurchenko, E.Z. Gusakov, A.Yu. Popov, L.V. Simonchik, M.S. Usachonak, EPL **126**, 15002 (2019)

[3] L. Simonchik, A. Altukhov, V. Arkhipenko, A. Gurchenko, E. Gusakov, A. Popov, M. Usachonak, Europhys. Conf. Abstracts **43C**, P2.4012 (2019)

[4] H.P. Laqua, V. Erckmann, H.J. Hartfuß, H. Laqua, W7-AS Team, ECRH Group, Phys. Rev. Lett. **78**, 3467 (1997)

[5] H.P. Laqua, Plasma Phys. Control. Fusion **49**, R1 (2007)

[6] S.K. Hansen, S.K. Nielsen, J. Stober, J. Rasmussen, M. Stejner, M. Hoelzl, T. Jensen, the ASDEX Upgrade team, Nucl. Fusion **60**, 106008 (2020)

[7] M. Gryaznevich, A. Nicolai, V. Chuyanov, Tokamak Energy Ltd. Team, Probl. At. Sci. Technol. Ser. Thermonucl. Fusion **44**, 107 (2021)

[8] S.K. Nielsen, M.P. Gryaznevich, A.S. Jacobsen, T. Jensen, M. Jessen, S.B. Korsholm, J. Rasmussen, M. Salewski, M.G. Senstius, V. Naulin et al., Fusion Eng. Des. **166**, 112288 (2021)

[9] R. Wilhelm, V. Erckmann, G. Janzen, W. Kasparek, G. Müller, E. Räuchle, P.G. Schüller, K. Schwörer, M. Thumm, the W VII-A team, Plasma Phys. Control. Fusion **26**, 1433 (1984)

[10] D.G. Bulyginskii, V.V. D'yachenko, M.A. Irzak, M.M. Larionov, L.S. Levin, G.A. Serebrenyi, N.V. Shustova, Sov. J. Plasma Phys. **12**, 77 (1986)

[11] Z.A. Pietrzyk, A. Pochelon, R. Behn, A. Bondeson, M. Dutch, T.P. Goodman, M.Q. Tran, D.R. Whaley, Nucl. Fusion **33**, 197 (1993)

[12] ITER Organization, Tech. Rep. ITR-18-003, ITER Organization, St. Paul-lez-Durance (2018), `https://www.iter.org/technical-reports`

[13] M. Porkolab, B.I. Cohen, Nucl. Fusion **28**, 239 (1988)

[14] S. Hiroe, H. Ikegami, Phys. Rev. Lett. **19**, 1414 (1967)

[15] B. Grek, M. Porkolab, Phys. Rev. Lett. **30**, 836 (1973)

[16] M. Okabayashi, K. Chen, M. Porkolab, Phys. Rev. Lett. **31**, 1113 (1973)

[17] F.S. McDermott, G. Bekefi, K.E. Hackett, J.S. Levine, M. Porkolab, Phys. Fluids **25**, 1488 (1982)

[18] V. Shevchenko, G. Cunningham, A. Gurchenko, E. Gusakov, B. Lloyd, M. O'Brien, A. Saveliev, A. Surkov, F. Volpe, M. Walsh, Fusion Sci. Technol. **52**, 202 (2007)

[19] E.Z. Gusakov, A.V. Surkov, Plasma Phys. Control. Fusion **49**, 631 (2007)

[20] A. Köhn, G. Birkenmeier, A. Chusov, P. Diez, A. Feuer, U. Höfel, H. Höhnle, E. Holzhauer, W. Kasparek, S. Merli et al., Plasma Phys. Control. Fusion **55**, 014010 (2013)

[21] S. Kubo, M. Nishiura, K. Tanaka, D. Moseev, S. Ogasawara, T. Shimozuma, Y. Yoshimura, H. Igami, H. Takahashi, T.I. Tsujimura et al., J. Instrum. **11**, C06005 (2016)

[22] A.V. Arefiev, I.Y. Dodin, A. Köhn, E.J. Du Toit, E. Holzhauer, V.F. Shevchenko, R.G.L. Vann, Nucl. Fusion **57**, 116024 (2017)

[23] S.K. Nielsen, P.K. Michelsen, S.K. Hansen, S.B. Korsholm, F. Leipold, J. Rasmussen, M. Salewski, M. Schubert, M. Stejner, J. Stober et al., Phys. Scr. **92**, 024001 (2017)

[24] S.K. Hansen, S.K. Nielsen, M. Salewski, M. Stejner, J. Stober, the ASDEX Upgrade team, Plasma Phys. Control. Fusion **59**, 105006 (2017)

[25] S.K. Hansen, S.K. Nielsen, J. Stober, J. Rasmussen, M. Stejner, the ASDEX Upgrade team, EPJ Web Conf. **203**, 02007 (2019)

[26] S.K. Hansen, S.K. Nielsen, J. Stober, J. Rasmussen, M. Salewski, M. Stejner, ASDEX Upgrade Team, Phys. Plasmas **26**, 062102 (2019)

[27] M.G. Senstius, S.K. Nielsen, R.G.L. Vann, EPJ Web Conf. **203**, 01010 (2019)

[28] M.G. Senstius, S.K. Nielsen, R.G. Vann, S.K. Hansen, Plasma Phys. Control. Fusion **62**, 025010 (2020)

[29] M.G. Senstius, S.J. Freethy, J. Allen, S.K. Nielsen, Rev. Sci. Instrum. **93**, 103522 (2022)

[30] J. Han, Z. Gao, S.K. Hansen, *Kinetic theory of parametric decay instabilities near the upper hybrid resonance in plasmas* (2023), accepted by Phys. Plasmas

[31] A.D. Piliya, *Decay Instability in Weakly Inhomogeneous Plasma*, in *Proc. 10th Int. Conf. Phenomena in Ionized Gases (Oxford)*, edited by R.N. Franklin (Donald Parsons & Co. Ltd., Oxford, 1971), p. 320

[32] M.N. Rosenbluth, Phys. Rev. Lett. **29**, 565 (1972)

[33] A.D. Piliya, Sov. Phys.–JETP **37**, 629 (1973)

[34] S.K. Hansen, Ph.D. thesis, Technical University of Denmark, Kgs. Lyngby (2019), `https://pure.mpg.de/rest/items/item_3182239/component/file_3182252/content`

[35] V. Shevchenko, M.R. O'Brien, D. Taylor, A.N. Saveliev, MAST Team, Nucl. Fusion **50**, 022004 (2010)

[36] V.F. Shevchenko, Y.F. Baranov, T. Bigelow, J.B. Caughman, S. Diem, C. Dukes, P. Finburg, J. Hawes, C. Gurl, J. Griffiths et al., EPJ Web Conf. **87**, 02007 (2015)

[37] E. Westerhof, S.K. Nielsen, J.W. Oosterbeek, M. Salewski, M.R. De Baar, W.A. Bongers, A. Bürger, B.A. Hennen, S.B. Korsholm, F. Leipold et al., Phys. Rev. Lett. **103**, 125001 (2009)

[38] S.K. Nielsen, M. Salewski, E. Westerhof, W. Bongers, S.B. Korsholm, F. Leipold, J.W. Oosterbeek, D. Moseev, M. Stejner, the TEXTOR team, Plasma Phys. Control. Fusion **55**, 115003 (2013)

[39] E.Z. Gusakov, V.I. Fedorov, Sov. J. Plasma Phys. **5**, 463 (1979)

[40] A.Yu. Popov, E.Z. Gusakov, Plasma Phys. Control. Fusion **57**, 025022 (2015)

[41] E.Z. Gusakov, A.Yu. Popov, Phys. Plasmas **23**, 082503 (2016)

[42] E.Z. Gusakov, A.Yu. Popov, Nucl. Fusion **59**, 104003 (2019)

[43] E.Z. Gusakov, A.Yu. Popov, P.V. Tretinnikov, Nucl. Fusion **59**, 106040 (2019)

[44] E.Z. Gusakov, A.Yu. Popov, Plasma Phys. Control. Fusion **62**, 025028 (2020)

[45] E.Z. Gusakov, A.Yu. Popov, Phys. Plasmas **27**, 082502 (2020)

[46] E.Z. Gusakov, A.Yu. Popov, Plasma Phys. Control. Fusion **63**, 125017 (2021)

[47] E.Z. Gusakov, A.Yu. Popov, Plasma Phys. Rep. **48**, 327 (2022)

[48] M.G. Senstius, S.K. Nielsen, R.G.L. Vann, Phys. Plasmas **27**, 062102 (2020)

[49] M.G. Senstius, S.K. Nielsen, R.G.L. Vann, Plasma Phys. Control. Fusion **63**, 065018 (2021)

[50] M.G. Senstius, E.Z. Gusakov, A.Yu. Popov, S.K. Nielsen, Plasma Phys. Control. Fusion **64**, 115001 (2022)

[51] A. Tancetti, S.K. Nielsen, J. Rasmussen, E.Z. Gusakov, A.Yu. Popov, D. Moseev, T. Stange, M.G. Senstius, C. Killer, M. Vecséi et al., Nucl. Fusion **62**, 074003 (2022)

[52] A. Tancetti, S.K. Nielsen, J. Rasmussen, D. Moseev, T. Stange, S. Marsen, M. Vecséi, C. Killer, G.A. Wurden, T. Jensen et al., Plasma Phys. Control. Fusion **65**, 015001 (2023)

[53] E.Z. Gusakov, A.Yu. Popov, A.N. Saveliev, E.V. Sysoeva, Plasma Phys. Control. Fusion **59**, 075002 (2017)

[54] E.Z. Gusakov, A.Yu. Popov, Phys. Plasmas **25**, 012101 (2018)

[55] E.Z. Gusakov, A.Yu. Popov, A.N. Saveliev, Plasma Phys. Control. Fusion **61**, 025006 (2019)

[56] E.Z. Gusakov, A.Yu. Popov, Plasma Phys. Control. Fusion **63**, 015016 (2021)

[57] A. Bruschi, E. Alessi, W. Bin, O. D'Arcangelo, B. Baiocchi, F. Belli, G. Calabrò, I. Casiraghi, V. Cocilovo, L. Figini et al., Nucl. Fusion **57**, 076004 (2017)

[58] A. Bruschi, E. Alessi, B. Baiocchi, W. Bin, O. D'Arcangelo, F. Fanale, L. Figini, C. Galperti, S. Garavaglia, G. Gittini et al., EPJ Web Conf. **203**, 02005 (2019)

[59] B. Baiocchi, W. Bin, A. Bruschi, L. Figini, U. Tartari, E. Alessi, O. D'Arcangelo, J. Instrum. **15**, C01046 (2020)

[60] E.Z. Gusakov, A.Yu. Popov, Phys. Rev. Lett. **128**, 065001 (2022)

[61] M. Bornatici, R. Cano, O. De Barbieri, F. Engelmann, Nucl. Fusion **23**, 1153 (1983)

[62] S.K. Hansen, S.K. Nielsen, J. Stober, the EUROfusion MST1 team, the ASDEX Upgrade team, *Relativistic analysis of upper hybrid wave propagation and trapping* (2023), submitted to Phys. Plasmas

[63] M. Greenwald, Plasma Phys. Control. Fusion **44**, R27 (2002)

[64] P.T. Lang, C. Angioni, M. Bernert, A. Bock, T.M.J. Engelhardt, R. Fischer, O.J.W.F. Kardaun, B. Ploeckl, M. Prechtl, W. Suttrop et al., Nucl. Fusion **60**, 092003 (2020)

[65] M. Giacomin, A. Pau, P. Ricci, O. Sauter, T. Eich, the ASDEX Upgrade team, JET Contributors, the TCV team, Phys. Rev. Lett. **128**, 185003 (2022)

[66] Y.R. Martin, T. Takizuka, ITPA CDBM H-mode Threshold Database Working Group, J. Phys.: Conf. Ser. **123**, 012033 (2008)

[67] V.S. Udintsev, S. Danani, G. Taylor, T. Giacomin, J. Guirao, S. Pak, S. Hughes, L. Worth, G. Vayakis, M.J. Walsh et al., EPJ Web Conf. **203**, 03003 (2019)

[68] Y. Liu, V.S. Udintsev, S. Danani, G. Paraiso, G. Taylor, M.E. Austin, A. Basile, J.H. Beno, B. Bunkowski, R. Feder et al., J. Instrum. **17**, C04019 (2022)

[69] C.M. Muscatello, C. Anderson, J. Anderson, A. Basile, R.L. Boivin, M. Duco, D.K. Finkenthal, A. Gattuso, J. Klabacha, G.J. Kramer et al., Nucl. Fusion **60**, 066005 (2020)

# Paraxial beams in fluctuating fusion plasmas: diffusive limit and beyond

*Hannes* Weber[1], *Omar* Maj[1], and *Emanuele* Poli[1,*]

[1]Max-Planck-Institut für Plasmaphysik, Boltzmannstr. 2, 85748 Garching bei München, Germany

**Abstract.** A paraxial expansion of the (ensemble-averaged) Wigner function in the relevant wave kinetic equation for electron cyclotron waves in fluctuating plasmas allows the derivation of phase-space equations similar to the equations for the Gaussian beam parameters in the paraxial WKB method [G.V. Pereverzev, Phys. Plasmas **5**, 3529 (1998)]. This is relatively straightforward when the scattering of the wave field by density fluctuations can be described by a diffusion operator in refractive-index space. The general case is rather more complicated, yet we could find a heuristic construction of a paraxial Wigner function. Here we use a simple model, which has an analytical solution, to test both the theoretical validity of the diffusion approximation and the heuristic paraxial approach beyond the diffusion approximation.

## 1 Introduction

The wave kinetic equation (WKE) provides a suitable tool to tackle the problem of the scattering of short-wavelength waves from density fluctuations in fusion plasmas [1, 2]. In this approach, the typical correlation length of the fluctuations is not required to be larger than the radiation wavelength. Just the spatial variation of the correlation length and of the root-mean-square density fluctuation is required to satisfy the WKB limit, and this is usually the case, since their magnitude is related to the equilibrium profiles. The solution of the WKE gives the lowest-order approximation of the statistically averaged Wigner function of the wave field in the short-wavelength limit, where the statistical average is taken over an ensemble of density fluctuations. Density fluctuations, typically generated by plasma turbulence, are predicted to lead to a significant beam broadening particularly in large machines, in which the beam path across the plasma can be much longer than in present-day fusion experiments [3, 4]. A direct numerical solution of the WKE for electron cyclotron (EC) waves in tokamak geometry has been implemented in the WKBeam code [5]. Comparisons with both full-wave codes [6] and experimental data [7] supported the applicability of the underlying scattering model (based on the Born approximation) to typical fusion-plasma parameters. The numerical approach implemented in WKBeam, although much more efficient than a direct solution of the wave problem (if this can be afforded at all), still leads to a much heavier computational burden as compared to beam-tracing codes like TORBEAM [8] or GRAY [9], in which, however, the effect of density fluctuations is not included.

A possibility to reduce the computational cost of the solution of the WKE may still come from a generalization of the paraxial approach used in TORBEAM to the averaged Wigner function. In fact, from the results of the WKBeam code, we observe a strong localization of the averaged Wigner function around a trajectory in phase space. In previous work [10], we showed how such a paraxial approximation of the Wigner function can be constructed in a simplified model. The result is a set of ordinary differential equations (ODEs), called phase-space beam-tracing equations. If the diffusion approximation for the wave scattering holds, the derivation of such phase-space beam-tracing equations is relatively straightforward (at least in this simplified model). In the general case, however, a different strategy for the evaluation of the scattering effects must be considered.

In this paper, we discuss the validity of the diffusive limit on the basis of an analytical solution of the WKE. We also use the analytical solution in order to assess the proposed paraxial approach in a particular scattering scenario, namely that of a thin fluctuation layer. In this context, "thin" means that the variation of the beam parameters due to refraction and diffraction can be neglected within the layer, while outside the layer the propagation occurs in the limit of vanishing fluctuations. This setup is similar to the phase-screen model employed in trans-ionospheric communication [11].

## 2 Model problem and analytical solution

The Wigner function description of EC wave beams is formulated in spatial coordinates $x$, normalized to the scale $L$ of variation of the background medium. Here, the scale length $L$ is defined with the equilibrium quantities, that are assumed independent of time, and therefore does not capture the scale of fluctuations, which can be arbitrary. Further, we have the refractive-index vector $N = ck/\omega$, where $c$ is the speed of light in free space, and $\omega$ the angular frequency of the wave field. The statistically averaged Wigner matrix of the wave electric field [1] is defined on the position-refractive-index phase space with coordinates

*e-mail: emanuele.poli@ipp.mpg.de

$(x, N)$ and depends on the parameter $\kappa := \omega L/c$. Since $L$ is the scale length of the equilibrium, for EC waves in plasmas we have $\kappa \gg 1$. In the limit $\kappa \to +\infty$, each eigenvalue $W$ of the leading-order term of Wigner matrix satisfies the constrained transport equation [1, 5]

$$\begin{cases} \{H, W\} = -2\gamma W + S(W), \\ HW = 0, \end{cases} \tag{1}$$

where $\{F, G\} = \nabla_N F \cdot \nabla_x G - \nabla_x F \cdot \nabla_N G$ denotes the canonical Poisson bracket of two smooth functions $F, G$ in the $(x, N)$ phase space, $H$ is the geometrical optics Hamiltonian, hence $H = 0$ gives the dispersion relation of the wave mode associated to $W$, $\gamma$ is the absorption coefficient, and

$$\begin{aligned} S(W) = \Big(\frac{\kappa}{2\pi}\Big)^d \int \eta^2(x)\Gamma(\kappa, x, N - N') \\ \times \big[2\pi\delta(H(x, N))W(\kappa, x, N') \\ - 2\pi\delta(H(x, N'))W(\kappa, x, N)\big]\mathrm{d}N', \end{aligned} \tag{2}$$

with $\eta$ a given function depending on the mode polarization (the precise definition [5] is not needed here), $\delta$ is the Dirac's distribution, and

$$\Gamma(\kappa, x, N) = \int e^{-i\kappa N\cdot s}\mathbb{E}(\delta n_e(x + \tfrac{s}{2})\delta n_e(x - \tfrac{s}{2}))\mathrm{d}s, \tag{3}$$

where $n_e(x) = n_{e,0}(x) + \delta n_e(x)/\sqrt{\kappa}$ is the total electron density split into the equilibrium part $n_{e,0}$, and small random fluctuations $\delta n_e/\sqrt{\kappa}$. The reasons for the scaling $1/\sqrt{\kappa}$ are discussed in earlier work [1, 5]. The expectation-value operator $\mathbb{E}$ is the average over an ensemble of fluctuations $\delta n_e$. Equation (1) complemented by appropriate boundary condition at the launching mirror is the WKE solved by the WKBeam code.

The simplified test problem addressed in this paper is obtained by choosing a two-dimensional domain ($d = 2$) with coordinates $x = (\mathrm{x}, \mathrm{y}) = (\tau, \mathrm{y})$, $N = (N_\mathrm{x}, N_\mathrm{y})$, and

$$\begin{aligned} H(x, N) = N_\mathrm{x}^2 + N_\mathrm{y}^2 - 1, \quad \gamma = 0, \quad \eta = 1, \\ \Gamma(\kappa, x, N) = \Gamma_0(\tau, N_\mathrm{y}) = 2\pi L_C^2 F^2 e^{-\frac{1}{2}\kappa^2 L_C^2 N_\mathrm{y}^2}, \end{aligned} \tag{4}$$

where $F = F(\tau) = \mathbb{E}(\delta n_e^2)^{1/2}/n_{e,0}$ is the root-mean-square of density fluctuations relative to the equilibrium density, and $L_C$ is the correlation length, here assumed to be constant. The dependence on $\kappa$ is understood in the rest of the paper.

Upon using $\delta(H) = (2N_{\mathrm{x},0})^{-1}\delta(N_\mathrm{x} - N_{\mathrm{x},0})$, and the approximation $N_{\mathrm{x},0} = (1 - N_\mathrm{y}^2)^{1/2} = 1 + O(N_\mathrm{y}^2)$ for small $N_\mathrm{y}$, we obtain $W = (1/2)\upsilon(\tau, \mathrm{y}, N_\mathrm{y})\delta(N_\mathrm{x} - N_{\mathrm{x},0})$ where $\upsilon(\tau, \cdot, \cdot)$ is a distribution defined on the reduced $(\mathrm{y}, N_\mathrm{y})$ phase space and satisfies the linear Boltzmann equation

$$\partial_\tau \upsilon + N_\mathrm{y}\partial_\mathrm{y}\upsilon = S_0(\upsilon), \tag{5}$$

with

$$\begin{aligned} S_0(\upsilon) = \frac{\kappa^2}{8\pi}\int \Gamma_0(\tau, N_\mathrm{y} - N_\mathrm{y}') \\ \times [\upsilon(\tau, \mathrm{y}, N_\mathrm{y}') - \upsilon(\tau, \mathrm{y}, N_\mathrm{y})]\mathrm{d}N_\mathrm{y}. \end{aligned} \tag{6}$$

One should note that simplification (4) is somewhat unphysical: On the one hand, one assumes that the frequency is so large that propagation can be described with the vacuum Hamiltonian; on the other hand, fluctuations are retained. Therefore model (4) should be regarded as a computational test. Nonetheless, such simplified model captures the essential parts of the WKE: Indeed, equation (5) is essentially the same as the more realistic model studied by Chellaï et al. [7], the only difference being simpler coefficients.

Particularly, we can obtain an analytical solution of (5) on the line of the derivation sketched in the appendix of the paper by Chellaï et al. [7]. This analytical solution is then written in Fourier space, that is,

$$\begin{aligned} \hat{\upsilon}(\tau, p, q) &= \int e^{-i(p\mathrm{y}+qN_\mathrm{y})}\upsilon(\tau, \mathrm{y}, N_\mathrm{y})\mathrm{d}\mathrm{y}\mathrm{d}N_\mathrm{y}, \\ &= e^{\int_0^\tau \varphi_0(\tau', q+p(\tau-\tau'))\mathrm{d}\tau'}\hat{\upsilon}_0(p, q + p\tau), \end{aligned} \tag{7}$$

where

$$\varphi_0(\tau, q) = \sqrt{\frac{\pi}{8}}\kappa L_C F^2\Big[e^{-q^2/(2\kappa^2 L_C^2)} - 1\Big], \tag{8}$$

and

$$\hat{\upsilon}_0(p, q) = \frac{2\pi}{k_0}\sqrt{\frac{\pi}{2}}w_0|A_0|^2 e^{-\frac{w_0^2}{8}(p+\frac{q}{R_0})^2 - v_0^2 q^2/2}. \tag{9}$$

is the Fourier transform of the initial condition, with $A_0$ being the initial electric-field amplitude, $w_0$ the initial Gaussian beam width, and $R_0$ is the initial phase-front radius of curvature. The energy densities in both real and refractive-index space are given by

$$\mathcal{E}_x(\tau, \mathrm{y}) = \frac{\kappa}{2\pi}\int \upsilon(\tau, \mathrm{y}, N_\mathrm{y})\mathrm{d}N_\mathrm{y},$$

and

$$\mathcal{E}_N(\tau, N_\mathrm{y}) = \int \upsilon(\tau, \mathrm{y}, N_\mathrm{y})\mathrm{d}\mathrm{y},$$

respectively, and their calculation requires the function $\upsilon$ in the $(\mathrm{y}, N_\mathrm{y})$-space. However, their moments can be evaluated directly from $\hat{\upsilon}$ [7]. In particular, from the second- and fourth-order moments we can compute the width and the kurtosis [12] of both distributions. The skewness is identically zero for both distributions. Upon defining the quantities

$$\mathbb{F}_n(\tau) := \frac{\sqrt{2\pi}}{L_C}\int_0^\tau F^2(\tau')(\tau - \tau')^n\mathrm{d}\tau',$$

for any integer $n \geq 0$, the results can be written as

$$\begin{aligned} &w_x^2(\tau) = w_{\mathrm{GB}}^2(\tau) + \mathbb{F}_2(\tau), \quad w_N^2(\tau) = (w_0^2/L_R^2) + \mathbb{F}_0(\tau), \\ &\mathrm{kurt}\,(\mathcal{E}_x(\tau)) = 3 + \frac{12}{k_0^2 L_C^2}\frac{\mathbb{F}_4(\tau)}{(w_{\mathrm{GB}}(\tau)^2 + \mathbb{F}_2(\tau))^2}, \\ &\mathrm{kurt}\,(\mathcal{E}_N(\tau)) = 3 + \frac{12}{k_0^2 L_C^2}\frac{\mathbb{F}_0(\tau)}{(w_0^2/L_R^2 + \mathbb{F}_0(\tau))^2}, \end{aligned} \tag{10}$$

where $w_x$ and $w_N$ are the widths in real and refractive-index space, respectively, whereas $\mathrm{kurt}(\rho)$ denotes the kurtosis of the density $\rho$. We recall that any value in excess of 3 indicates that the distribution decays to zero more slowly than a Gaussian, and thus have "fat tails". In addition, $w_{\mathrm{GB}}(\tau)^2 = w_0^2[1+\frac{2\tau}{R_0}+\frac{\tau^2}{L_R^2}]$ and $w_0^2/L_R^2$ are the squared beam width and spectral width of a standard Gaussian beam, with $1/L_R^2 = 1/R_0^2 + 4/(\kappa^2 w_0^4)$.

## 3 Diffusive limit and diffusive scattering

For propagation distances longer than the transport mean-free path [13, section 5.1 and references therein] the energy density $\mathcal{E}_x$ in *physical space* can be described by a diffusion equation. This regime is referred to as diffusive limit of the linear Boltzmann equation (5). A validity criterion for the diffusive regime has been discussed in [4] and can be re-derived here on the basis of kurt($\mathcal{E}_x$). For instance, with a piecewise-constant profile $F^2(\tau) = F^2_{\max} > 0$ for $0 \le \tau \le \tau_F$ and $F^2(\tau) = 0$ otherwise, from (10) one finds that as $\tau \to +\infty$,

$$\mathrm{kurt}(\mathcal{E}_x(\tau)) \sim \mathrm{kurt}(\mathcal{E}_N(\tau)) \sim 3 + \frac{6}{\lambda}\frac{(\xi\lambda/2)^2}{(1+\xi\lambda/2)^2} \tag{11}$$

where

$$\lambda = \sqrt{\pi/2}F^2_{\max}(k_0 L_C)(k_0\tau_F), \tag{12}$$

is the parameter introduced by Snicker et al. [4] and

$$\xi = (4L_R^2)/(\kappa^2 w_0^2 L_C^2). \tag{13}$$

It is remarkable that the kurtosis of both $\mathcal{E}_x$ and $\mathcal{E}_N$ have the same asymptotic behavior. We deduce that both distribution observed at large $\tau$, i.e., well after propagation through the layer of fluctuations, have negligible excess kurtosis if either $\lambda \gg 1$, in agreement with the criterion established earlier [4], or $\xi \ll 1$.

We shall now show that the parameter $\xi$ controls the regime in which the scattering operator (6) can be approximated by a diffusion operator acting in *refractive-index space* only. This should not be confused with the diffusive limit [13] in which one finds a diffusion operator in position space. It is also somewhat stronger than the condition of negligible kurtosis, since a diffusion operator preserves the Gaussian form of the initial condition. Formally this "diffusive scattering approximation" is obtained by a Taylor expansion of the solution,

$$S_0(v) = \frac{\kappa^2}{16\pi}\Big[\int (N_y')^2\Gamma_0(\tau,\tau,N_y')\mathrm{d}N_y'\Big]\partial^2_{N_y}v + S_r(v),$$

where $S_r(v)$ involves the fourth-order remainder term in Taylor's formula for $v$ (odd-order terms vanish due to symmetry of the integrand). Neglecting $S_r$ in equation (5) leads to

$$\partial_\tau v + N_y\partial_y v = \mathcal{D}(\kappa,\tau)\partial^2_{N_y}v(\kappa,\tau,y,N_y), \tag{14}$$

with refractive-index diffusion coefficient

$$\mathcal{D} = \frac{\sqrt{2\pi}}{8}\frac{F^2(\tau)}{\kappa L_C}. \tag{15}$$

The validity of (14) can be established on noting that the Taylor's expansion of $\varphi_0$ for $q^2 \ll 2\kappa^2 L_C^2$ gives, cf. (8),

$$\varphi_0(\tau,q) \approx -\mathcal{D}(\kappa,\tau)q^2, \quad (q^2 \ll 2\kappa^2 L_C^2),$$

and the exact solution (7) of equation (5) reduces to the exact solution of the diffusive-scattering approximation (14). Therefore, we expect that (14) yields a good approximation of the solution of (5) if the initial condition $\hat{v}_0$ is concentrated in a region of its domain where $q^2 \ll 2\kappa^2 L_C^2$. For a Gaussian beam, the width in $q$ of the initial condition (9) is given by $8L_R^2/w_0^2$ and thus the condition amounts to $\xi \ll 1$ as anticipated. In the special case of flat initial phase front $1/R_0 \to 0$, we have $L_R = \kappa w_0^2/2$ and this reduces to $\xi = w_0^2/L_C^2 \ll 1$. We note that in realistic application, typically $\xi > 1$ and the full scattering operator (6) should be used.

## 4 Paraxial solution of the WKE

The Fokker-Planck equation (14), which holds for $\xi \ll 1$, admits exact solutions of the form of a Gaussian distribution

$$v(\kappa,\tau,y,N_y) = c(\tau)e^{-\kappa G^{rr}(\tau)y^2 - 2\kappa G^{rN}(\tau)yN_y - \kappa G^{NN}(\tau)N_y^2}, \tag{16}$$

where the amplitude $c$ and the Gaussian parameters $G^{rr}$, $G^{rN}$ and $G^{NN}$ at the launch position can be obtained in terms of electric field amplitude, beam width and focusing, cf. (30) of [14]. Substitution into (14) gives ordinary differential equations for the parameters of the Gaussian, that is, [14]

$$\frac{\mathrm{d}c}{\mathrm{d}\tau} = -2\kappa\mathcal{D}G^{NN}c, \tag{17a}$$

$$\frac{\mathrm{d}G^{rr}}{\mathrm{d}\tau} = -4\kappa\mathcal{D}(G^{rN})^2, \tag{17b}$$

$$\frac{\mathrm{d}G^{rN}}{\mathrm{d}\tau} = -G^{rr} - 4\kappa\mathcal{D}G^{rN}G^{NN}, \tag{17c}$$

$$\frac{\mathrm{d}G^{NN}}{\mathrm{d}\tau} = -2G^{rN} - 4\kappa\mathcal{D}(G^{NN})^2. \tag{17d}$$

The general form (5) of the WKE is more complicated since the operator $S_0$ does not preserve the Gaussian form of the function $v$. However, equation (5) can be written as an abstract evolution equation of the form [14]

$$\frac{\mathrm{d}v}{\mathrm{d}\tau} = Av + Bv, \tag{18}$$

where $Av = -N_y\partial_y v$ and $Bv = S_0(v)$. A time-discretized solution in terms of the Lie-Trotter operator splitting method [15] is proposed [14, 16]. If $0 = \tau_0 < \tau_1 < \cdots < \tau_k < \cdots$ is a sequence of points in $\tau$, the exact solution at $\tau = \tau_k$ formally can be written as

$$v(\tau_k) = \Phi^{A+B}(\tau_k,\tau_{k-1})v(\tau_{k-1}), \tag{19}$$

with the exact evolution operator $\Phi^{A+B}(\tau_k,\tau_{k-1})$ accounting for the effect of both operators $A$ and $B$ during the time interval $\Delta\tau_k = \tau_k - \tau_{k-1}$ and $v(\tau_{k-1})$ the solution before this time step. After Fourier transform, $\Phi^{A+B}$ can be constructed in the same way as the exact solution (7). Analogously, let $\Phi^A$ and $\Phi^B$ be the evolution operators for the two sub-problems

$$\frac{\mathrm{d}v}{\mathrm{d}\tau} = Av, \quad \frac{\mathrm{d}v}{\mathrm{d}\tau} = Bv, \tag{20}$$

respectively. A first-order approximation of $\Phi^{A+B}(\tau_k, \tau_{k-1})$ is then obtained by the Lie-Trotter formula

$$\Phi^{A+B}(\tau_k, \tau_{k-1}) = \Phi^A(\tau_k, \tau_{k-1})\Phi^B(\tau_k, \tau_{k-1}) + O(h_\tau^2),$$

where $h_\tau = \max_k \Delta\tau_k$ is the maximum step in $\tau$.

For solutions of the form (16) the evolution operator $\Phi^A$ is exactly constructed by solving equations (17) with $\mathcal{D} = 0$ for the parameters of the Gaussian. For the evolution operator $\Phi^B$ associated to the full scattering operator, we apply the explicit Euler scheme which gives $\Phi^B(\tau_k, \tau_{k-1})v(\tau_{k-1}) = \Phi^B_{\mathrm{Euler}}(\tau_k, \tau_{k-1})v(\tau_{k-1}) + O(h_\tau^2)$ where

$$\begin{aligned}\Phi^B_{\mathrm{Euler}}(\tau_k, \tau_{k-1})v(\tau_{k-1}) &= (1 - \Delta\tau_k\Sigma(\tau_{k-1}))v(\tau_{k-1}) \\ &+ \frac{\kappa^2\Delta\tau_k}{8\pi}\int \Gamma_0(\tau_{k-1}, N_y - N'_y)v(\tau_{k-1}, y, N'_y)\mathrm{d}N'_y, \end{aligned} \tag{21}$$

with

$$\Sigma(\tau) = \frac{\kappa^2}{8\pi}\int \Gamma_0(\tau, N_y)\mathrm{d}N_y = \sqrt{\frac{\pi}{8}}\kappa L_C F^2(\tau).$$

When applied to a Gaussian, the result of the action of this operator is an amplitude decay of the original Gaussian function (first term) plus the convolution of the (Gaussian) fluctuation spectrum with the Gaussian initial Wigner function. The result of such computation, again, has the Gaussian shape as in (16) with the parameters $(c, G^{rr}, G^{rN}, G^{NN})$ to be determined by evaluation of the convolution integral.

As a result of the procedure described above, starting from a Gaussian Wigner function after one time step we have the sum of two Gaussian functions. Owing to the linearity of equation (21) the strategy can be applied on a superposition of Gaussians as well, doubling the number of such Gaussian components of the beam after each time step. Thus, at time $\tau_k$ the obtained paraxial approximation $v_{\mathrm{Euler}}$ of the solution can be written as superposition

$$v_{\mathrm{Euler}}(\tau_k, y, N_y) = \sum_{i=0}^{m} v_i(\tau_k, y, N_y), \tag{22}$$

with an exponentially growing number of scattered contributions $m = 2^k$ (starting with one mode at $\tau_0$). Here, each Gaussian mode $v_i$ is described by ansatz (16), with a set of parameters $(c_i, G_i)$. The development of these modes is illustrated in figure 1.

However, the fast growth of the number of Gaussian modes might require a large numerical effort when the solution is constructed in practice and makes it impossible to provide an analytical interpretation of the results. On the other hand, many effects may be retained when a simplifying model is considered. Here, we introduce the model of a thin fluctuation layer, described by the piecewise-constant profile introduced above, with $F^2_{\max} = D/\tau_F$. The overall effect of fluctuations for $\tau > \tau_F$ is determined by the integral of the fluctuation amplitude function $\int_0^{\tau_F} F^2(\tau)\mathrm{d}\tau = D$ and, thus, in this model measured by the parameter $D$. We consider layers so thin that the variation of the parameters in each Gaussian component $v_i$ of (22) due to the operator $\Phi^A$ can be neglected, hence $\Phi^A(\tau_k, \tau_{k-1}) \approx 1$. Within the layer, this simplification allows us to replace the paraxial solution $v_{\mathrm{Euler}}$ with an adapted simpler approximation given by a linear combinations of functions that are constructed by a repeated application of the convolution operator in (6). This construction of the functions is reminiscent of the Krylov subspace method. The first element is chosen to be the initial Gaussian launched at $\tau = 0$ with parameters

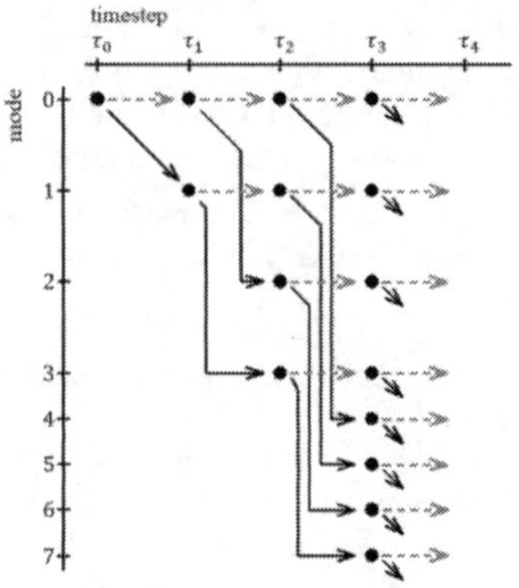


**Figure 1.** Propagation of Gaussian modes: Bullets symbolize propagating Gaussian modes. Red dashed arrows pointing to the right indicate mode propagation due to operator A, black arrows indicate the effect of the scattering operator.

$$G_0^{rr} = \Phi, \quad G_0^{rN} = 0, \quad G_0^{NN} = \Phi^{-1}, \tag{23}$$

where $\Phi = 2/\kappa w_0^2$ computed from the initial beam width $w_0$ and the phase front is flat $(1/R_0 \to 0)$ for simplicity. Then the functions $\psi_\alpha = \psi_\alpha(y, N_y)$ for $\alpha = 0, 1, \ldots$ are defined recursively by

$$\begin{aligned}\psi_0 &= e^{-\kappa G_0^{rN} y^2 - \kappa G_0^{NN} N_y^2}, & \alpha &= 0, \\ \psi_\alpha &= K(\psi_{\alpha-1})/\gamma_{\alpha-1}, & \alpha &\geq 1,\end{aligned} \tag{24}$$

with constant $\gamma_\alpha$ determined by the normalization condition $\psi_\alpha(0, 0) = 1$ and with $K$ being the convolution in $N_y$ by the Gaussian $\exp[-\kappa^2 L_C^2 N_y^2/2]$ so that

$$B\psi = (\kappa^2 L_C^2 F^2/4)K(\psi) - \Sigma\psi.$$

Explicitly, (24) yields

$$\psi_\alpha(y, N_y) = e^{-\kappa G_\alpha^{rN} y^2 - \kappa G_\alpha^{NN} N_y^2},$$

with parameters defined recursively by

$$G_\alpha^{rr} = G_{\alpha-1}^{rr} = G_0^{rr}, \quad G_\alpha^{rN} = G_{\alpha-1}^{rN} = 0,$$

$$G_\alpha^{NN} = G_{\alpha-1}^{NN}\Big(1 + \frac{2G_{\alpha-1}}{\kappa L_C^2}\Big)^{-\frac{1}{2}}, \quad \gamma_\alpha = \frac{\sqrt{2\pi}}{\kappa L_C}\Big(1 + \frac{2G_\alpha}{\kappa L_C^2}\Big)^{-\frac{1}{2}}.$$

The recurrence relation for $G_\alpha^{NN}$, in particular, leads to the closed equation

$$G_\alpha^{NN} = G_0^{NN}\Big(1 + 2\alpha\frac{G_0^{NN}}{\kappa L_C^2}\Big)^{-1}, \qquad (25)$$

which can be proven by induction.

The solution within the layer is then approximated by a linear combination of such Gaussian functions,

$$v^{(m)}(\tau, y, N_y) = \sum_{\alpha=0}^{m} C_\alpha(\tau)\psi_\alpha(y, N_y), \quad 0 \le \tau < \tau_F, \quad (26)$$

where $C_\alpha = C_\alpha(\tau)$ are coefficients to be determined. Outside the layer, each Gaussian component is propagated with the exact evolution operator $\Phi^A(\tau, \tau_F)$, i.e., by solving (17) with $\mathcal{D} = 0$.

The subspace spanned by a finite number of $\psi_\alpha$ is not closed under the action of the operator $K$, or equivalently $B$. Therefore a truncation is applied, that is,

$$\begin{aligned} Bv^{(m)} &= \tfrac{1}{4}\kappa^2 L_C^2 F^2 K(v^{(m)}) - \Sigma v^{(m)} \\ &= \tfrac{1}{4}\kappa^2 L_C^2 F^2 \sum_{\alpha=0}^{m} \gamma_\alpha C_\alpha \psi_{\alpha+1} - \sum_{\alpha=0}^{m} \Sigma C_\alpha \psi_\alpha \\ &\approx \tfrac{1}{4}\kappa^2 L_C^2 F^2 \sum_{\alpha=1}^{m} \gamma_{\alpha-1} C_{\alpha-1}\psi_\alpha - \sum_{\alpha=0}^{m} \Sigma C_\alpha \psi_\alpha, \end{aligned}$$

where the term containing $\psi_{m+1}$ has been dropped. The error of this approximation depends on how fast the coefficients $C_\alpha$ decay to zero for $\alpha \to +\infty$.

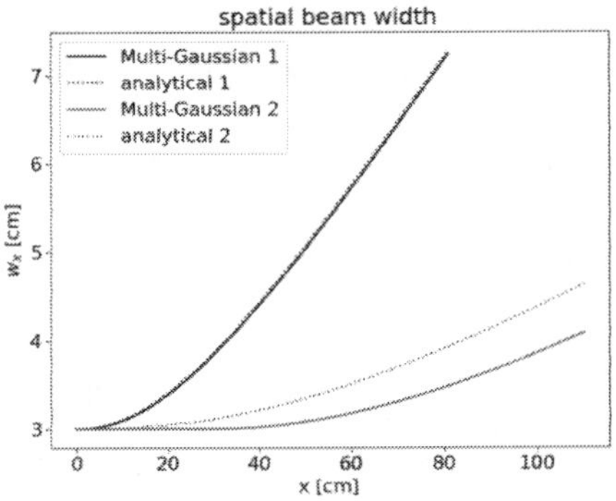


**Figure 2.** Evolution of the spatial width $w_x$ of the beam along the propagation axis $x = \tau$. Comparison of the multi-Gaussian solution (26) and the analytical reference (10) for two sets of fluctuation parameters. Beam parameters: initial width $w_x(0) = 3$ cm, initial curvature radius $1/R_0 \to 0$, beam frequency $f = 140$ GHz. First (second) set of fluctuation parameters: fluctuation layer width $\tau_F = 1$ cm (30 cm), fluctuation strength $F = 0.05$ (0.003), correlation length $L_C = 1$ cm (1 cm).

The substitution of (26) into the equation for $\Phi^B$ in (20) together with the truncated form of $Bv^{(m)}$ leads to

$$\frac{dC_\alpha}{d\tau} = \sigma_{\alpha-1}C_{\alpha-1}(\tau) - \Sigma C_\alpha(\tau), \qquad (27)$$

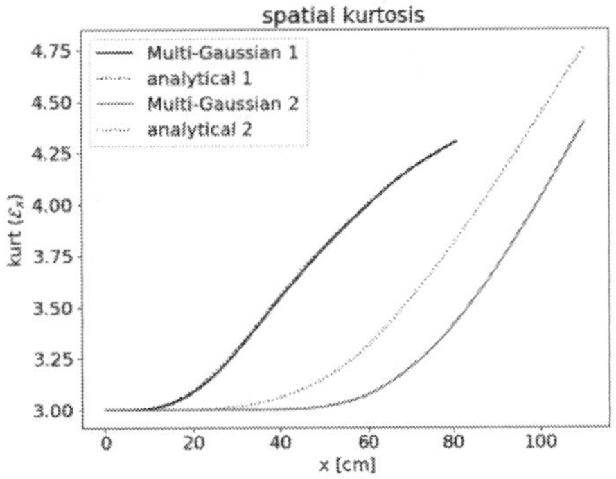


**Figure 3.** Evolution of the spatial kurtosis $\mathrm{kurt}(\mathcal{E}_x)$ of the beam along the propagation axis $x = \tau$. Comparison of the multi-Gaussian solution (26) and the analytical reference for two sets of fluctuation parameters. Parameters cf. fig. 2.

with the coefficients

$$\sigma_\alpha = \frac{1}{4}\kappa^2 L_C^2 \gamma_\alpha = \Sigma \cdot \Big(1 + 2\frac{G_\alpha^{NN}}{\kappa L_C^2}\Big)^{-\frac{1}{2}}, \qquad (28)$$

and $\sigma_{-1} = 0$ by definition (no contribution to the non-scattered Gaussian function). Equation (27) can be readily integrated with the result that

$$C_\alpha(\tau) = \frac{\tau^\alpha}{\alpha!}\prod_{\mu=0}^{\alpha-1}\big(\sigma_\mu\big)e^{-\Sigma\tau}C_0(0). \qquad (29)$$

When (25), (28), and the definition of $D$ are inserted, again by induction one can show that

$$C_\alpha(\tau_F) = \frac{\Big(\sqrt{\frac{\pi}{8}}\kappa L_C D\Big)^\alpha}{\alpha!}e^{-\sqrt{\frac{\pi}{8}}\kappa L_C D}\sqrt{\frac{G_\alpha^{NN}}{G_0^{NN}}}C_0(0). \qquad (30)$$

The projection onto the configuration space of the Wigner function (26) can be computed from (30), with the result

$$\mathcal{E}_x(\tau_F, y) = \sqrt{\frac{\kappa}{4\pi}}\sum_{\alpha=0}^{m}\frac{C_\alpha(\tau_F)}{\sqrt{G_\alpha^{NN}}}e^{-\kappa G_0^{rr}y^2}. \qquad (31)$$

After inserting the amplitudes $C_\alpha(\tau_F)$ and letting $m \to +\infty$ the sum becomes a series which can be computed analytically, with the result

$$\mathcal{E}_x(\tau_F, y) = \sqrt{\frac{\kappa}{4\pi G_0^{NN}}}C_0(0)e^{-\kappa G_0^{rr}y^2}, \qquad (32)$$

which is equivalent to the initial beam cross section in configuration space $\mathcal{E}_x(0,\cdot)$. The conservation of $\mathcal{E}_x$ is a property of the exact evolution operator $\Phi^B$, since $\int S_0(v)dN_y = 0$, and we find that it is preserved by the approximated solution (26) in the limit $m \to +\infty$.

As for the spectrum $\mathcal{E}_N$, with the approximated solution $v^{(m)}$ and in the limit $m \to +\infty$, we obtain

$$\mathcal{E}_N(\tau_F, N_y) = \sqrt{\frac{\pi}{\kappa\Phi}}e^{-\lambda/2}\sum_{\alpha=0}^{\infty}\frac{(\lambda/2)^\alpha}{\alpha!}\frac{C_0(0)}{\sqrt{1+\alpha\xi}}e^{-\kappa G_\alpha^{NN}N_y^2}, \qquad (33)$$

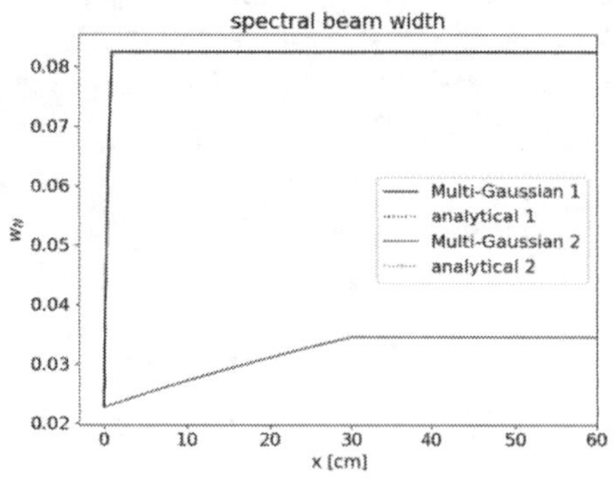


**Figure 4.** Evolution of the spectral width $w_N$ of the beam along the propagation axis x = $\tau$. Comparison of the multi-Gaussian solution (26) and the analytical reference (10) for two sets of fluctuation parameters. Parameters cf. fig. 2.

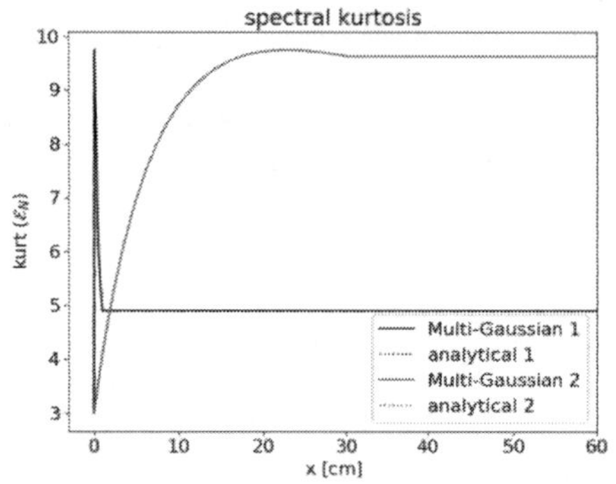


**Figure 5.** Evolution of the spectral kurtosis kurt($\mathcal{E}_N$) of the beam along the propagation axis x = $\tau$. Comparison of the multi-Gaussian solution (26) and the analytical reference (10) for two sets of fluctuation parameters. Parameters cf. fig. 2.

where the parameters $\lambda$ and $\xi$ are defined in section 3, cf. equations (12) and (13) with $L_R = \kappa^2 w_0^2/2$ for an initially flat phase-front, and $G_\alpha^{NN} = \Phi^{-1}/(1+\alpha\xi)$. This result reflects the fact that the scattering operator $\Phi^B$ has a direct impact on the spectrum of the beam only, leaving its spatial structure unaffected. In turn, application of $\Phi^A$ leads to a spatial evolution of the beam which preserves its spectrum. The spectrum $\mathcal{E}_N$ computed with $v^{(m)}$, equation (33), becomes a superposition of Gaussian functions of different widths. With every application of the convolution part of the scattering operator, $G_\alpha^{NN}$ decreases leading to eventually broad tails in the spectrum arising from higher $\alpha$ modes. We can compute the kurtosis of the distribution (32). The even-order moments are

$$\int N_y^{2n}\mathcal{E}_N(\tau_F, N_y)\mathrm{d}N_y = \frac{\pi}{\kappa}\Phi^n\frac{(2n-1)!!}{(2\kappa)^n}C_0(0) \times e^{-\lambda/2}\sum_{\alpha=0}^{\infty}\frac{(\lambda/2)^\alpha}{\alpha!}(1+\alpha\xi)^n .$$

The remaining sums can be traced back to the Taylor series of the function $f(\lambda) = e^{\lambda/2}$ and evaluated analytically, with the result

$$\mathrm{kurt}\,(\mathcal{E}_N(\tau_F)) = 3 + 3\frac{1}{\frac{\lambda}{2}+\frac{2}{\xi}+\frac{2}{\xi^2\lambda}} .$$

This result is identical to the exact asymptotic limit obtained in equation (11).

The evolution of the spatial beam width, spatial kurtosis, spectral beam width and spectral kurtosis is shown versus the spatial position x = $\tau$ along the beam axis for two scenarios in the figures 2 to 5. For scenario 1, the parameters represent a relatively thin fluctuation layer ($x_F = \tau_F = 1$ cm), whereas scenario 2 includes a broader fluctuation layer with $x_F = \tau_F = 30$ cm. As expected, in all figures multi-Gaussian solution and the analytical solution coincide very well when a thin fluctuation layer is considered. In figures 2 and 3 one can see that within the fluctuation layer the quantities describing the spatial distribution are exactly conserved by the multi-Gaussian solution. As discussed above, the effect of a disturbed spectrum (due to the action of $\Phi^B$) on the spatial distribution arises from the evolution of the beam via the transport operator $\Phi^A$. This evolution is deferred in the multi-Gaussian solution for which the operator $\Phi^A$ is applied only after the fluctuation layer. In turn, the transport operator leaves the spectral distribution unchanged as can be seen in figures 4 and 5. Broadening of the spectrum inside the fluctuation layer can be observed in figure 4. The kurtosis in figure 5 starts from the Gaussian value kurt = 3 and, after reaching a maximum, matches the exact value analytically if $m \to +\infty$, as shown above.

## References

[1] S. McDonald, Phys. Reports **158**, 337 (1988).
[2] S. McDonald, Phys. Rev. A **43**, 4484 (1991).
[3] C. Tsironis *et al.*, **16**, 112510 (2009).
[4] A. Snicker *et al.*, Nucl. Fusion **58**, 016002 (2018).
[5] H. Weber *et al.*, EPJ Web of Conf. **87**, 01002 (2015).
[6] A. Koehn *et al.*, Plasma Phys. Control. Fusion **60**, 075006 (2018).
[7] O. Chellai *et al.*, Nucl. Fusion **61**, 066011 (2021).
[8] E. Poli *et al.*, Comp. Phys. Comm. **225**, 36 (2018).
[9] D. Farina, Fus. Sci. Tech. **52**, 154 (2007).
[10] H. Weber *et al.*, J. Phys. Conf. Series **1125**, 012022 (2018).
[11] J. B. Uscinski, *The Elements of Wave Propagation in Random Media*, McGraw-Hill, 1977.
[12] E. G. Westfall, The American Statistician **68**, 191 (2014).
[13] L. Ryzhik *et al.*, Wave Motion **24**, 327–370 (1996).
[14] H. Weber *et al.*, Journal of Comp. Electronics **20**, 2199 (2021).
[15] E. Faou, *Geometric Numerical Integration and Schrödinger Equations*, EMS, 2012.
[16] M. Spivack and B. J. Uscinski, J. Comput. Appl. Math. **27**, 349 (1989)

# New modelling capabilities to support the ITER EC H&CD System optimisation and preparation of plasma operation

*Melanie* Preynas[1*], *Mireille* Schneider[1], *Jose Manuel* Arroyo[1], *Florent* Beaumont[1,4], *Natalia* Casal[1], *Munseok* Choe[3], *Giuseppe* Carannante[1], *Franco* Gandini[1], *Mark A.* Henderson[2], *Toshimichi* Omori[1], *Sandrine* Pascal[4] and the ITER Scientist Fellow Network

[1]ITER Organization , Route de Vinon sur Verdon, 13067 Saint Paul Lez Durance Cedex, France
[2]UKAEA, Culham Science Centre, Abingdon, OX14 3DB, United Kingdom
[3]Korea Institute of Fusion Energy (KFE), Daejeon 34133 Republic of Korea
[4]DAES, Genève, Switzerland

**Abstract.** The ITER Electron Cyclotron Resonance Heating (ECRH) & Current Drive (ECCD) system is planned to be progressively installed and commissioned following the four stage approach of the ITER Research Plan. Starting with an injection of up to 5.8 MW from one Upper Launcher (UL) in the vacuum vessel to assist the plasma breakdown during First Plasma (FP) operation, the system will then be extended to achieve a capability of 20 MW injected power in Pre-Fusion Plasma Operation (PFPO) and Fusion Power Operation (FPO) phases. Development of optical modelling was required to characterize the optical performance of the FP configuration with the so-called First Plasma Protection Components. An optical 3D model using Zemax OpticStudio® has been developed and extended to the UL. Effects of higher order modes, thermal deformations and tolerances on the UL functionality have been characterized and are presented. Finally, in preparation of plasma operation and in the frame of the EC system upgrade layout optimisation, ECRH-ECCD modelling is being undertaken within the ITER Integrated Modelling and Analysis (IMAS) suite.

## 1 The EC system within the ITER Research Plan

### 1.1 The four stage approach

The ITER Research Plan defines the strategy to achieve its mission goals through the scientific and technical exploitation of the tokamak and its ancillary systems [1]. It will unfold in four phases, as summarised in Table 1:

- **First Plasma** (FP) for which only one upper launcher of the Electron Cyclotron (EC) system will be operational, associated with 7 gyrotrons injecting 170 GHz waves, leading to a total injected power of up to 5.8 MW into the plasma. The main goal of First Plasma operation is to demonstrate the integrated commissioning of tokamak core components.
- **The 1st Pre-Fusion Power Operation phase** (PFPO-1) for which the full EC system will be operational and inject 170 GHz waves from 24 gyrotrons, providing up to 20 MW inside the plasma (with a possible early 10-20 MW power upgrade under assessment). The 24 beams will be launched from 1 equatorial launcher (EL) and/or 4 upper launchers (UL). The main goal of PFPO-1 is to commission the main plant systems and to determine the L-H power threshold for ITER.
- **The 2nd Pre-Fusion Power operation phase** (PFPO-2) for which all the H&CD auxiliary systems will be operational: 20 MW of ECRH as described above, 20 MW of ICRH launched via 2 antennas injecting waves of tuneable frequency in the range [40–55] MHz, and 33 MW of 870 keV NBI launched via 2 injection boxes injecting hydrogen beams. The main goal of PFPO-2 is to raise the plasma current and power to the nominal values of 15 MA and 73 MW respectively, and to increase the pulse duration. Qualification of all control systems for MHD instability control is to be proven prior DT operation.
- **The Fusion Power Operation phase** (FPO) in which the NBI system will be the only heating and current drive system modified within the tokamak building. NBI system will inject deuterium beams with an energy of up to 1 MeV and a power of up to 33 MW. The main goal of FPO is to demonstrate Q=10 operation (ITER baseline 15MA/5.3T scenario), long pulse operation (ITER hybrid 12.5MA/5.3T and steady-state 10MA/5.3T scenarios) and to study burning plasma physics.

Accurate EC resonance heating (ECRH) and current drive (ECCD) predictive modelling over the large range of toroidal magnetic fields [1.8T – 5.3T] foreseen in the

* Corresponding author: melanie.preynas@iter.org

ITER Research Plan is crucial for the preparation of the EC system operation and would allow to optimise the use of the upgraded power.

**Table 1.** The four stage approach of the ITER Research Plan; for First Plasma the duration includes Integrated Commissioning (11 months), First Plasma operation (1 month) and Engineering Operation (6 months).

| Phase | FP | PFPO-1 | PFPO-2 | FPO |
|---|---|---|---|---|
| **Duration** | 18 months | 18 months | 21 months | n x 16 months |
| Main ion species | H | H, $^4$He | H, $^4$He | D, DT |
| H&CD power (MW) | ECH - 5.8 | ECH - 20 | ECH – 20<br>ICRH - 20<br>NBI – 33 | ECH - 20<br>ICRH - 20<br>NBI - 33 |

Hence, an important aspect of the modelling activities described in this paper is to assess the optimal conditions and identify the potential issues for the commissioning and operation of the EC system in the various phases of the ITER Research Plan. It implies an optimisation of the design of the EC system itself but also of the scenarios foreseen in all phases of the ITER Research Plan.

## 1.2 Description of the ITER EC system

### *1.2.1 First Plasma*

The first plasma breakdown in ITER will be assisted by ECRH. Up to 5.8 MW will be injected via one UL located in the upper port 16 across the magnetic null. The UL optics has been designed for minimizing the power required to stabilise the Neo-classical Tearing Modes (NTMs), which requires a well-focused beam at the resonance in the region of 0.6<ρT<0.88 (where ρT is the normalised radius with respect to the toroidal flux). This region is high in the vacuum chamber far from the magnetic null located below the plasma axis. Furthermore, focusing the beams at the resonance results in a large beam expansion upon arrival at the central column. This necessitates a set of mirrors on the inner wall of the vacuum chamber to collect the beams and redirect them across the magnetic null and into a dump in equatorial port 17 (EP-17) (see fig. 1). The FP beam configuration is divided into two sets of beams:

- the lower beams (L-beams) that are launched by the lower steering mirror (LSM) of the UL and then bounce off the LA and LB focusing mirrors
- the upper beams (U-beams) that are launched by the upper steering mirror (USM) of the UL and then bounce of the UA focusing mirror

Estimates of the plasma breakdown performance have been carried out using the Hamiltonian adiabatic formalism [2], helping the definition of EC requirements for ITER First Plasmas.

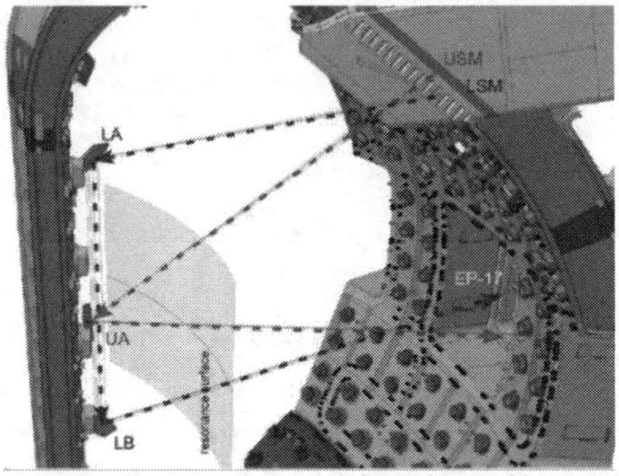


**Fig. 1.** First Plasma EC beam configuration. In blue are the lower beams launched by the LSM of the UL-16 and redirected to the resonance layer below the plasma null thanks to the LA and LB focusing mirrors. In red are the upper beams launched by the USM and refocused by the UA mirror to cross the resonance layer above the plasma null. The residual non-absorbed power of the beams will be attenuated in the beam dump located in the EP-17.

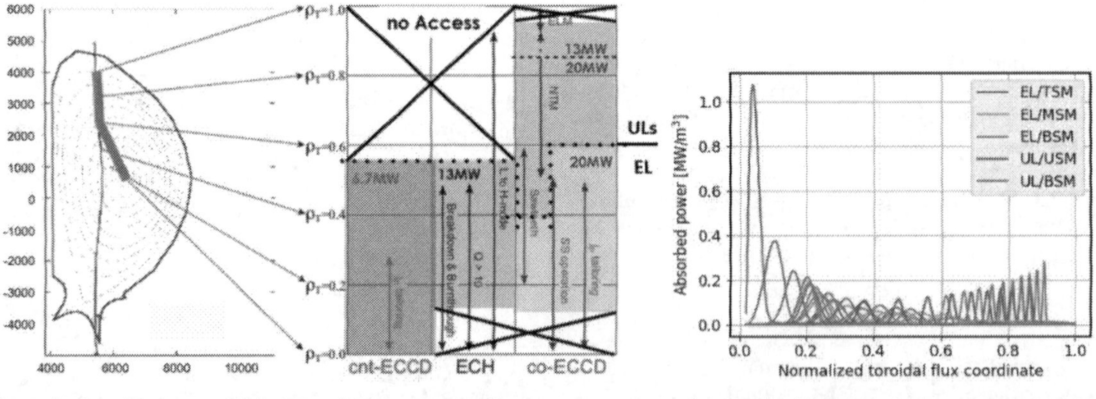


**Fig. 2.** Left: deposition access of the EL and ULs across the plasma cross section and categorized based on driving counter (cnt) ECCD (shaded red), co-ECCD (shaded green) or pure heating (shaded blue), considering 1MW gyrotrons. Right: radial coverage of EC system when steering the poloidal angle for each mirror of the EL and ULs, as simulated with the TORBEAM wave code [3].

### *1.2.2 . PFPO and FPO phases*

The system will have a capability of 20 MW injected (from 24 1MW gyrotrons) via a set of 1 Equatorial Launcher (EL) and 4 Upper Launchers for the first phase of the Pre-Fusion Plasma Operation (PFPO-1) [4].

The EL launcher will be used for central heating and current drive accessing the inner half of the plasma volume, as shown in fig. 2. The EL can launch up to 24 beams that are distributed over 3 rows of 8 beams, i.e. 3×6.7MW injected to the plasma (see fig. 3). Each beam set is incident on a single fixed mirror and then a steering mirror which steers the beam in the poloidal plane. The upper set of EL beams is orientated for counter-ECCD injection, whereas the middle and lower sets are oriented for co-ECCD injection. This effectively decouples the current drive from the heating of the EC beams. Indeed, balancing the co and counter contribution provide a net 'zero' driven current (or pure heating).
The four upper launchers are identical and can launch up to 8 beams each, i.e. 6.7 MW, distributed between the lower and upper rows. The ULs will be used in particular for stabilizing MHD activity, which occurs in the outer half of the plasma volume. Each beam is incident on three fixed mirrors, which are used to regroup and focus the beams such that the absorption in the plasma is very localized. As shown in fig. 3. the last mirror is a steering mirror, which directs the beams in a vertical plan with roughly a 20° toroidal injection angle (optimized for peak $j_{CD}$ injecting in the co-current drive direction). The two steering mirrors, upper (USM) and lower (LSM) cover different ranges in the plasma. The resulting heating and current drive access is illustrated in fig. 2. UL access is distinguished from the EL access. As represented in the plot on the right, each steering mirror has its own range of radial coverage.
The ITER EC system has some flexibility in the way the 24 gyrotrons can be operated and how their 24 beams are distributed among the 24 entries of the EL and 4×8 entries of the ULs. As shown in fig. 4, five sets of switches along the transmission lines enables the beam distribution. The 24 gyrotrons can be distinguished in three different groups. The gyrotrons #1 to #8 can feed either the top row of the EL or the UL 12 and 13. Indeed, each of these two upper launchers is connected to four gyrotrons only. An additional switch redirects the beams either to the top or the bottom row of the UL. The gyrotrons #9 to #16 can feed the middle row of the EL or the UL-15. Finally, the gyrotrons #17 to #24 are connected to the EL or the UL-16 in the same way.
The EC system is designed to accommodate an upgrade up to 20 MW, and the activation of this potential is being considered. This upgrade aims at enhanced physics capabilities combined with an optimized layout to distribute the additional available power among the 56 launcher entries of the five launchers. The potential operational limitations of the system are considered as well as upgraded gyrotrons (>1MW). This requires thermal and structural analysis to define the operational conditions that are acceptable for the system.

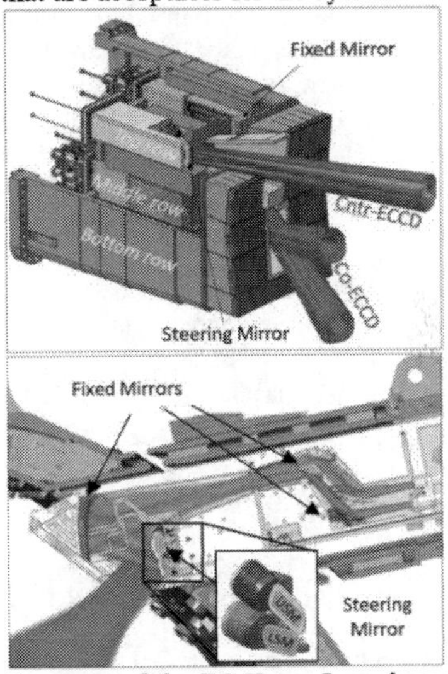


**Fig. 3:** Geometry of the EC Upper Launchers with the static and steering mirrors. Top: Equatorial Launcher; Bottom: Upper Launcher. In red are represented the beams that propagate inside the launchers.

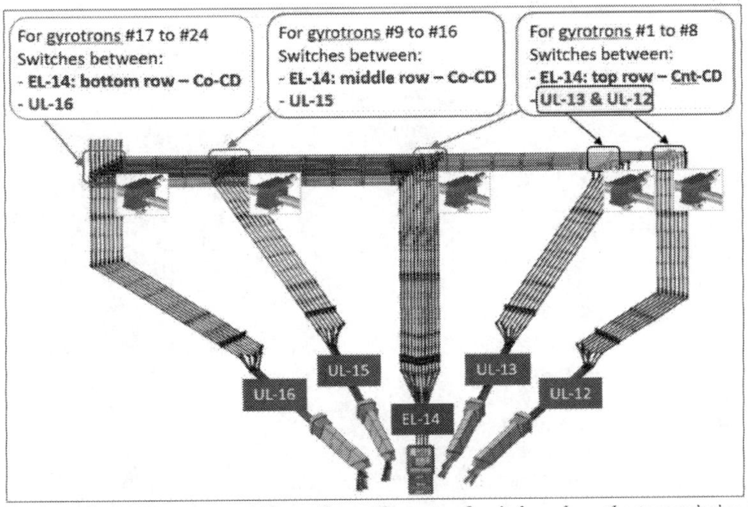


**Fig. 4**: Gyrotron distribution of 24MW system – 56 beamlines. Five sets of switches along the transmission lines bring flexibility to the EC system operation.

## 2 Zemax Optic Studio simulations of the EC launching system

### 2.1 Purpose

In the frame of the design development of the FP configuration of the EC launching system, i.e. UL-16 and LA, LB and UA mirrors described in section 1.2.1, an optical 3D model using Zemax OpticStudio® has been developed. Characterization of the propagation losses was requested to insure a safe level of stray radiation during first plasma operation, as well as a sophisticated model of the beam dump. Presentation of this development and associated results are not yet published and will not be presented in this report. This initial model has been extended to the misalignment assessment in the UL considering several beam deviation contributors: Higher Order Modes (HOM), thermal deformations and mechanical tolerances. Besides the quantification of the beam deviation at the mirrors, the consequences in terms of stray radiation, local hot spots in the launcher and propagation efficiency of the launcher were estimated. Those results associated to the plasma performance analysis performed by GRAY [7] brought to the conclusion in 2021 that a further optimisation of the optical design was necessary before proceeding to the manufacturing. Some modifications of the optical design to mitigate beam misalignment has been tested in Zemax and their impact on the FP launching configuration evaluated.

### 2.2 System definition

#### *2.2.1 Definition of Gaussian beam source*

The Gaussian beams are modelled using the source dll « gaussiansource.dll » available in Zemax. The source behavior have been verified against quasi-optical gaussian beam propagation theory: both beam divergence and ray distribution along beam propagation showed a proper behavior but a minimum of $1.10^6$ rays is needed. The Gaussian beams are launched at the tip of the in-vessel waveguides.

#### *2.2.2 Geometry*

The optical system (mirrors) and port plug are defined using objects which consist of CAD import (STEP file) from ENOVIA model. A comparison of the interception points of the chief rays on mirrors between Catia and Zemax shows that rays propagate in the same way. The global coordinate system is TGCS (ITER Tokamak Global Coordinate System). Thus, the exact real geometry of both mirrors and surrounding port plug is considered in the ray trace analysis, which is key for complete and realistic beam propagation analysis. Proper absorption coefficient of the different parts has been defined in the model, in particular for the reflective surface of the mirrors.

### 2.3 Ray trace analysis: evaluation of the beam propagation losses

Ray trace analysis combined with the definition of detectors at different positions in the launchers allows to extract two kind of propagation losses, at the mirrors and in other regions of the UL.
Spillover losses at each mirror for each of the eight beams were calculated using a set of detectors defined just before and just after the mirror reflective surface. Percentage of beam losses and mapping of the losses can be easily extracted from the analysis. Example of the mapping of losses at M4 considering some HOM is illustrated in fig. 5 by plotting the irradiance on the detector placed behind the mirrors. 8 beams of 0.86MW were considered in the model and the pixel size of the detector is $1mm^2$. A peak irradiance of $9MW/m^2$ is obtained, a value 50% much higher than without any HOM considered. This value is extremely high and can damage an impacted surface that is non-actively cooled. Please note that the HOM in the model are simply modelled by a tilt of the beam k-vector and/or an offset at the launching position.
Another type of propagation losses can occur in the launcher if sources of beam deviation are considered in the model. Some critical regions were identified using Zemax. One of them is one L-beam interfering with the steering mechanism between M2 and M3 that leads to local overheating.
Propagation losses generate stray radiation everywhere in the launcher, a load that needs to be quantified carefully and to be taken properly into account in the structural analysis of the port plug and sub-components. A particular attention to sensitive components such as the springs and the bellows of the M4 steering mechanism is requested since their cooling is intrinsically limited by their geometry. It has to be stressed that the degradation of the global efficiency of the launcher due to a certain category of beam deviation was characterised. For instance, losses go from 0.3% for the nominal configuration up to 1.3% when thermally deformed mirror surfaces are considered, with an increase from 0.27% to 1.2% for the losses at M4 mirror. This Zemax model has many other capabilities like the detection and characterisation of ghost (that is a portion of a beam badly reflected by a surrounding sub-mirror).

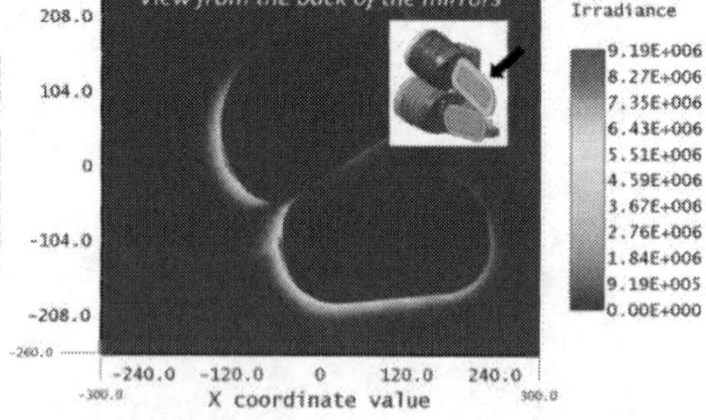


**Fig. 5:** Irradiance in $W/m^2$ of the detector placed right after the M4 mirrors to characterise the spillover considering beam deviation from HOM.

For example, part of L-beams are reflected by the upper M3 mirror instead of the lower mirror. This has been quantified and needs to be mitigated in the up-coming optimization of the optical design. All the analysis performed with Zemax were complemented by some parallel analysis performed by the European Fusion Laboratory ISTP-CNR using GRASP and by the University of Stuttgart using PROFUSION [5].

## 3 EC simulations within the ITER Integrated Modelling & Analysis Suite

### 3.1 The Integrated Modelling & Analysis Suite

The ITER Integrated Modelling and Analysis Suite IMAS is the collection of physics software that will be used to support ITER operation and research as defined in the ITER Integrated Modeling Programme [6].
It uses standard Interface Data Structures (IDS) to represent experimental and simulated data. IDSs are used to get/put data from/to storage, they are applicable to any fusion device, and they are passed between IMAS physics models to build up integrated modelling workflows. The global set of IDSs represent the IMAS Data Dictionary. IMAS will be capable of high physics fidelity predictive simulations of ITER plasmas and will be used for ITER data processing and analysis. IMAS is only used for off-line analysis so it will not be used as part of the Plasma Control System.
The use of a standard data model such as IMAS offers a great level of modularity and flexibility, ensuring the use of identical input data when changing model, making it ideal also for code benchmarking: for example, excellent agreement was recently found in between TORBEAM [3] and GRAY [7] in IMAS [6].

### 3.2 EC modelling in IMAS

Modelling in IMAS relies on the execution of models that exclusively exchange IDSs. IMAS EC modelling can be undertaken in various contexts like standalone EC simulations, or in physics workflows within a suite of codes for transport modelling.

#### *3.2.1 Input and output data for EC modelling*

In IMAS, EC wave codes always receive a minimal set of input IDS data, which are:

- *equilibrium* to describe the magnetic configuration of a given scenario
- *core_profiles* which provides the kinetic profiles (densities, temperature, etc.) for a given scenario
- *ec_launchers* which provides the geometry of the EC system together with the configured waveforms, i.e. the scheduled time evolution of the applied power, polarization and steering angles.

The results of EC codes in IMAS involve specific output IDSs. The evaluation of ECRH and ECCD is often performed using a wave code only. However, for more accurate results, wave codes can be associated to a Fokker-Planck calculation to simulate the evolution of the distribution of EC-accelerated electrons inside the plasma. These two types of code provide their results using the following IDSs:

- *waves* from EC wave propagation codes
- *distributions* from EC Fokker-Planck codes

As an illustration of EC wave modelling, Fig. 6 shows the ray tracing generated by the TORBEAM code [3] when taking its input from an ITER DT baseline 15MA / 5.3T scenario of the IMAS database (130012/3/public/ITER) and the EC launcher geometry from the Machine Description database (120000/2/public/ITER_MD), applying power on all mirrors of the EL and all mirrors of one of the ULs.

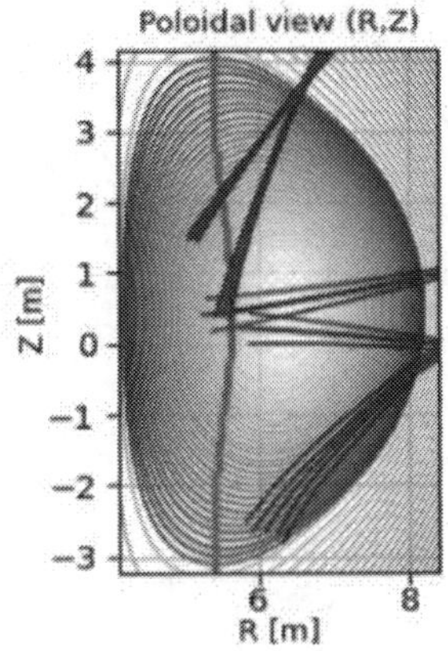


**Fig. 6.** EC rays read from the *waves* IDS generated by the TORBEAM code [3], for an ITER DT baseline 15MA/5.3T scenario; the cold resonance layer is displayed in red.

#### *3.2.2 EC in standalone and within Integrated Modelling contexts*

In IMAS, EC modelling can be carried out 1) as a standalone program, or 2) within the IMAS H&CD workflow where all H&CD sources can be combined together including synergetic effects [8], or 3) within a transport suite of codes. The best way to insure modularity and self-consistency of a scenario computation is to run EC modelling inside the H&CD workflow within a transport solver. This is the case with the JINTRAC transport suite of codes [9] into which the H&CD workflow has been integrated to provide the *core_sources* IDS used as heating sources for the transport equations. The ultimate goal is to combine such a transport suite with a free-boundary equilibrium code, including control algorithms to adjust the H&CD waveforms (e.g. the EC power allocation among launchers and mirrors and their steering angles, e.g. for NTM control) according to the signals (simulated and later measured during ITER operation) from the (synthetic) diagnostics. Such a suite will enable high fidelity plasma simulations which is one of the ultimate goals of IMAS.

When EC modelling is carried out as a standalone simulation, it takes its input from an existing scenario and from the machine description. In ITER, we have developed and populated two databases to provide the necessary input:

- **The Scenario database**, from which *equilibrium*, *core_profiles* and possibly other IDSs can be read; note that all the key scenarios of the ITER Research Plan for the various values of magnetic field foreseen during ITER operation are available in the IMAS scenario database.
- **The Machine Description (MD) database** which contains the geometry of the various ITER plant systems. The geometry of the ITER EC system is stored in the *ec_launchers* IDS which is read from this database.

As an example, Fig. 7 shows ECCD profiles from X2 resonance, applying 13.3 MW on the EL and 6.7 MW on one of the ULs, for an ITER 7.5 MA / 2.65 T scenario in hydrogen as simulated by the JINTRAC transport suite (scenario 104103/22/public/ITER). Note that following the IMAS sign convention, negative ECCD means co-current. Fig. 7 then shows that EL/TSM is the only mirror injecting in the counter-current direction.

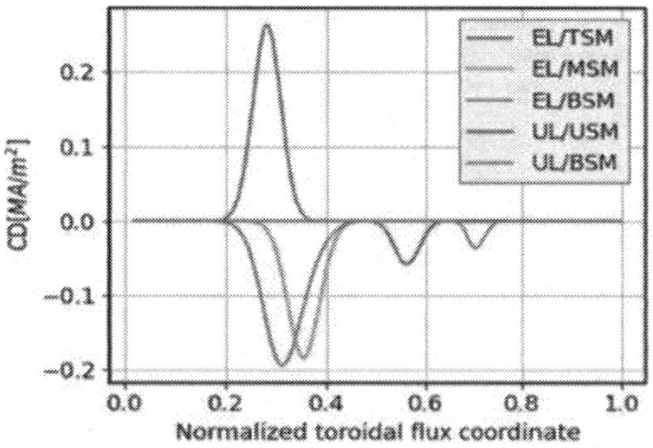


**Fig. 7.** ECCD profiles from X2 absorption simulated by TORBEAM for a JINTRAC 7.5 MA / 2.65 T H scenario.

The EC waveforms, as mentioned earlier, are part of the *ec_launchers* IDS, which contains both the static data (geometry of the system) and dynamic data (time evolution of the power, polarization, steering angles, etc.) configured for a given scenario. For the time being, this dynamic data is configured in the *ec_launchers* IDS via a waveform editor, but it will ultimately be read from the *pulse_schedule* IDS, including the system latency time along with some operational variations, as illustrated in Fig. 8.

In its current version, the geometry of the EC launchers stored in the MD database is described at the level of the steering mirrors. However, to model the whole flexibility and capabilities of the EC system, each of the 56 beamlines of the five launchers will soon be defined in the MD database instead. Another aspect to later consider is to account for the coordinate change of the beam launching points at the steering mirrors w.r.t. to the chosen steering angles. In ITER, as illustrated in Fig. 3, from the exit of the waveguides the beams first encounter three static mirrors (M1, M2 and M3), then the M4 steering mirror.

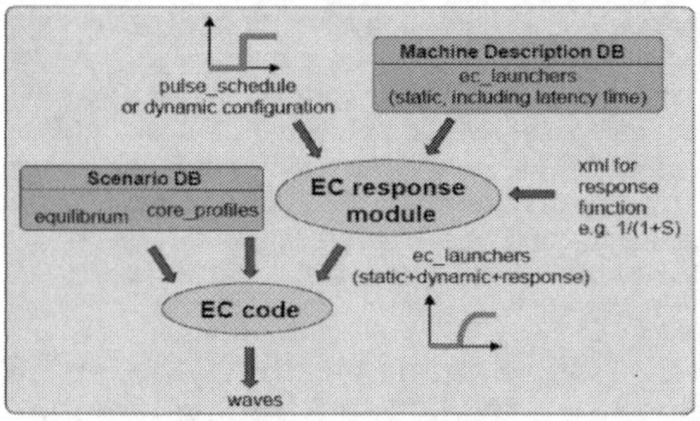


**Fig. 8.** EC modelling using Scenario and Machine Description data combined with dynamic waveforms configured by a pulse scheduler.

So far the geometry of the EC system was stored considering a fixed position of the launching point of the beams at M4, but this approximation will be replaced by a proper calculation of the actual position of the beams with respect to the steering angle configuration.

## 4 Conclusion

The installation of the EC system has started but part of it is still under design finalisation. Zemax OpticStudio modelling supports the optimisation of optical design, but will also be used during the preparation of plasma operation. Several EC codes (beam and ray tracing codes) are available within the IMAS Integrated Modelling and Analysis suite and are used to assess the ECRH and ECCD capabilities for the various scenarios of the ITER Research Plan. Those modelling capabilities are under permanent development and will support the definition of an EC system upgrade if required.

## Acknowledgments

ITER is a Nuclear Facility INB-174. The views and opinions expressed herein do not necessarily reflect those of the ITER Organization. This publication is provided for scientific purposes only. Its contents should not be considered as commitments from the ITER organization as a nuclear operator in the frame of the licensing process. This work has been supported by the ITER Scientist Fellow Network.

## References

1. ITER Research Plan, Tech. Rep. ITR-18-03 (2018)
2. D. Farina, Nucl. Fusion 58 (2018) 066012
3. E. Poli et al, Comp. Phys. Com. (2018) 225 36–46
4. M. A. Henderson, Phys. Plasmas 22 (2015) 021808
5. B. Plaum et al. this workshop (2022)
6. F. Imbeaux et al, Nucl. Fusion 55 (2015) 123006
7. D. Farina Fusion Sci. Technol. 52 (2007) 154
8. M. Schneider et al, Nucl. Fusion 61 (2021) 126058
9. M. Romanelli et al, Pl. Fus. Res. 9 (2014) 3403023

EPJ Web of Conferences **277**, 01005 (2023)
*EC21*
https://doi.org/10.1051/epjconf/202327701005

# Calculations for the optical system for the first ITER plasma

*Burkhard* Plaum[1*], *Mélanie* Preynas[2], and *Munseok* Choe[2]

[1]Institut für Grenzflächenverfahrenstechnik und Plasmatechnologie, Universität Stuttgart, Pfaffenwaldring 31, D-70569 Stuttgart, Germany
[2]ITER Organization, Route de Vinon-sur-Verdon - CS 90 046 - 13067 St Paul Lez Durance Cedex - France

**Abstract.** A dedicated optical system is designed for the breakdown of the first ITER plasma, when the machine will not be equipped with the equatorial ECRH-launcher. Seven beams from the upper launcher will be transmitted via several in-vessel mirrors through the resonance layer to a beam-dump. The whole system was simulated with the PROFUSION package to obtain the beam parameters, estimate spill-over effects and determine the required polarization in the launcher in order to have X-mode polarization in the resonance.

## 1 Introduction

For the first plasma experiments in ITER, the machine will not be equipped with all components. In particular the equatorial launcher, which is foreseen for central EC-heating, will not be available. This means that for the initial plasma breakdown the upper launcher [1] has to be used.

Since the resonance is near the plasma axis, where the magnetic field lines are mostly in toroidal direction, the microwave needs to cross the resonance with vertical polarization mostly in the horizontal plane in order to heat in X-mode (See Fig. 1). This requires a dedicated optics system [2,3]. In the first version, it contained a grating, which was, however, not feasible due to the extreme angles (shallow incidence, almost perpendicular reflection). A new design consisting only of smooth mirrors was simulated and analysed in this paper.

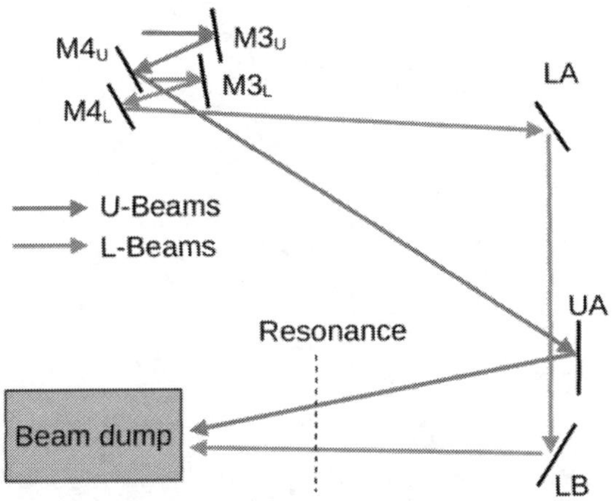


**Fig. 1.** Beam path of the from the launcher to the beam dump

For the initial plasma breakdown the beams of 7 gyrotrons (1 MW each) will be used. The 3 beams launched by the upper row of the launcher are denoted U1, U2 and U3 respectively. The beams coming from the lower row are denoted L1-L4.

For the upper beams, the steering mirror (M4) in the launcher will be modified with a cylindrical surface, and a common in-vessel mirror (UA) is used to direct the beams towards the resonance.

The lower beams are directed via one common ellipsoidal in-vessel mirror (LA) and four individual hyperboloid mirrors (LB) towards the resonance.

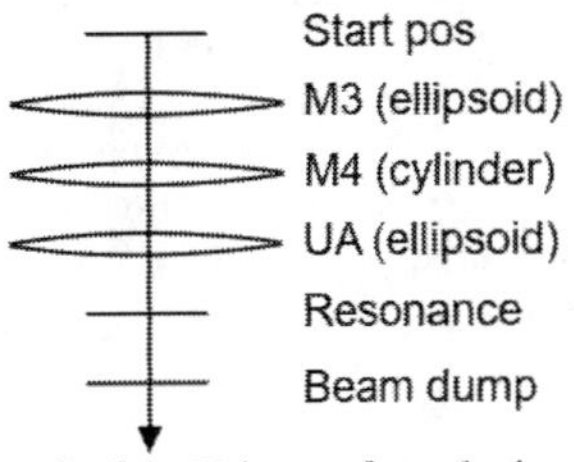


**Fig. 2.** Beam path of the U-beams from the launcher to the beam dump

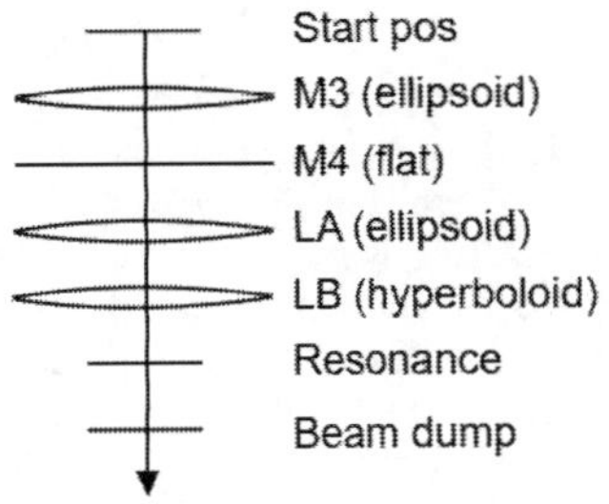


**Fig. 3.** Beam path of the L-beams from the launcher to the beam dump

* Corresponding author: burkhard.plaum@igvp.uni-stuttgart.de

Figures 2 and 3 show the calculation schemes for the U- and L-beams respectively, where focusing elements are drawn as lenses.

The calculations were done with the PROFUSION code [4], which was extended for this project.

## 2 Calculation methods

### 2.1 Physical Optics

The calculation starts inside the upper launcher at a location after the mirror M2, where the beams have a nearly circular shape with a known waist radius. The analytical description is used to generate the incident field on the first mirror.

The reflection is calculated using a physical optics algorithm. It divides the mirror surface into small ($< \lambda/2$) elements. On each of the elements the surface current $J$ is calculated from the incident magnetic field and the local normal vector $n$ of the mirror surface under the assumption of a perfect electric conductor:

$$\vec{J} = 2\,\vec{n} \times \vec{H_i} \quad (1)$$

From the surface currents, the magnetic field $H_r$ of the reflected beam can be calculated using the Green function of free space:

$$\vec{H_r} = \iint \left(-jk - \frac{1}{R}\right)\vec{J} \times \frac{\vec{R}}{R}\frac{e^{-jkR}}{4\pi R}\,dxdy \quad (2)$$

where $R$ is the vector from the mirror point to the location, where $H_r$ is calculated, and $k$ is the free space wave vector.

The reflected field is calculated on a plane, which is perpendicular to the axis of the beam. In contrast to simpler methods the Physical Optics solver also considers mode conversion effects due to non-perpendicular incidence on the mirror as well as a small cross polarised component, which is generated when straight field lines are reflected by a curved mirror surface.

### 2.2 FFT Propagation

For the free space propagation after the last mirror, an FFT-based algorithm is used. It is based on the fact, that the Fourier transform of a spatial field distribution is equivalent to a spectrum of plane waves, which propagate under different angles. After passing a distance $d$, each of these waves is multiplied with a phase factor, which corresponds to the k-component in the main propagation direction. The spatial field distribution can then be obtained with an inverse FFT. To ensure the maximum speed and accuracy, the forward transform is only applied to the initial field and the inverse transform is calculated for each point, where the field is of interest.

Since the 2D Fourier transform implies, that the field distribution is periodic in x- and y-directions, some artefacts can occur in the propagated field, which are interferences from the virtual neighbour beams. They can be avoided by padding the field with zeros, which increases the FFT-size and thus the calculation time.

### 2.3 Gaussian beam analysis

To compare the simulated results with design values or results from other software packages, the beam profiles, which are available only numerically, need to be fitted to Gaussian beam parameters. This is done with a multidimensional optimizer, which was already used successfully for different optimization tasks [5].

In each iteration, the fitting routine generates a synthetic Gaussian beam and calculates the overlap integral with the given field pattern. From this, the power fraction of higher order modes can be calculated. The optimizer varies the Gaussian beam parameters such, that the power of the higher modes becomes as small as possible. Start values for the beam parameters are obtained by analysing the field pattern and calculating some statistical parameters. From a field pattern, the following parameters can be obtained:

- The position of the beam centre
- The angle of the beam axis with respect to the z-axis of the coordinate system
- The ellipse radii
- The ellipse orientation

For beams with a simple astigmatism it is also possible to obtain the positions and radii of the beam waists. The beams in this paper, however, have a general astigmatism, which was not supported by PROFUSION, when the analysis was done (see section 4.1).

### 2.4 Tracking of the polarisation

As mentioned above, the linear polarisation in the resonance should be vertical (i.e. in z-direction of the global torus coordinate system). In order to determine the correct polarisation vector in the launcher, a tool was written, which tracks the polarisation over a sequence of mirrors. For one reflection on one mirror, the relevant vectors are shown in Fig. 4.

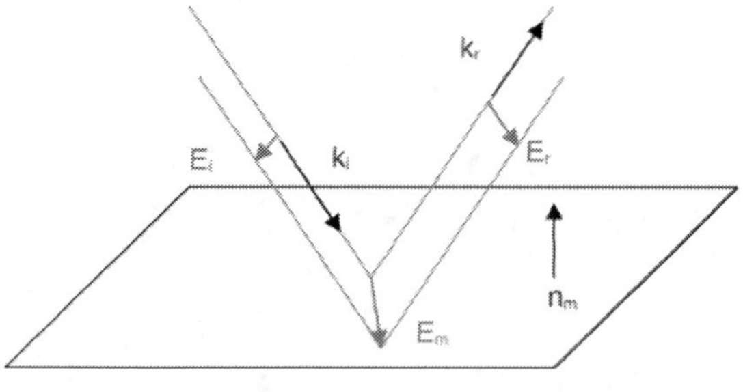


**Fig. 4.** E-field vectors for a reflection on a flat mirror

The incident beam has a wave vector $k_i$ and an electrical field vector $E_i$. Both vectors define the *polarization plane* of the beam. The simple rule is that the polarisation planes of the incident and reflected beams intersect the mirror plane in the same line, which is equivalent to the E-field vector $E_m$ projected onto the mirror.

Mathematically the following formalism can be used to calculate $E_r$. Note that the lengths of the vectors are not relevant here and are assumed to be normalised. Also, the sign of the vector products does not need to be considered, because a multiplication with -1 results only in an inverted phase but keeps the polarisation unchanged.

The normal vector $n_i$ of the polarisation plane of the incident beam is:

$$\vec{n}_i = \vec{E}_i \times \vec{k}_i \quad (3)$$

The projected field $E_m$ on the mirror can then be calculated with the normal vector $n_m$ of the mirror:

$$\vec{E}_m = \vec{n}_i \times \vec{n}_m = (\vec{E}_i \times \vec{k}_i) \times \vec{n}_m \quad (4)$$

Together with the wave-vector $k_r$ of the reflected beam the normal vector $n_r$ of the polarisation plane of the reflected beam becomes:

$$\vec{n}_r = \vec{E}_m \times \vec{k}_r = ((\vec{E}_i \times \vec{k}_i) \times \vec{n}_m) \times \vec{k}_r \quad (5)$$

Finally, the electric field vector $E_r$ can be calculated from $n_r$ and $k_r$

$$\vec{E}_r = \vec{n}_r \times \vec{k}_r\,, \quad (6)$$

which leads to the following expression for $E_r$:

$$\vec{E}_r = (((\vec{E}_i \times \vec{k}_i) \times \vec{n}_m) \times \vec{k}_r) \times \vec{k}_r \quad (7)$$

Note that $n_m$ can be calculated using the reflection law from $k_i$ and $k_r$. This means that the only necessary parameters for a sequence of mirrors are the incident and reflected wave vectors and the direction of the initial E-vector.

Ideally the field vector in the resonance should be in X-mode, which corresponds to the $z$-direction of the global torus coordinates. This condition can, however, be fulfilled only approximately, because the field vector is always perpendicular to the $k$-vector (in an isotropic medium) and the latter is slightly tilted with respect to the x-y-plane (i.e. the equatorial plane of the vessel). The best physically possible E-direction was obtained by defining an auxiliary vector a, which is perpendicular to the k-vector and lies in the x-y plane:

$$\vec{a} = \begin{pmatrix} 0 \\ 0 \\ 1 \end{pmatrix} \times \vec{k}_{res} \quad (8)$$

The best possible E-vector $E_{res}$ then becomes:

$$\vec{E}_{res} = \vec{a} \times \vec{k}_{res} \quad (9)$$

From this field vector, the described method was applied in backwards direction to get the E-vector at the exit of the waveguides in the upper launcher. This vector was then used to do the full simulation with the physical optics solver in forward direction.

## 3 Simulation Framework

All solvers are part of the PROFUSION code package [4], which was developed over more than two decades and contains a large number of routines tools for microwave optics, both for free space and in waveguides.

The PROFUSION tools are commandline programs, which are written in C and run on Linux computers. This has the advantage that a calculation is fully defined by the data files (if needed) and the command line. This improves the reproducibility and allows to control even complex simulations (e.g. including parameter scans) by simple shell scripts. PROFUSION also contains a large number of small tools for doing standard calculation (like 3D vector arithmetic, coordinate transformations, evaluation of mathematical expressions), which make it possible to convert the parameters obtained from IO to the values needed by the solvers. The calculation scheme works basically as follows:

- Use redundant parameters provided by the ITER Organization (IO) to make some consistency checks
- Transform the parameters from the ones provided by IO to the ones used by the solvers where necessary
- Run the solvers to calculate the field patterns
- Run the analysis to obtain the Gaussian parameters and other data
- Generate the field plots

Since we always start with the data provided by ITER, potential errors from manual conversion could be avoided.

The most time-consuming routine is the physical optics solver, which calculates the output field in a 2D plane according to Eq. (2), which results in 4 nested loops. Therefore, it was parallelized to run on multiple CPU cores (up to 40 on the available machines). In addition, different beams could be calculated by different servers in parallel. This made it possible to simulate a whole set of beams (U or L) including the analysis in a few hours. During the project, the PROFUSION tools were upgraded with some new features, most notably:

- Support for mirrors with irregular (i.e. numerically given) boundaries
- Support for passing all coordinates in an arbitrary global coordinate system (TGCS in the ITER case).
- Determination of the ellipse rotation angle of the Gaussian fit tool
- Implementation of a tool for tracking the polarisation across a sequence of mirrors with Eq. (7)
- A tool for generating discretized datasets of mirror surfaces from geometry parameters (cylinder, ellipsoid or hyperboloid).

# 4 Results

## 4.1 Field patterns

The fields reflected by the mirrors were calculated at different reference position, which were chosen such, that they are as close as possible to the following mirror without the field plane intersecting the mirror surface.

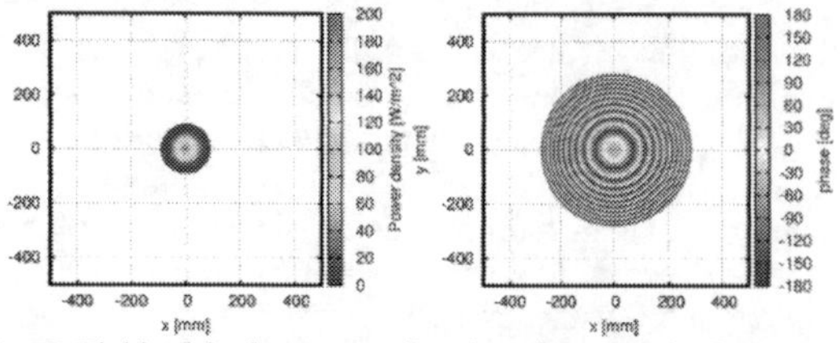


**Fig. 5.** Field of the beam U1 after the mirror M2 in the upper launcher.

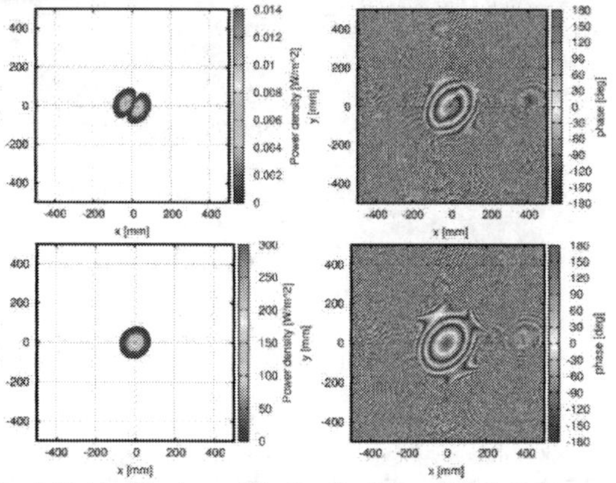


**Fig. 6.** Field of the beam U1 after the Mirror M3. Left: Power density in [W/m²] for a 1 W beam, right: Phase in degrees.

Fig. 5 shows the initial field of the beam U1 between M2 and M3 in the upper launcher, which is a circular Gaussian beam. The left image shows the power density of the field in W/m² for a total beam power of 1 W, which is the standard normalization in PROFUSION. The right image shows the phase in degrees. Fig. 6 shows the fields of U1 after the mirror M3. The upper images show the cross polarization, the lower image shows the main polarization. The coordinate systems for the field plots after M3 are chosen such, that the y-axis is aligned to the main polarization direction, which is required for a maximum power in the X-mode polarization. Therefore, the very small cross polarization becomes visible. For a beam power of 1 MW the power in the cross polarization is less than 0.05% and is considered uncritical since the only effect is a slightly decreased heating efficiency.

Figure 7 shows the field of the beam U1 after the mirror M4. Figure 8 shows the field of U1 after the mirror UA. Some stray radiation is visible especially in the cross polarization, which comes from beam truncation due to the limited size of the mirror UA. It will further be analysed in section 4.3. The effect also occurs for the co-polarised component, but it is invisible in the field plot due to the different scale.

The beams have a general astigmatism [6], where the amplitude and phase ellipses are not aligned. Furthermore, the amplitude ellipse changes its orientation along the propagation axis. Such beams can be described with a complex rotation angle of the ellipse, which was not supported by PROFUSION at the time the analysis was done. Therefore, the beam phase was set to zero before the analysis and the most important parameters, the ellipse radii, were determined by assuming a flat phase front.

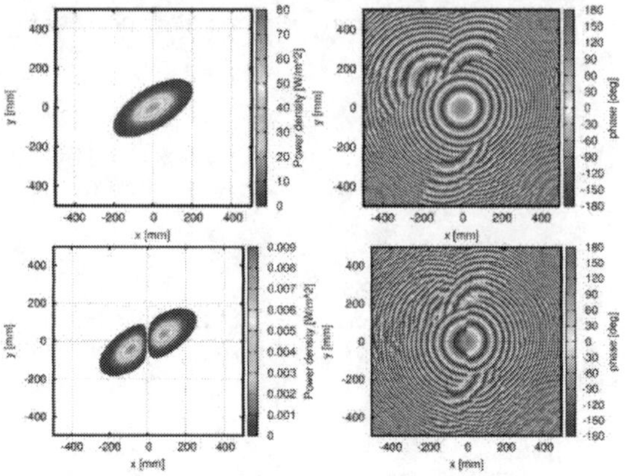


**Fig. 7.** Field of the beam U1 after the mirror M4

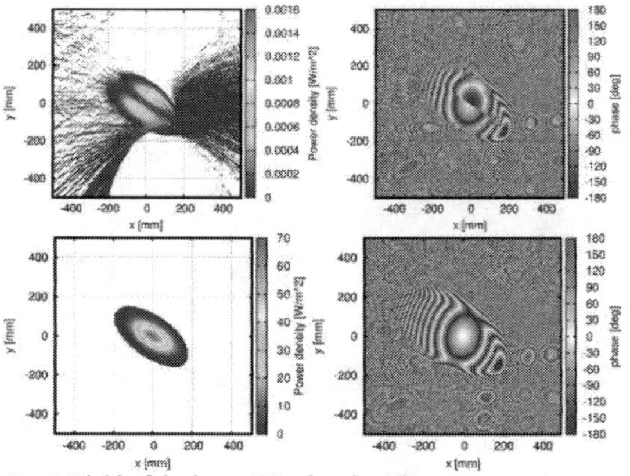


**Fig. 8.** Field of the beam U1 after the mirror UA

## 4.2 Beam profiles

The profiles of the U1-beam are shown in Fig. 9. They are plotted over a distance of 7 m from the last reference plane, which includes the way through the resonance. Again, we see a slight disturbance coming from the truncation by the mirror UA. It is, however, very small in terms of spurious mode power and only visible in a logarithmic plot.

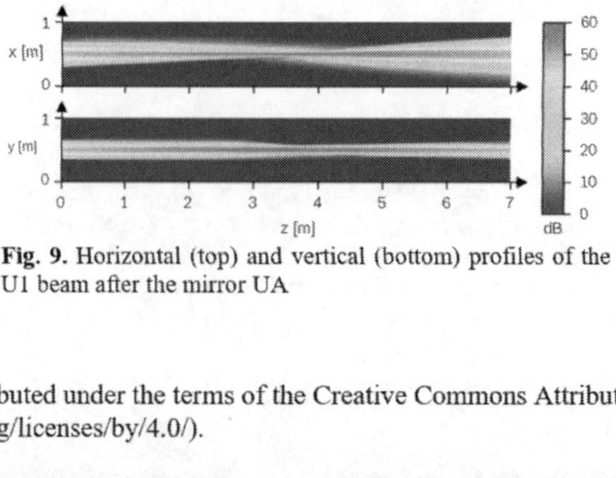


**Fig. 9.** Horizontal (top) and vertical (bottom) profiles of the U1 beam after the mirror UA

The power density, which is relevant for the plasma breakdown and the heat load in the beam dump, is shown in Fig. 10. It is, again, the power density along the beam axis in $W/m^2$ for a total beam power of 1 W.

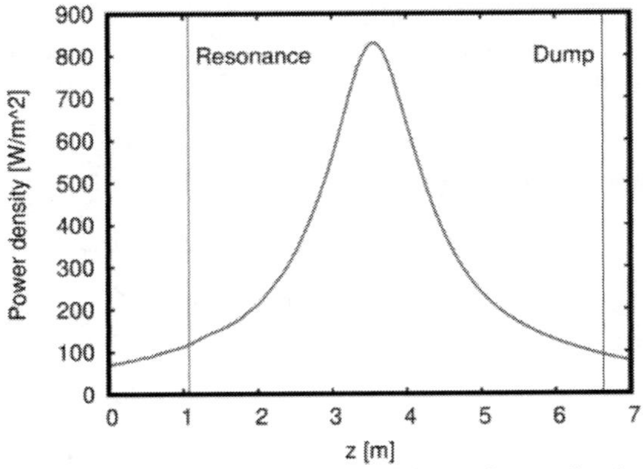


**Fig. 10.** Power density at the axis of the U1 beam after the mirror UA for a total beam power of 1 W.

### 4.3 Beam truncation by the mirror UA

The physical optics solver allows to handle the beam truncation by a limited mirror size by simply assuming the surface currents to be zero for points outside the mirror contour. Together with a routine for the point-in-polygon problem, it is possible to consider irregular mirror contours, which are given in terms of a polygon.

The most critical situation is the beam U1 on the mirror UA. The mirror UA is located very close to the plasma and therefore needs to be as small as possible. Fig. 11 shows the beam footprint on the mirror and the mirror contour.

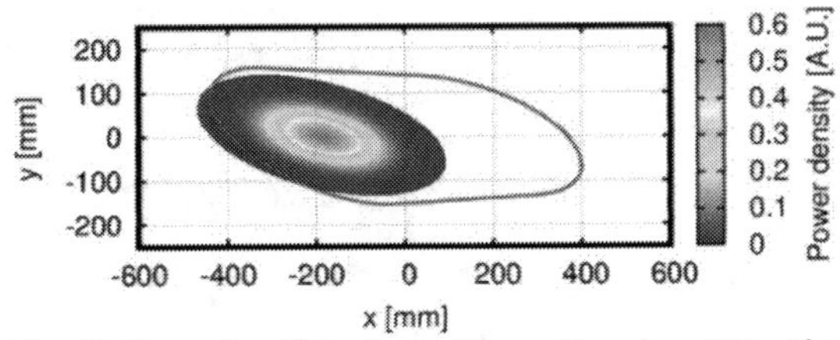


**Fig. 11.** Footprint of the beam U1 on the mirror UA. The contour is shown in red

With the current configuration the power outside of the mirror is 0.48 %. To estimate the effects of the spill-over on sensitive in-vessel components, the field outside of the mirror contour was calculated (Fig. 12) and propagated by 100 mm (Fig. 13) to estimate the power density near the vessel wall. Note that the coordinate system of Fig. 11 is a mirror-local system of UA, the coordinate system of Figs 12 and 13 is a beam-local coordinate system of the beam U1.

The estimated power density was checked against the coordinates of some sensitive in-vessel components, which are located near UA and no critical values could be identified.

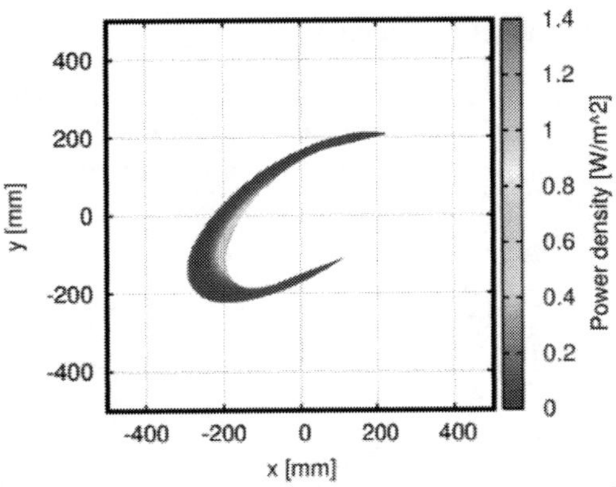


**Fig. 12.** Spill-over power of the beam U1 in the plane of the mirror UA.

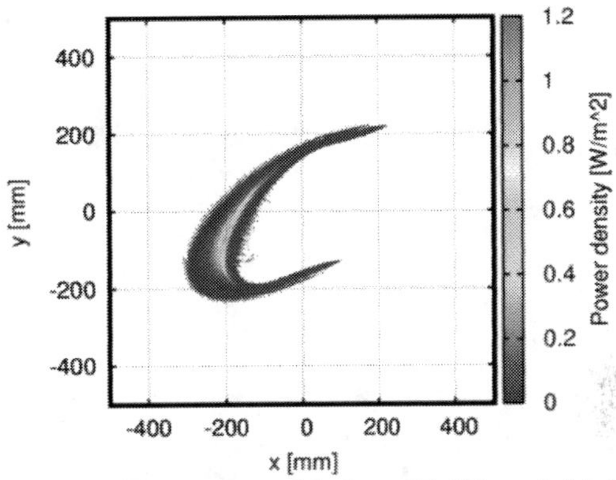


**Fig. 13.** Spill-over power of the beam U1 100 mm behind the plane of the mirror UA.

### 4.4 Comparison with the design values

To verify the calculation, the positions of the beam centres were calculated with the Gaussian fit tool (see section 2.3) and compared with the design values (obtained with the Catia software). Table 1 shows the result for the U-beams at the location of the beam dump. At this location the beam passed 3 mirrors and more than 13 m of free space propagation, so the errors are expected to be the largest there.

We can see a very good agreement. The maximum error is 1.702 mm, which is much smaller than the minor beam radius (57.309 mm) and therefore negligible compared to other expected errors, e.g. from spurious waveguide modes.

**Table 1.** Coordinates (in mm) of the U-Beams in the beam dump

| Beam | Reference | Calculation | Error [mm] |
|---|---|---|---|
| U1 | 9397.991<br>-5614.515<br>277.215 | 9398.183<br>-5614.194<br>278.000 | 0.870 |
| U2 | 9523.528<br>-5387.328<br>252.132 | 9523.031<br>-5388.469<br>253.293 | 1.702 |
| U3 | 9639.352<br>-5174.561<br>220.869 | 9639.284<br>-5174.834<br>221.843 | 1.014 |

## 5 Comparison with raytracing

In parallel to the PROFUSION simulation, the system was simulated with the commercial Zemax OpticStudio software. Fig. 14 shows the field of the beam U2 after the mirror UA calculated with PROFUSION and Zemax.

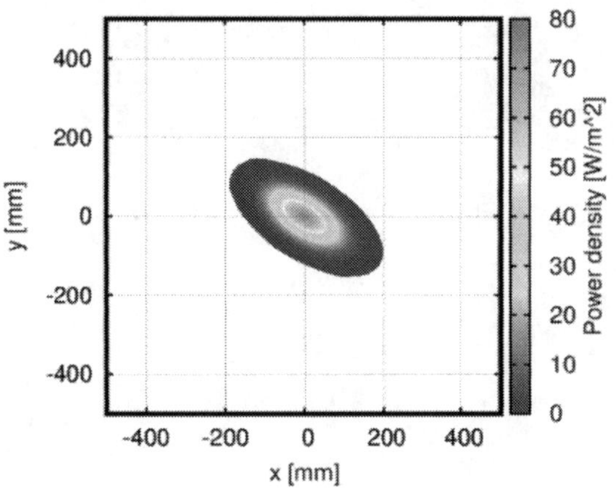


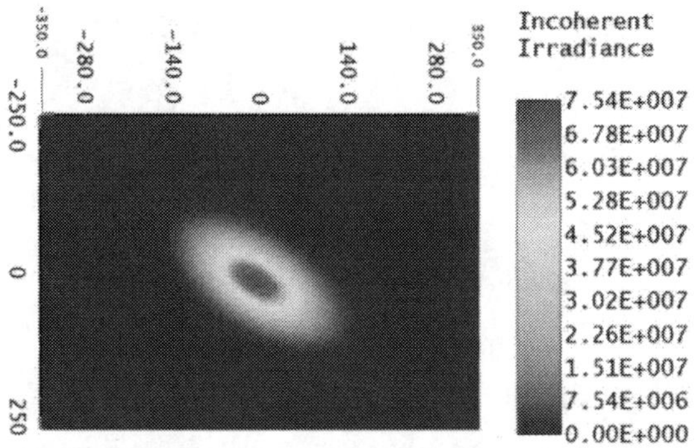


**Fig. 14.** Field of the beam U2 after the mirror UA calculated with PROFUSION (top) and Zemax (bottom)

We can see a good agreement of the shape and orientation of the beam ellipse. Furthermore, the peak value of the field (which is 73.6 $W/m^2$ for the PROFUSION result) is in reasonable agreement. The factor of approximately $10^6$ is due to the fact, that the PROFUSION result is for a beam power of 1 W, while the Zemax calculation is for 1 MW.

## 6 Modifications of the system

During the project, some modifications of the system were suggested by ITER and the simulation was re-run. Due to organizational issues and the ending of the contract, a comprehensive analysis of the modified design could not be done but is planned for the future. All results presented in this paper are for the initial design, where all necessary data is available.

## 7 Conclusion

The simulation and analysis of the in-vessel optics for the first ITER plasma could verify both a feasible design and the accuracy as well as the flexibility of the simulation framework, which was built around the PROFUSION tools.

The design fulfils the requirements in terms of minimum power density in the resonance and maximum power density in the beam dump. The most critical component is the mirror UA, which has a limited size due to its position near the plasma and suffers from a small fraction of the power reaching the vessel wall and a truncation of the reflected beam. A detailed analysis using the coordinates of sensitive in-vessel components could not identify any critical issues, also the effects of the beam truncation on the reflected beam are uncritical.

Comparisons with the initial design of the U-beams and raytracing calculations with Zemax showed a very good agreement.

This work was funded by the ITER organization under the service contract No. 43-2080.

## References

1. D. Strauss et.al., *Nearing final design of the ITER EC H&CD Upper Launcher*, Fusion Engineering and Design **146**, 2019, 23–26, https://doi.org/10.1016/j.fusengdes.2018.11.013
2. F. Fanale et.al., *Design validation of in-vessel mirrors and beam dump for first plasma operations in ITER*, Fusion Engineering and Design, **172**, 2021, 112717, ISSN 0920-3796, https://doi.org/10.1016/j.fusengdes.2021.112717 .
3. A. Moro, et al., Design of Electron Cyclotron Resonance Heating protection components for first plasma operations in ITER, Fusion Engineering and Design, 154, 2020, 111547, ISSN 0920-3796, https://doi.org/10.1016/j.fusengdes.2020.111547 .
4. B. Plaum, *Simulation of microwave beams with PROFUSION*, 2019 Edition, http://dx.doi.org/10.18419/opus-10252
5. B. Plaum, *Optimization of Broadband Smooth-Wall Circular Horn Antennas*. J Infrared Milli Terahz Waves 39, 984–995 (2018). https://doi.org/10.1007/s10762-018-0510-6
6. J. A. Arnaud and H. Kogelnik, "*Gaussian Light Beams with General Astigmatism,*" Appl. Opt. *8*, 1687-1693 (1969)

# ECH and ECCD modelling studies for DTT

*Benedetta* Baiocchi[1*], *Lorenzo* Figini[1], *Alessandro* Bruschi[1], *Francesco* Fanale[2], *Saul* Garavaglia[1], *Gustavo* Granucci[1], and *Afra* Romano[2,3]

[1]Institute for Plasma Science and Technology – CNR, via Cozzi 53, 20125 Milano, Italy
[2]EDP ENEA, Fusion and Nuclear Safety Department, C. R. Frascati, via E. Fermi 45 00044, Frascati, Italy
[3]DTT S.c. a r.l., C. R. Frascati, via E. Fermi 45 00044, Frascati, Italy

**Abstract.** In this work the Electron Cyclotron (EC) physics performances of the EC system foreseen for the new Divertor Tokamak Test facility (DTT) are investigated using the beam tracing code GRAY on the flat top phase of the most recent DTT full power scenario. The whole core plasma region can be reached by EC beams with complete absorption, assuring bulk heating and core current drive (CD) for profile tailoring, and NTM mitigation in correspondence of the rational surfaces. A detailed analysis regarding modifications of the EC propagation, absorption and CD location due to density fluctuations caused by pellet injection is performed. The compatibility between the EC system and the pellet injection system is verified: the density variations due to pellet injection are foreseen to negligibly influence the EC performances, allowing the EC beams to reach the plasma central region for bulk heating and to drive current on the rational surfaces for NTM mitigation. Finally, the polarization variations originated by the angle steering foreseen for the operational and physics tasks accomplishment during the flat top phase of the discharge are assessed. Negligible power losses have been found keeping fixed polarization during the needed steering.

## 1 Introduction

The new Divertor Tokamak Test facility (DTT [1]), aimed to perform studies regarding power exhaust and divertor load and currently under construction, will be provided with a mix of additional heating systems. The electron cyclotron (EC) system is foreseen to be the main heating system. It will contribute to assure high power flow, and it has the aim of accomplishing several operational and physics tasks: plasma current ramp-up & ramp down, L- to H-mode transition, core heating & current drive (CD), control of MHD modes as sawteeth (ST) and neoclassical tearing modes (NTM), plasma current density and electron temperature profile tailoring, mitigation of impurity accumulation and wall cleaning. Then proper different radial localizations and combination of heating and non-inductive current drive must be provided, taking into account the peculiar physics requirements and the engineering constraints of such new machine. To this purpose, two types of launcher have been proposed [2,3]: two different antennas are foreseen to be hosted in the equatorial and in the upper port of each ECRH sector, with 6 and 2 single launchers respectively. The 8 launchers of a cluster are independently steerable to improve the flexibility, allowing EC absorption over a broad range of plasma locations. In order to investigate and to optimize launchers performances for the foreseen EC tasks, the beam-tracing code GRAY [4] has been used to perform a comprehensive study regarding the propagation, absorption and current drive of the EC beams, as reported in Section 1. Such study has been applied to the flat top phase of the reference DTT Single Null full power scenario [5], characterized by $B_t$=5.85 T, $I_p$=5.5 MA, R=2.19 m, a=0.7 m, by additional power coupled to the plasma $P_{add}$=45 MW (28.8 MW from EC), and by the kinetic profiles reported in fig. 1-left. Fig. 1-right shows the main baseline scenario equilibrium features (the poloidal projection of the flux surfaces, with rational surfaces represented with green lines) and the EC launcher parameters, i.e. the resonance layer in the plasma correspondent to the selected frequency of 170 GHz, and examples of the beams paths launched from the Upper and from the top and bottom Equatorial ports launcher positions.

Then sensitivity studies regarding the EC performances variation have been carried out, considering the characteristics of the DTT full power scenario which can be critical for proper EC system operation. As shown in fig. 1-left, the reference DTT full power scenario is characterized by high density plasma ($n_{e0}$~2.5e20 m$^{-3}$, which means a plasma frequency $f_p$~142 GHz), requiring then pellet injection as fuelling system [5]. Perturbations due to pellet injection on high density profiles can imply high beam refraction and then issues can rise regarding beam localization in correspondence of the rational surfaces, which are externally located in the considered scenario (see fig. 1-right). Section 2

* Corresponding author: benedetta.baiocchi@istp.cnr.it

reports evaluations regarding EC beam propagation deviations and (if needed) steering corrections for the DTT full power scenario with injection of pellets with different sizes. In order to fulfill all the physics tasks foreseen for the EC system during the plasma discharge, wide angles steering is required. The correct injected beam polarization has to be preserved for achieving an optimal coupling to Ordinary mode when operating at the first harmonic resonance, as foreseen for the full power DTT scenario. A detailed investigation of polarization variation due to angles steering is reported in Section 3. Finally, a short summary follows in Section 4.

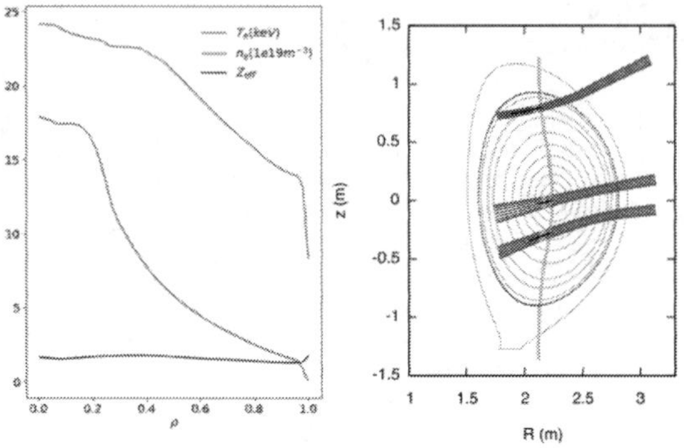


**Fig. 1.** (left) Electron density (green), electron temperature (red) and effective Z (black, with average $Z_{eff}$~1.6) profiles of the DTT full power scenario, used as input in beam tracing simulations. ρ is calculated as the squared root of the normalized toroidal flux. (right) Poloidal projection of the flux surfaces and of the path of different EC beams varying the launching position/angles. In yellow the resonance layer for 170 GHz. In (dark) green the poloidal projection of the q=3/2 (q=2/1) surface, located at ρ=0.701 (ρ=0.814).

## 2 EC propagation, absorption and CD analysis

In order to investigate and to optimize the EC performances, parametric studies have been performed, varying the injection angles (α=poloidal, β=toroidal), considering the DTT full power scenario reported in fig.1 and the EC launcher specifications as [2] (injection frequency of 170 GHz in Ordinary mode, and 3 launcher positions: top and bottom equatorial (EQT, EQB), and upper (UP)).

Fig. 2 shows the driven current per unit of injected power with varying α and β, for the top and bottom equatorial launchers. Complete absorption is found for large angles intervals, narrower limits are given by engineering constraints [3]. The allowed ranges of angles (-10°<α<25° for EQT, -25°<α<10° for EQB, -25°<β<25°) permit to reach the plasma in the central region (ρ~0.1) as the peripheral core plasma (ρ<~0.7-0.8). β<0 gives co-driven current, with the maximum value of 25 kA/MW in correspondence of ρ~0.1. Core heating and CD are well performed by equatorial launchers, then allowing the accomplishment of the task of bulk heating. Such results are promising also in order to achieve the additional tasks of profile tailoring and sawteeth instabilities stabilization. Very recently, theoretical studies and integrate modelling works have been started to perform a detailed analysis of the quantitative requirements for EC to fulfill such tasks in the most recent reference full power DTT scenario [6]. For NTM mitigation, which is carried out by driving current in correspondence of the rational surfaces q=3/2 and q=2, the Upper launcher is preferable for launching geometry.

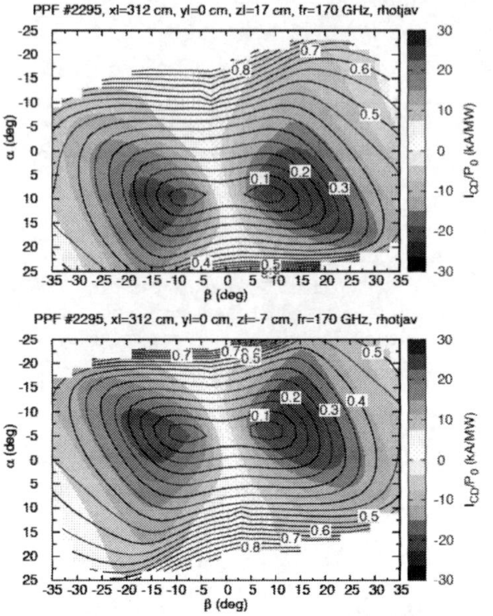


**Fig. 2.** Calculated driven current per unit power for the top (top) and bottom (bottom) equatorial launchers with varying poloidal (α) and toroidal (β) angles. Driven current average location is represented by black contours.

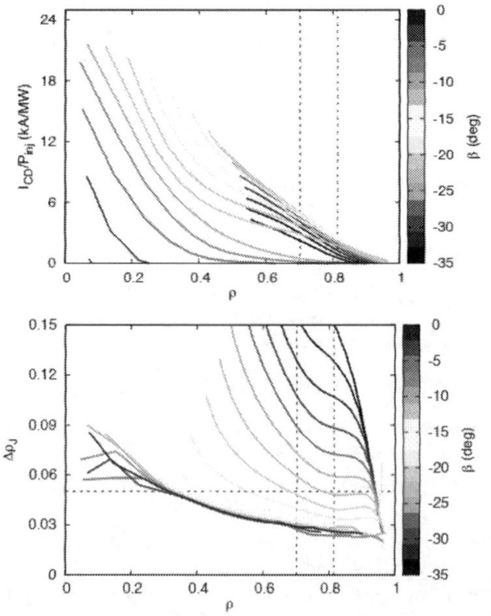


**Fig. 3.** (top) Calculated driven current per unit power and (bottom) average current density profile width as functions of the driven current average location. Colours refer to different β values. Every point of every line corresponds to a different

$\alpha$ value (10°-50°). Locations of the rational q surfaces ($\rho$=0.701 for q=3/2 and $\rho$=0.814 for q=2) are shown by vertical dashed black lines. The horizontal dashed line in figure on the bottom represents the value $\Delta\rho$=0.05

Fig. 3 shows the driven current per unit of injected power and the width of the relative current density profiles as function of the driven current average location, with varying $\alpha$ and $\beta$, and launching from the upper position. The access to the plasma in the internal core region ($\rho$<0.4) is reached for $|\beta|$<15°. In the outer core region, where q=3/2 and q=2 are located, the maximum value of the driven current per unit of injected power is obtained for $|\beta|$ between 20° and 22°, obtaining $I_{cd}/P_0$~5 kA/MW for q=3/2 and $I_{cd}/P_0$~2.5 kA/MW for q=2. The narrower current density profile is found for the minimum $|\beta|$ of such interval, ($|\beta|$=20°), and it satisfies $\Delta\rho \leq 0.05$ (i.e. ~3.5 cm when measured at the plasma midplane), which is the expected requirement for NTM control for DTT [1].

## 3 EC analysis in presence of density fluctuation due to pellets injection

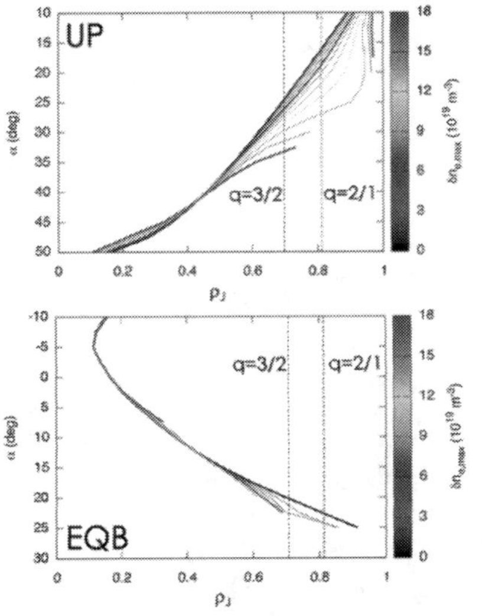


**Fig. 4**. Poloidal angles $\alpha$ and relatives current density average location for upper (top: with toroidal angle $\beta$=15°) and bottom equatorial launchers (bottom: with toroidal angle $\beta$=5°). Colours refer to different density perturbation amplitude. Location of the rational q surfaces are shown by vertical dashed black lines.

Among the particle fuelling methods, the pellets injection system has been foreseen for DTT. In order to evaluate the compatibility of the EC system with the density fluctuations due to pellet injection, first an analysis regarding the maximum allowed density perturbation has been performed. We have considered as density perturbation a Gaussian deposition profile given by $\delta n_e(\rho)$~$\delta n_{e,max}\exp[-(\rho-\rho_{dep})^2/(2\Delta\rho^2_{dep})]$, where $\rho_{dep}$=0.8 is the central location, $\Delta\rho_{dep}$ = 0.09 is the width at half height, and $\delta n_{e,max}$ is the amplitude of the considered perturbation. The full power scenario density profiles with adding the above described density perturbation have been used as GRAY inputs, with adjusting the temperature profiles in order to have a constant pressure profile. The analysis performed with varying the amplitude of the density perturbation gives the results shown in fig. 4. Higher values of the perturbation density could cause strong beam refraction, and then the reduction of available poloidal steering range, which, as shown in fig. 4, does not cover anymore the whole core region with complete absorption. Regarding the upper launcher performances, the beam is not negligibly affected for $\delta n_{e,max}$>6e19 m$^{-3}$, and even large angle adjustments (2°-20°) can be required for off-axis aiming, i.e. to reach the rational surfaces. The core is still reachable despite large density perturbations. However the prediction and the control of the power deposition and the ECCD location could be scarcely reliable.

After such analysis, which gives general indication about the EC compatibility with density perturbations in the peripheral plasma core for the DTT full power scenario, a detailed investigation has been performed with the GRAY code, using as input the kinetic profiles and the equilibrium data of the time-dependent full power DTT scenario simulations [7]. They have been modelled with the JETTO transport code [8]: the pellet ablation and deposition have been predicted by the code HPI2 [9], which has been used self-consistently with the code suite. QuaLiKiz [10] has been used as transport model, while the ECRH has been prescribed. Although the current DTT reference configuration is based on a plasma radius equal to R=2.19 m, without loss of significance the time-dependent simulations, and then the analysis here illustrated, have been carried out on a previous configuration characterized by R=2.14 m.

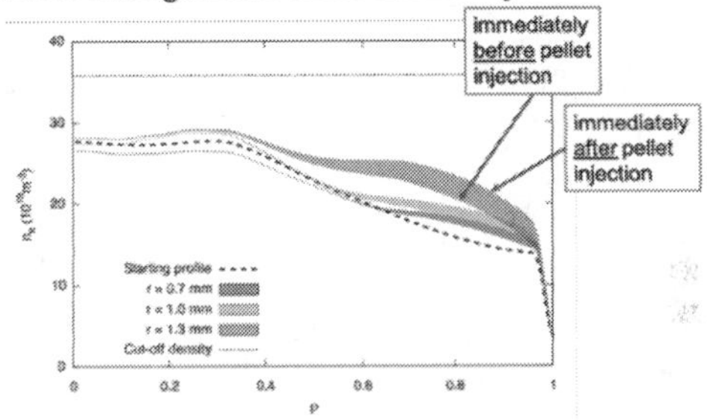


**Fig. 5**. Density profiles of the JINTRAC simulations without (dashed) and with pellets (coloured: blue for radius of 0.7 mm, green for 1 mm, red for 1.3 mm. The bottom and top sides of every band represent the profile immediately before and after one pellet). Density cut-off for Ordinary mode is represented by red dashed line.

The pellet injection geometry (oblique high field side) and the reference pellet parameters (radius 1 mm and velocity 516 m/s) have been chosen according to the previous DTT pellet injection modelling work performed with stand-alone HPI2 [11]. The pellet frequency injection (16 Hz) has been selected in order to sustain the desired density pedestal. Fig. 5 shows the resulting density profiles of the time-dependent simulations using the reference pellet parameters (green), and two realistic variations reported in [11] (blue: radius=0.7 mm, red: radius=1.3 mm). The lower and the higher sides of every colored band represent the density profile just before and just after the pellet injection for every 'pellet cycle' in stationary conditions. The dashed line represents the starting density profile of the time-dependent simulations, and it has been obtained using gas puffing as fuelling method. The density perturbation due to pellet ablation can involve a broad region of the plasma ($0.4 \leq \rho \leq 0.95$). The perturbed density profile keeps lower values than the Ordinary mode density cut-off for all the considered cases. The local density increase compared to the initial profile which spans from $\delta n_{e,max}$=1e19 $m^{-3}$ (15%) to $\delta n_{e,max}$=7e19 $m^{-3}$ (40%) depending on the pellet dimensions. However, in all the cases, the variation of the density values between two consecutive pellets does not exceed 10%.

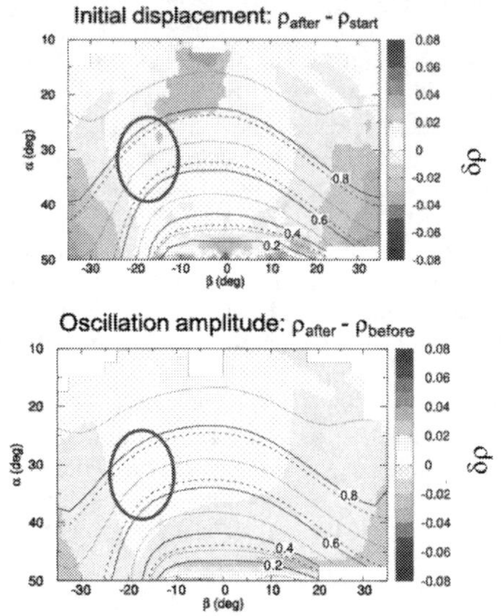


**Fig. 6**. Displacement δρ of the ECCD location due to pellets $n_e$ modifications by varying toroidal and poloidal angles for reference pellets (1 mm) (top: between the case just after pellet and without pellet, bottom: between the case just after pellet and just before pellet). Black contours: reference ECCD location (before pellet injection). Dashed lines represent the rational q surfaces for the DTT full power scenario (R=2.14 m configuration). Violet ellipse highlights the optimized intervals of injection angles which reach the rational q surfaces 2 and 3/2.

In order to investigate in detail the effects that these more realistic density profiles variations due to pellet injection have on EC propagation, absorption and CD, stand-alone GRAY simulations have been performed varying the injection angles and using as input the profiles of the different phases of the pellet cycle. We limit the analysis to the upper launcher case, because, as it has been found from the initial investigation regarding the maximum allowed perturbation, the inner region of the plasma core is reached even in presence of large density deviation.

Fig. 6 shows the radial displacement of the ECCD location due to density modifications induced by pellets for the reference case of radius=1 mm. With respect to the case without pellets, the radial deviation in correspondence of the location of the rational surfaces reached by the optimized injection angles (see Section 1) is $\delta\rho \sim 0.02$, and $\delta\rho < 0.02$ during a pellet cycle. Such displacement is lower than the width of the current density profile, then it should not require any adjustment of the steering angles.

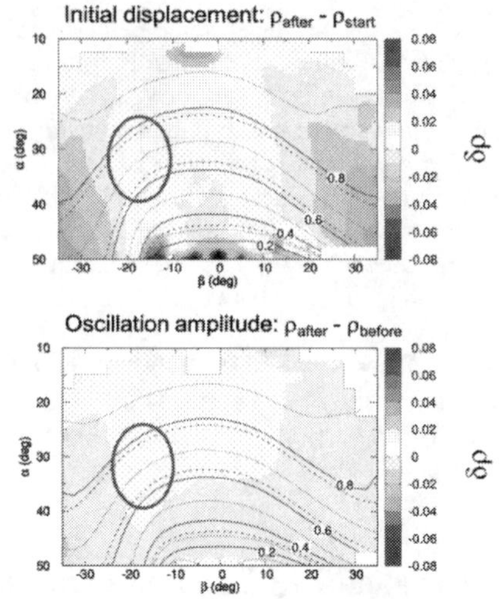


**Fig. 7**. Displacement δρ of the ECCD location due to $n_e$ modifications due by pellets varying toroidal and poloidal angles for pellets of 0.7 mm (top: between the case just after pellet and without pellet, bottom: between the case just after pellet and just before pellet). Black contours: reference ECCD location (before pellet injection). Dashed lines represent the rational q surfaces (R=2.14 m configuration). Violet ellipse highlights the optimized intervals of injection angles which reach the rational q surfaces 2 and 3/2.

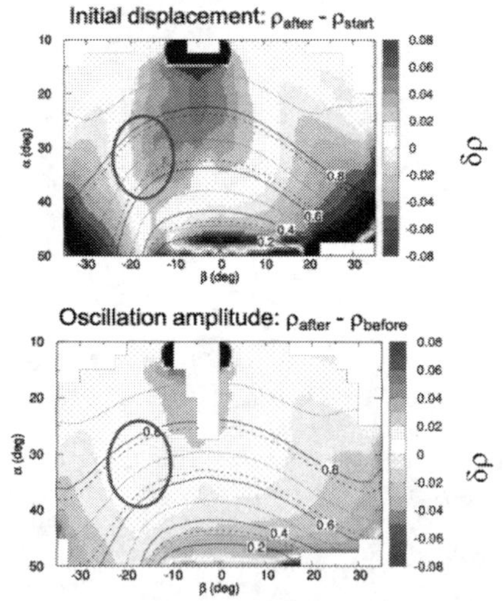


**Fig. 8**. Displacement δρ of the ECCD location due to $n_e$ modifications due by pellets varying toroidal and poloidal angles for pellets of 1.3 mm (top: between the case just after pellet and without pellet, bottom: between the case just after pellet and just before pellet). Black contours: reference ECCD location (before pellet injection). Dashed lines represent the rational q surfaces (R=2.14 m configuration). Violet ellipse highlights the optimized intervals of injection angles which reach the rational q surfaces 2 and 3/2.

As fig. 7 shows, in the case with pellets of radius 0.7 mm the deviation is almost negligible. The radial displacement for larger pellets can instead require some modification: for pellets with radius=1.3 mm (see fig. 8) the radial deviation with respect to the case with no pellets reaches values $\delta\rho\sim 0.04$ for the optimized angles for NTM control, while in a pellet cycle it is $\delta\rho\sim 0.02$. $\delta\rho=0.04$ is comparable to the ECCD profile width, then, in order to correctly localize the current on the rational surfaces, a modification of the poloidal angle of 2° is needed. We can conclude that, for realistic pellet sizes, the 'real-time' corrections of the steering angles during the plasma discharge, if needed, are of the order of the degree. The density variations due to pellet injection as predicted by HPI2 are then highly lower with respect to the values which prevent the EC beams reaching the outer plasma region (see Fig. 4). Such perturbations are then foreseen not really problematic for EC performances in realistic DTT scenarios.

## 4 Polarization analysis

In order to couple the EC launched wave to the desired mode in the plasma, the wave polarization must be properly imposed by the launcher. However, incident wave ellipticity ($\chi$) and ellipse inclination ($\psi$), which characterize the polarization, depend respectively on the magnitude of the magnetic field and the angle of incidence, and on the magnetic pitch angle at the plasma boundary. The angles steering, which could be needed during the plasma discharge for the achievement of the operational tasks, can have a not negligible effect on the ellipticity and the ellipse inclination, requiring to consequently adjust the injected wave polarization. Fig. 9 shows the values obtained for ellipticity and ellipse inclination varying the steering angles for the upper launcher. As expected, larger changes of the ellipticity are due to toroidal angle variation, while the ellipse orientation depends both by toroidal and poloidal steering.

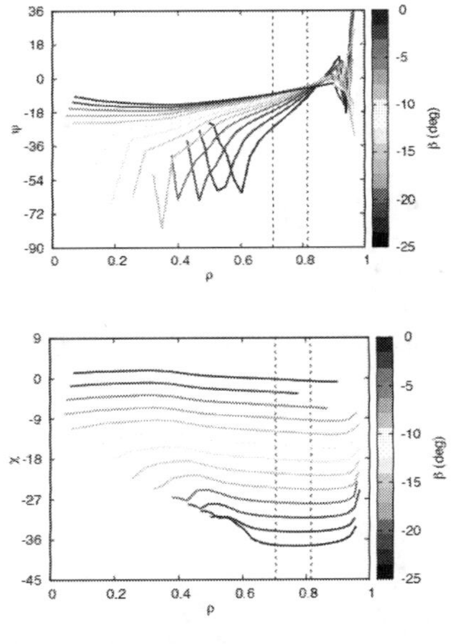


**Fig. 9**. (top) $\psi$ and (bottom) $\chi$ as functions of the average current density location. Colours refer to different $\beta$ values. Every point of every line corresponds to a different $\alpha$ value (10°-50°). Location of the rational q surfaces are shown by vertical dashed black lines.

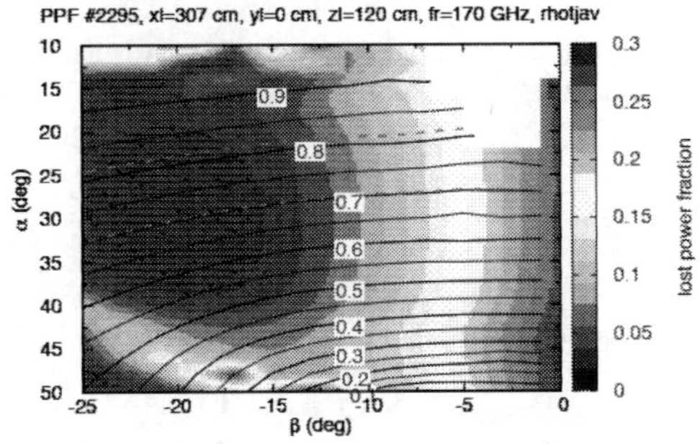


**Fig 10.** Lost power fraction using optimal $\psi$ and $\chi$ for q=3/2 for the upper launcher, varying toroidal and poloidal angles. Red dashed lines represent the rational surfaces locations.

An estimate of the power which could be lost during the angle steering, if the polarizer is not properly adjusted, is shown in Fig. 10. It has performed using as antenna polarizer parameters the optimal values for a selected angle couple ($\alpha$=30°, $\beta$=-20°, for reaching the rational surface q=3/2) and calculating for every other couple of angles the power which is not coupled to the ordinary mode. Considering the couple of angles reaching the rational surface q=2 ($\alpha$=23°, $\beta$=-20°), the fraction of lost power is very low, nearly 0.7%. The same analysis has been performed using as reference for the antenna polarizer parameters the ellipticity and the ellipse direction values of the couples of angles reaching q=2 in the case without pellets ($\alpha$=28°, $\beta$=-20°). Without variations of the reference polarization values, the lost power calculated for the couples of angles reaching q=2 for the cases with pellets is less than 0.6%, considering both profiles at the two sides of the pellet cycle.

## 5 Summary

The physics performances of the EC system have been investigated for the DTT reference full power scenario with the beam tracing code GRAY. The accessibility to the whole plasma core region with complete absorption is assured by both Equatorial and Upper EC launchers, with maximum current drive efficiency of 25 kA/MW located in $\rho\sim 0.1$. Regarding the optimization analysis for reaching the rational surfaces q=3/2 and q=2, the best compromise between maximum CD and enough narrow current density profile has been found for the toroidal angle $\beta$=20°.

The impact of pellet deposition perturbation on EC beam propagation has been studied, and EC performances have been assessed in time-dependent scenarios with pellet injection. The inner plasma core is always reached without any noticeable deviation, although lower prediction reliability is expected. In the region of the rational surfaces, for realistic pellets, the radial deviation of the EC localization with respect to the case without pellet profiles requires a maximum poloidal angle correction of few degrees. However, if the density profiles corresponding to the initial and final time of a pellet cycle are considered, no corrections are

required. Then the compatibility between the EC system and the pellet injection system has been verified for DTT. Additional calculations are needed to check such results for the time-dependent simulations of the DTT full power scenario with the present configuration R=2.19 m, however similar conclusions are expected.
Finally, the investigation of the wave polarization modifications due to injection angles steering has been performed. For the angle steering required for typical foreseen aims (i.e. sweeping between rational surfaces locations, or pointing the same rational surface during pellet injection) the power which is not be coupled with the ordinary mode has been found negligible.
The studies here reported have been performed on the flat-top of the reference DTT full-power scenario. As future work, the performances and optimization analysis is going to be focused on the dynamical phases of the DTT scenarios.

1. R. Martone *et al*, DTT Divertor Tokamak Test facility. Interim Design Report, ENEA (ISBN 978-88-8286-378-4), April 2019 ("Green Book")
2. S. Garavaglia *et al*, FED, 168, 112678 (2021)
3. A. Romano *et al*, Poster 1.17 at this conference
4. D. Farina, *et al*, Fusion Sci. Tech. **52**, 154 (2007)
5. I. Casiraghi, *et al*, Nucl. Fusion **61**, 116068 (2021)
6. I. Casiraghi *et al*, 48th EPS Conference on Plasma Physics (Virtual Edition, June 27-July 1 2022), O2.108
7. B. Baiocchi *et al*, to be submitted to Nucl. Fusion
8. G. Cenacchi *et al JETTO: A Free-boundary Plasma Transport Code* (JET Joint Undertaking, 1988)
9. B. Pegourie´ *et al* Nucl. Fusion 47 44, (2006)
10. C. Bourdelle *et al* Phys. Plasmas 14, 112501 (2007)
11. B.Pégourié and E.Geulin, 2020 Final Report of EFDA_D_2PEJWH - PMI-2.3-T011-D001 EXTRA - Common milestones PEX/DTT, Topic 4: - CEA-01 Neutrals and impurities model and pumping http://idm.euro-fusion.org/?uid=2PEJWH

# Refinement of the EC stray radiation estimates for ITER

*Franco Gandini*

ITER Organization, Route de Vinon sur Verdon - CS 90 046 - 13115 Saint Paul Lez Durance, France

**Abstract.** The ITER ECRH&CD system is composed by 24 gyrotrons at 170 GHz that will deliver 20 MW at the plasma. Up to 6.7 MW will be injected in the empty vacuum vessel at the beginning of each plasma discharge to provide the gas breakdown. In that phase and when the plasma absorption is non ideal, a certain level of EC non-absorbed power, usually addressed as stray radiation, will be present. The EC stray radiation interaction with ITER first wall and diagnostics has been described in preliminary works. A more refined assessment is here described, following update of the diffuse stray radiation model and update of the launchers optics parameters. The optical design of the EC equatorial launcher has been entirely redesigned to optimize the power deposition and minimize interaction with the launcher structures. The updated parameters for the 24 launched beams are now available and have been used to estimate the interaction of the beams to be used for the breakdown phase with the tokamak structures. The preliminary stray radiation model described every opening of the tokamak as a "black" hole, that is a perfect power sink. Refining this crude description, using for the openings a "grey" hole model, provide a better agreement with benchmarks from other alternative models. Examples of stray radiation estimates performed for various ITER structures, systems and diagnostics are discussed.

## 1 Introduction

The ITER tokamak is equipped with an ECRH&CD (Electron Cyclotron Resonance Heating & Current Drive) system composed by 24 gyrotrons at 170 GHz, each one with 1 MW unit power (baseline value). Each source is connected to an equatorial launcher (EL, installed in Port 14) and one of the upper launchers (UL, installed in Ports 12, 13, 15 and 16). The EL is used for central heating and current drive applications with a deposition range covering 0< $\varrho_T$ <0.6 (where $\varrho_T$ is the square root of the toroidal flux). The UL is used for control of magnetohydrodynamic (MHD) activity such as the neoclassical tearing mode (NTM) and sawtooth oscillation with a deposition range covering 0.4< $\varrho_T$ <0.88. The access ranges of each steering mirror are different for all launchers. This is particularly the case for the UL in which the upper steering mirror (USM) is used to access the inner region of the plasma for controlling sawtooth and q=3/2 NTMs, while the lower steering mirror (LSM) is used to control the q=3/2 and 2 NTMs. For what concern the equatorial launcher, three steerable mirrors collecting the RF beams delivered from eight waveguides are installed at the bottom of the shield slots. The beams are injected in a fixed toroidal direction and can sweep vertically from -30° to +30° by rotation of the steerable mirror. One mirror directs its 8 beams in the counter-current drive direction (cnt-ECCD, clockwise looking from the top, from port 14 towards ports 13, 12, 11, etc.) while the other two mirrors direct their 8+8 beams in the co-current drive direction (co-ECCD, counter-clockwise looking from the top, from port 14 towards ports 15, 16, 17, etc.). About the upper launchers, each one has 8 beams in input, 4 routed to the upper steering mirror and 4 routed to the lower steering mirror. The beams can sweep vertically from about -23° to about -68° by rotation of the steerable mirror. The precise angle range depends on the launching mirror (see table 1 for a summary of the functions and the steering capability of both launchers type).

**Table 1.** Summary of launcher design parameters

| Type | |
|---|---|
| Equatorial Launcher | Upper Launcher |
| **Functions** | |
| 1. Plasma heating and access to H-mode.<br>2. On- & off-axis CD for steady state operation.<br>3. Initial breakdown for plasma start up. | 1. Plasma heating and access to H-mode.<br>2. Off-axis CD mainly for NTM stabilization.<br>3. Initial breakdown for plasma start up. |
| **Ports** | |
| Equatorial 14 | Upper 12, 13, 15 & 16 |
| **Injected power** | |
| Up to 20 MW/port | Up to 6.7 MW/port |
| **Frequency** | |
| 170 GHz | |
| **beam steering capability** (α: poloidal, β: toroidal) | |
| Top mirror<br>$\alpha = \pm 10^\circ$<br>$\beta = 20^\circ$ cnt-ECCD<br><br>Middle mirror<br>$\alpha = -5^\circ \div -30^\circ$<br>$\beta = 25^\circ$ co-ECCD<br><br>Bottom mirror<br>$\alpha = 10^\circ \div 30^\circ$<br>$\beta = 25^\circ$ co-ECCD | Upper Steering Mirror<br>$\alpha = -28^\circ \div -68^\circ$<br>$\beta = 15^\circ \div 22^\circ$ co-ECCD<br><br>Lower Steering Mirror<br>$\alpha = -23^\circ \div -60^\circ$<br>$\beta = 17^\circ \div 21^\circ$ co-ECCD |

### 1.1 Operating modes

To accomplish the design scopes of the EC system there are critical phases characterized by low absorption and therefore high levels of stray radiation that could harm tokamak structures:

#### *1.1.1 Breakdown*

Up to 6.7 MW of EC power are going to be injected in the "empty" ITER chamber at the beginning of discharge in order to assist gas ionization and plasma formation (breakdown). In this phase there is no RF power absorption from the plasma, only absorption from the wall at the points where the beam is reflected and, after some bounces, diffuse radiation that can escape from the chamber openings. The wavelength in vacuum of the EC radiation is 1.76 mm, therefore any aperture larger than 0.8 mm can be considered as an "escaping opening".
The expected duration of this phase is no longer than few seconds, with the EC power injected by burst of no more than 300 ms.

#### *1.1.2 Burn through (total EC power injected: up to 6.7 MW)*

It is the phase after initial ionization and breakdown, when a current channel forms and the discharge is heated. In this phase the radiation from low Z impurities must be lower than the injected power to successfully evolve the discharge to the current ramp up phase. Moreover, in this phase the power injected to compensate the radiation from impurities is not significantly absorbed by the plasma and the power loading for the chamber wall is of the same kind described at the previous point.

The expected duration of this phase is few seconds (nominally 5.5 seconds). The amount of power to be used will be assessed during the PFPO1 operational phase.

#### *1.1.3 HCD with non-optimum polarization (total EC power injected: up to 20 MW)*

The absorption of EC waves is optimized only when injected with the correct polarization. Any error in polarization would mean additional stray radiation. The fraction injected with the wrong polarization will undergo strong diffraction and/or reflection at the cut off layer. The correct description of this effect requires a beam tracing calculation. A qualitative estimate of the power deposition can be obtained evaluating the beam size of the cross polarized fraction of the injected beam at the possible incidence point and the power density dividing the power in the wrong polarization by the projected area of the beam.
Without the control of the effective injected polarization, a conservative assumption is that the stray power coming from non-optimum polarization will be present through the whole discharge.

## 2 Description of the residual Power

When the plasma absorption is poor or negligible, it is assumed that the structure of the injected astigmatic Gaussian beams is maintained at least through the first 3 bounces: this assumption is a simplification to over-estimate the peak power deposition and the average power loading on the chamber wall from the collimated beam (shine through). Also, the wall structure and shape are not taken into account: the wall is described as a plane perfect reflector. After some reflections we assume that the radiation inside the tokamak chamber cannot be anymore described as a collimated beam but, on the contrary, it may be described as diffuse (stray) radiation.

### 2.1 Shine through

The breakdown and burn through assistance require up to 6.7 MW injected from the equatorial launcher. The following results have been obtained considering the injection of 8 beams (0.8375 MW each) from the top steering mirror of the equatorial launcher (port 14). As zero order assumption, the exact vertical locations of the mirror (z = 0.04 m) is neglected, and the beams are considered to come from the equatorial plane z = 0 and propagate in the horizontal plane (see Fig.1).

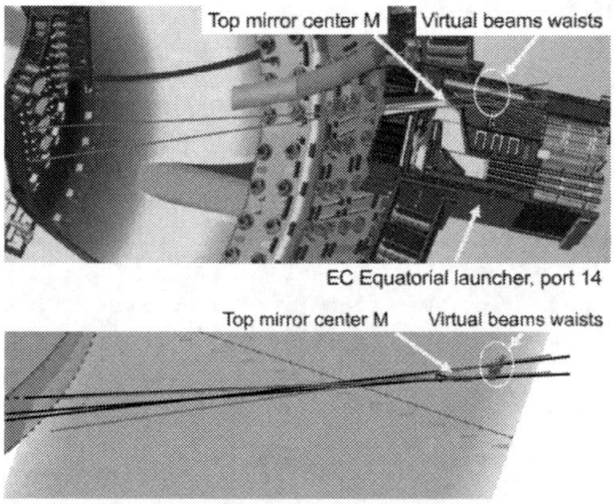


**Fig. 1**. Launching scheme from the top mirror of port equatorial 14: above, cross section of launcher and vacuum vessel; below, geometrical representation of the beams.

The following assumptions are considered:

- Pure gaussian propagation from virtual waists locations (Fig. 2)
- Central column represented as cylindrical surface at R=4.0 m
- Low field side wall (LFSW) represented as cylindrical surface at R=8.5 m
- Mirror-like reflections at tangential plane at beam axis incident points (Fig. 3)
- Merged beams described as gaussian (conservative assumption, Fig. 5)

The eight beams are followed individually. The beams are considered merged (to simplify the analysis) when, at the reflection's locations, they are close enough.

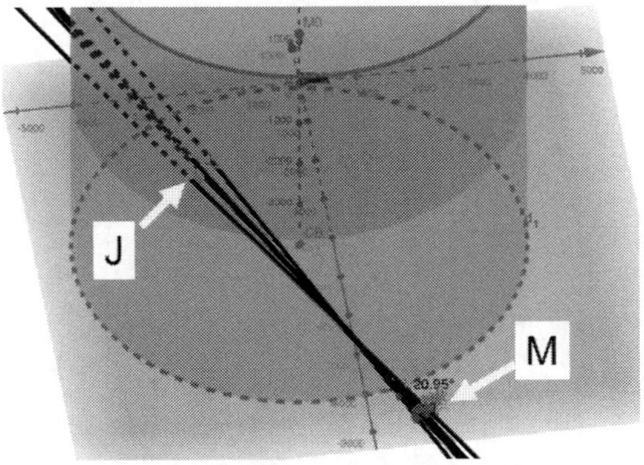


**Fig. 2**. Beams propagated from their virtual waist location along their **k** vectors, estimated path length from waists area to the central column (7.3 m), identified point "J" at 6.6 m from "M"

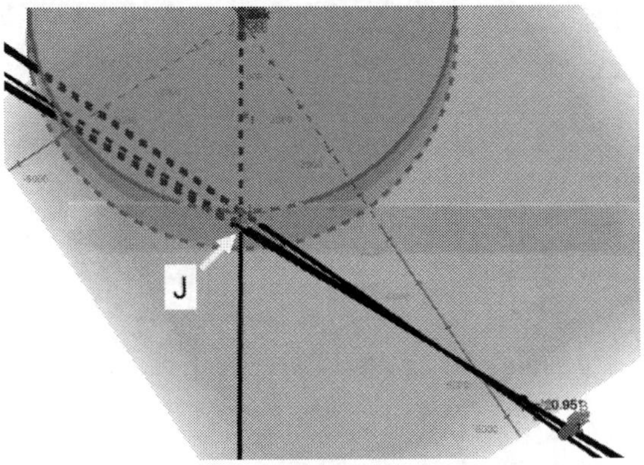


**Fig. 3**. Built plane in "J" tangential to the central column

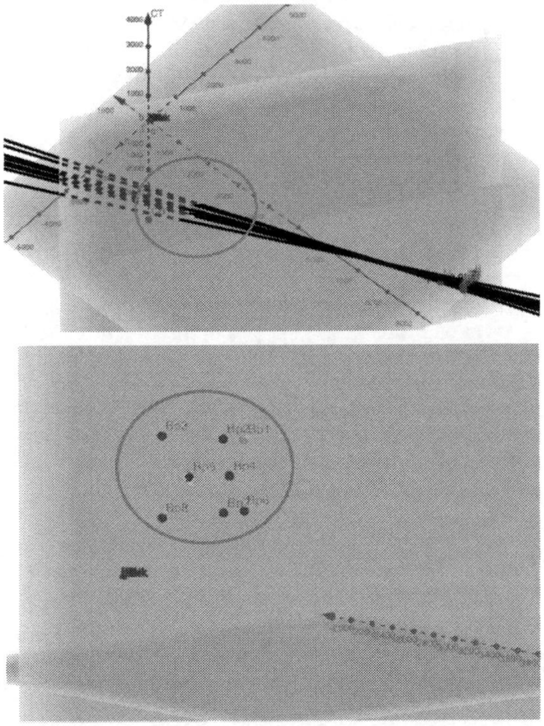

**Fig. 4**. Incident points of the 8 beams **k** vectors

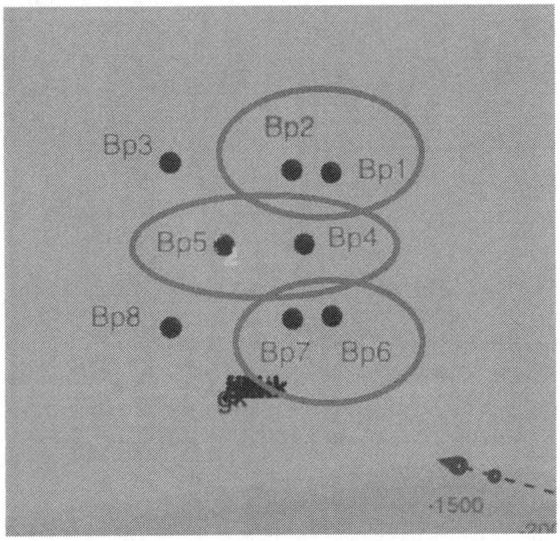


**Fig. 5**. Incident points of the 8 beams. The beams considered merged are the ones incident at (Bp1-Bp2), (Bp4-Bp5) and (Bp6-Bp7)

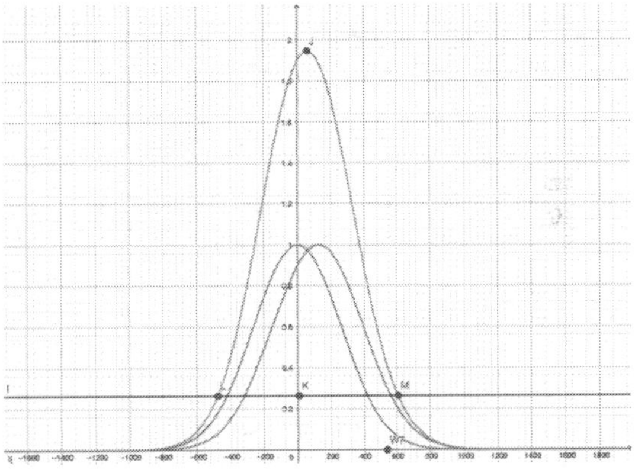

**Fig. 6**. Beams at (Bp1-Bp2) and (Bp6-Bp7) superposed in the horizontal direction, The distance (Bp1-Bp2) and (Bp6-Bp7) is 124 mm

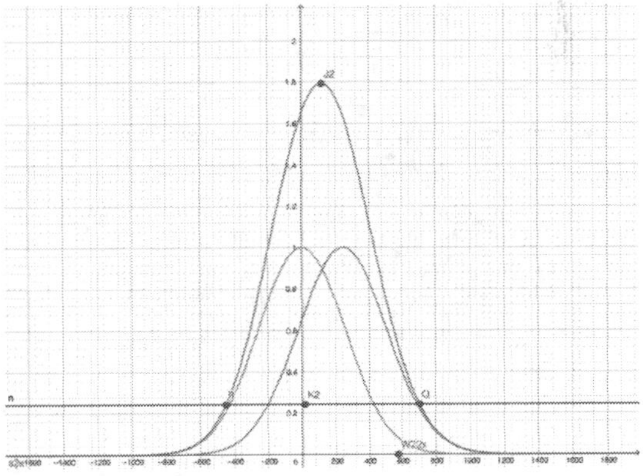

**Fig. 7**. Beams at (Bp4-Bp5), distance 244 mm, superposed in the horizontal direction

The six beams (Bp1+Bp2), (Bp6+Bp7) (see Fig. 6) and (Bp4+Bp5) (see Fig. 7), once merged horizontally and described as gaussian beams, are then further merged vertically. The beam sizes (radius at 1/e, projected on the plane tangential to the central column) are: (Bp1+Bp2) and (Bp6+Bp7) 543 mm horizontally and

224 mm vertically; (Bp4+Bp5) 577 mm horizontally and 214 mm vertically.

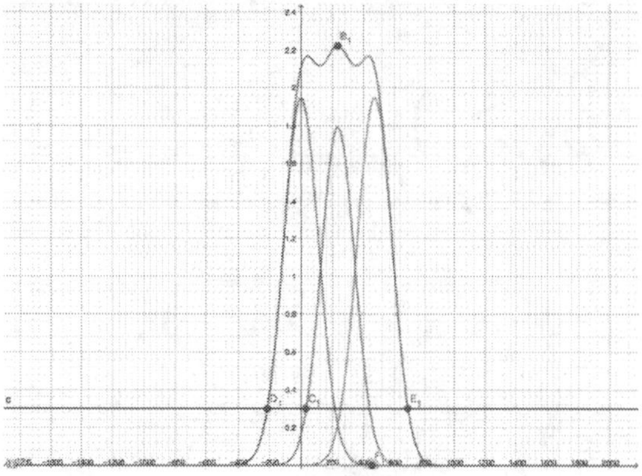

**Fig. 8**. Beams (Bp1+Bp2), (Bp6+Bp7) and (Bp4+Bp5) superposed in the vertical direction

Finally, the six beams are described as a superposed astigmatic gaussian beam with sizes $W_x$ = 565 mm horizontally and $W_y$ = 451 mm vertically (see Fig. 8). The peak intensity is calculated as

$$P_{axis}=(2\,P_{in})/(\pi\,W_x\,W_y) \qquad (1)$$

where $P_{in}$ is the six beams merged input power and $P_{axis}$ is the power density on the axis. The average power density $P_{aw}$ is estimated considering the whole transmitted power distributed over the projection of the merged beam on the chamber wall at the reflecting points

$$P_{aw}=P_{in}/(4\pi\,W_y\,W_x/\cos\alpha) \qquad (2)$$

where $\alpha$ is the merged beam incident angle. The same analysis can be extended to the next reflection at the low field side (Fig. 9)

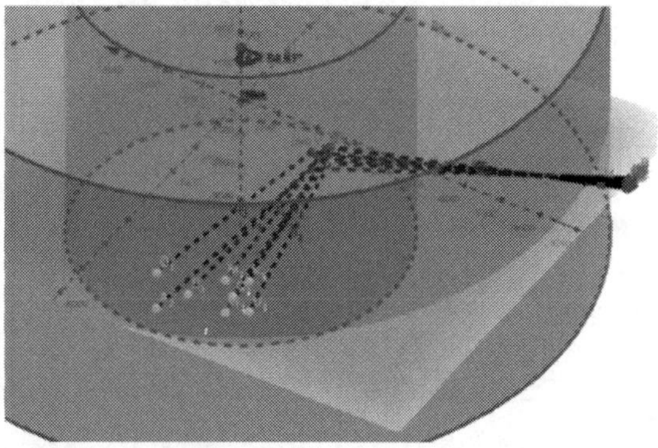

**Fig. 9**. Incident points at low field side, second reflection

Considering EC-assisted breakdown from the ECH Equatorial launcher (8 beams injected from the same mirror for a total power of 6.7 MW), the expected incident peak power density at the first bounce location on the HFS wall is about 13 MW/m$^2$ (but just on the beam axis), with absorption of about 1% in the wall itself (absorbed peak power: 0.13 MW/m$^2$). The average incident power loading is 1.6 MW/m$^2$ (absorbed average power: 0.016 MW/m$^2$). The second bounce is in the area between port 10 and port 11 (see Fig. 10) and the expected peak power is about 3.8 MW/m$^2$ (again, to be considered only on the beam axis), while the average power loading is 0.5 MW/m$^2$. To be considered that the beams footprint at the second bounce is quite large, with a projected horizontal beam size $W_x$ = 767 mm and a vertical beam size $W_y$ = 928 mm, for a 4w beam area of nearly 9 m$^2$.

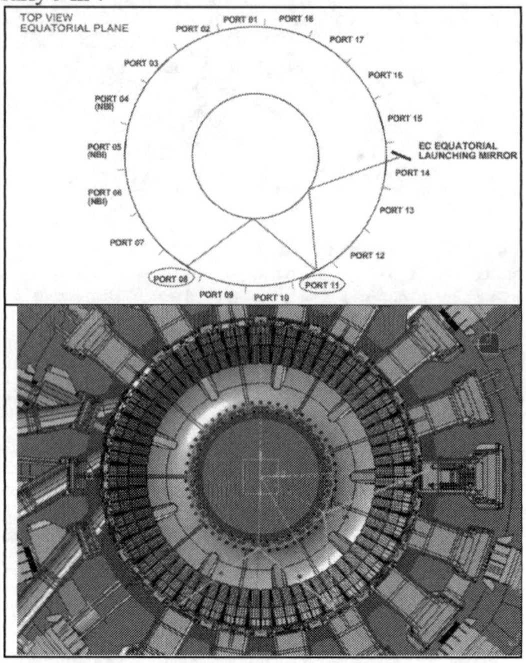


**Fig. 10**. First reflection locations for beams injected from equatorial port 14: above, geometrical representation of the beams; below, top view of tokamak equatorial plane section.

These results have been obtained considering the refined equatorial launcher beams parameters and significantly relax the previous loads that were calculated using the preliminary beams parameters [1]: the previous $P_{axis}$ at the first bounce was 334 MW/m$^2$, while the $P_{aw}$ was 25 MW/m$^2$.

**Table 2.** Summary of incident peak power density $P_{axis}$ and average power density $P_{aw}$ for the first two bounces

| 1$^{st}$ bounce | | 2$^{nd}$ bounce | |
|---|---|---|---|
| $P_{axis}$ *MW/m$^2$* | $P_{aw}$ *MW/m$^2$* | $P_{axis}$ *MW/m$^2$* | $P_{aw}$ *MW/m$^2$* |
| 13 | 1.6 | 3.8 | 0.5 |

## 2.2 Diffuse stray radiation

After few reflections on the tokamak walls, it is assumed that the injected power loses its beam structure and a diffuse, more or less isotropic background level of stray radiation is established. A simplified description of the behaviour of the stray radiation in the chamber can be developed from the integrating sphere theory, a concept that has been developed to describe the input/output radiation ratio in a sphere with finite reflectivity $\rho$ and multiple openings [2, 3]. The simplest version of the model considers the reflective surfaces having all the same reflectivity and all the openings as perfect absorbers (black holes):

$$\tau_{BH}=\frac{A_i}{A_s}\frac{\rho}{1-\rho\left(1-\frac{\sum A_j}{A_s}\right)} \qquad (3)$$

where $\tau_{BH}$ is the throughput of the sphere ("black hole" case), i.e., the ratio of the exiting flux from a test opening (the "exit opening") to that entering the sphere. $A_i$ is the exit opening area; $A_j$ is the summation of all the openings areas and $A_s$ is the area of the sphere. However, most of the large opening considered (equatorial and upper ports openings, divertor gap, NBI ports) have a complex structure and describe them as "black" hole is not completely justified. The results from the preliminary work [1] ("black hole" model) have been compared with alternative modeling [4, 5], and a good agreement can be achieved including a term in the equation and treat the openings as "grey", attributing them a reflection coefficient $\rho_G$:

$$\tau_{GH} = \frac{A_i}{A_s}\frac{\rho}{1-\rho\left(1-\frac{\Sigma A_j}{A_s}\right)-\rho_G\frac{\Sigma A_j}{A_s}} \quad (4)$$

$A_s$ = Total surface area = 858.3 $m^2$
$\Sigma A_j$ = Total opening surfaces = 43.5 $m^2$
$\rho$ = Wall reflectivity = 0.99 average

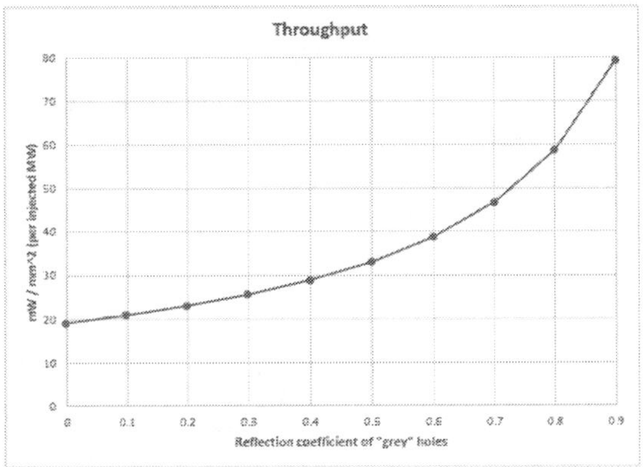


**Fig. 11**. Output flux (throughput) from a small opening as function of "grey" hole's reflection coefficient

A conservative preliminary assumption is to consider an average reflection coefficient $\rho_G = 0.6$ for the openings and therefore a throughput of about 40 mW/mm² per injected MW (Fig. 11).
In the tables 3 and 4 are reported few examples of expected output flux from some openings and behind the blankets during the breakdown phase (6.7 MW injected in the empty chamber).

**Table 3.** expected output flux from some openings during the breakdown phase (6.7 MW injected in the empty chamber)

| | Opening surface $m^2$ | Output flux (kW) black holes 20 mW/mm² | Output flux (kW) grey holes 40 mW/mm² |
|---|---|---|---|
| NBI large port | 1.19 | 160 | 319 |
| Reflectometer (LFS, EqP11) ϕ31.75 mm | 0.00079 | 0.10 | 0.21 |
| Reflectometer (LFS, EqP11) ϕ50.8 mm | 0.0020 | 0.27 | 0.54 |
| Reflectometer (LFS, EqP11) ϕ63.5 mm | 0.0032 | 0.43 | 0.86 |
| Reflectometer (LFS, EqP11) ϕ88.9 mm | 0.0062 | 0.83 | 1.66 |

**Table 4.** power to be expected diffused behind the blanket modules

| | mW/mm² per injected MW Black holes | mW/mm² per injected MW Grey holes |
|---|---|---|
| HFS – mod 1÷5 – vertical and horizontal gaps | 4.5 | 9.0 |
| HFS – modules 6÷7 – vertical gaps | 8.6 | 17.2 |
| HFS – modules 6÷7 –horizontal gaps | 6.1 | 12.2 |
| LFS – vertical and horizontal gaps | 16.7 | 33.4 |

## 3 Conclusion

The model here described is a simplification of reality with the purpose of providing a conservative assessment of the expected level of EC incident power on the central column and on the Low Field Side Wall (LFSW).
The real surface of the Blanket panels at the central column is different from an ideal cylinder and even more different from a plane: the reflected beams from such shaped surface will diverge more than considered here, lowering the power density at LFSW.
The merged beams are described as they would have gaussian shape, as consequence both the peak power and the average power density are overestimated.
For a more realistic assessment of the wall loading is needed a beam tracing and accurate model of the reflecting wall shape.
The updated parameters for the 24 launched beams have been used to estimate the interaction of the beams to be used for the breakdown phase with the tokamak structures.
The preliminary stray radiation model described every opening of the tokamak as a "black" hole, that is a perfect power sink. Refining this crude description using for the openings a "grey" hole model provides a better agreement with benchmarks from other alternative models.
*The views and opinions expressed herein do not necessarily reflect those of the ITER Organization*

## References

1. F. Gandini, M. Henderson, C. Nazare, C. Darbos, T. Gassmann, D. Purohit, and T. Omori , "Stray RF Power Estimates From EC Exploitation During ITER Plasma Operations", AIP Conference Proceedings 1406, 173-176 (2011) https://doi.org/10.1063/1.3664955
2. D. G. Goebel, "Generalized Integrating-Sphere Theory", *App. Opt.* **6**, n.1, 125-128 (1967).
3. D. J. Lovell, "Theory and Application of Integrating Sphere Technology", *Laser focus/Electro-optics*, 86-96 (1984)
4. D. Moseev, J.W. Oosterbeek, A. Sirinelli, Y. Corre, M. Houry, S.B. Korsholm, H.P. Laqua, S. Marsen, M. Preynas, J. Rasmussen, M. Salewski, T. Stange, V. Udintsev, *Stray radiation energy fluxes in ITER based on a multiresonator model*, Fusion Eng. Des., **172**, (2021), 112754, ISSN 0920-3796, https://doi.org/10.1016/j.fusengdes.2021.112754
5. J. Oosterbeek, V.S. Udintsev, F. Gandini, M. Hirsch, H.P. Laqua, N. Maassen, Y. Ma, A. Polevoi, A. Sirinelli, G. Vayakis, M.J. Walsh, *Loads due to stray microwave radiation in ITER*, Fusion Eng. Des., **96–97**, 553-556 (2015), ISSN 0920-3796, https://doi.org/10.1016/j.fusengdes.2015.05.068

# ECRF stray radiation studies in preparation of the operations of JT-60SA

*Carlo* Sozzi[1*], *Ken* Kajiwara[2], *Takayuki* Kobayashi[2], *Lorenzo* Figini [1], *Luca* Garzotti[3], *Alessandro* Moro[1], *Silvana* Nowak[1], *David* Taylor[3].

[1]Institute for Plasma Science and Technology (ISTP-CNR), Italy
[2]National Institutes for Quantum and Radiological Science and Technology, QST, Naka, Ibaraki 311-0193, Japan
[3]CCFE, Culham Science Centre, Abingdon, UK

**Abstract.** JT-60SA tokamak is equipped with an ECRF system since the beginning of its operational phase. Starting from two gyrotrons units during the Integrated Commissioning, applicable for core heating, assisted breakdown and assisted Wall Conditioning, the system capabilities will be progressively extended from the Initial Research phase for wider applications. The development of the full current plasma H mode scenario 2 (inductive, type I ELM, $I_p$=5.5 MA, $B_T$=2.25 T, $q_{95}$=3) is among the first scientific objectives of the research program. In preparation of this, predictive modelling of the current ramp-up in scaled versions of scenario 2 is being done, based on parameters previously published. In this scenario the ECRF power is injected from an early phase of the discharge. Such modelling provides the kinetic profiles giving the opportunity to estimate the expected amount of EC stray radiation during the ramp-up phase when the EC power absorption might be less than 100% and consequently the potential risk of damage of the in-vessel components is higher.

## 1 Introduction

JT-60SA is the large superconducting tokamak built under the Broader Approach agreement jointly by Europe and Japan [1] and now under its commissioning phase. It is designed to address many areas of fusion science in preparation of the burning plasma era of ITER and DEMO, in particular the ones related with the control of high β steady state plasmas and the confinement of high energy particles. Several different main operation scenarios (plus variants) are foreseen for JT-60SA [1, 2, 3] in the research phases, with toroidal magnetic field of 2.25-2.28 T (inductive scenarios) and 1.62-1.72 T (non-inductive).

At its full development, the ECRF system is designed for 9 gyrotrons able to inject up to 7 MW to the plasma from 4 toroidally distributed upper oblique launchers with extended poloidal and toroidal steering capability and pulse duration up to 100 s. Each gyrotron can provide EC power at three different frequencies, namely 82, 110 and 138 GHz [4, 5]. Before full performance will be made available, gradually increasing capabilities are foreseen for the Integrated Commissioning (IC) phase (~2023) and for the Initial Research phase (IR, ~2025-2026).

For the IC 2 gyrotron of about 1 MW power each (at source) are available, one at fixed 110 GHz frequency, and one at selectable frequency of 82, 110 and 138 GHz. In the latter, the maximum power depends on the selected frequency (1.5 MW at 110 GHz, 0.75 at 138 and 0.6 at 82). The two gyrotrons have separated power supplies and then can be pulsed with different waveforms, in particular at different starting times. In the IC the ECRF power is launched with a fixed angle two-waveguide launcher parallel to the port axis (and perpendicular to the toroidal field) at 35.5° with respect to the equatorial plane. Both X and O polarization (and their combinations) are available. The ECRF pulse duration is up to 5 sec for 110 and 138 GHz and up to 1 sec for 82 GHz. The launching settings available during the IC, with beam injection towards the centre of the vessel, allow the application of the ECRF power for core heating, ECRF assisted breakdown and ECRF assisted Wall Conditioning.

For its IR phase, the JT-60SA tokamak will be equipped with four gyrotrons units delivering a maximum of 3 MW at 110 GHz to the plasma, or 1.5 MW at 110 GHz plus 1.5 MW at 138 GHz. The available additional power also includes 6 MW of P-NBI and 10 MW of N-NBI.

The development of the so-called scenario 2 (inductive H mode scenario, lower single null CFC divertor plasma configuration, type I ELMs, $I_p$=5.5 MA, $B_T$=2.25 T, $q_{95}$=3) is among the first scientific objectives of the IR phase. In preparation of this goal, predictive modelling of the current ramp-up is being performed. This was first analysed with a fast integrated tokamak modelling tool (METIS+FEEQS) in [6]. A similar current ramp-up in fully predictive way was simulated using the JINTRAC suite of codes [7]. In these simulations the plasma current evolution was imposed as boundary condition, the equilibrium was calculated

* Corresponding author: carlo.sozzi@istp.cnr.it

with the ESCO equilibrium solver and updated periodically throughout the simulation. The transport model adopted for the simulation is the Bohm/gyro-Bohm transport model. The density was controlled by a numerical feed-back mechanisms adjusting the prescribed gas puff to deliver the required density ramp. The power deposition was calculated with GRAY (ECRF) and PENCIL (NBI) [7].

Safe approach to operations at full plasma current suggests a gradual development path to reach the full performance scenario, like the one described in the JT-60SA Research Plan [1]. For this purpose, variations of scenario 2 at reduced plasma current and magnetic field are being considered in this paper. In particular a range of ramp-up cases are analysed. The ECRF power is injected from an early phase of the discharge when the EC power absorption is generally less than 100%. Consequently, the potential risk of damage of the in-vessel components may be higher. The JINTRAC modelling provides $T_e$ ad $n_e$ profiles from which is possible to estimate the expected EC absorption and, conversely, the amount of EC stray radiation during the ramp-up. Fixed ECRF launching settings are assumed, with the beams directed towards the plasma core. Such launching settings are representative either of the basic configuration for core heating in the IR phase and of the one available during the IC phase.

In all the cases except one the rise of the plasma current is limited to 2.1 MA. For the IC plasma currents ranging from few hundred kA up to about 2.5 MA [8] are being considered in single null configuration with the X-point close to the top of the vessel. However, since the absorption and beam trajectory mainly depend on plasma temperature and density in the plasma core, the results here obtained for the reduced scenario 2 are also relevant for the IC provided that similar axial density and temperature are realized.

The study of the current ramp-up phase complements the analysis of the low absorption scenarios being considered for the design of the EC stray detection system for the IR phase [9]. To this purpose, the adaptation to the JT-60SA parameters of the differential bolometers being developed for ITER is in progress. No direct measurements of the EC stray radiation are available in the IC.

The expected EC stray power loads on PFCs due to shine-through of the EC beams for some specific case is also reported in the paper.

## 2 Heating scenarios during the plasma current ramp-up

One possible option for the development of a plasma scenario up to full performance is the progressive increase of the flat top plasma current and of the toroidal field while keeping constant the edge safety factor $q_{95}$. This approach is motivated by the opportunity of minimizing the operation risks related to high plasma current while keeping similar magnetic equilibria and plasma shape, also reducing the amount of additional heating needed to sustain the scenario.

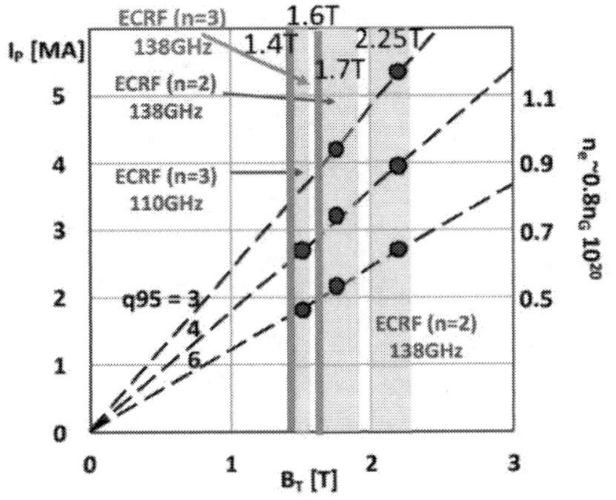


**Fig. 1.** Possible operational path for the rise to 5.5MA in the Initial Research phase. Adapted from [1]

Alternatively, the plasma current can be increased at a given toroidal field progressively lowering the $q_{95}$. The range of possible $I_p$, $B_T$, EC frequency combinations that allow the use of the ECRF heating is shown Fig. 1 for three values of $q_{95}$.

**Table 1.** ECRF heating cases during the current ramp-up

| case | $B_T$ [T] | Ip [MA] | Te [keV] | $n_e$ [$10^{19}$ $m^{-3}$] | t[s] | Injected $P_{EC}$ [MW] |
|---|---|---|---|---|---|---|
| 1a | 2.25 | 0.81 | 0.55 | 0.81 | 2.9 | 0.5 |
| 1b | 2.25 | 2.86 | 6.7 | 1.18 | 8.3 | 3.5 |
| 1c | 2.25 | 5.5 | 10 | 3.1 | 15 | 7 |
| 2a | 2 | 2.1 | 1.55 | 1.6 | 8.2 | 0.75 |
| 2b | 2 | 2.1 | 3.32 | 1.75 | 9.2 | 1.5 |
| 3 | 2 | 2.1 | 1.57 | 1.61 | 8.2 | 1.5 |
| 4a | 1.8 | 1.08 | 0.56 | 0.9 | 3.3 | 0.215 |
| 4b | 1.8 | 1.9 | 1.31 | 1.07 | 5.3 | 0.428 |
| 4c | 1.8 | 2.1 | 2.33 | 1.6 | 8.3 | 0.75 |
| 5a | 1.8 | 2.1 | 1.52 | 1.58 | 8.2 | 0.75 |
| 5b | 1.8 | 2.1 | 2.85 | 1.74 | 9.2 | 1.5 |
| 6a | 1.7 | 2.1 | 1.37 | 1.5 | 8.25 | 0.75 |
| 6b | 1.7 | 2.1 | 2.27 | 1.87 | 9.3 | 1.5 |
| 7 | 1.7 | 2.1 | 1.32 | 1.5 | 8.2 | 1.5 |
| 8 | 1.6 | 2.1 | 1.6 | 1.7 | 8.2 | 0.75 |
| 9a | 1.4 | 1.06 | 0.41 | 1.2 | 3.3 | 0.215 |
| 9b | 1.4 | 1.92 | 1.07 | 1.17 | 5.3 | 0.428 |
| 9c | 1.4 | 2.1 | 1.39 | 1.67 | 8.3 | 0.75 |

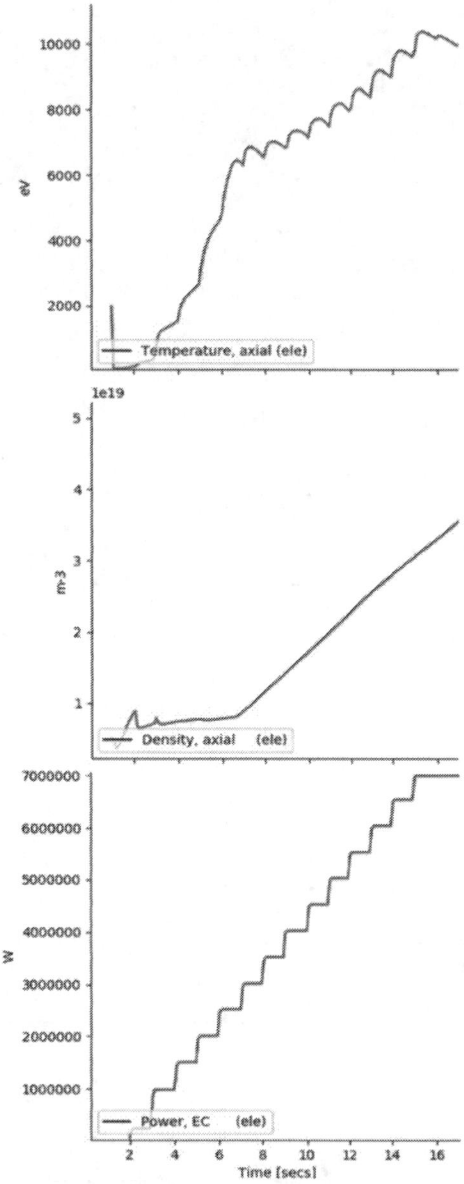


**Fig. 2.** Time evolution of the axial electron temperature (top), of the axial density (mid) and of the absorbed EC power (bottom) for case 1 simulation in Table 1.

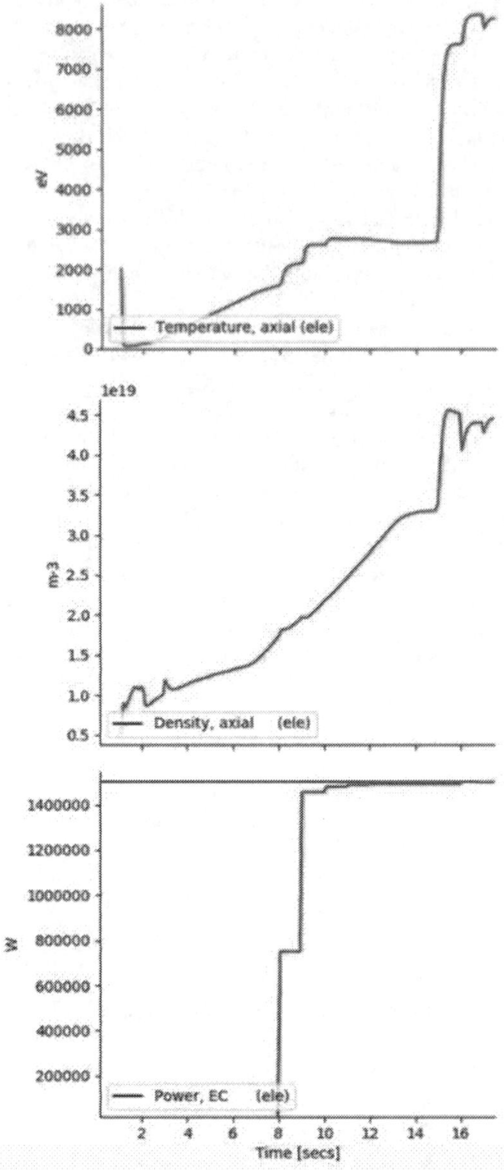


Fig. 3. Time evolution of the axial electron temperature (top), of the axial density (mid) and of the absorbed EC power (bottom) for case 6 simulation in Table 1.

A reduced subset of heating cases in current ramp-up is listed in Table 1 in which the main parameters at the time and radial location of the ECRF injection are listed, along with the power at the corresponding time.

The case label (first left column) corresponds to a given simulation run (1,2…) at different times (a,b,…). Cases 1, 4 and 9 are power ramps with injection starting 2 s after the breakdown (see Fig. 2), while the others are power steps with injection starting at 8 s (see Fig. 3).

During the ramp-up of the discharge ECRF is the only active additional heating system.

Case 1 refers to the nominal scenario 2 at full current (5.5 MA) and full ECRF power (7 MW to the plasma). The time evolution of the electron temperature, of the density at the magnetic axis and of the absorbed ECRF power is shown in Fig. 2. In all the other cases 2-9, the power of one or two gyrotrons only is considered. For these cases the current ramp ends at 6 s, then in cases 2, 3, 5, 6, 7 and 8 the injection occurs 1 in 1 or 2 power steps, separated by 1 s, at the beginning of the current flat top during the rise of the density and of the temperature as shown in Fig. 3 for case 6. The sudden rise of density and temperature at a later time in the simulation (15 s) with respect to the ECRF power steps is related with the switch on of the Neutral Beam power which is not discussed in this paper. The effect of the ECRF power is visible in density and temperature time traces. The resonance layers for the two EC frequencies in this case are located at 20-30% of the minor radius.

In cases 4 and 9 the ECRF injection starts in the early phase of the current ramp, at a temperature of few hundred eV and density of about $1 \cdot 10^{19}$.

In some cases (4, 5, 6, 7, 9) two different EC frequencies are injected, 110 and 138 GHz. The combination of two different frequencies provides some flexibility in the heating radial profile prior of the availability of the steerable ECRF antenna.

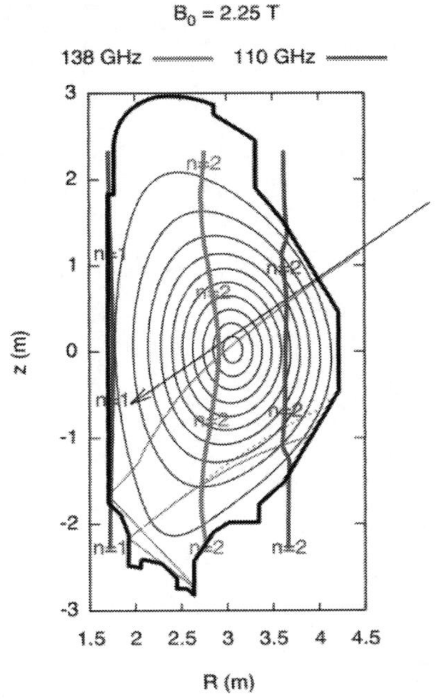


**Fig. 4.** Beam trajectory (red lines) and EC resonances for the harmonic n for scenario 2 at nominal magnetic field. The black arrow represents the initial injection direction. Applies to case 1.

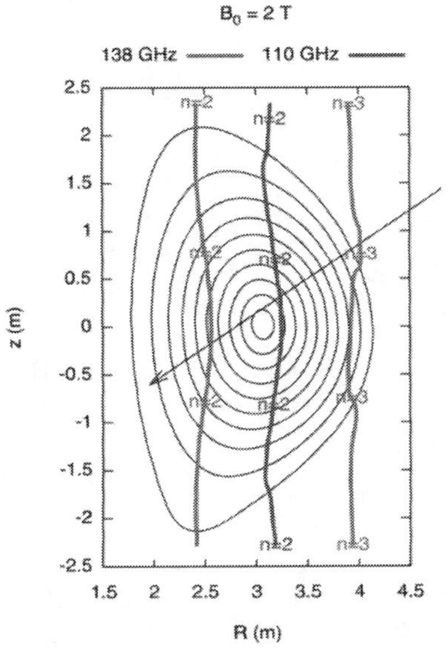


**Fig. 5.** EC resonances for $B_0$=2T (cases 2, 3).

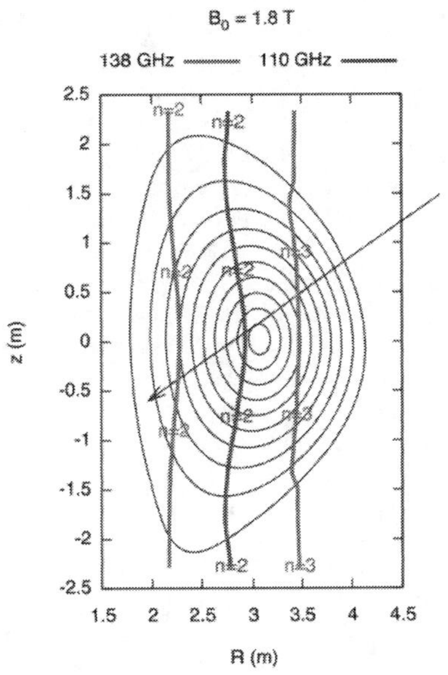


**Fig. 6.** EC resonances for $B_0$=1.8T (cases 4, 5)

Example applications in which this may be used is for extending the volume of the region of EC assisted breakdown, or for tailoring the current profile during the ramp.

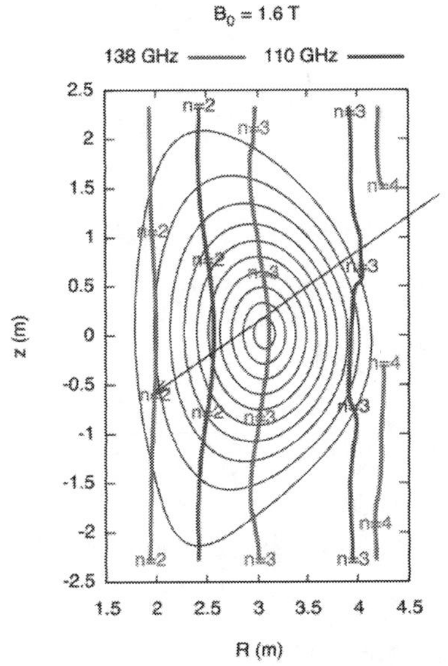


**Fig. 7.** EC resonances for $B_0$=1.6 T (case 8).

Figures 4-8 show the position of the EC resonances for harmonics n=2,3,4, X polarization, for both 110 GHz and 138GHz frequencies for the heating scenarios of Table 1. In all the cases the beam is launched towards the plasma centre (-35.5° with respect to the horizontal plane). Fig. 4 also shows the beam trajectory computed for the reference scenario (volume averaged electron temperature and density respectively 6.3 keV and 0.56 $10^{20}/m^3$ in the current flat top). The beam trajectory is extended showing the reflection points in the vessel in the case in which the absorption is not total.

In some of the heating scenarios (e.g. Figures 5 and 7) absorption layers enter the outer region of the plasma, either in the low or high field side, resulting in the broadening of the absorption region and in the depletion of the power delivered to the plasma centre. This is for example visible in Fig. 9 that shows density (top), temperature (mid) and power deposition profiles (bottom) for case 6b, where the appearance of the 2nd harmonic absorption layer at the plasma edge subtracts power to the 3rd harmonics absorption layer in the plasma centre. This occurs to a much lesser extents when the edge absorption layer is of a higher harmonic as for example cases 2 and 3 in Fig. 5. However, higher harmonic outer absorption may become detrimental when the scenario is fully developed and the pedestal temperature increases. In the ramp-up cases here discussed, third harmonics absorption become effective, and apparent in the power deposition profile, when the temperature at the absorption region is above ~1 keV.

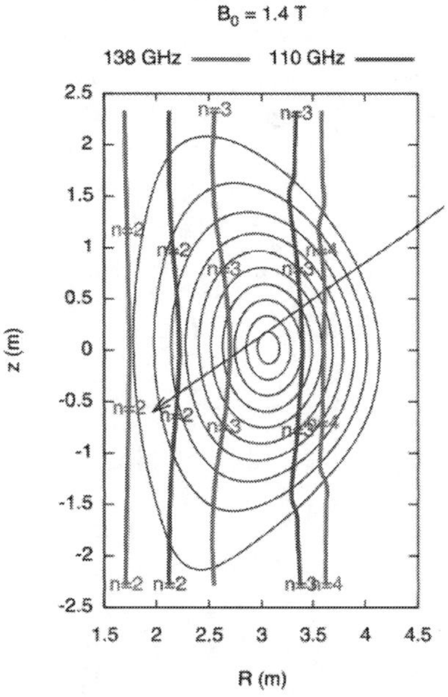


Fig. 8. EC resonances for $B_0$=1.4T (case 9)

# 3 Evaluation of the stray EC radiation

The results of the simulations for what concerns the evaluation of the EC stray radiation are summarized in Table 2 for the same cases listed in Table 1. For each heating case, along with the EC frequencies and harmonics, the position of the absorption layer, the percentual absorption (abs%) and the corresponding amount of EC stray power ($P_{EC}$ stray) are listed in the two last right-hand columns. The absorption is evaluated at the time of the switch on assuming perfect matching of the polarization of the injected beam (100% X in this case). The amount of EC stray power (last right column) is estimated as follows. For the given value of the non

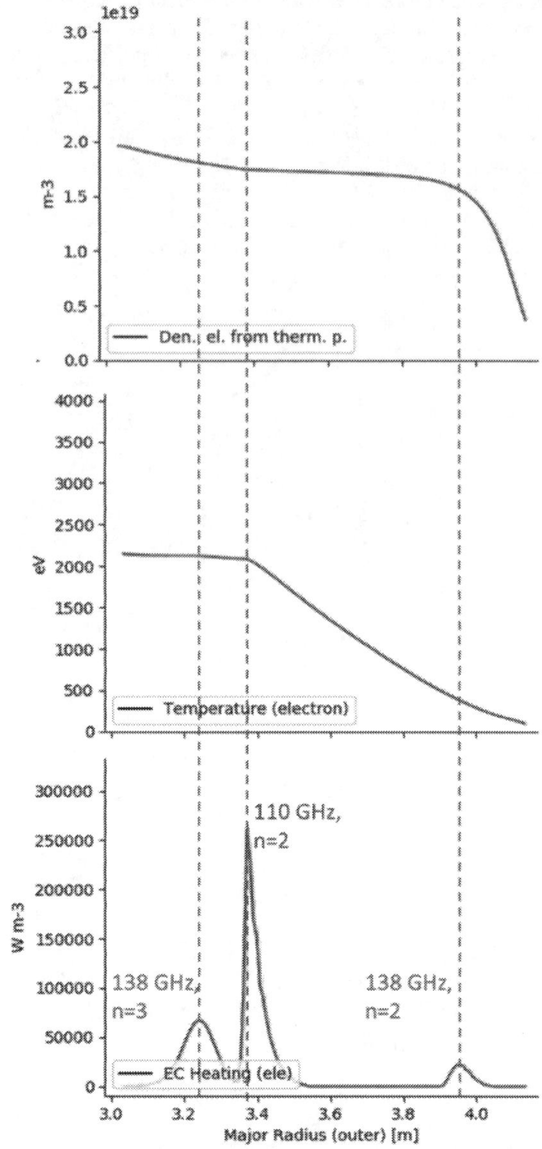


**Fig. 9.** Density, temperature and absorption profiles for case 6b

absorbed power as for column abs%, a further amount is added. This takes into account 5% of spurious cross-polarized component with respect to the main (X) polarization. This is a conservative, reference value from literature, justified by the effect of High Orders Modes [10] in the transmission line and by cross-polarization scattering due to turbulent fluctuations in the tokamak edge [11]. Within the range of the studied cases, relatively low values of absorption occur for injection times before about 5 s when the temperature is about 1 KeV or less (cases 1a, 4a, 4b, 9a, 9b) and when 3rd harmonic is used without a sufficiently long period of 2nd harmonic pre-heating. This can be seen for example comparing cases 6b and 7, with respectively 97% and 93.3% of absorption, that only differ for the time of the 3rd harmonic switch-on with respect to the

2nd harmonic (simultaneous in case 7). In all the other cases (10 out of 18) the absorption is higher than 99%. Only 4 cases out of 18 have absorption minor than 90%.

Non absorbed EC radiation may cause additional thermal load on specific locations of the first wall or in other components exposed to the EC waves, in particular in the area in front of the ECRF launchers. The beam propagation in cases of low plasma absorption has been studied for the steerable launchers in [12] using a realistic antenna modelling [4, 13].

**Table 2.** EC power stray for the cases of Table 1. Symbol (…) in the $R_{abs}$ means that the absorption corresponding to the frequency and harmonic is negligible

| case | $f_{EC}$[Ghz](N) | $R_{abs}$ [m] | abs% | $P_{EC}$ stray [kW] |
|---|---|---|---|---|
| 1a | 138(2) | 3.31 | 48.30 | 261.00 |
| 1b | 138(2) | 3.29 | >99 | 17.50 |
| 1c | 138(2) | 3.06 | >99 | 35.00 |
| 2a | 110(2) | 3.15 | >99 | 3.75 |
| 2b | 110(2) | 3.17 | >99 | 7.50 |
| 3 | 110(2) | 3.15 | >99 | 7.50 |
| 4a | 110(2) +138(3) +138(2) | 3.30+ (...)+ 3.84 | 69.77 | 66.08 |
| 4b | 110(2) +138(3) +138(2) | 3.26 +(...) +3.82 | 92.10 | 35.95 |
| 4c | 110(2) +138(3) +138(2) | 3.23 +3.42 +3.83 | 99.07 | 10.75 |
| 5a | 110(2) | 3.21 | >99 | 3.75 |
| 5b | 110(2) +138(3) +138(2) | 3.23 +3.43 +3.85 | 99.56 | 14.16 |
| 6a | 110(2) | 3.35+ | >99 | 3.75 |
| 6b | 110(2) +138(3) +138(2) | 3.37 +3.24 +3.95 | 97.00 | 52.50 |
| 7 | 110(2) +138(3) +138(2) | 3.36 +3.23 +3.94 | 93.33 | 107.50 |
| 8 | 138(3) +138(2) | 3.02 +4.04 | 62.14 | 287.73 |
| 9a | 110(3) +138(3) +110(2) | 3.24 +3.24 +3.88 | 50.29 | 107.96 |
| 9b | 110(3) +138(3) +110(2) | 3.29 +3.29 +3.85 | 91.06 | 40.39 |
| 9c | 110(3) +138(3) +110(2) | 3.40 +3.40 +3.85 | 99.93 | 4.29 |

The peak power density incident in and the peak absorbed power by the plasma facing components corresponding to the aforementioned cross-polarization are summarized in Table 3 where the EC stray load due to the residual cross polarized component (OM2) at launch is reported. This unwanted component is partially transmitted through the plasma even at relatively high plasma temperature. The range 0-8 keV has been considered using beam tracing [12] and values for 5 keV temperature are listed for reference in the table. Depending on the plasma parameters and on the position in the vessel this effect may lead to relatively localized power deposition in the range of 3-50 kW/m$^2$.

**Table 3.** Residual cross polarized fraction (5%) at 5 keV plasma temperature for scenario 2, 138 GHz. Estimated per 1 MW power injected beam

| (kW/m$^2$) | peak incident power density | | peak absorbed power density on CFC | | dura-tion |
|---|---|---|---|---|---|
| poloidal angle | bounce | | bounce | | |
| | 1st | 2nd | 1st | 2nd | |
| -35.5 | 334.5 | 41.4 | 29.8 | 3.7 | pulse length |

Similarly, Table 4 summarizes the same quantities for the case of a plasma having low absorption characteristics, as for example at the start-up. Values in the table corresponds to a plasma with 50 eV central temperature. A more extensive analysis for a range of ECRF launching settings and plasma parameters is reported in [12]. Specific evaluations for the launching settings of the IC phase are instead reported in [9]. The last column of Tables 3 and 4 contain an indication of the duration of the conditions causing the EC stray effect. For the cross-polarization case such effect is expected to last for all the duration of the ECRF heating. The EC stray due to the low absorption phase is expected to last from few ms to 2 s, and it can in any case be limited setting an appropriated control loop within the Plasma Control System.

**Table 4.** Low absorption phase scenario 2, 138 GHz at low plasma temperature (50 eV). Estimated per 1 MW power injected beam

| (kW/m$^2$) | peak incident power density | | peak absorbed power density on CFC | | dura-tion |
|---|---|---|---|---|---|
| poloidal angle | bounce | | bounce | | |
| | 1st | 2nd | 1st | 2nd | |
| -35.5 | 669 | 0.28 | 59.54 | 0.02 | 1-2000 ms |

## 4 Summary and perspectives

In this work we address the application of ECRF during the ramp-up of the plasma current and the early phase of the current flat top, with fixed ECRF launching settings, assuming a few hypothetical plasma scenarios scaled from the reference scenario 2. Since the absorption and beam trajectory in this early phase of the discharge mainly depend on plasma temperature and density, the results are relevant for both the IC and for the development of scenario 2 in IR.

This paper is a starting point in the exploration of the application of ECRF during the ramp-up from the standpoint of ECRF operation and in particular for what concerns the EC stray power. The reported results indicate good perspectives for the use of ECRF also in plasma scenarios with reduced magnetic field according with the development path described in the JT-60SA Research Plan.

Use of mixed ECRF frequencies at different harmonic, typically 110Ghz n=2+138Ghz n=3, appears feasible with the caveat that the absorption of the 3rd harmonic is favored by pre-heating with the second one.

An extension of the present analysis will include a wider range of the start-up times of the ECRF and of the launching settings keeping into account the steering capabilities of the ECRF launcher after the Integrated Commissioning phase.

This work has been carried out within the framework of the EUROfusion Consortium, funded by the European Union via the Euratom Research and Training Programme (Grant Agreement No 101052200 - EUROfusion). Views and opinions expressed are however those of the author(s) only and do not necessarily reflect those of the European Union or the European Commission. Neither the European Union nor the European Commission can be held responsible for them.

## References

1. JT-60SA Research Unit, JT-60SA Research Plan, Version 4.0, September 2018, *http://www.jt60sa.org/pdfs/JT-60SA_Res_Plan.pdf* Chapter 3, "Operation Regime Development"
2. L. Garzotti et al., Nucl. Fusion 58 026029 (2018)
3. M Yoshida et al 2022 Plasma Phys. Control. Fusion 64 054004
4. T. Kobayashi et al., Fusion Engineering and Design 96–97 (2015)
5. T. Kobayashi et al., Nucl. Fusion 62 026039 (2022)
6. V. Ostuni et al., Nucl. Fusion 61 026021 (2021)
7. L. Garzotti et al., report SA-SE.CM.M.02-T001-D001 EFDA_D_2P4FS5, 2021 (private communication)
8. Y. Kamada et al., Nucl. Fusion 62 042002 (2022)
9. A. Moro et al, Proc. Of SOFT Conference 2022
10. E. Kowalski et al., IEEE Transactions on Microwave Theory and Techniques Vol. 58, Oct 2010
11. L. Guidi et al 2016 J. Phys.: Conf. Ser. 775 012005
12. C. Sozzi et al., P4.1089 46th EPS Conference on Plasma Physics EPS 2019, Milan Italy
13. P. Platania et al, AIP Conference Proceedings 1689, 090010 (2015); https://doi.org/10.1063/1.4936547

# Nonlinear degradation of O-X-B mode conversion in MAST Upgrade

*Mads Givskov* Senstius[1,*], *Simon James* Freethy[2], and *Stefan Kragh* Nielsen[1]

[1]Department of Physics, Technical University of Denmark, Fysikvej, DK-2800 Kgs. Lyngby, Denmark
[2]Culham Centre for Fusion Energy, UKAEA, Abingdon, OX14 3DB, United Kingdom

**Abstract.** Spherical tokamaks like the MAST Upgrade device are often operated in an overdense regime. As a consequence, conventional electron cyclotron resonance heating (ECRH) and current drive (ECCD) are typically not possible. MAST Upgrade is planned to investigate a mode coupling scheme known as O-X-B at high power, which may allow gyrotrons to heat and drive current in overdense plasmas by coupling electromagnetic waves to electrostatic electron Bernstein waves (EBWs) at the upper hybrid (UH) layer. However at the gyrotron beam intensities planned for MAST Upgrade, several nonlinear effects may degrade the linear mode coupling into EBWs. Using particle-in-cell simulations, parametric decay instabilities (PDIs) and stochastic electron heating (SEH) are investigated in the region near the UH layer. It is found that nonlinear effect could have a substantial impact on the O-X-B scheme in MAST Upgrade at high gyrotron intensities.

## 1 Introduction

Spherical tokamaks are often operated in a highly overdense regime, which means that their core is unavailable to electron cyclotron (EC) resonance heating (ECRH) and current drive (ECCD) at several of the lowest harmonics. However, a mode coupling scheme known as O-X-B[1] may enable the use of high power microwaves for heating and current drive in such plasmas. The O-X-B scheme couples electromagnetic waves from e.g. a gyrotron to electrostatic waves known as electron Bernstein waves (EBWs)[2] which can propagate without a high density cutoff and are strongly damped at the electron cyclotron resonances. The use of EBWs is considered a key enabling technology at MAST Upgrade which has acquired two 28/34.8 GHz gyrotrons chosen specifically for O-X-B instead of regular ECCD. Although the mode coupling scheme is promising at low power densities[3], a number of nonlinear effects may degrade its performance when unprecedented gyrotron power levels are used for O-X-B operation. Parametric decay instabilities (PDIs) and stochastic electron heating (SEH) are considered likely nonlinear loss mechanisms in upcoming MAST Upgrade EBW experiments. Even though the conversion rate of O-X-B is favorable at low power, it may scale unfavorably with an increase in power above a nonlinear threshold. The follow up device to MAST Upgrade will be the prototype fusion power plant STEP which is to be a microwave driven machine. If the promising low power O-X-B results remain at higher gyrotron power levels, STEP may employ O-X-B as its main current drive scheme.

The EBWs are excited at the upper hybrid (UH) layer where PDIs are known to take place even at comparably low power densities. PDIs are unstable nonlinear wave interactions which may cause power to be diverted away from the gyrotron beam into waves at undesired frequencies. In addition to the PDIs, EBWs can also experience strong damping due to SEH. The EBWs are carried by gyrating electrons but the electron gyromotion gets warped if the EBW amplitude is too large. Instead of an energy conserving orbit, the electrons may then follow stochastic trajectories with periods of substantial acceleration. Both nonlinear effects could lead to great losses in the outer parts of the plasma.

In this contribution, we investigate the frequencies that are likely to be excited in PDIs during O-X-B in MAST Upgrade as well as a threshold for SEH to set in. We compare with particle-in-cell simulations and look at the power scaling of O-X-B. Although both PDIs and SEH can be considered loss mechanisms, they have different impacts on how a gyrotron beam interacts with a plasma. We discuss how both impact mode coupling schemes involving EBWs.

## 2 Theory

### 2.1 The O-X-B mode coupling scheme

For the conventional microwave based heating and current drive schemes, a high power microwave source injects an electromagnetic wave, typically O- or X-mode, into the plasma and is absorbed at a low EC harmonic near the core of the plasma. In overdense plasmas such as the MAST Upgrade and other spherical tokamaks, the electron cyclotron frequency is small compared to the plasma frequency and the confined core becomes unavailable to O- and X-mode. However by angling an O-mode beam optimally to the magnetic field, it can couple to slow X-mode at the plasma frequency inside the plasma, which in

*e-mail: mgse@fysik.dtu.dk

turn can couple to an electrostatic EBW at the UH layer[1]. Figure 1a shows an O-X-B trajectory calculated with a ray tracer, dividing it into segments associated with the O-, X- and EBW. Where exactly the X-mode ends and the EBW start is debatable as the transition is smooth. Near the UH layer, X-mode becomes longitudinal and the wave characteristics become similar to that EBWs under an electrostatic approximation, but we instead define the forward propagating wave as X-mode, the backward propagating wave as the EBW and the turning point as the UH layer. For typical MAST Upgrade parameters, this occurs between the 2nd and 3rd EC resonances and the turning point is then outside the UH frequency layer as can be seen in figure 1a.

This mode coupling scheme is called O-X-B and can be utilized to get microwave radiation absorbed at the core of an overdense plasma as EBWs can propagate without a high density cutoff. At low power, the O-X-B scheme has shown to drive current more efficiently than conventional ECCD. However, the UH layer where X is coupled to EBW is a resonant region where nonlinear effect can become significant. The impact of nonlinear effects on O-X-B is not well known at the high gyrotron beam intensities that the MAST Upgrade gyrotron system will be capable of delivering to the UH layer.

### 2.2 Parametric decay instabilities (PDIs)

One of the nonlinearities commonly associated with the mode coupling at the UH layer is three-wave interactions. One of the most well documented three-wave interaction for X to EBW conversion is a decay event where the X-mode gyrotron pump wave excites an EBW and a lower hybrid (LH) wave[2]. The process conserves frequency and wave vector which leads to the following selection rules

$$\omega_0 = \omega_1 + \omega_2, \quad \mathbf{k}_0 = \mathbf{k}_1 + \mathbf{k}_2, \tag{1}$$

where $\omega_i$ is the angular frequency of wave $i$, $\mathbf{k}_i$ its wave vector, and index 0 refers to the pump wave, while 1 and 2 refer to the daughter waves produced in the decay process. Because of the conservation of frequency, the EBW daughter wave will be produced at frequency downshifted from the pump by that of the LH daughter wave, which would be approximately the LH frequency for an interaction between these waves. The strength of the interaction between the waves depends on the amplitude of the pump wave. If the interaction is weak, damping from collisions and other effects may prevent the daughter waves from growing beyond their background level. However above a pump wave amplitude threshold, growth due to the three-wave interactions exceed the losses and the interactions manifest themselves as a parametric decay instability (PDI)[4–7]. Using [8], the threshold is $\approx 2 \times 10^5\,\mathrm{W/m^2}$ for the L-mode parameters presented later in section 3. It is also possible for three waves to interact in a combination event where typically a large amplitude pump wave combines with a natural wave of the plasma to produce a third one. Again, this process requires conservation of frequency and wave vector. Often, both decay and combination events occur in a system, producing side bands to the pump frequency, shifted by a characteristic frequency such as the LH frequency around it. The region around the UH layer is associated with PDIs because the pump wave encounters a turning point, causing its group velocity to decrease and consequently its amplitude to swell[9] in order for the flow of power to be conserved. The PDIs may be of concern because they allow for the flow of power to change characteristics. If a gyrotron beam is intended to be absorbed at a target EC resonance in the center of the plasma, a PDI may cause a substantial fraction of the power to go into waves which resonate in other parts of the plasma and propagate in entirely different directions. What is more, low power experiments do not necessarily reflect high power experiments due to the nonlinear dependence on pump wave amplitude. Upcoming O-X-B experiments at MAST Upgrade will also investigate PDIs and a radiometer is being developed to allow for signatures of nonlinear wave interactions with the gyrotron beam to be observed[10].

### 2.3 Stochastic electron heating (SEH)

The EBWs are longitudinal waves which propagate as coherent electron density perturbations. As the electrons gyrate around the magnetic field lines in and out of phase, they can carry an electric field which points in the same direction as it propagates. At low electric field amplitudes, the electrons are not particularly affected by the electric field of the EBW, but above a wave amplitude threshold, the electron trajectories transition from harmonic orbits around the magnetic field line into a seemingly stochastic motion. This is illustrated in figure 1b where a simple Boris solver[11] is used to calculate particle trajectories in the presence of a magnetic field and an EBW at different electric field amplitudes. Whilst the green line appears unaffected by the EBW, the orange line displays visible modifications to the radius of gyration and the blue line appears stochastic. In the stochastic regime, the electrons experience a strong acceleration and their kinetic energy can rapidly increase as they ride the waves in the electric field at the expense of the power carried by the EBW. For this reason, the term stochastic electron heating (SEH) is often used. In the literature, different conditions have been suggested for the SEH threshold[12–15]. Following [16], we introduce the normalized quantity

$$A \equiv -\frac{m_e}{eB^2}\partial_x E_x, \tag{2}$$

where $m_e$ is the electron mass, $e$ is the elementary charge, $B$ is the magnitude of the background magnetic field and $E_x$ is the electric field of a longitudinal wave propagating in the $x$-direction. The threshold condition for SEH is then $|A| \geq 1$ and this should lead to strong increases in electron temperature. The derivation of this expression is based on the assumption that only one wave is present but near the UH layer, both a large amplitude X- and EBW are present at the same frequency but propagating in different directions with quickly varying wave vectors and amplitudes. This makes the $\partial_x E$ tedious to calculate analytically but is easy to evaluate numerically based on data from full

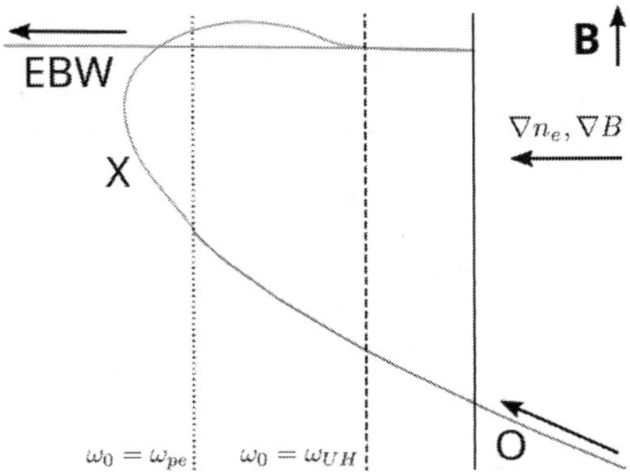


(a) Illustration of the O-X-B mode coupling scheme using a ray tracer and parameters relevant to MAST Upgrade. The injected O-mode (blue) is coupled to X-mode (orange) at the plasma frequency (dotted) by angling it correctly. X-mode then couples an EBW (green) near the UH frequency (dashed), in our definition at the turning point (solid black).

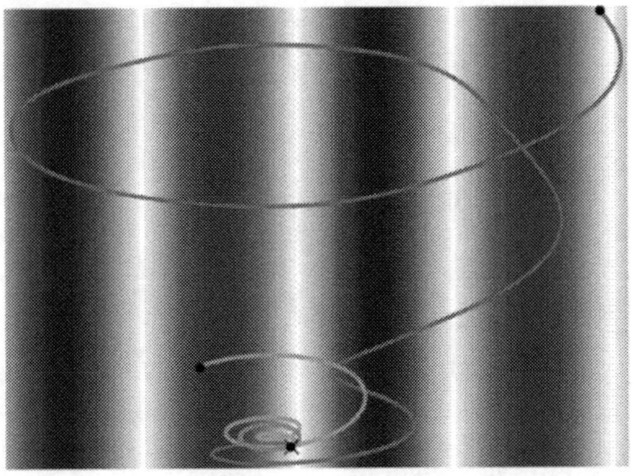

(b) Examples of electron trajectories at different EBW electric field amplitudes, an instance of the wave field is shown in the red-black heat map. The lines are for $|A| \ll 1$ (green), $|A| \approx 1$ (orange) and $|A| \gg 1$ (blue). The cross marks the starting position and circles are ending positions after 2 cyclotron periods. Trajectories are calculated using a Boris update scheme.

Figure 1: Figures illustrating basic concepts of the theory section.

wave simulations. Both the presence of two large amplitude waves as well as wave swelling near the turning point makes the UH layer a region where SEH is more likely to occur.

## 3 Simulations

### 3.1 Domain parameters

We study the impact of PDIs and SEH on O-X-B numerically using the fully kinetic particle-in-cell (PIC) code EPOCH[17, 18], focusing on the X- to EBW conversion in the vicinity of the UH layer. Whilst the conversion is a three dimensional problem, we here simplify the geometry to 1D as a first look into the problem for MAST Upgrade parameters. The gyrotron pump wave is mainly longitudinal near the UH layer and propagates close to perpendicular to the magnetic field and parallel to the direction of the density and temperature gradients. Making the 1D approximation does not affect the trajectory of the pump wave much nor should it affect SEH if the inhomogeneous direction is chosen to be perpendicular to the magnetic field. However, the small wave number component parallel to the magnetic field does lead to damping, in particular of EBWs. This may suppress some PDIs but PDIs may still excite EBWs that propagate perpendicularly to the magnetic field. In the 1D geometry, we also neglect the transverse beam profile and assume a uniform field in the homogeneous directions. In an experimental situation, the pump wave would likely have a Gaussian profile with a strong field at the center, which decreases away from the center. Gyrating and drifting particles therefore do not experience the strongest field everywhere along their trajectory. In particular, electrons spend a limited amount of time in the interaction region as they propagate across the beam profile due to their parallel velocity, and the assumption of a uniform transverse profile may therefore lead to a lower threshold[19]. We note that the electron Larmor radius is much smaller than the focused gyrotron beam at the UH layer.

The 1D approximation in terms of the EPOCH code means simply that there is only one inhomogeneous spatial direction. Macroparticles have three velocity components, which are affected by forces as expected, but only one position component. The full 6D phase space is realized by constructing 1D position space along with 3D velocity space from the macroparticles and then assuming perfect translational symmetry in the remaining position dimensions. This allow for particle orbits and drifts to be modeled even in 1D.

The code EPOCH allows for particle species to be made immobile by taking them out of the particle pusher step of the update scheme. This way, they still contribute to the background but cannot generate waves. This will be utilized to test if certain spectral features later can be explained as the result of PDIs involving ion waves.

The parameters to be investigated in this papers correspond to a high density L-mode plasma in MAST Upgrade. For a 37 GHz gyrotron beam, the region of focus between the UH layer and the L-cutoff is on the order of a centimeter long and gradients are approximately constant throughout. The UH layer is found in the pedestal, between the second and third EC resonances. The domain size is chosen to be $0.0\,\mathrm{mm} < x < 7.5\,\mathrm{mm}$ with $N_{\mathrm{grid}} = 750$ grid points and $N_{\mathrm{part}}/N_{\mathrm{grid}} = 2000$ superparticles per grid point. The initial density and temperature are chosen to be linear profiles satisfying $n_e(0\,\mathrm{mm}) = 2.16\times10^{19}\,\mathrm{m}^{-3}$, $n_e(7.5\,\mathrm{mm}) = 1.41\times10^{19}\,\mathrm{m}^{-3}$, $T_e(0\,\mathrm{mm}) = 63\,\mathrm{eV}$, and

$T_e(7.5\,\text{mm}) = 48\,\text{eV}$. The initial ion temperature is $T_i(x) = T_e$, the magnetic field is $B = 0.46\,\text{T}$ and an $\omega_0/(2\pi) = 37\,\text{GHz}$ is injected at the $x = 0\,\text{mm}$ in X-mode polarization. The particle boundary conditions replace escaping particles with thermally distributed particles entering the domain. The X-mode pump wave is excited at the left boundary at intensities ranging from $10^3\,\text{W/m}^2$, near the background level where nonlinear effects are of little importance, up to $10^9\,\text{W/m}^2$, which is the theoretical upper limit that the new MAST Upgrade gyrotron system will be able to generate.

### 3.2 Spectral development

As the pump wave intensity is increased from the background level, the longitudinal electric field spectrum near the pump wave develops. Before discussing nonlinear spectral features, a linear regime is established. Figure 2a shows the spectra for two low intensity simulations in the linear regime. The spectra are very similar with the main difference being the magnitude of the $\omega_0/(2\pi) = 37\,\text{GHz}$ pump wave peak which increases by approximately one order of magnitude as expected.

As the pump wave intensity is increased to $I_0 = 10^5\,\text{W/m}^2$, the spectrum begins to develop more broadly. Figure 2b shows the spectrum from two simulations, one where the ions have been made immobile (orange) and a regular simulation with mobile ions (blue). When ions are immobile, they provide a charge background but are taken out of the particle pusher and ion waves therefore cannot be excited. Comparing the spectra, there is a development of the spectrum around 37.25 GHz when ions are mobile. Furthermore, there is some development close to the pump frequency and figure 2c shows a zoomed spectrum near the pump. Here, symmetric sidebands shifted by $\sim 20$ MHz, or a few ion cyclotron frequencies, around the pump frequency has appeared when ions are mobile only, suggesting this could be the result of PDIs involving low IBW branches.

At even higher pump intensities in figure 2d, the spectrum starts to develop also at frequency shifts of approximately 0.2 GHz, which is close to the LH frequency. Two prominent symmetric peaks can be seen in the figure, only when ions are mobile, suggesting that they are the result of PDIs involving LH waves.

### 3.3 Indications of SEH

Around pump intensities of $10^7 - 10^8\,\text{W/m}^2$, the injected pump wave reaches a magnitude where the normalized parameter $A$ exceeds the SEH threshold $|A| \gtrsim 1$. The plots in figure 3 compare the parameter $|A|$ in the top row and the electron temperature in the bottom row. The electron temperature is calculated through the second moment of the distribution function which need not be Maxwellian. For $I_0 = 10^7\,\text{W/m}^2$, the threshold is not exceeded and although a wave pattern is visible in the electron temperature, no noticeable heating is observed in particular at $x \approx 6$ mm which is near the UH layer. For $I_0 = 10^8\,\text{W/m}^2$, the threshold is exceeded and a substantial heating occurs, which continues beyond the 12 ns shown in the plots. Now, the electron heating could be caused by some other process, e.g. hot electrons generated through PDIs, but the onset of rapid heating appears to agree with the SEH threshold and not the onset of PDIs which happens at much lower intensities. As the UH layer is found in the pedestal region of the MAST Upgrade plasma, SEH appears to constitute an unfavorable loss mechanism which may prevent the power of an injected beam from reaching the core of the plasma.

### 3.4 Impact on the linear conversion

With indications of both PDIs as well as SEH, the obvious question is how it affects the linear X- to EBW conversion. To investigate this, we use an FFT on the intervals $0.5\,\text{mm} < x < 2.5\,\text{mm}$ and $10\,\text{ns} < t < 260\,\text{ns}$ to get a spectrum in $(k, \omega)$-space. We then identify and integrate the peaks associated with the incoming X-mode and with the returning EBW at the same frequency and sign of $k$. We refer to this EBW as the linearly converted EBW as we expect that it mostly comes from linear coupling with the X-mode pump at the UH layer. We do this for different pump intensities, normalize to the lowest intensity simulation and show the results in figure 4. As a sanity check, the injected X-mode pump is seen to scale close to linearly with pump intensity as expected. The returning EBW, on the other hand, scales linearly at low pump intensity but stalls around $I_0 = 10^5\,\text{W/m}^2$ which is also around the point where indications of PDIs were observed in figures 2b-2c and similar to the PDI threshold predicted using [8], although this was for interactions involving LH waves. However, this particular PDI alone may not be the sole cause for the unfavorable intensity scaling. We note that there is an actual dip in conversion around where the SEH sets in, but it cannot be clearly attributed to pump losses due to SEH. Other spectral features outside the LH range also appear and may account for some of the lost pump power. These features will be the focus of later investigations. The figure predicts poor scaling of the O-X-B conversion efficiency in the intensity range that the MAST Upgrade is expected to be able to operate at, however, heating and current drive of the plasma may still occur if PDIs simply excite waves at shifted frequencies, which also propagate inwards.

Lastly, changing parameters such as pump frequency and magnetic field profile can have profound impact on the spectral development and the extend of the linear regime. Still, the theoretical upper operational limit of $I_0 = 10^9\,\text{W/m}^2$ is in the nonlinear regime in all simulations so far. Even if a parallel component of the wave vector in higher dimensions may help suppress PDIs, SEH should not be affected much by this.

## 4 Conclusion

First considerations of nonlinear losses in connection to O-X-B operation in MAST Upgrade were presented. The

https://doi.org/10.1051/epjconf/202327701009

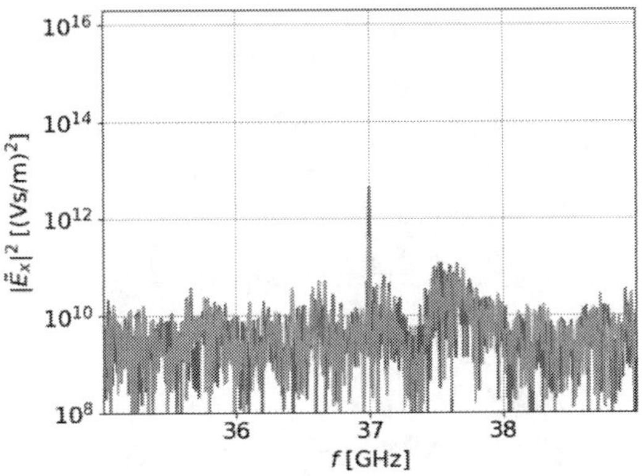


(a) Two spectra in the linear pump intensity regime. The orange line is for $I_0 = 10^3\,\mathrm{W/m^2}$ and the blue line is for $I_0 = 10^4\,\mathrm{W/m^2}$. Ions are mobile in both.

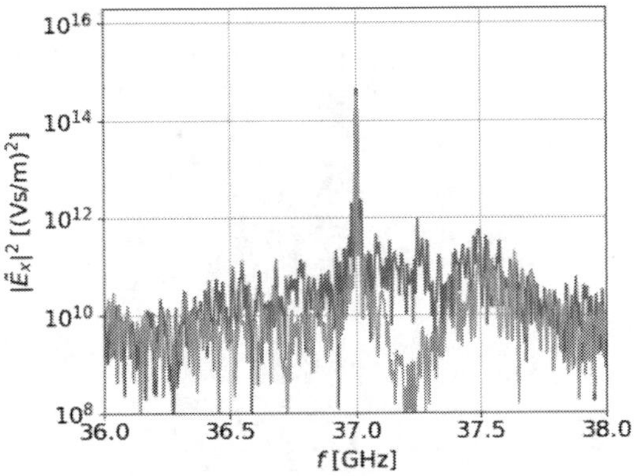


(b) The pump wave intensity is $I_0 = 10^5\,\mathrm{W/m^2}$, the orange line is for immobile ions and the blue line is for mobile ions. Notice the difference near 37.25 GHz and near the pump, see the zoom in figure 2c.

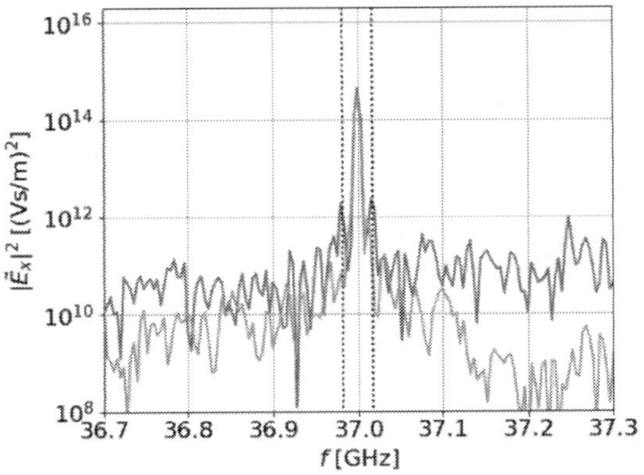


(c) Same as figure 2b but zoomed in near the pump frequency. Notice the symmetric peaks shifted by ~ 20 MHz, marked by vertical dotted lines and visible only when ions are mobile.

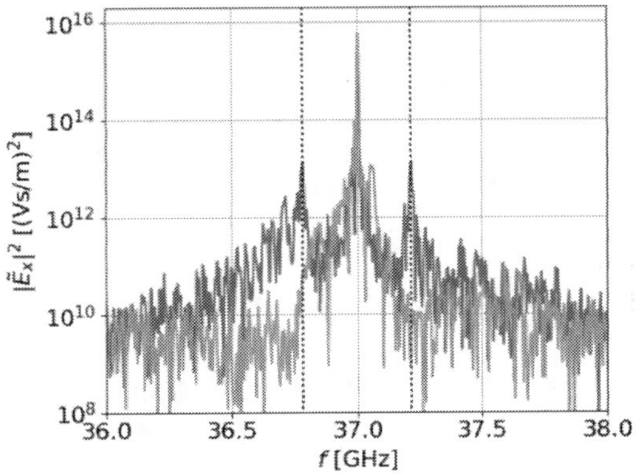


(d) The pump wave intensity is $I_0 = 10^7\,\mathrm{W/m^2}$. The orange line is for immobile ions and the blue line is for mobile ions. Notice the development ~ 0.2 GHz from the pump when the ions are mobile.

Figure 2: Frequency spectra of the longitudinal electric field spectrum near the $\omega_0/(2\pi) = 37$ GHz pump wave.

region around the UH layer is of particular concern for both nonlinear wave interactions as well as SEH in the X- to EBW mode coupling step. PIC simulations in a simplified 1D geometry at various pump wave intensities within the operational limits of the MAST Upgrade gyrotrons show indications of both PDIs and SEH. Although a regime exists where the coupling into an EBW scales linearly with pump intensity, a fairly low intensity threshold for sublinear scaling is found around $I_0 = 10^5\,\mathrm{W/m^2}$. Whilst this first look into nonlinear degradation of O-X-B in MAST Upgrade is limited and may underestimate nonlinear thresholds due to the 1D approximation, it does justify a more thorough investigation. In particular, there were indications of spectral features outside the LH range from the pump wave which will be investigated in the future.

## 5 Acknowledgments

This work has been supported by research grant 15483 from VILLUM FONDEN, as well as the Enabling Reseach grant ENR-MFE19.DTU-03 and the EUROfusion Researcher Grant AWP22-ERG-DTU/Senstius from the EUROfusion Consortium. This work has been carried out within the framework of the EUROfusion Consortium, funded by the European Union via the Euratom Research and Training Programme (Grant Agreement No 101052200 — EUROfusion). Views and opinions expressed are however those of the author(s) only and do not necessarily reflect those of the European Union or the European Commission. Neither the European Union nor the European Commission can be held responsible for them. The EPOCH code used in this work was in part funded by the UK EPSRC grants EP/G054950/1, EP/G056803/1, EP/G055165/1, EP/ M022463/1 and EP/P02212X/1.

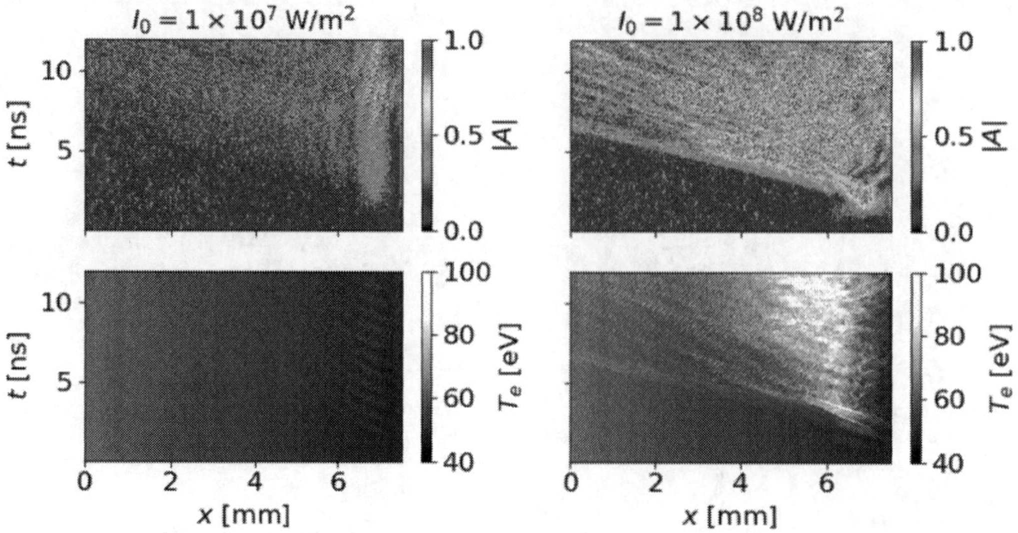


Figure 3: Comparisons of the normalized SEH parameter $|A|$ (top row) and the electron temperature (bottom row) for pump intensities around the SEH threshold of $|A| \approx 1$. The left column is below the threshold while the right column is above it.

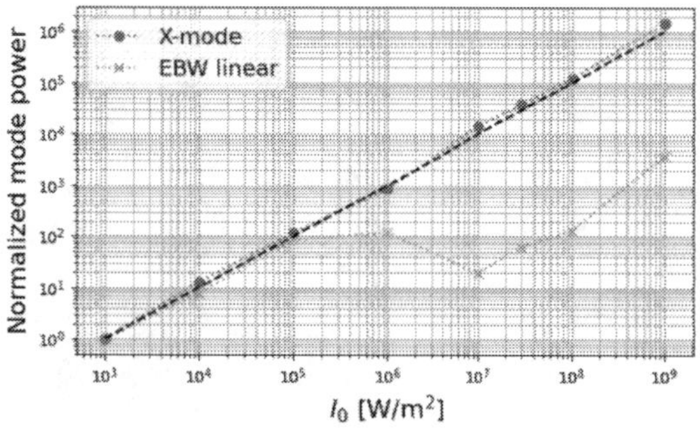


Figure 4: Scaling of X- and linearly converted EBW power with pump intensity. The points are normalized to that of the lowest intensity simulation and the dashed black line is a linear reference line.

## References

[1] J. Preinhaelter and V. Kopecký, *J. Plasma Physics*, **10**, 1–12 1973.

[2] H. P. Laqua, *Plasma Physics and Controlled Fusion* **49** R01 2007

[3] V. Shevchenko, G. Cunningham, A. Gurchenko, E. Gusakov, B. Lloyd, M. O'Brien, A. Saveliev, A. Surkov, F. Volpe, and M. Walsh. *Fusion Science and Technology*, 52:2, 202–215, 2007.

[4] A.D. Piliya. Decay instability in weakly inhomogeneous plasma. *Proc. 10th Int. Conf. Phenomena in Ionized Gases (Oxford) ed Franklin R.N. (Oxford: Donald Parsons & Co. Ltd.)*, page 320, 1971.

[5] M.N. Rosenbluth. Parametric instabilities in inhomogeneous media. *Phys. Rev. Lett.*, 29:565–7, 1972.

[6] A.D. Piliya. Nonstationary theory of decay instability in a weakly inhomogeneous plasma. *Sov. Phys.–JETP*, 37:629–32, 1973.

[7] M. Porkolab and B. I. Cohen, *Nuclear Fusion*, **28**, 239, 1988.

[8] E. Z. Gusakov and A. V. Surkov, *Plasma Physics and Controlled Fusion*, **49**, 631-639, 2007

[9] R. B. White and F. F. Chen, *Plasma Physics* **16** 565–587, 1974

[10] M. G. Senstius, S. J. Freethy, J. Allen and S. K. Nielsen, *Review of Scientific Instruments* **93** 103522, 2022

[11] J. P. Boris, *Proceedings of 4th Conference on Numerical Simulation of Plasmas ( Naval Research Laboratory, Washington D. C., 1970)*, pp. 3–67.

[12] J. M. McChesney, R. A. Stern and P. M. Bellan, *Physical Review Letters* **59** 1436, 1987

[13] M. Balikhin and M. Gedalin, *Physical Review Letters* **700**, 9, 1993

[14] K. Stasiewicz, R. Lundin and G. Marklund, *Physica Scripta* **T84**, 60–63, 2000

[15] C. F. F. Karney and A. Bers, *Pysical Review Letters* **39**, 9, 1977

[16] D. C. Speirs, B. Eliasson and L. K. S. Daldorff, *Journal Geophysics Research Space Physics*, **122**, 10,638–10,650, 2017

[17] T. D. Arber *et al*, *Plasma Physics and Controlled Fusion*, **57**, 113001, 2015

[18] M. G. Senstius, S. K. Nielsen, R. G. Vann and S. K. Hansen, *Plasma Physics and Controlled Fusion*, **62**, 025010, 2020

[19] R. Kamendje, S. V. Kasilov, W. Kernbichler, I. V. Pavlenko, E. Poli and M. F. Heyn, *Physics of Plasmas*, **12**, 012502, 2005

# Benchmarking full-wave codes for studying the O-SX mode conversion in MAST Upgrade

*Alf* Köhn-Seemann[1,*], *Bengt E.* Eliasson[2], *Simon J.* Freethy[3], *Lou A.* Holland[4], and *Roddy G.L.* Vann[4]

[1]Institute of Interfacial Process Engineering and Plasma Technology, University of Stuttgart, Stuttgart, Germany
[2]SUPA, Department of Physics, University of Strathclyde, Glasgow, U.K.
[3]Culham Centre for Fusion Energy, Culham, U.K.
[4]York Plasma Institute, University of York, York, U.K.

**Abstract.** Three full-wave codes for simulating microwave propagation and O-SX mode conversion in magnetized plasma are described and compared. Their feasibility to investigate mode conversion processes and obtain conversion efficiencies for parameters relevant for a potential MAST Upgrade 28 GHz electron Bernstein wave heating scenarios is explored.

## 1 Introduction

Electromagnetic waves in the microwave range of frequencies play an indispensable role in fusion plasmas for heating [1] and diagnostic purposes [2]. One of their advantages are the small space requirements for in-vessel components compared to other heating sources which is of special importance in a potential reactor where the in-vessel wall will be covered with Tritium-breeding blankets [3].

In conventional tokamaks & stellarators the microwave energy is usually deposited where the angular frequency of the injected microwave $\omega_0$ equals the electron cyclotron frequency $\omega_{ce} = eB_0/m_e$, with $e$ the elementary charge, $B_0$ the background magnetic field, and $m_e$ the electron mass. If the electron plasma frequency $\omega_{pe} = \sqrt{n_e e^2/(\epsilon_0 m_e)}$ (with $n_e$ the electron plasma density and $\epsilon_0$ the vacuum permittivity) exceeds $\omega_{ce}$, the wave is in cut-off and an injected O-mode can no longer reach the electron cyclotron resonance (ECR) layer. Such a plasma is often referred to as *overdense*. One solution is to double the frequency of the injected microwave and heat at harmonics of the ECR. Second harmonic heating is a standard scheme in fusion plasmas but higher harmonics are rarely used as the absorption decreases with increasing harmonics number.

Compared to conventional tokamaks, spherical tokamaks exhibit a higher plasma-$\beta$ [4], resulting in highly overdense plasmas. Heating at harmonics becomes ineffective due to the high harmonics number. A solution is provided by the electrostatic electron Bernstein wave (EBW) which has no high-density cut-off, is very well absorbed at the ECR even at high harmonics and can very efficiently drive toroidal net currents [5]. This makes them an attractive candidate for spherical tokamaks, where non-inductive current drive techniques are mandatory due to a small (or completely absent) central solenoid.

*e-mail: koehn@igvp.uni-stuttgart.de

The paper gives an overview of a set of numerical tools to be used to explore the feasibility of a MAST Upgrade 28 GHz EBW heating scenario. The numerical set-up used in the simulations is described in Section 2, including a brief description of the different numerical codes. Section 3 contains the benchmark studies, Section 4 investigates the cases of very steep plasma density profiles. A brief summary concludes the paper.

## 2 Numerical set-up

### 2.1 The simulation domain

A Gaussian beam with a frequency of $f_0 = 28\,\mathrm{GHz}$ in O-mode polarization is considered in the simulations. Following e.g. Ref. [2], its electric field distribution reads

$$E(x,z) = E_0 \frac{w_0}{w} \exp\left(\frac{-x^2}{w^2} - ikz - ik\frac{x^2}{2R} + i\phi_0\right), \qquad (1)$$

with $x$ the radial distance to the beam axis, $z$ the axial distance to the beam waist, $w_0$ the size of the beam waist (here $w_0 = 4\,\lambda_0$), $R$ the radius of curvature of the wavefront, and $\phi_0$ the Gouy phase shift. Note that $w$, $R$ and $\phi_0$ are all functions of $z$.

A linear plasma density profile is used with a normalized density scale length of $k_0 L_n = 25$, corresponding to

$$n_e(z) = (z - z_{\mathrm{n,start}}) \cdot \frac{2\pi}{\lambda_0}\frac{1}{k_0 L_n} \quad \left[= \frac{z - z_{\mathrm{n,start}}}{L_n}\right], \qquad (2)$$

with $z_{\mathrm{n,start}} = 0.15\,\mathrm{m}$. The background magnetic field is set to a constant magnitude of $|\mathbf{B}_0| = 0.85\,\mathrm{T}$, pointing into the $y$ direction: $\mathbf{B}_0 = (0, B_0, 0)$. Note that $z$ is understood here as the direction into which the microwave beam is primarily injected (which looses of course its meaning when considering O-SX mode conversion which requires an oblique injection with respect to the background magnetic field).

| scenario | background | $z_0$ | $\theta$ |
|---|---|---|---|
| #1 | vacuum | 0 | 90° |
| #2 | vacuum | 0.2 m | 90° |
| #3 | vacuum | 0 | 60° |
| #4 | plasma (no B-field) | 0 | 60° |
| #5 | magnetized plasma | 0 | 60° |
| #6 | magnetized plasma | 0 | $\theta_{\text{opt}}$ |

**Table 1.** Scenario configurations in the toroidal plane, $z_0$ corresponds to the position of the beam waist, $\theta$ refers to the injection angle with respect to the $y$-axis, where $\theta_{\text{opt}}$ is the angle resulting in maximum mode conversion efficiency, Eq. (10).

The simulation box is set to a size of $N_y = 80\,\lambda_0$ and $N_z = 30\,\lambda_0$. The different scenario configurations are summarized in Table 1: they increase in complexity with scenario #6 finally corresponding to a case of optimized O-SX mode conversion. Note that those scenarios all treat propagation in a "toroidal plane", i.e. $y$ corresponds to the toroidal direction assuming a conventional tokamak whose magnetic field points primarily into the toroidal direction.

To compare results between the different numerical codes, we will use cross sections of the squared wave electric field summed up over a few wave periods (i.e. time-averaged) along the $y$ direction. Cross sections at $z = \{0, 5, 10, 15, 20, 25, 30\}$ cm will be considered. For the cases where O-SX mode conversion takes place, we will also compare the conversion efficiencies.

### 2.2 The codes

#### 2.2.1 IPF-FDMC

IPF-FDMC is a 2.5D finite-difference time-domain (FDTD) code [6] solving Maxwell's equations and the fluid equation of motion of the electrons on a 2D Cartesian grid. The code is written in C and uses OpenMP for parallelization on shared memory systems. It allows to simulate propagation of electromagnetic waves in a cold magnetized plasma. The evolution equations considered for the magnetic field **B**, the electric field **E**, and the current density **J** of the wave in a plasma equilibrium are:

$$\frac{\partial}{\partial t}\mathbf{B} = -\nabla \times \mathbf{E} \tag{3}$$

$$\frac{\partial}{\partial t}\mathbf{E} = c^2\nabla \times \mathbf{B} - \mathbf{J}/\epsilon_0 \tag{4}$$

$$\frac{\partial}{\partial t}\mathbf{J} = \epsilon_0\omega_{pe}^2\mathbf{E} - \omega_{ce}\mathbf{J} \times \hat{\mathbf{B}}_0 - \nu_e\mathbf{J} \tag{5}$$

with $c$ the speed of light and $\hat{\mathbf{B}}_0$ the unit vector into the direction of the background magnetic field. An electron collision frequency $\nu_e$ is included in Eq. (5) as a dissipation mechanism. The code has been successfully benchmarked against cold plasma theory [7] and against the full-wave code EMIT-3D [8]. For the implementation of Eqs. (3)-(5), the standard FDTD scheme has to be complemented by a discretization scheme for the current equation (5). IPF-FDMC uses a "straightforward" way: first advance $J_x$ using the old values of $J_y$ and $J_z$ (in the cross product with $\mathbf{B}_0$), then advance $J_y$ using the updated value of $J_x$ and the old value of $J_z$, and finally advance $J_z$ using the updated values of $J_x$ and $J_y$. As has been shown in previous works, this method is completely sufficient for a rather large set of problems [9, 10]. More advanced methods exist for situations with extreme density gradients [11, 12].

A Gaussian beam is injected into the grid as a soft source [6] with the electric field amplitude, as defined in Eq. (1), specified in an antenna plane. The beam waist $w_0$ and the axial distance $z$ to the beam waist are input parameters. According to e.g. Ref. [2], $w$ and $R$ in the antenna plane are given in vacuum by the following equations:

$$w(z) = w_0\left[1 + \left(\frac{z}{z_R}\right)^2\right]^{1/2}, \quad R(z) = z\left[1 + \left(\frac{z_R}{z}\right)^2\right], \tag{6}$$

with $z_R = \pi w_0^2/\lambda_0$ the Rayleigh range. Some algebra yields the expressions for the antenna input used in IPF-FDMC:

$$w_0 = \frac{w\lambda_0 R}{\left(\lambda_0^2 R^2 + w^4\pi^2\right)^{1/2}}, \quad z = \frac{w^4\pi^2 R}{w^4\pi^2 + \lambda^2 R^2}. \tag{7}$$

The simulation domain considered here is surrounded by non-radiating boundaries. IPF-FDMC uses a physical absorbing boundary layer which damps the wave electric field by multiplying it with a damping function $d(x)$ whose strength depends on the position $x$ in the absorber:

$$d(x) = 1 - \frac{10}{\lambda_{\text{grid}}}\left(\frac{x - d_{\text{absorb}}}{d_{\text{absorb}}}\right)^2, \tag{8}$$

with $\lambda_{\text{grid}}$ the wavelength in grid points, set to 64 grid points per vacuum wavelength, $d_{\text{absorb}} = 3\,\lambda_0$ the thickness of the absorbing layer. This method also yields the amount of wave energy damped in the absorber and thus indirectly the O-SX conversion efficiency as the converted SX-wave is dissipated in the vicinity of the upper-hybrid resonance (UHR) and not in the absorber.

#### 2.2.2 EMIT-2D

EMIT-2D is a 2D full-wave, cold plasma code parallelised using OpenMP. It is a novel code, developed from EMIT-3D [8] using the FDTD method to solve Maxwell's Eqs. (3) and (4). The plasma response is accounted for via Eq. (5) (with $\nu = 0$). Equations (3) - (5) are solved on a Yee grid with **J** being defined on the same spatial points as **E**. A leap frog scheme is used, i.e. **E** and **B** are calculated on alternate time-steps (with **J** being calculated on the same time-step as **B**) using the previous value of the other. As EMIT-2D is a 2D code, derivatives in the third dimension (chosen as the $y$ dimension arbitrarily) vanish.

Boundary conditions are required to solve the update equations on the boundary grid points. In EMIT-2D, the electric field at the boundary is set at zero. In order to avoid reflections from the boundary grid points, an absorbing boundary layer is implemented. The boundary layer is three vacuum wavelengths thick, and in it all components of the electric field are multiplied by a cubic function:

$$D(d) = 1 + \frac{13}{\tau}\left(\frac{d - d_{\text{bound}}}{d_{\text{bound}}}\right)^3 \tag{9}$$

where $\tau$ is the wave period divided by the timestep $\Delta t$, $d_{\text{bound}}$ is the boundary thickness and $d < d_{\text{bound}}$ is the distance from the edge of the simulation domain. This ensures that power returned into the simulation domain is diminished by a factor of less than $10^{-8}$ [13]. The spatial resolution is set to 50 Yee cells per vacuum wavelength.

A Gaussian beam is launched to propagate in the $z$-direction using a soft antenna, meaning the electric field is added to whatever is already present. The Gaussian beam launched by the antenna is described by Eq. (1) multiplied with a time-harmonic variation, $\sin(\omega t)$.

#### 2.2.3 Fourier Full-wave simulation code

The Fourier Full-Wave (FFW) simulation code solves Maxwell's equations coupled with the electron fluid equations, using a Fourier method in the plane normal to the density gradient. Separation of scales in time of the unknown variables is used leading to a system governing the slowly varying in time complex envelope of the wave electric field. Such separation of time scales are frequently used when studying quasi-monochromatic waves with slowly varying amplitudes and phases, leading to, e.g., the non-linear Schrödinger equation in non-linear optics and plasmas, and to the Zakharov equation when studying the non-linear coupling to ion waves [14–17]. Since we are studying the linear behaviour of microwaves, the non-linearity has been "switched off" keeping the ions immobile. Exploiting that the amplitude and phase are slowly varying, it is possible to take much longer time-steps in time than the oscillation period of the microwave.

An implicit trapezoidal (Crank-Nicholson) scheme advances the solution in time, with the advantage of being unconditionally stable with respect to the time-step and second-order accurate. In the simulations performed, one time-step corresponds to approximately 60 oscillation periods of the microwave, and for our purposes it suffices to take 50 – 100 time-steps to obtain the desired solution.

By Fourier transforming the system of equations in the plane perpendicular to the density gradient, the system separates to one 1D system per Fourier component. Simple out-flow boundary conditions are used at the antenna plane $z = 0$, where waves with the desired polarization are injected [16]. The remaining spatial variable parallel to the density gradient is resolved by 40 grid points per vacuum wavelength. To resolve the small-scale electrostatic waves resulting from mode conversion and resonant absorption of the EM wave a region around the resonant layer is resolved by 2000 grid points per wavelength. After solving each of the 1D systems in space and time, the resulting solutions are superimposed with appropriate weights and phase factors to obtain the full multi-dimensional solution.

## 3 Benchmark results

### 3.1 Scenario 1

A vacuum, no plasma and $B_0 = 0$, is considered in this scenario. A Gaussian beam, see Section 2.1, is injected from the bottom, with the waist at the antenna plane, propagating parallel to the $z$-axis. To compare the results of the full-wave codes, cross sections along various $z$-positions are taken, see Sec. 2. These signals will be referred to as *detector antenna signals* in the following. As shown in Fig. 1, one can see that with increasing distance from the antenna plane at $z = 0$, the amplitude of the beam cross section decreases and the width increases (as expected).

For a more quantitative comparison, the detector antenna signals of the different codes are subtracted from each other. This requires to map the results of one code to the spatial grid of another code. As an interpolation technique, *Piecewise Cubic Hermite Interpolating Poynomial* (a 1D monotonic cubic interpolation) from Python's `SciPy` module is used [18]. The resulting differences are shown in Fig. 2. The differences between IPF-FDMC and EMIT-2D are very small which is not surprising as both codes are based on the same numerical method. A further reduction can be expected if the spatial resolution of one of the codes were to be increased. The difference between FFW and the other codes is larger by an order of magnitude which is, however, still small (and thus acceptable).

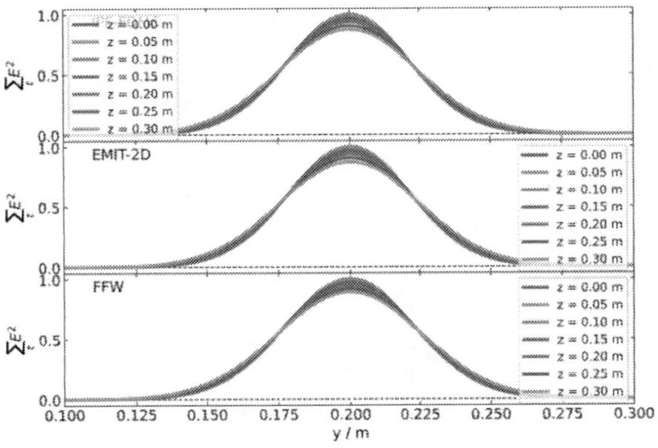


**Figure 1.** Full-wave simulation results obtained for scenario #1, shown is the time-averaged squared wave electric field at various cuts along the $y$-axis.

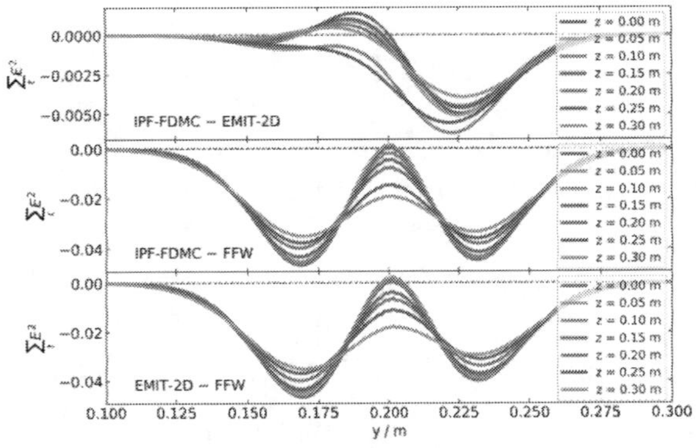


**Figure 2.** Difference of the detector antenna signals obtained from the full-wave simulations for scenario #1.

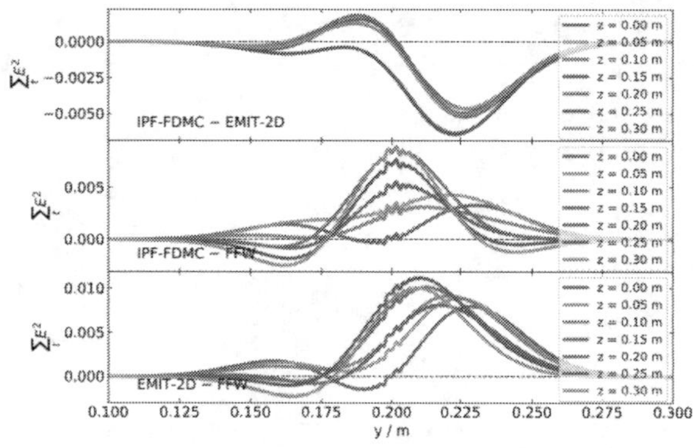


**Figure 3.** Difference of the detector antenna signals obtained from the full-wave simulations for scenario #2.

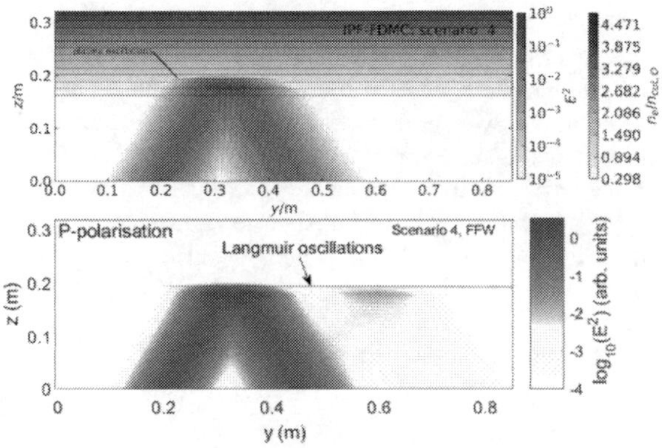


**Figure 5.** Full-wave simulation results for scenario #4, shown is a snapshot of the squared wave electric field.

### 3.2 Scenario 2

In scenario #2, the position of the beam waist is shifted upwards to $z = 0.2$ m, resulting in a converging beam. The quantitative comparison is plotted in Fig. 3: the differences between IPF-FDMC and EMIT-2D are again very small. The differences between FFW and the two FDTD codes are again somewhat larger but still small.

### 3.3 Scenario 3

In scenario #3, the waist is located at $z = 0$ and the beam is injected at an angle of $\theta = 30°$ with respect to the $z$-axis. The quantitative comparison of the detector antenna signals gives a similar picture as for scenario #1: as shown in Fig. 4, the differences between IPF-FDMC and EMIT-2D are very small and can be expected to be further reduced by increasing the spatial resolution. Although the difference between FFW and the two FDTD-codes is larger by an order of magnitude, it is still small and, looking at the spatial distribution of the differences, seems to be mostly due a small shift (or rather an offset) of the spatial grid of FFW with respect to EMIT-2D and IPF-FDMC.

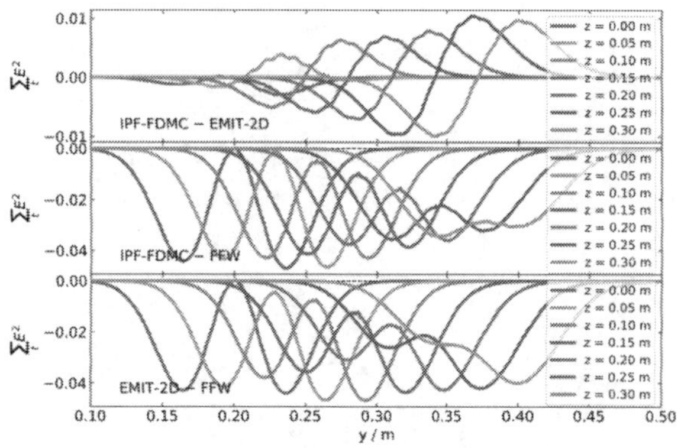


**Figure 4.** Difference of the detector antenna signals obtained from the full-wave simulations for scenario #3.

### 3.4 Scenario 4

In this scenario we have introduced the electron plasma density profile as described by Eq. (2), but still have $B_0 = 0$. A microwave beam with the waist located at the antenna plane is emitted at an angle of $\theta = 30°$. Since there is no magnetic field here, no O-SX conversion can happen (for $B_0 = 0$, the two solutions of the dispersion relation degenerate and O- and X-mode are indistinguishable). There is, however, coupling to plasma oscillations, also known as Langmuir waves, possible if the wave electric field has a component along the density gradient [19].

Figure 5 shows a snapshot of the wave electric field obtained from the full-wave simulations. Small scale oscillations can be seen at the plasma cut-off layer: these are the aforementioned plasma oscillations. In the codes, such small-scale oscillations are eventually dissipated via a collisional damping term. Note that for the case of FFW there seems to be a second wave injected at a position of roughly $y = 0.5$ m. This corresponds to reflections at the bottom absorber from the "main" wave leaving the grid at that position. While this might look like an issue, it is only visible due to the logarithmic color scale and is negligible when calculating power absorption in the grid. It is, nevertheless, planned to further improve the bottom absorber to suppress these spurious reflections.

FFW yields an absorption at the plasma frequency layer of approximately 4 %, IPF-FDMC of approximately 2.1 %. A formula by Mjølhus [20] allows to estimate this value to 5 % considering a plane wave description and thus serving as an upper boundary which would be approached if beams with increasing waist would be used in our simulations. The difference between FFW and IPF-FDMC is due to the different implementation of the beam, see the corresponding discussion in Sec. 3.6.

The differences between the detector antenna signals of the codes are shown in Fig. 6 (note that there is no signal for $z \geq 0.2$ m as the wave is in cut-off at the corresponding densities). While for most detector antenna positions the difference is small, it is more pronounced when being close to the cut-off. The reason is probably a small mis-

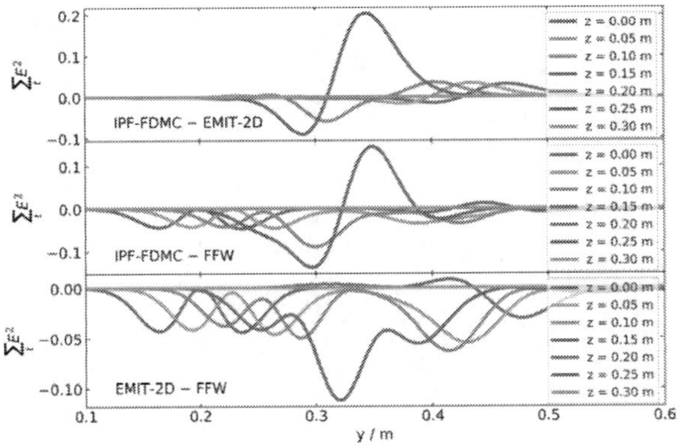


**Figure 6.** Difference of the detector antenna signals obtained from the full-wave simulations for scenario #4.

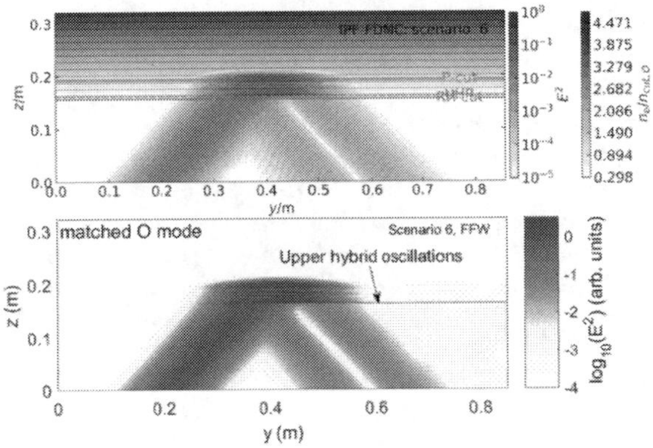


**Figure 8.** Full-wave simulation results for scenario #6, shown is a snapshot of the squared wave electric field.

alignment of the numerical grids along the $z$-axis. When being very close to the cut-off, the shape of the horizontal cross sections strongly depends on that position.

### 3.5 Scenario 5

In this scenario, a homogeneous background magnetic field $B_0 = 0.85\,\mathrm{T}$ is introduced pointing into the $y$-direction. The microwave beam is injected at an angle of 30° with respect to the $z$-axis. A pure O-mode is injected requiring an elliptically polarized beam [21]. Despite being far off the optimum injection angle for O-SX mode conversion, conversion is actually found in the simulations: a value of 1.1 % is obtained from IPF-FDMC, 1.9 % from EMIT-2D, and FFW yields 0.2 %, which gives an idea about the error-bars when comparing the codes.

Analyzing the differences between the detector antenna signals, see Fig. 7, one sees small differences between IPF-FDMC and EMIT-2D. The differences between FFW and the two FDTD codes are larger by a factor 2 but looking at their spatial distribution it seems to be due to a small misalignment of their spatial grids.

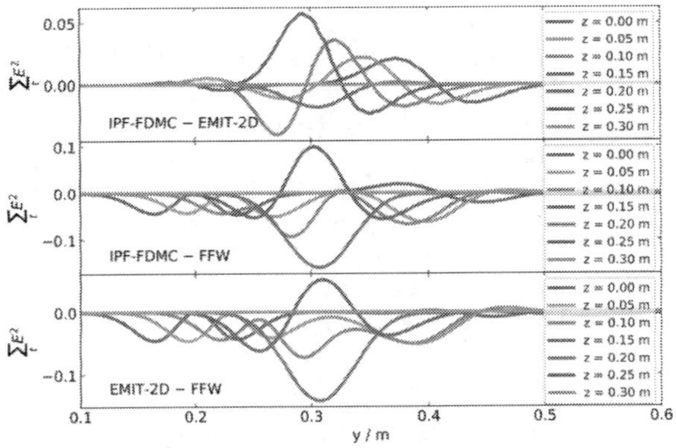


**Figure 7.** Difference of the detector antenna signals obtained from the full-wave simulations for scenario #5.

### 3.6 Scenario 6

The microwave beam is now injected at the optimum angle $\theta_{\mathrm{opt}}$ for achieving maximum mode conversion efficiency,

$$\theta_{\mathrm{opt}} = \arcsin\left\{\left[\frac{Y}{1+Y}\right]^{1/2}\right\}, \tag{10}$$

corresponding to $\theta_{\mathrm{opt}} \approx 43°$. A snapshot of the resulting wave electric fields is shown in Fig. 8. An enhancement in the vicinity of the UHR can be clearly seen. Another important feature is the "hole" in the reflected, non-converted part of the beam: the SX-mode content is "filtered" by the UHR (eventually dissipated via collisional damping in the codes), it cannot pass it. Only the tails of the Gaussian beams, which have non-optimum k-components, can pass the UHR as they cannot couple to the SX-mode. An O-SX conversion efficiency of 86.4 % is obtained from IPF-FDMC, 86.7 % from EMIT-2D, and 88 % from FFW. The higher value obtained from FFW is due to its implementation of a Gaussian beam: it is represented by a spectrum of $k$ components where each component is in pure O-mode polarization. In the two other codes the Gaussian beam is defined by appropriate spatial wave electric field distributions with the tails having non-optimum polarization. Conversion efficiencies from FFW are therefore to be understood as an upper bound.

The difference between the detector antenna signals, shown in Fig. 9, reveal a noticeable mismatch only in the vicinity of the O-SX conversion layer. This is likely due to the different dissipation mechanisms in the codes. Note that the conversion efficiencies agree, however, very well.

## 4 Steep density profiles

Plasma scenarios in MAST Upgrade relevant for a potential 28 GHz EBW heating system include L- and H-mode scenarios and different antenna positions. The resulting value of $k_0 L_n$ defines the sensitivity of the O-SX conversion on angular mismatch: higher values of $k_0 L_n$ (corresponding to more shallow profiles) are more sensitive to angular mismatch. An estimation based on a high elongation, low plasma-$\beta$ scenario yields a range of $k_0 L_n = 2 \ldots 25$. Figure 10 shows the O-SX conversion efficiencies obtained from full-wave simulations for this range, showing good agreement between the different codes. Furthermore, one can clearly see a peak in the conversion efficiency at around $k_0 L_n \approx 7$. For smaller values (thus steeper profiles), the efficiency starts to decrease as now the SX-mode can tunnel through the evanescent layer defined by the UHR and the right-hand cut-off and couple to the FX-mode which simply leaves the plasma.

## 5 Summary

Three full-wave codes have been benchmarked against each other, where good agreement was found. All were shown to be able to simulate the O-SX mode conversion yielding very similar results. This makes them suitable tools for a feasibility study of an EBW heating scenario in MAST Upgrade, to be published in a subsequent paper. It was also shown that for too steep plasma density profiles, the O-SX conversion efficiency starts to decrease which might pose a threat to in-vessel components.

## 6 Acknowledgments

The authors are indebted to the efforts of the open-source software community.

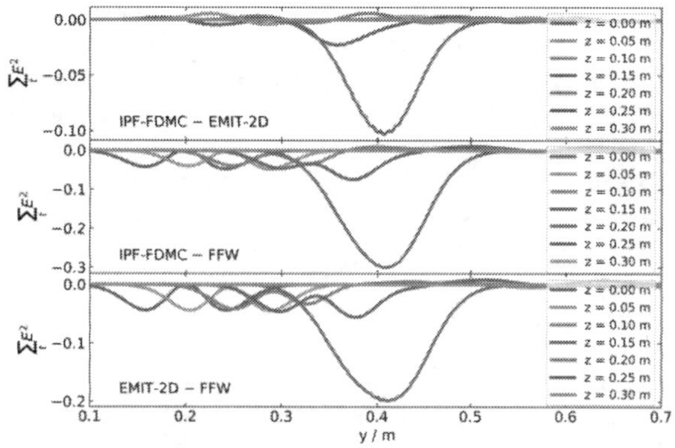


**Figure 9.** Difference of the detector antenna signals obtained from the full-wave simulations for scenario #6.

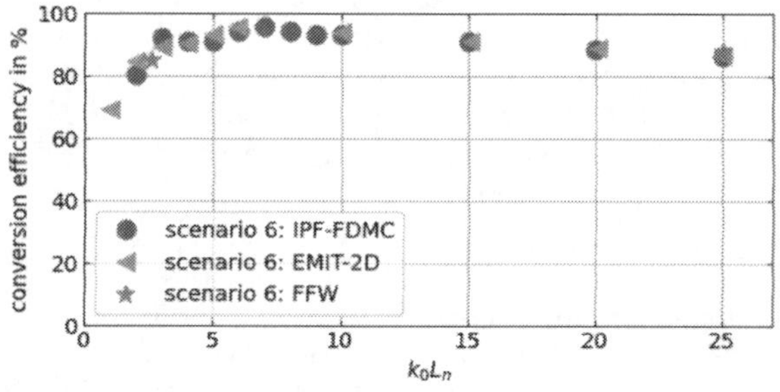


**Figure 10.** O-SX conversion efficiency obtained from full-wave simulations as a function of $k_0 L_n$ .

## References

[1] V. Erckmann, U. Gasparino, Plasma Physics and Controlled Fusion **36**, 1869 (1994)
[2] H.J. Hartfuß, T. Geist, *Fusion Plasma Diagnostics with Mm-Waves* (Wiley-VCH Verlag GmbH & Co. KGaA, Weinheim, Germany, 2013), ISBN 978-3-527-67627-9 978-3-527-41105-4
[3] C.L. Smith, S. Cowley, Philosophical Transactions of the Royal Society A: Mathematical, Physical and Engineering Sciences **368**, 1091 (2010)
[4] Z. Gao, Matter and Radiation at Extremes **1**, 153 (2016)
[5] H.P. Laqua, Plasma Physics and Controlled Fusion **49**, R1 (2007)
[6] A. Taflove, S.C. Hagness, *Computational Electrodynamics: The Finite-Difference Time-Domain Method*, Artech House Antennas and Propagation Library, 2nd edn. (Artech House, Boston, 2000), ISBN 978-1-58053-076-7
[7] A. Köhn, T. Williams, R. Vann, E. Holzhauer, J. Leddy, M. O'Brien, M. Ramisch, EPJ Web of Conferences **87**, 01003 (2015)
[8] T.R.N. Williams, A. Köhn, M.R. O'Brien, R.G.L. Vann, Plasma Physics and Controlled Fusion **56**, 075010 (2014)
[9] A. Köhn, Á. Cappa, E. Holzhauer, F. Castejón, Á. Fernández, U. Stroth, Plasma Physics and Controlled Fusion **50**, 085018 (2008)
[10] C. Lechte, IEEE Transactions on Plasma Science **37**, 1099 (2009)
[11] S. Heuraux, F. da Silva, T. Ribeiro, B. Despres, M. Campos Pinto, J. Jacquot, E. Faudot, S. Wengerowsky, L. Colas, L. Lu, Journal of Plasma Physics **81**, 435810503 (2015)
[12] F. da Silva, M.C. Pinto, B. Després, S. Heuraux, Journal of Computational Physics **295**, 24 (2015)
[13] M. Thomas, Ph.D. thesis, University of York, York (2018)
[14] V. Zakharov, Soviet Physics JETP **35**, 908 (1972)
[15] E. Kuznetsov, Soviet Physics JETP **39**, 1003 (1974)
[16] B. Eliasson, Modern Physics Letters B **27**, 1330005 (2013)
[17] B. Eliasson, A. Senior, M. Rietveld, A.D.R. Phelps, R.A. Cairns, K. Ronald, D.C. Speirs, R.M.G.M. Trines, I. McCrea, R. Bamford et al., Nature Communications **12**, 6209 (2021)
[18] SciPy 1.0 Contributors, P. Virtanen, R. Gommers, T.E. Oliphant, M. Haberland, T. Reddy, D. Cournapeau, E. Burovski, P. Peterson, W. Weckesser et al., Nature Methods **17**, 261 (2020)
[19] R.B. White, F.F. Chen, Plasma Physics **16**, 565 (1974)
[20] E. Mjølhus, Radio Science **25**, 1321 (1990)
[21] F.R. Hansen, J.P. Lynov, C. Maroli, V. Petrillo, Journal of Plasma Physics **39**, 319 (1988)

# Electron Bernstein Wave (EBW) current drive profiles and efficiency for STEP

*Thomas* Wilson[1], *Simon* Freethy[1], *Mark* Henderson[1], *Alf* Köhn-Seeman[2], *Ivan* Konoplev[1], *Samuli* Saarelma[1], *David* Speirs[3], *Roddy* Vann[4], and the STEP team[1]

[1] UKAEA, Culham Science Centre, Abingdon, OX14 3DB, United Kindgom
[2] University of Stuttgart, IGVP, Pfaffenwaldring 31, 70569 Stuttgart, Germany
[3] University of Strathclyde, 16 Richmond St, Glasgow, G1 1XQ, United Kingdom
[4] University of York, York, YO10 5DD, United Kingdom

**Abstract.** GENRAY and CQL3D were used to estimate the Electron Bernstein Wave (EBW) current drive profiles and normalised current drive efficiency $\zeta_{CD}$ (6) for several STEP reactor concepts with varying temperature, density, geometry and magnetic field. $\zeta_{CD} > 1.0$ was readily found for $\rho = 0.65 - 0.9$ while $\zeta_{CD} > 0.5$ was found for $\rho \geq 0.5$. Okhawa is found to be the most efficient current drive mechanism due to the high trapped fraction in STEP. Optimal current drive was found for 2nd harmonic absorption for $\rho \leq 0.8$ due to the higher $v_\perp$ at the wave-particle resonance. 1st harmonic absorption is required for $\rho > 0.8$ as there is no access to the 2nd harmonic at the launch frequencies examined. Rays with negative starting $N_\parallel$ were found to penetrate furthest. Due to the high electron temperature in STEP the relativistic downshift of the harmonic becomes comparable to the Doppler shift, increasing access to the 2nd harmonic.

## 1 Introduction

The UK's Spherical Tokamak for Energy Production (STEP) reactor design program is now exclusively investigating concepts using microwave based heating and current drive (HCD) systems. Electron Bernstein Wave (EBW) HCD is a relatively immature technology compared to Electron Cyclotron (EC) HCD but is of interest due to the promise of high current drive efficiency and access to dense plasmas at low magnetic fields or high density where EC is cut off.

The general ray tracing code GENRAY [1] and bounce-averaged Fokker-Planck code CQL3D [2] were used to estimate the EBW current drive profiles and the current drive efficiency for several STEP Prototype Reactor (SPR) concepts. We aimed to find estimates of the optimal EBW current drive efficiency and gain understanding of how EBW compares to EC and what parameters limit EBW access. This is used to dimension the total required microwave power and guide design of plasma scenarios with high current drive efficiency.

The need for quick estimates and large parameter ranges on launch location and frequency necessitated a simplified modelling approach where effects such as coupling and finite beam width were ignored. This work should be therefore considered a first step used to generate an initial design for the STEP EBW scenario, which will then be further analysed using more realistic EBW models.

We find the predicted EBW current drive efficiency is nearly triple that for EC in all SPRs studied, with normalised efficiency $\zeta_{CD} > 1.0$ (6) found for $\rho = 0.65 - 0.9$. Okhawa was the most efficient current drive mechanism for all SPRs. Optimal current drive is found for 2nd harmonic absorption where it is accessible.

The high electron temperatures in the STEP concepts (Table 1) produce a large Doppler broadening of the cyclotron resonance, limiting EBW access to $\rho \geq 0.5$. The relativistic downshift is also seen to be comparable with the Doppler shift. Combined with the large well in the magnetic field this allows 2nd harmonic access for an increased range of frequencies.

## 2 Physics Background

STEP requires a completely non-inductive current drive system which will be provided by purely microwave based HCD systems. These systems must operate as efficiently as possible to maximise net electrical power and commercial viability. We are interested in waves in between the 1st and 2nd harmonic of the electron cyclotron frequency $\omega_{ce}$. EBWs are electrostatic electron waves, sustained by electron cyclotron motion, propagating across the magnetic field in a hot plasma. It is convenient here to define the Stix parameters $X$ and $Y$ for a wave at frequency $\omega$

$$X = \left(\frac{\omega_{pe}}{\omega}\right)^2 \quad Y = \frac{\omega_{ce}}{\omega} \tag{1}$$

Compared to O and X mode, EBWs have two properties which make them appealing for a HCD system

1) The wave-particle interaction is stronger due to the electrostatic nature of the EBW compared to the electromagnetic O/X mode. This increases

the efficiency of current drive processes as energy is transferred to higher energy electrons.

2) EBWs have no density cut off and can propagate into the 'overdense' region $X > 1$.

High theoretical current drive efficiency is the primary reason for research into EBW HCD systems. STEP is currently being designed with an option for EBW heating though there is currently little experimental data for EBW current drive on tokamaks.

## 3 EBW Excitation

EBWs can be excited from microwaves via mode conversion of slow X mode at the Upper Hybrid (UH) layer. This can be achieved in three ways

1) Launching slow X mode from the high field side (HFS) (XB conversion).
2) Launching fast X mode from the low field side (LFS) and tunnelling onto the slow X mode (direct XB tunnelling).
3) Launching O mode from the LFS at a critical angle and converting to slow X mode at the $X = 1$ layer (OXB conversion).

In this paper we only consider LFS launch and OXB conversion as option 1 suffers from absorption of the X-mode at the fundamental harmonic and option 2 requires high density gradients that STEP will not achieve.

### 3.1 EBW Absorption

EBWs are absorbed very strongly compared to O and X mode due to their electrostatic nature [3]. The finite temperature of the plasma and large $N_\parallel$ leads to a large Doppler shift away from the cyclotron harmonics, where the EBW is completely absorbed on high energy electrons at the tail of the distribution far from the cold resonance.

The relativistic Doppler shifted resonance condition is [4]

$$1 = \frac{nY}{\gamma} + N_\parallel \beta_\parallel \quad (2)$$

where $\gamma$ is the Lorentz factor, $N_\parallel$ is the component of the wavevector parallel to the magnetic field $\boldsymbol{B}$ and $\beta_\parallel = \frac{v_\parallel}{c}$ is the velocity of the resonant electrons parallel to $\boldsymbol{B}$ normalised to the speed of light.

For $N_\parallel^2 < 1$ (2) traces an ellipse in ($\beta_\parallel$, $\beta_\perp$) velocity space [4]. It is useful to compare $\beta$ to the normalised thermal velocity $\beta_T$ including relativistic corrections [5]. As a simple estimate we assume the EBW is only absorbed for $|\beta_\parallel| < \alpha\beta_T$, where $\alpha = 3 - 3.5$ [6]. As the EBW absorption is so strong this effectively means absorption occurs at $\beta_\parallel = \alpha\beta_T$ [7,8]. Substituting into (2) we get an estimate for the resonance condition

$$\mathrm{nY} = \frac{1 \pm \alpha N_\parallel \beta_T}{\sqrt{1 - \alpha^2\beta_T^2 - \beta_\perp^2}} \quad (3)$$

However, the absorption of the wave is a finite Larmor radius effect. The absorption on the $n$-th harmonic is proportional to $\frac{J_n(\mu)}{\mu}$ [4] where $J_n$ is the $n$-th Bessel function of the first kind and $\mu = k_\perp r_L$.

For the 1st harmonic this has maximum at $\beta_\perp = 0$ while for the 2nd harmonic this is zero at $\beta_\perp = 0$. Therefore, for the 1st harmonic in we set $\beta_\perp = 0$ and for the 2nd harmonic $\beta_\perp = \beta_T$. However, the value of $\beta_\perp$ has little effect on (2) as it is small compared to $3 - 3.5\beta_T$.

It should be noted for STEP the high temperature means $\beta_T \geq 0.1$ for $\rho > 0.8$ and $\beta_T \geq 0.15$ for $\rho > 0.55$, resulting in a relativistic downshift comparable with the Doppler shift.

### 3.2 EBW Current Drive

When the EBW is absorbed, the wave-particle interaction results in a transfer of energy to electrons, creating a perturbation in the distribution function away from pure Maxwellian, called quasilinear diffusion.

Resonant electrons can gain both perpendicular and parallel momentum from the wave. From considering conservation of energy and momentum we get [4]

$$\frac{dp_\perp}{dp_\parallel} = \left(\frac{1}{N_\parallel \beta_\parallel} - 1\right)\frac{\beta_\parallel}{\beta_\perp} \quad (4)$$

In our cases $|N_\parallel \beta_\parallel| \sim 0.1$, meaning momentum transfer is almost completely perpendicular. This means two dominant mechanisms for current drive are the Fisch-Boozer effect [4] and the Okhawa effect [4], neither of which require parallel momentum transfer.

### 3.3 Current Drive Efficiency

GENRAY and CQL3D output the total toroidal driven current $I_{CD}$ by an EBW of input power $P$, which allows calculation of the absolute current drive efficiency $\eta$

$$\eta = \frac{I_{CD}[A]}{P[W]} \quad (5)$$

Current drive efficiency is typically analysed using a normalised parameter $\zeta_{CD}$ defined as [9]

$$\zeta_{CD} = 32.7\frac{\eta[A/W]n_e[10^{20}m^{-3}]R[m]}{T_e[keV]} \quad (6)$$

where $n_e$ is the plasma density, $R$ is the radius of the magnetic axis and $T_e$ is the electron temperature. This parameter allows comparison of different plasma scenarios by removing the effect of collisionality and geometry from the absolute current drive efficiency.

## 4 Modelling Codes and Input Data

### 4.1 GENRAY

GENRAY is a general ray tracing code for the calculation of electromagnetic wave propagation and absorption in the geometrical optics approximation. This code calculates the trajectory of the wave and evolution of wave parameters such as $N_\parallel$ and power.

All GENRAY calculations had two stages. The first is to scan over the poloidal and toroidal launch angles to find the most efficient OX conversion. There are always two solutions for positive and negative $N_\parallel$. The second is ray tracing at the optimal OX conversion angles. Wave power was fixed at 1MW. To speed up calculations only electron dynamics were included and we used a single ray representation (implying 100% efficient OX conversion). Our ray tracing procedure follows a similar method to [10]. The Hamiltonian used (id=6) is described here [1].

### 4.2 CQL3D

CQL3D is a multi-species, 2D-in-momentum-space, 1D in radial coordinate, fully relativistic, bounce-averaged, collisional-quasilinear Fokker-Planck equation solver. It calculates the resulting heating and current drive from the GENRAY ray tracing calculations.

CQL3D uses output from GENRAY to calculate the quasilinear diffusion coefficient along the ray trajectory, then calculates the flux surface averaged and bounce averaged electron distribution function resulting from the balance of wave heating, electric field effects and collisional relaxation. This allows calculation of the current driven on each flux surface.

In our cases, the Fokker Planck equation was solved on a grid of 26 flux surfaces. CQL3D outputs the total driven toroidal current and current density, which was used to calculate the current drive efficiency and the localisation of the current drive. It also outputs the electron distribution function on the midplane, which can be scaled to the magnetic field at the resonance location by transforming the pitch angle, preserving energy and magnetic moment [2], and discarding particles which fail to reach the resonance.

### 4.3 STEP Prototype Reactors (SPRs)

Important geometrical and plasma parameters for the considered 4 SPRs are given in Table 1. SPR-039 and SPR-045 are the same size but have different kinetic profiles. The density and temperature profiles were generated from Europed runs [10]. Z effective was set as a flat profile at 1.8.

All SPRs considered have a well in the total magnetic field on the low field side (LFS, with the minimum magnetic field occurring on midplane at ρ~0.6, as can be seen for SPR-046 in Figure 1. This is important for the interpretation of the current drive results as launching towards the core causes $Y$ to drop.

| SPR | $R_{maj}$ [m] | $R_{min}$ [m] | $B_{geo}$ [T] | $\langle n_e \rangle$ $[10^{19}\text{m}^{-3}]$ | $\langle T_e \rangle$ [keV] |
|---|---|---|---|---|---|
| 008 | 3.64 | 2.18 | 2 | 13.2 | 8.7 |
| 011 | 2.5 | 1.5 | 2.4 | 18.9 | 8.3 |
| 039 | 3.4 | 2 | 3.2 | 12.3 | 10 |
| 046 | 3.4 | 2 | 3.2 | 14.7 | 10 |

**Table 1.** Important geometrical and plasma parameters for SPRs considered in the scans.

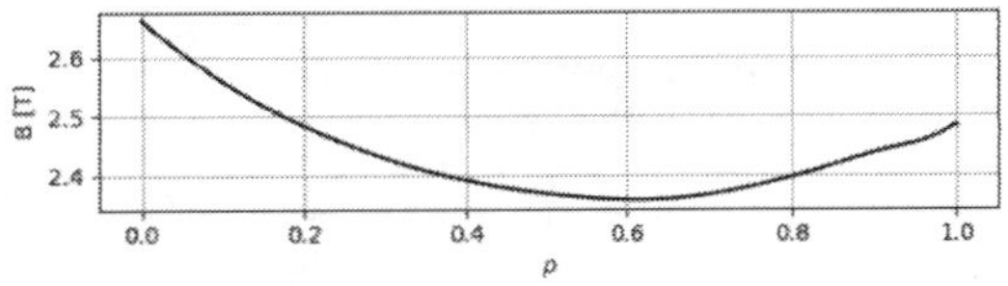


**Figure 1.** Magnetic field strength on midplane low field side for SPR-046. Note the minimum at ρ~0.6 and that the field is below the value at $\rho = 1$ for $0.2 < \rho < 1.0$.

### 4.4 Scan Parameters

For a launch frequency $f$ and launch height $Z_L$ we ran GENRAY and CQL3D to find the total driven current for 1MW input power. The launcher radius was set 1m outboard from the last closed flux surface at $Z_L$. For each launcher height we analysed both OX solutions for positive and negative optimal $N_\parallel$.

For each SPR, $f$ and $Z_L$ were scanned over the ranges given in Table 2. The height was scanned in steps of 0.25m for $Z_L$ and 1 GHz for $f$. The launcher height was only scanned above the midplane and we took the absolute value of the current when calculating efficiencies. We assume we can recover current drive of the same magnitude but with the opposite sign by launching from the opposite side of the midplane with opposite launch angles.

| SPR | $Z_L$ [m] | $f$ [GHz] |
|---|---|---|
| 008 | 0 – 3.0 | 66 - 88 |
| 011 | 0 – 2.0 | 66 - 88 |
| 039 | 0 – 3.5 | 73 - 97 |
| 046 | 0 – 4.0 | 73 - 97 |

**Table 2.** Scan range for launch height $Z_L$ [m] and frequency f [GHz] for the EBW scans for each SPR.

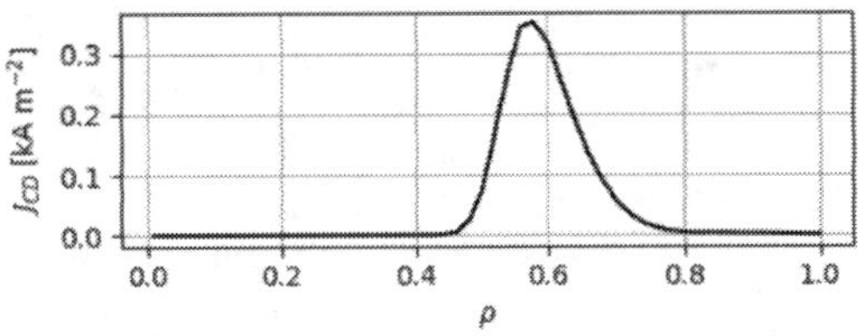


**Figure 2.** Driven current density profile giving maximum current drive at $\rho = 0.55$ for SPR-046. Current is driven for approximately $\rho = 0.1$ each side of the peak.

We associated the current drive efficiency with the rho value of the peak in the driven current density. The current drive profiles are not completely localised at a given $\rho$ value though typically the bulk of the current was within $\rho = 0.1$ of the peak value, as seen in Figure 2. If the peak driven current density was less than 1 kA $m^{-2}MW^{-1}$ the efficiency was set to 0.

## 5 Current Drive Results

### 5.1 Maximum Current Drive Efficiency

The maximum $\zeta_{CD}$ for each SPR considered is shown in Figure 3. $\zeta_{CD} > 0.5$ was readily achieved for all scenarios for $\rho > 0.5$ and $\zeta_{CD} > 1$ is found for $0.6 < \rho < 0.9$. This is almost 3 times larger than the values found for EC [*L. Figini, this workshop*]. The EBW fails to penetrate the plasma beyond $\rho = 0.4$, though current drive becomes inefficient for $\rho < 0.5$.

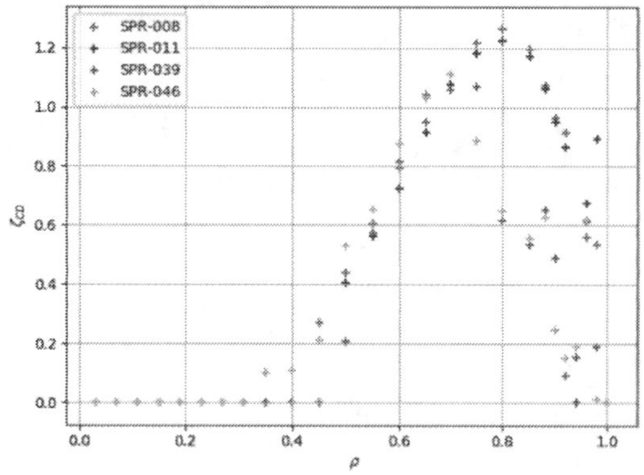


**Figure 3.** Maximum $\zeta_{CD}$ vs ρ of the peak in the driven current density for each SPR.

There is good agreement between all SPRs apart from between $0.7 < \rho < 0.9$ for SPR-039 and SPR-046. There were some numerical issues with these SPRs causing GENRAY to exit as the OX conversion failed. This can likely be resolved via adjustment of numerical parameters and so SPR-039 and SPR-046 are expected to agree with SPR-008 and SPR-011 in this region.

### 5.2 Analysis of results

To understand these results we need to study the wave particle resonance. Figure 4 shows the power damped on the 1st and 2nd harmonic from CQL3D for all SPRs.

We see a transition 1st harmonic to 2nd harmonic absorption when moving towards the plasma core. We also see the optimal current drive occurs for the positive $N_{\parallel}$ solution of the OX conversion for $\rho > 0.6$ and for negative $N_{\parallel}$ for $0.4 \leq \rho \leq 0.6$.

From (2) we see resonance $N_{\parallel}\beta_{\parallel} > 0$ for a 1st harmonic and $N_{\parallel}\beta_{\parallel} < 0$ for a 2nd harmonic resonance. From the sign of $N_{\parallel}$ we get the sign of $\beta_{\parallel}$, which can be compared to the sign of $I_{CD}$ to figure out if the current drive is Fisch-Boozer or Okhawa. The total current drive multiplied by the sign of the resonant electron velocity is plotted in Figure 5.

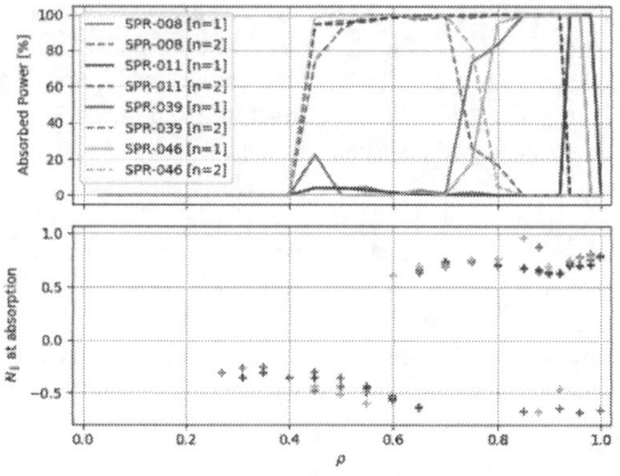


**Figure 4.** Percentage of total power absorbed on the 1st and 2nd harmonic (top) and $N_{\parallel}$ at absorption for each SPR. There is a transition from 1st harmonic absorption to 2nd harmonic. This transition is further inboard SPR-039 and SPR-045. Rays with best access have negative $N_{\parallel}$. Note the SPR-008 and SPR-011 traces overlay each other (top).

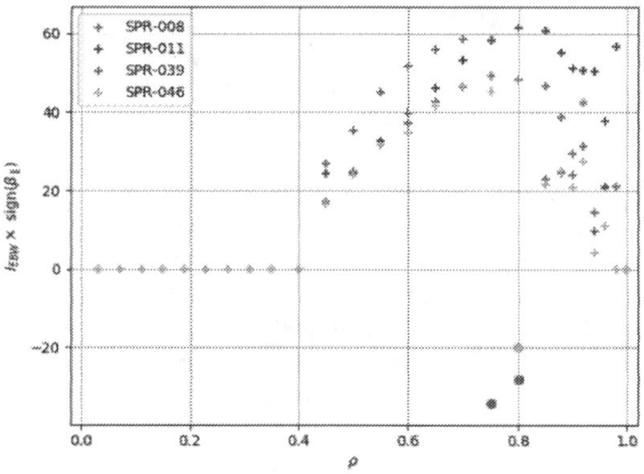


**Figure 5.** Total EBW driven current multiplied by the sign of the resonant electron velocity. Almost all runs show positive values implying the Okhawa effect is the dominant current drive mechanism.

Almost all SPRs show positive values meaning Okhawa is the dominant current drive mechanism. The only Fisch-Boozer dominant points occur in the region where SPR-039 and SPR-046 had numerical issues.

As STEP is a spherical tokamak there is a large trapped fraction due to the large variation in magnetic field strength between the low and high field side. This means the trapped-passing (TP) boundary occurs at small $\beta_{\perp}$ so the resonant interaction is likely to drive particles close to the TP boundary. We can investigate by looking at the deviation of the electron distribution function from Maxwellian after the absorption of the

EBW. The perturbation in the electron distribution function for SPR-008 at $\rho = 0.88$ for run giving the best current drive efficiency in Figure 6.

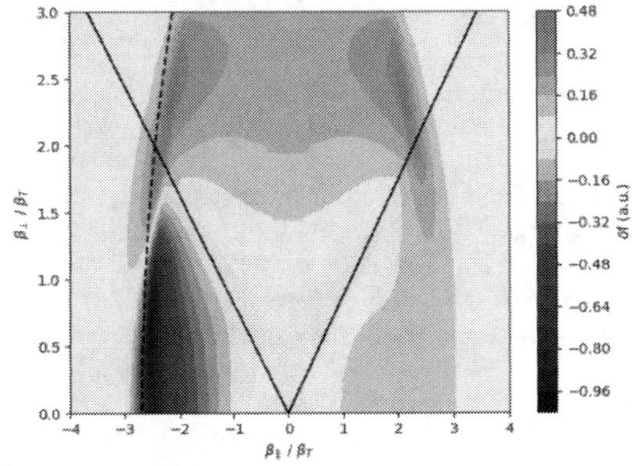

**Figure 6.** Perturbation in distribution function [a.u.] from Maxwellian at $\rho = 0.88$ for run giving the best current drive efficiency for SPR-008. $I_{CD} = -38.8$ kA, $N_\parallel = 0.64$ and absorption is on 2nd harmonic so Okhawa current drive is dominant. The resonance curve (dashed black) and trapped-passing boundary (solid black) are also shown.

The wave particle interaction has created an increase in density of electrons at higher $\beta_\perp$ near the downshifted resonance predicted in (3). Okhawa current dominates as the resonance is close to the TP boundary and large numbers of particles are trapped. Bounce averaging forces the distribution function to be symmetric in $\beta_\parallel$ inside the TP boundary, resulting in zero current. The fast bounce time of trapped electrons means momentum gets rapidly transferred across the TP boundary. This causes some de-trapping at the opposite sign of the resonant velocity which enhances the Okhawa effect [6].

### 5.3 Analysis of Absorption

The simple estimate of absorption (3) gives a prediction for the $Y$ value where the EBW will be absorbed. This value is plotted against the value from GENRAY ray tracing in Figure 7. $\alpha = 3.35$ gave best fit.

There is reasonable agreement between the estimate and the predictions from GENRAY, with a mean absolute error of 0.02. This cannot replace GENRAY ray tracing calculations as it requires $N_\parallel$ and the magnetic field along the ray trajectory. We can use it to explain the switch from 1st to 2nd harmonic absorption and the change of sign in $N_\parallel$.

$N_\parallel$ increases over the ray trajectory, meaning $|N_\parallel|$ grows for rays starting with positive $N_\parallel$ and shrinks for rays starting with negative $N_\parallel$ as seen in Figure 8. The Doppler broadening of the harmonic therefore grows and shrinks during ray propagation for positive and negative starting $N_\parallel$ respectively.

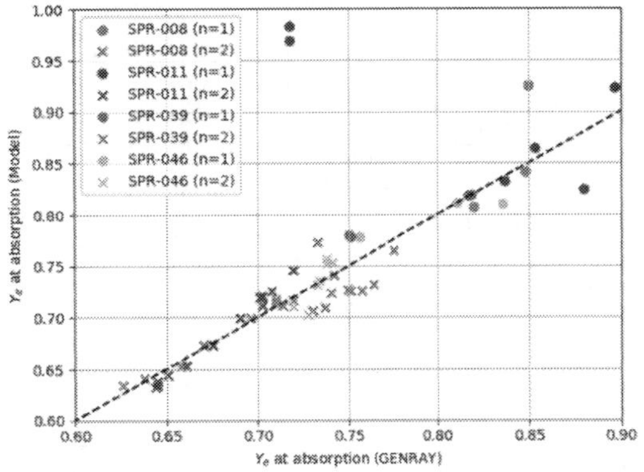


**Figure 7.** Predicted Y value at absorption using $\alpha = 3.35$ (y axis) vs value from GENRAY, for rays giving optimal current drive. The dashed black line is y=x. There is reasonable agreement with a mean absolute error of 0.02.

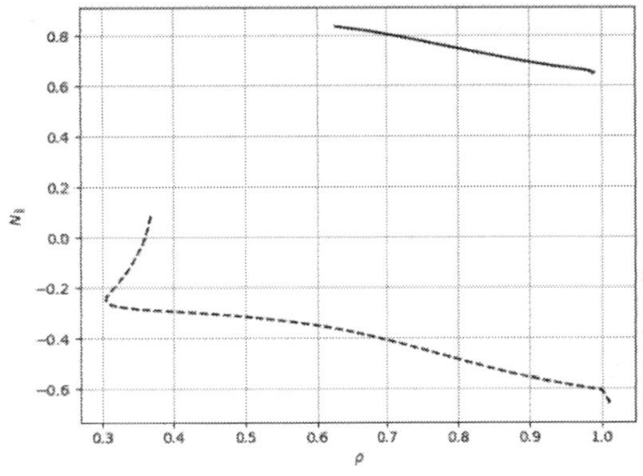

**Figure 8.** $N_\parallel$ evolution across the EBW ray trajectory for two rays giving maximum current drive at $\rho = 0.45$ (dashed) and $\rho = 0.75$ (solid) for SPR-008. Both rays start with $|N_\parallel| \sim 0.65$. The ray traces continue beyond the point of absorption but there is no power left in the ray.

When a ray meets the resonance condition it will be rapidly absorbed, so it is not surprising that rays with smaller $|N_\parallel|$ travel further into the plasma before being absorbed. Figure 9 shows the predicted absorption location vs launch frequency for SPR-008 and SPR-046 for $N_\parallel = 0.65$, a typical value after OXB conversion.

We see absorption on the 1st harmonic is only possible for $\rho < 0.9$ for a small range of frequencies unless $N_\parallel$ increases significantly. The magnetic well and the relativistic downshift of the resonance frequency reduces the possibility for 1st harmonic absorption and increases the possibility of 2nd harmonic absorption.

This explains the transition from 1st to 2nd harmonic absorption. We also see for SPR-008 the 1st harmonic is downshifted much further compared to SPR-046 due to the deeper magnetic well. This explains why 1st harmonic absorption was only possible for $\rho > 0.75$ at large $N_\parallel$ for SPR-046 but only seen for $\rho > 0.9$ in SPR-

008. In this region there is also no access to the 2[nd] harmonic at the launch frequencies examined.

The last observation to explain is that 2[nd] harmonic absorption gave higher current drive efficiency than 1[st] harmonic absorption in regions both were possible. This can be explained by the resonant interaction of the 2[nd] harmonic occurring at higher $\beta_\perp$ than the 1[st] harmonic, resulting in more efficient trapping as the resonant interaction is closer to the TP boundary.

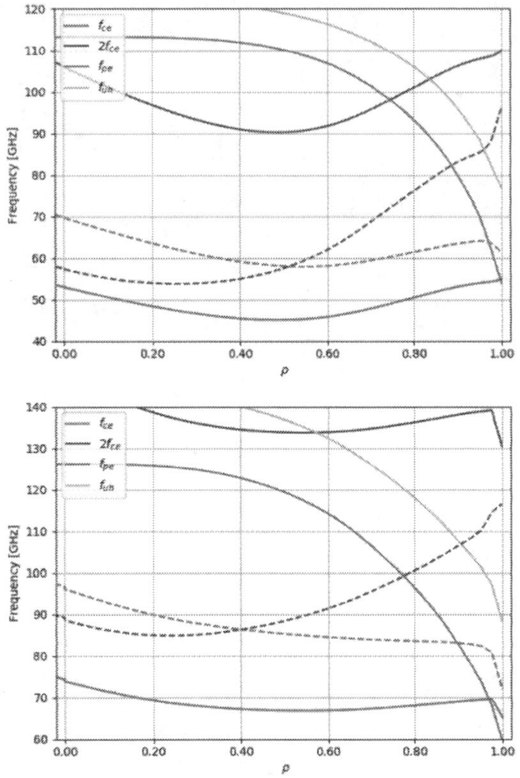


**Figure 9.** Estimate of EBW absorption on midplane for SPR-008 (top) and SPR-046 (bottom) for $N_\parallel = 0.65$. The cold 1[st] (red) and 2[nd] (purple) cyclotron harmonic frequencies and predicted 1[st] harmonic upshift / 2[nd] harmonic downshift from (3) (dashed) are shown. The plasma frequency (green) and upper hybrid frequency (orange) are also shown.

By estimating $N_\parallel$ and the ray trajectory we could also use (3) to identify frequencies and launch heights likely to give optimal access. It also can estimate the effect of changes in temperature on absorption which can be used to estimate EBW access for different plasma scenarios.

## 6 Summary

GENRAY and CQL3D were used to scan the EBW current drive efficiency for 4 STEP prototype reactor concepts. $\zeta_{CD} > 0.5$ was achieved for all scenarios for $\rho > 0.5$ and $\zeta_{CD} > 1$ is found for $0.6 < \rho < 0.9$. This is almost 3 times larger than the values found for EC. The EBW fails to penetrate the plasma beyond $\rho = 0.4$, though current drive becomes inefficient for $\rho < 0.5$.

The Okhawa effect is the most efficient mechanism for current drive in STEP. As STEP is a spherical tokamak the absorption of the EBW pushes electrons near or across the trapped-passing boundary preventing efficient Fisch-Boozer current drive.

EBW current drive occurs deepest in the plasma for rays starting with negative $N_\parallel$ and absorb on the 2[nd] harmonic. This is because $N_\parallel$ increases as the ray propagates inwards, reducing $|N_\parallel|$ for rays starting with negative $N_\parallel$ and reducing the size of the Doppler broadening of the cyclotron harmonics.

A simple estimate for the Doppler broadened EBW resonance is given. Due to the high temperature of STEP plasmas, relativistic downshifts of the resonant frequencies due to mass increase are comparable with the Doppler shift. This means 1[st] harmonic absorption is only possible for $\rho > 0.75$ at large $N_\parallel$.

All current drive efficiencies should be considered optimistic as we have assumed 100% OXB coupling efficiency and neglected any effects due to finite beam width. In quoting current drive efficiencies, we have also assumed current drive is linear with power and ignored any non-linear effects of instabilities which occur at high power.

## References

1. Smirnov, A. P., and R. W. Harvey. *CompX Report CompX-2000-01* (2001).
2. Harvey, R. W., and M. G. McCoy. *In Proceedings of the IAEA Technical Committee Meeting on Simulation and Modeling of Thermonuclear Plasmas*, pp. 489-526. 1992.
3. Decker, Joan, and Abhay K. Ram. *Physics of Plasmas* 13, no. 11 (2006): 112503.
4. Prater, Ronald. *Physics of Plasmas* 11, no. 5 (2004): 2349-2376.
5. Karney, Charles FF. *Computer Physics Reports* 4, no. 3-4 (1986): 183-244.
6. Decker, Joan. "Electron Bernstein wave current drive modeling in toroidal plasma confinement." PhD diss., Massachusetts Institute of Technology, 2005.
7. Preinhaelter, J., et al. In *AIP Conference Proceedings*, vol. 787, no. 1, pp. 349-352. American Institute of Physics, 2005.
8. Volpe, F., H. P. Laqua, and W7-AS Team. *Review of scientific instruments* 74, no. 3 (2003): 1409-1413.
9. Luce, T. C., et al. *Physical review letters* 83, no. 22 (1999): 4550.
10. Taylor, G., et al. *Physics of Plasmas* 11, no. 10 (2004): 4733-4739.
11. Saarelma, S., et al. *Physics of Plasmas* 26, no. 7 (2019): 072501.

# Induced microwave scattering in the ITER edge transport barrier at ECRH and possibility of its modeling at ASDEX-Upgrade

*Evgeniy* Gusakov, *Alexei* Popov, *Michael* Irzak

Ioffe Institute, 194021 Saint-Petersburg, Russia

**Abstract.** The induced scattering parametric decay instability of O-mode polarized microwaves leading to excitation of lower hybrid or electron plasma waves localized within a tokamak edge transport barrier is investigated. Numerical estimates of the instability threshold and its growth rate for this scenario are given for the conditions of O2-mode ECRH experiments at ASDEX-Upgrade, where experimental investigation of these phenomena is possible.

## 1 Introduction

Electron cyclotron resonance heating (ECRH) is widely used in current toroidal devices. The O1-mode ECRH technique is also considered for the local electron heating providing the neoclassical tearing mode control in ITER. Until very recently the propagation and absorption of O-mode microwaves were believed to be well-described by the linear theory and assumed to be predictable in detail. However, as it was shown in [1], in ITER they can suffer from a low threshold induced scattering parametric decay instability (PDI) in the edge transport barrier (ETB). The presence of a large density gradient have a significant impact on the properties of waves in the lower hybrid (LH) frequency range, leading to new transparency windows that are absent in a homogeneous plasma [2,3]. These new modes can be 2D localized along the direction of a plasma inhomogeneity due to gradient effects and along the magnetic field due to magnetic ripples. The instability power threshold leading to 2D localized LH or electron plasma (EP) wave excitation under the ITER conditions is much less than 1 MW. They are likely to be present in future O1-mode ECRH experiments at ITER and at DEMO, leading to broadening of the power deposition profile and therefore decreasing the neoclassical tearing mode suppression efficiency. Thus, the urgent experimental investigation of this parametric decay instability seems important for the ITER experiment planning. In the present paper we demonstrate a possibility to investigate the O-mode induced scattering PDI and its consequences in O2-mode ECRH experiments on ASDEX-Upgrade. The instability threshold is shown to be well below 0.5 MW. Its dependence on the density gradient in the ETB, magnetic field ripple and on the scattering angle is studied. The instability growth rate is also determined.

## 2 Lower hybrid and electron plasma waves in the edge transport barrier

To describe the behavior of LH and EP waves in the ETB with a proper account of magnetic field ripples we introduce the local Cartesian coordinate system $(x,y,z)$ with its origin located at the inflection point of the density profile and the local minimum of the magnetic field between two nearby coils. The coordinate $x$ is related to the flux surface label, $y$ is the coordinate perpendicular to the magnetic field line on the magnetic surface and $z$ is the coordinate directed along the magnetic field line. We consider an electrostatic wave propagating mainly across the magnetic field on a magnetic surface, whose potential can be represented by means of the WKB approximation as $\phi=\Phi(x,z)\exp(iq_y y-i\omega_L t)+c.c.$ with its amplitude $\Phi$ obeying the Poisson equation

$$\hat{D}(\omega_L)\Phi=\varepsilon(\omega_L)\left(\frac{\partial^2}{\partial x^2}-q_y^2\right)\Phi+\frac{\partial g(\omega_L)}{\partial x}q_y\Phi+\eta(\omega_L)\frac{\partial^2}{\partial z^2}\Phi=0 \quad (1)$$

where $\varepsilon$, $g$, $\eta$ are the cold-plasma dielectric tensor components; $\omega_{pi}$ and $\omega_{pe}$ - the ion and electron plasma frequencies; $\omega_{ci}$ and $\omega_{ce}$ - the ion and electron cyclotron frequencies. Then, we introduce a function $Q=\varepsilon(\omega_L)^{-1}\partial g(\omega_L)/\partial x$ which has a local maximum at the inflection point of the density profile in the ETB region leading to a localized solution around $x=0$. Figure 1 shows the typical density profile (thick solid curve) at an ETB in the ASDEX-Upgrade. The solid curve shows the wave number $q_x=q_x(q_y,q_z=0)$ of the LH wave ($f_L=2.8$ GHz, $q_y=Q_0/2=2.4\text{ cm}^{-1}$, $Q_0=Q(0,0)$). To describe this localized solution analytically, we expand the magnetic field in powers of $z$ around the magnetic field minimum position $B\approx\bar{B}\left(1-\delta(x,y)(Nz/R)^2/2+...\right)$, where $\bar{B}$ is the magnetic field on a given magnetic surface, $N$ is the number of toroidal field coils, $R$ is the tokamak major radius; and $\delta$ is the amplitude of magnetic ripples. This allows approximating the function $Q$ with a parabola over both coordinates $Q\approx Q_0\left(1-\frac{x^2}{2l_x^2}+\frac{z^2}{2l_z^2}\right)$ and reducing equation (1) to

$$\left(\varepsilon(\omega_L,0)\left(\frac{\partial^2}{\partial x^2}+Q_0q_y-q_y^2-K_x^4x^2\right)-\left|\eta(\omega_L,x_m)\right|\left(\frac{\partial^2}{\partial z^2}-K_z^4z^2\right)\right)\Phi=0 \quad (2)$$

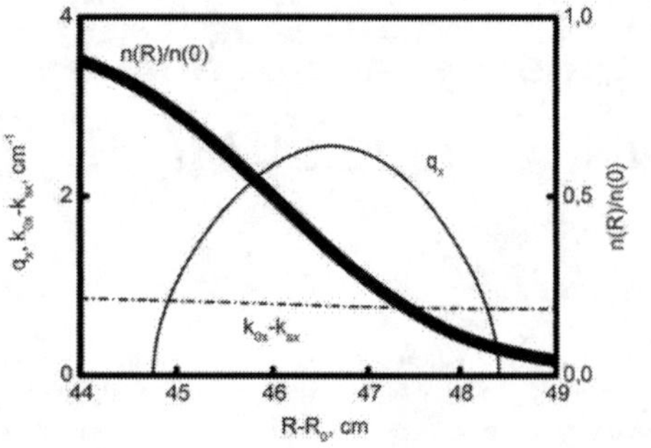


**Fig. 1.** Dispersion curves of the interacting waves. The thick solid curve shows the dimensionless density. The dashed dotted curve is the difference between the numbers of the pump and scattered O-mode waves $k_{0x}-k_{sx}$. The solid curve is the EP wave number $q_x = q_x(q_{ym}, q_z = 0)$, $q_{ym} = 2.4$ cm$^{-1}$. The pump and scattered wave frequencies are $f_0 = 140$ GHz, $f_s = 137.02$ GHz; the scattering angle is $\vartheta = 5.3^0$; $\Delta_{\text{ETB}} = 2$ cm.

where $K_x = \left(Q_0 q_y / \left(2l_x^2\right)\right)^{1/4}$, $K_z = \left(Q_0 q_y / \left(2l_z^2 \left|\eta(\omega_L)\right|\right)\right)^{1/4}$. Equation (2) describes a wave localized both in the radial direction and along the magnetic field and has a solution $\Phi(\mathbf{r}) = \Phi_{p,r} f_p(K_x x) f_r(K_z z)$, expressed in terms of the Hermite polynomials $f_p(Kx) = \sqrt{K / (\sqrt{\pi} 2^p p!)} \exp\left(-K^2 x^2 / 2\right) H_p(Kx)$ with the amplitude $\Phi_{p,r}$ taken constant. The eigenfrequency of the 2D trapped wave is determined by the following equation

$$\kappa^2\left(\omega_L^{p,r}\right) = (2r+1)\left|\eta\left(\omega_L^{p,r}\right)\right| K_z^2\left(\omega_L^{p,r}\right) + \varepsilon\left(\omega_L^{p,r}\right)\left(Q_0\left(\omega_L^{p,r}\right) q_y - q_y^2 - (2p+1) K_x^2\left(\omega_L^{p,r}\right)\right) = 0 \quad (3)$$

If the density gradient is small, or even absent, the plasma is evanescent for such a LH wave propagating almost perpendicular to the external magnetic field. However, in the case of strongly inhomogeneous plasmas the new transparency regions for LH waves appear first discovered in [2]. Later, it was demonstrated numerically [3] that at large density gradients these oscillations, propagating strictly across the magnetic field, can be localized in the plasma inhomogeneity direction. It should be noted that this wave has a remarkable property. In a certain range of wave numbers $q_y$, it changes the sign of its group velocity $\upsilon_{gy}$ in the $y$ direction within the localization region along the direction of the plasma inhomogeneity. When $q_{ym} = Q_0 / 2\left(1 - (2p+1) / \left(4K_x^2 l_x^2\right)\right)$, the average convection velocity along the $y$ coordinate is zero $u_{Ly} = \int_{-\infty}^{\infty} dx \left|f_p(K_x x)\right|^2 \upsilon_{gy}(x) = 0$. For such a wave, the convective loss along the $y$ direction is suppressed and the only sink of its energy from the decay region is due to diffraction. The latter is a slower process than convection. The LH or EP wave possessing a $y$ wave number component close to the optimal value $q_{ym}$ is the most unstable when excited due to a PDI.

## 3 O-mode induced scattering off LH or EP waves 2D localized in an ETB

In inhomogeneous plasmas the parametric decay occurs in a narrow layer in the close vicinity of a point at which the decay resonance for the frequencies and numbers of coupled oscillations are fulfilled. We assume that the decay point is within the ETB and consider a pump O-mode wave propagating perpendicular to the magnetic field along $x$ towards the plasma core. The pump wave can decay into a localized LH wave and an O-mode wave. The potential of the LH wave without the nonlinear coupling is described by equation (2). Its frequency and wave number obey the quantization condition (3). We assume that the $y$ component of the LH wave number is equal to the optimal value. As decay begins, the amplitude $\Phi_{p,r}$ of the non-linearly excited LH or EP wave starts to grow in the pump localization zone and experiences diffraction along $y$. According to [1], this phenomenon is described by the following equation

$$\left(\frac{\partial}{\partial t} + i < \Lambda_y > \frac{\partial^2}{\partial y^2}\right)\Phi_{p,r} = \gamma_0 \exp\left(-\frac{y^2}{\tilde{w}^2}\right)\Phi_{p,r} \quad (4)$$

where $< \Lambda_y > = \int_{-\infty}^{\infty} \left|f_p(K_x x)\right|^2 \left|f_p(K_z z)\right|^2 \Lambda_y dx dz$ is the diffraction coefficient averaged over the wave localization area, $\tilde{w}^2 = |w^2|^2 / w_0^2$, $w = \sqrt{w_0^2 - iR_f c / \omega_0}$ is the local complex width of a beam, $w_0$ is the beam width at the focus position $x_f$, the parameter $R_f$ when it obeys the inequality $w_0^2 >> R_f c / \omega_0$ has the meaning of a wavefront curvature radius at the antenna,

$$\gamma_0 = \frac{A_0^2}{8B(0)^2} \frac{\omega_{pe}^4(0)}{\omega_0^2 \omega_{ce}^2(0)} \frac{c}{\omega_s} \frac{1}{< \partial D(\omega_L) / \partial\omega >} \times \frac{\left(q_{ym}^2 + K_x^2(2p+1)\right)^2}{n_{sx}(0) n_{0x}(0)} \int_{-\infty}^{\infty} dz \exp\left(-\frac{z^2}{\tilde{w}^2}\right)\left|f_r(K_z z)\right|^2 \times \int_{-\infty}^{\infty} dx f_p(K_x x)^* \int_{-\infty}^{x} dx' \exp\left(i\int_{x'}^{x} \Delta K dx''\right) f_p(K_x x') \quad (5)$$

is the pumping rate; $\omega_0$, $\omega_s = \omega_0 - \omega_L^{p,r}$ are the pump and scattered O-mode frequencies, $\Delta K = (\omega_s n_{sx} - \omega_0 n_{0x}) / c$; $n_{0,sx} = n_x\left(\omega_{0,s}, n_{0,sy}\right)$, $n_x\left(\omega_0, n_y\right) = \sqrt{\eta(\omega_0) - n_y^2}$, $P_0$ and $A_0 = \sqrt{8P_0 / (c\tilde{w}^2)}$ are the pump wave's power and amplitude. If the pump power significantly exceeds the power threshold value, *i.e.* $P_0 >> P_0^{th}$, then we can approximate solution to (4) analytically (see [4]) $\Phi_{p,r}(t, y) = \exp\left(\gamma_{ins}^s t + i\delta\omega_{ins}^s t\right) f_s\left(K_y y\right)$, $s \in Z$, where

$$K_y^{-1} = \frac{< \Lambda_y >^{1.4} \tilde{w}^{1/2}}{\sqrt[4]{\gamma_0}} \exp\left(-i\frac{\pi}{8} - i\arg\left(\frac{\gamma_0}{4}\right)\right), \quad (6)$$

$$\delta\omega_{ins}^s = \gamma_0'' + (2s+1)\sqrt{\frac{|\gamma_0| < \Lambda_y >}{\tilde{w}^2}} \cos\left(\frac{\arctan(\gamma_0'' / \gamma_0')}{2} + \frac{\pi}{4}\right)$$

and

$$\gamma_{ins}^s = \gamma_0' - (2s+1)\sqrt{\frac{|\gamma_0| < \Lambda_y >}{\tilde{w}^2}} \sin\left(\frac{\arctan(\gamma_0'' / \gamma_0')}{2} + \frac{\pi}{4}\right)$$

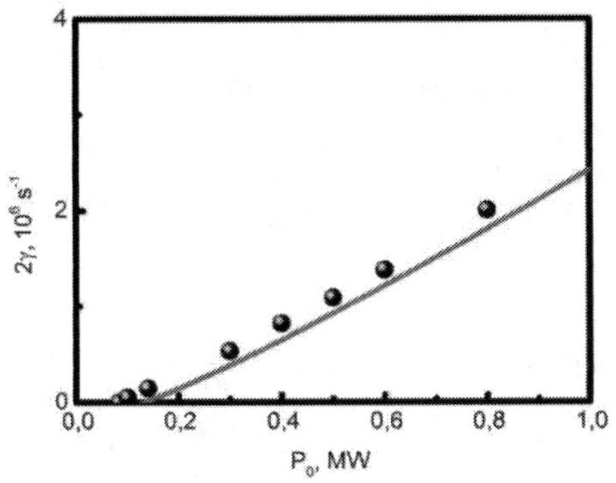


**Fig. 2.** Dependence of the growth rate on the pump power. The solid curve is defined by the analytical dependence (6). The scattered circles are the results of numerical solution of equation (4). The numerical solution gives an estimate for the power threshold of $P_0^{th}=0.094$ MW at $\delta=0.5\%$, $w=2$ cm and the same parameters of the coupled waves as in figure 1.

is the LH wave growth rate. Though equation (6) is no more valid if the pump power only slightly exceeds the threshold value $P_0 \geq P_0^{th}$, we can use it to derive a rough estimate of the PDI threshold

$$\gamma_0'\left(P_0^{th}\right)=(2s+1)\sqrt{\frac{\left|\gamma_0\left(P_0^{th}\right)\right|<\Lambda_y>}{\tilde{w}^2}}\times \sin\left(\frac{\arctan\left(\gamma_0''\left(P_0^{th}\right)/\gamma_0'\left(P_0^{th}\right)\right)}{2}+\frac{\pi}{4}\right) \quad (7)$$

## 4 Induced scattering PDI for O2-mode ECRH in ASDEX-Upgrade

The induced scattering PDI can be investigated in O2-mode ECRH experiments at the ASDEX-Upgrade tokamak ($R_0=165$ cm, a central magnetic field is up to 3T, $N=16$) proposed for central heating of dense plasma [5]. For the parameters of O2- mode ECRH at ASDEX-Upgrade with 140 GHz gyrotrons the single pass absorption is incomplete [6] which provides a chance of measuring of the induced forward-scattering signal at the high magnetic field side of the device after crossing the resonance layer. The magnetic ripple amplitude at the separatix is 0.2 - 1%, depending on its position [7]. Thus, if such experiments were carried out with ETB parameters similar to those discussed in [8], the phenomenon analyzed in this article could occur. To prove this, we plot in figure 1 the dispersion curves of the LH wave (thin solid curve) along with a difference of the pump and forward scattered O-mode wave numbers (dashed dotted curve). The pump and scattered O-mode wave frequencies are $f_0=140$ GHz and $f_s=137.02$ GHz. The $y$ component of the wave number is $q_{ym}=2.4$ cm$^{-1}$, that corresponds to a local scattering angle of $\vartheta=\tan(q_{ym}/k_{sx})^{-1}=5.3^0$. Since the O2-mode does not experience strong absorption in the optically thin ECR layer [6], the scattering signal can be detected on the high magnetic field side.

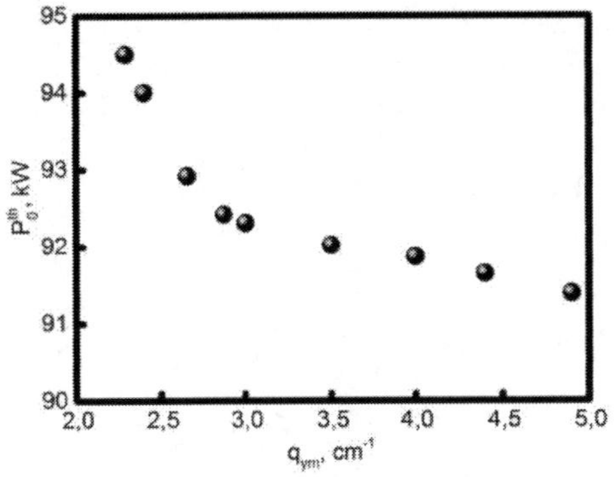


**Fig. 3.** Dependence of the instability threshold, which is determined numerically by solving equation (4), on the parameter $q_{ym}$ at an eigenfrequency corresponding to this value. The other parameters are the same as in figure 2.

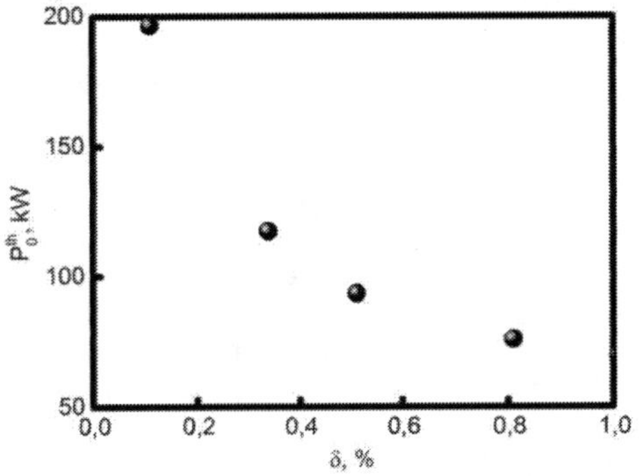


**Fig. 4.** Dependence of the power threshold on the ripple amplitude. The other parameters are the same as in figure 2.

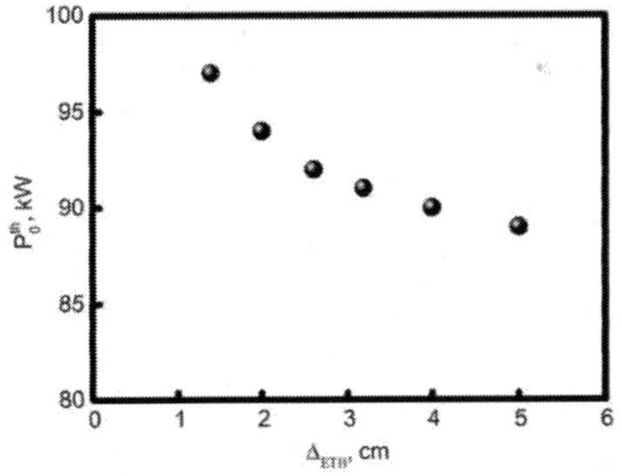


**Fig. 5.** Dependence of the instability threshold, which is determined numerically, on the inhomogeneity scale $\Delta_{ETB}$ at the density profile inflection point $x_{inf}$ and at $q_{ym}=2.4$ cm$^{-1}$. The other parameters are the same as in figure 2.

Figure 2 shows the dependence of the growth rate on the pump power. The solid curve is given by (6). The scattered circles are the results of numerical solution of (4). According to (8), the predicted instability threshold is $P_0^{th}=0.131$ MW, while the numerical solution gives a value of $P_0^{th}=0.094$ MW for it. The ripple amplitude here was assumed to be $\delta = 0.5\%$, while the pump beam width was assumed to be $\tilde{w} = 2$ cm. Figure 3 shows a weak dependence of the instability threshold, which is determined numerically, on the parameter $q_{ym}$ at the eigenfrequency (3) corresponding to this optimum value. The other parameters are the same as in figure 1. Figure 4 shows the dependence of the instability threshold calculated numerically on the ripple amplitude. Increasing the ripple amplitude reduces the LH wave localization area along the magnetic field. The latter increases the coupling efficiency and decreases the instability threshold. Figure 5 shows the dependence of the instability threshold, which is determined numerically, on the inhomogeneity scale $\Delta_{\mathrm{ETB}}$ at the density profile inflection point $x_{\mathrm{inf}}$ in the ETB and at $q_{ym} = 2.4\ \mathrm{cm}^{-1}$. The dependence in figure 5 on the scale of inhomogeneity $\Delta_{\mathrm{ETB}}$ is also weak. It remains to be noted that all figures 1-5 are plotted for the parameters at the inflection point of the density profile as follows $B_0 = 2.5$ T, $n(x_{\mathrm{inf}}) = 0.7\times10^{14}\ \mathrm{cm}^{-3}$, $T_{e,i}(x_{\mathrm{inf}}) = 200$ eV.

## 5 Conclusions

In the present work we have shown that an absolute induced scattering parametric decay instability, leading to the excitation of a lower hybrid wave localized in the edge transport barrier, may be investigated in the proposed O2-mode ECRH experiments on the ASDEX-Upgrade tokamak. We have estimated the instability threshold and shown that it is significantly lower than the power available in a single microwave beam. The growth rate of instability and spectral characteristics of the scattering signal in the proposed experiments were also discussed. However, in order to predict the power level of induced scattering, a study of the instability saturation is needed, which is planned by the authors and will be implemented in the future.

The analytical treatment is supported under the RSF 22-12-00010 grant, the numerical modelling is performed under the Ioffe Institute state contract 0040-2019-0023 whereas the code for the PDI modeling was developed under the Ioffe Institute state contract 0034-2021-0003.

## References

1. E.Z. Gusakov, A.Yu. Popov, Phys. Rev. Lett. **128**, 065001 (2022)
2. E.Z. Gusakov, M.A. Irzak, A.D. Piliya, JETP Letters **65**, 25 (1997)
3. E.Z. Gusakov, V.V. Dyachenko, M.A. Irzak, O.N. Shcherbinin, S.A. Khitrov, Plasma Phys. Control. Fusion **52**, 075018 (2010)
4. E.Z. Gusakov, A.Yu. Popov, Phys. Usp. **63**, 365 (2020) https://doi.org/10.3367/UFNe.2019.05.038572
5. M. Schubert et al. and the ASDEX Upgrade Team, Proceedings of 43rd EPS Conference on Plasma Physics 4 - 8 July 2016, Leuven, Belgium Vol **40A** (ISBN: 2-914771-99-1), P1.026
6. V. Erckmann, U. Gasparino, Plasma Phys. Control. Fusion **36**, 1869 (1994)
7. M. Garcia-Munoz et al. and the ASDEX Upgrade Team, Plasma Phys. Control. Fusion **55**, 124014 (2013)
8. W. Tierens et al. and the ASDEX Upgrade Team, Nucl. Fusion **57**, 116034 (2017)

# Development of ECRH-based methods for assisted discharge burn-through: experiment and simulation

*D.* Ricci[1,*], *J* Stober[2], *R.* Dux[2], *L.* Figini[1], *T.* Wauters[3], *E.* Lerche[4], *G.* Granucci[1], the ASDEX Upgrade Team, and Eurofusion WPTE Team

[1]Istituto per la Scienza e Tecnologia dei Plasmi, CNR, Milan, Italy
[2]Max-Planck-Institut fur Plasmaphysik, Garching, Germany
[3]ITER Organization, Building 72/2026, SCOP, SCOD, Science Division
[4]UKAEA Culham Centre for Fusion Energy, Abingdon

**Abstract.** Electron Cyclotron (EC) waves will be routinely used in future reactors not only for plasma heating and/or non-inductive current drive during the flat top but also to assist the plasma start-up phase in large tokamaks with superconductive coils. In ITER, for example, EC start-up is foreseen since first plasma operation. To limit the level of stray radiation, ECRH can be used after ohmic breakdown, as a robust solution to successfully sustain the plasma burn-through in the presence of pre-filling gas and impurity influx from the wall.

On ASDEX Upgrade (AUG), a series of dedicated experiments have been performed using EC heating (X2) with a controlled Ne impurity injection in the prefill phase, to mimic non-favourable burn-through conditions such as would be expected in a discharge following a disruption event. The time for EC heating onset has been optimised to assist the early burn-through and a scan of the Ne concentration has been performed to find the threshold for successful burn-through conditions for two ECH power levels (0.7 and 1.4 MW). The toroidal magnetic field flexibility has been also documented, with the cold resonance position being shifted up to 13 % in major radius to match the ITER condition. These experiments showed that optimised settings of ECH power (onset and duration of the pulse) have a key role in making feasible the early Ne burn-through (with Ne concentration up to 14% and EC power of 1.4 MW). Successful pulses will be extended to study stationarity and clean up properties.

For an efficient and robust use of such a technique, it is essential to develop appropriate models capable of describing present experiments and of extrapolating (or predicting) to future scenarios. In this work, the predictive 0D model for the burn-though phase BKD0 [1] has been used to reproduce experimental results and estimate the power required for a successful burn-through as a function of the impurity concentration, finding that ECH power of 1.4 MW is required to sustain burn-through with more than 20% of Ne. The scalability of the model has been also tested on TCV [2] and its implication for ITER will be discussed.

## 1 Introduction

In the past, several experimental studies have been performed to define the operation scenario especially in view of ITER Electron Cyclotron (EC) assisted start-up. Plasma initiation assisted by EC has been applied successfully on different tokamaks. This method provides additional heating during the pre-ionization of the prefill gas and also during the burn-through, although the influence of EC is different on the two phases [1][3]. On ASDEX Upgrade [3] it was observed that EC heating applied after ohmic breakdown is a valuable way to overcome the radiation barrier, although a systematic study with controlled impurity injection has not yet been performed. In case of ITER, for the burn-through, modelling suggests that the power absorption at the second harmonic of the cyclotron frequency (X2) is not efficient at relatively low temperatures and densities that characterise this phase, both in single-pass scheme foreseen for First Plasma (FP), as well as in multi-pass configuration [4]. In order to separate the effect and confirm modelling prediction, a series of experiments have been carried out on ASDEX Upgrade in presence of controlled impurity injection. To limit the level of stray radiation, EC operating at X2 has been used after a robust ohmic breakdown to successfully sustain the burn-through of pre-filling deuterium with addition of small amount of Ne. By keeping constant the initial parameters for ohmic breakdown (in terms of toroidal electric field and deuterium density) and varying the Ne content it is possible to mimic non-favourable conditions such as it would be expected in case of impurity influx from the wall. We explore the feasibility of a cleaning discharge by means of EC heating and extrapolate results from various devices to ITER.

*e-mail: daria.ricci@istp.cnr.it

EPJ Web of Conferences **277**, 02001 (2023) https://doi.org/10.1051/epjconf/202327702001
*EC21*

## 2 Experimental set up

In ASDEX Upgrade [5], EC assisted burn-through experiments were performed using X2 heating at 140 GHz with power level at 0.7 and 1.4 MW. The database collected consists of 22 discharges, most of them with toroidal magnetic field $B_{tor}$ on-axis of 2.4 T and radial position of the cold resonance slightly on the high field side, while it was varied from 2.2 to 2.6 T in 4 dedicated discharges to test the field flexibility. Deuterium was injected in the vacuum vessel 100 ms before t = 0 s, together with a controlled Ne puffing, injected at -50 ms. Glow discharge was performed in between discharges to avoid Ne accumulation and get reproducibility. The amount of Ne was comparable to what remains in the machine after a killer gas pulse. Figure 1 shows an overview of typical discharge, here performed with 7% of Ne, $B_{tor}$: 2.4 T, and EC input power at 1.4 MW. EC wave is injected with a toroidal angle of 3° from equatorial launchers (EL), configuration that typically gives reflection all over the cross section.

Parametric scan in $P_{EC}$ and Ne/D was performed to understand the effectiveness, together with EC timing and toroidal field flexibility, that were additionally explored for optimisation purpose.

In this initial phase of the discharge it is difficult to diagnose the plasma accurately because most of the diagnostics are optimised to measure high $n_e$ and $T_e$. Moreover, ECE emission is missing, since the level of stray radiation is not compatible with ECE system on AUG. The effect of Ne was analysed using VUV spectroscopy (SPRED) for evidencing Ne penetration into the plasma. Measurement of stray radiation was used to characterise the absorption, which is qualitatively compared with simulation results discussed below.

It is worth to note that Ne burn-through results completed in a time interval between 0.5-1 s, depending on the initial concentration, power level, pulse length etc. In order to characterise the EC effectiveness on the early burn-through, different times are considered for the analysis described in the following sections: T1 = 40 ms (chosen after optimisation discussed below) corresponds to the EC onset after the start of the current rise, that is usually around 30 ms. Because different EC pulse durations were tested, the first 30 ms are considered (T2 = 70 ms) in order to compare results from the entire database. Moreover, at T3 = 90 ms the feedback control on the plasma current starts to run. This time is sufficiently accurate to define criterion for successful Ne burn-through, that is achieved if plasma current reaches 200 kA at T3 (see figure 2).

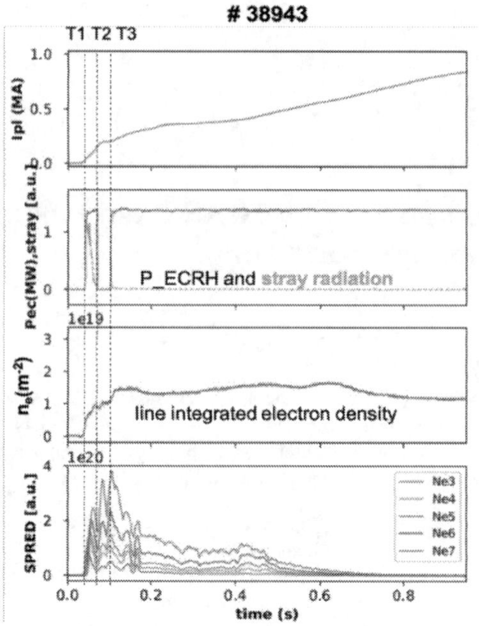


**Figure 1.** AUG #38943: Overview of EC-X2 assisted burn-through experiment: plasma current, EC power from 40 ms (T1) to 70 ms (T2) and stray radiation, line averaged electron density $n_e$ by interferometry and SPRED signal of NeIII up to NeVII levels. Ne/D: 7%; $B_{tor}$: 2.4 T

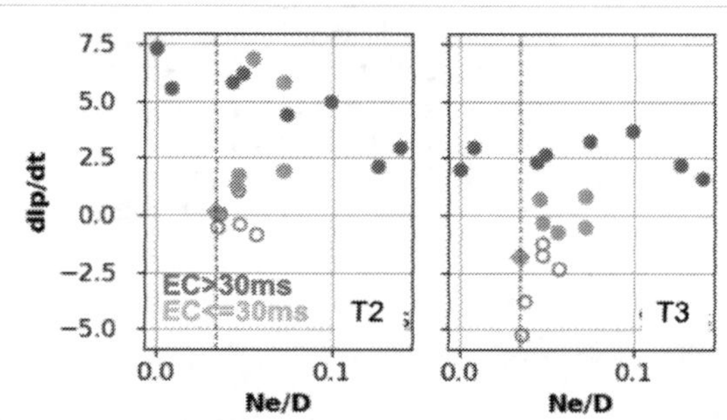


**Figure 2.** Plasma current development for the present database ($B_{tor}$: 2.4 T): (left) after the first 30 ms of EC pulse (T2) and (right) when the control of the ohmic coils takes place (T3). Dark blue bullets correspond to successful burn-through for EC blip longer than 30 ms. Light Blue bullets correspond to successful burn-through when EC pulse is up to 30 ms, empty ones to failures. Vertical dashed line is the limit for ohmic case.

## 3 Result and discussion

We choose to characterise the scenario as a function of the concentration of Ne injected before the t = 0 s, calculated by the integral of the corresponding calibrated gas fluxes. A minimum EC pulse duration of 30 ms allows to achieve successful Ne burn-through, at least for an initial concentration below 7%. Figure 2 (right) shows in light blue the developing of the plasma current after 30 ms of ECH power (at T2), giving a clear indications that a longer EC pulse would have helped if compared with the increase at T3 (figure 2, left), burning high level of impurity and making an effective cleaning discharge.

### 3.1 Optimization

As already outlined in the introduction, focus of the experiments is on the understanding the impact of EC on the burn-through phase only. One of the key point in the scenario developing was the definition the EC timing: not too early because of the need to separate the effect from previous pre-ionization phase, characterised here by a robust ohmic breakdown, and not too late, when the current initially raises and before Ne burn-through starts, to give the wave time to be absorbed in low $n_e$ and $T_e$ conditions.
Two different EC onsets were tested (at 40 and 60 ms, with discharge parameters reported in table 1) aiming to limit the stray radiation level, and making the additional heating effective with and without Ne (see figure 3). Discharge

**Table 1.** Discharge parameters used to optimise the EC onset

| # | $P_{EC}(MW)$ | $t_{EC,on}(ms)$ | Ne(%) |
|---|---|---|---|
| 37949 | 0.75 | 60 | 0 |
| 38936 | 0.75 | 60 | 4.7 |
| 38940 | 0.75 | 40 | 4.5 |

without impurity in prefill (figure 3, blue, $P_{EC}$: 0.7 MW at 60 ms) achieved a sustained burn-through, as expected and reported for comparison. At ECH power of 0.7 MW and Ne 4.5%, the successful case (figure 3, green) is when EC starts at 40 ms instead of 60 ms, which fails. However, the plasma current does not exceed 100 kA, an indication that higher current values could be achieved by increasing the power. The radiation from SPRED (figure 3, bottom panel) reveals that, in this experimental condition, the power is sufficient to create higher level of ionisation up to NeVII, from 55 ms, when absorption starts to improve, as seen from stray radiation signal.
The $B_{tor}$ flexibility was also documented, varying the cold resonance position up to 13% in major radius to match the ITER First Plasma condition [6]. Figure 4 (top) shows the position of the current center $R_{curr}$ and the time evolution of stray radiation up to T3 (bottom). The discharges were performed under identical experimental conditions (prefill, ECH power, Ne/D ), except for the $B_{tor}$ on axis, that has been varied from 2.2 T up to 2.6 T. On AUG, current forms close to the inner heat shield, and poloidal field currents are programmed to be identical for all the $B_{tor}$ scan. The radial position of the cold resonances at different fields are indicated by horizontal lines. The *EC* power level was 1.4 MW. The stray radiation increase significantly in all cases during the first 20 ms of EC pulse. Then it drops for 2.3 T, 2.4 T and 2.6 T in the next 10 ms, leading to sustained burn-through. The 2.2 T case failed, the stray radiation level remains high and tends to move inwards. Heating is more or less in the center of the current, as soon as variation of resonance location $R_{EC}$ is within 10%. Scanning $B_{tor}$ is different from other factors such as ECH power and Ne concentration. It affects the deposition location of ECH power, also with the poloidal magnetic field around it. Many variables make it difficult to analyse the toroidal-field-dependence of the burn-through conditions and give a strong conclusion. The best (rather than optimised) toroidal field should be determined with a specific poloidal field (not a fixed poloidal field for all shots). This would require much more detailed experimental scans, which have not been performed yet.

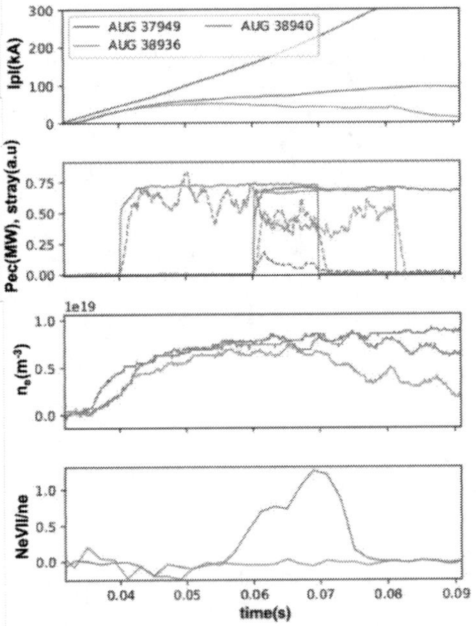


**Figure 3.** Effect EC onset at 40 or 60 ms, in conditions summarised in Table1. The stray radiation signals are plotted in arbitrary unity. High stray level in #38936 (and consequently low EC absorption) between 60 and 90 ms does not indicate that longer pulse would have helped.

### 3.2 Ne burn-through

Ne level was measured by SPRED following lines up to NeVII (77.4 nm). As the Ne concentration increases, the integral of NeVII line-emission from SPRED between T1 and T2 decreases (Figure 5), as the derivative of the plasma current does, as already shown in Figure 2(left). This results in keeping the temperature low and making the burn-through of NeVII ion to start with increasing delay. Here different colours correspond to different EC input power: in red no EC power (ohmic); in blue $P_{EC}$: 0.7 MW; in orange $P_{EC}$: 1.4 MW. Only few pulses at 0.7 MW are collected due to difficulty in controlling the valve aperture for this rather low Ne flux. Different power thresholds have been identified to achieve sustained burn-through: ohmic heating is sufficient when the Ne concentration is up to 3.5%, $P_{EC}$ = 0.7 MW are necessary to sustain burn-through when Ne is < 4.7% and 1.4 MW makes discharge with Ne impurity up to 14% effective.

Figure 6 shows the temporal evolution of the 1.4 MW stray radiation of EC for different Ne/D values. As the Ne concentration increases, the stray radiation measured by the sniffer probes also increases due to the low temperature and the delay in NeVII emission line is evident.

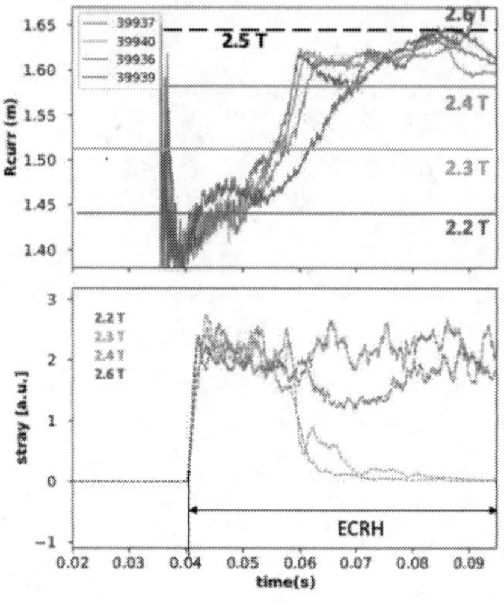


**Figure 4.** Variation of $B_{tor}$ in the range 2.2-2.6 T, ECH power 1.4 MW, Ne/D= 7%. Top: Position of the current center during EC pulse. Horizontal lines correspond to radial position of cold resonance at corresponding fields. Bottom: stray radiation measurements

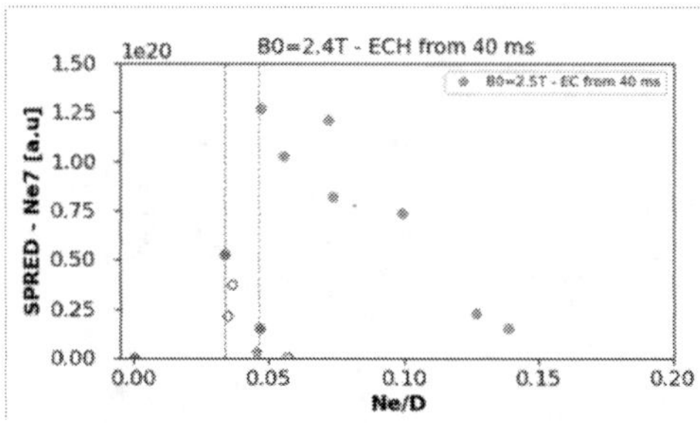


**Figure 5.** As the Ne/D increases, the integral of NeVII calculated from SPRED in the range 40-70 ms decreases. Power thresholds are (red) Ohmic: Ne < 3.5%, (Blue) EC 0.7 MW: Ne < 4.7%, (Orange) EC 1.4 MW: Ne up to 14%. Hollow circles correspond to failure.

To estimate $T_e$, unavailable from direct measurements at this early stage of the discharge, linear theory of absorption was applied: since sniffer probes exhibit similar temporal behaviour, their signals were used to interpolate between zero and full single pass absorption, which, for X2, scales linearly as the electron pressure as long as the $n_e$ is far below cut-off. The absorption, from the upper panel in figure 7, remains below 30% for the first 10-25 ms of the EC pulse, depending on the Ne concentration, and then rise up. The line integrated $n_e$ was used to determine the $T_e$ (Figure 7 (bottom panel)). This can be considered an upper limit for $T_e$ since a flat $n_e$ profile has been assumed. A peaked $n_e$ profile corresponding to the local electron pressure at the cold resonance location, would reduce $T_e$.

Considering that the NeVI ion stage lives in the range of 0.01-0.4 keV (emission peaks at 31 eV), while NeVII ion stage lives in temperature range 0.02-1keV (emission peaks at 42 eV) (from STRAHL [7], courtesy of M. Cavedon), the central temperature is in good agreement with estimation above.

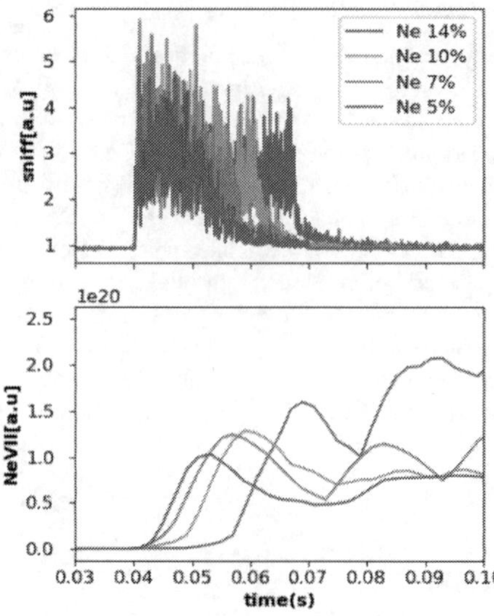


**Figure 6.** Top: Stray radiation increases with increasing Ne/D. Bottom: the derivative of the plasma current decreases (keeping the temperature low) and the burn-through of NeVii line starts with increasing delay. $B_{tor}$: 2.4 T; $P_{EC}$: 1.4 MW.

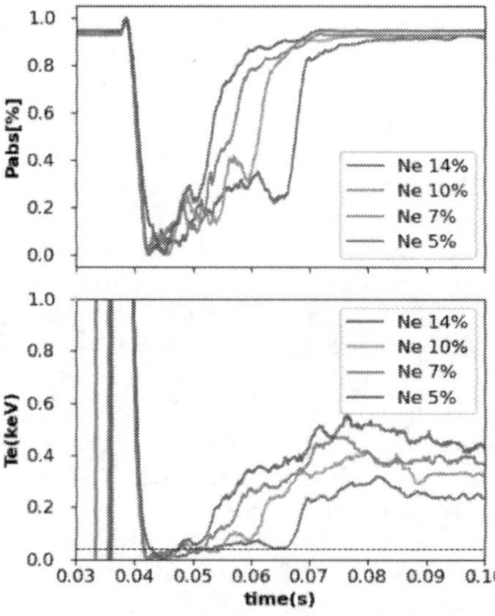


**Figure 7.** Top: Absorption calculated interpolating non calibrated sniffer probe signals between 0 (no absorption) and 1 (full single pass absorption). Bottom: $T_e$ derived using line integrated $n_e$ and assuming flat density profile.

## 4 Simulation

To address the question of how much heating power is needed to recover the discharge and assist the Ne burn-through, simulations were carried out with the code BKD0 [1]-[3] to reproduce the experimental data. The model is based on a two-step analysis: a 0D simulation of time evolution of main plasma parameters which contains energy balance equations for electrons and ions, particle balance equations for all the species involved in the simulations (electrons, ions and neutrals as D2 and impurities) and electric circuit equation. This is coupled with the self-consistent calculation of the ECH power absorption done with code GRAY[9]. Figure 8 shows similar trend between measurement and simulation of absorption done with BKD0, and absorbed power calculated by GRAY. The absorption will be discussed below.

An example of simulation is shown in figure 9, based on #37949. The loop voltage measured is used as input. The initial neutral density is derived from in-vessel pressure measurement by midplane gauge at t=0. The $n_e$ is measured by interferometry. The simulated volume-averaged $n_e$ approaches similar values with interferometry data, showing reasonable evolution. The time evolution of the main plasma quantities at different Ne/D and ECH power has been carried out. The effect of Ne was compared with simulation results with the following procedure: for each simulation, given the density of the charged stages of Ne, $n_e$ and $T_e$, a synthetic diagnostic has been set up. The photon emissivity coefficients (pec), obtained from OPEN-ADAS database [8] were used to perform the interpolation of the data onto the $n_e$, $T_e$ of the model and calculate the local emissivity via:

$$eps = \frac{1}{4\pi}(pec_{exc}n_e n_z + pec_{rec}n_e n_{z+1}). \quad (1)$$

Here, $eps$ is the emissivity in $[photons/m^3/s/sr]$, $pec_{exc}$ is the photon emissivity coefficient for excitation and $pec_{rec}$ is the photon emissivity coefficient for recombination $[m^3/s]$.

Finally, integrating $eps$ over the path length of the SPRED (or multiply by 1m because there is no profiles) and adding up all the lines of one-charge stage, SPRED data have been compared with the synthetic diagnostic. Results have been summarised in figure 10 where simulations reproduce quite successfully the experimental results, finding the power threshold for successful burn-through condition and predicting more than 20% of Ne to be injected with 1.4 MW of power. The baseline version of the GRAY code, already available in the ITER Integrated Modelling & Analysis Suite (IMAS), models the EC beam propagation, absorption and current drive for a single pass of the beam into the plasma. This approach is adequate for modelling scenarios with strong EC absorption, in which the plasma is optically thick and the beam is fully absorbed before reaching the central column of the tokamak, but a description of the beam propagation and absorption after its reflection off the inner wall is necessary for a more accurate modelling of the break-down and burn-through phases, when single-pass EC absorption is still low. A simple model for the

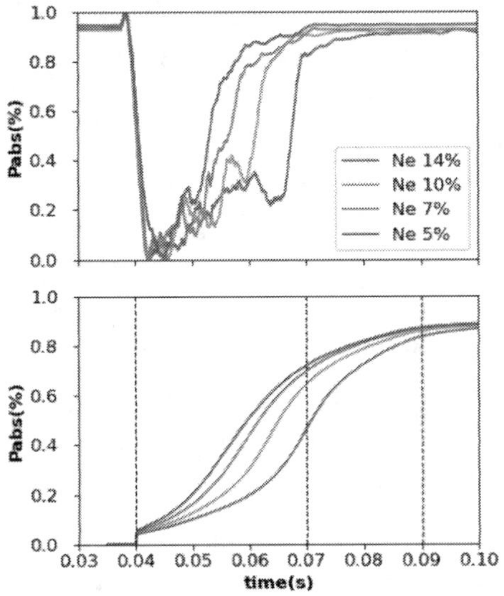


**Figure 8.** Comparison between absorption derived by sniffer probe signals using the procedure described in section 3.2 (Top) and the absorption done with BKD0 and GRAY codes (Bottom)

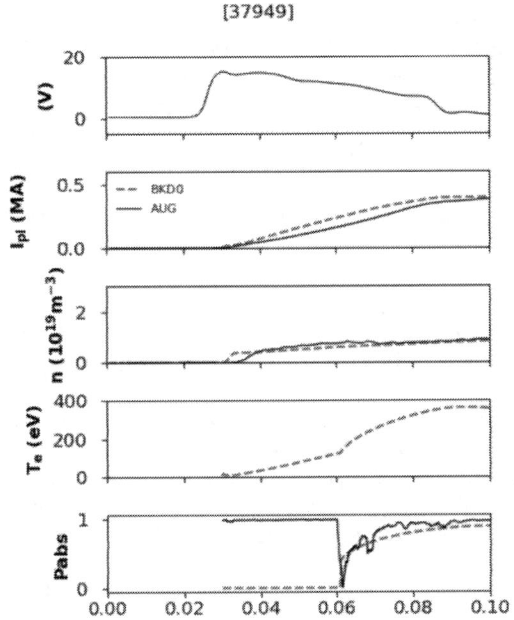


**Figure 9.** Comparison between BKD0 simulation (dashed red) and experimental time traces (black line) of #37949: Loop voltage, plasma current, volume-averaged $n_e$, $T_e$ and absorption. $P_{EC}$: 0.7 MW, $t_{EC,ON}$: 60 ms, neutral pressure: 4mPa, no impurity

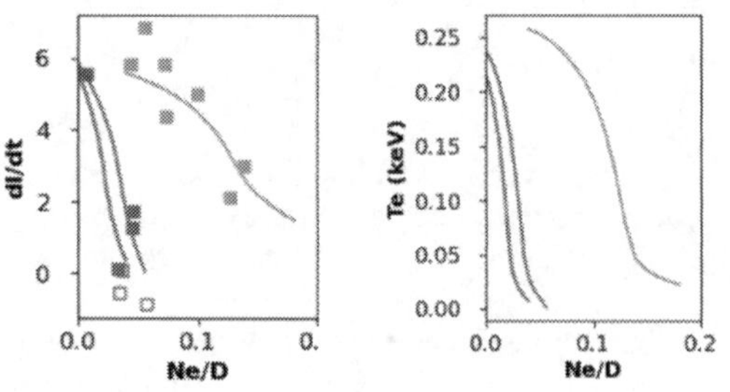


**Figure 10.** After 90 ms of simulation, red: Ohmic; Blue: EC: 0.7 MW; Orange: EC 1.4 MW. (Left) Comparison between plasma current development as a function of Ne concentration: experimental data (squares), simulation (line). (Right) simulation of $T_e$ as a function of Ne.

reflection of the EC beam on the first wall was included in GRAY, and it has now been implemented in IMAS. The inclusion of the wave reflection model required a change to the code interface: the first wall shape is now needed as additional input in order to compute the geometry of the beam reflection. The reflection model assumes a loss-less ideal conductor with a smooth surface at the wall, so that the full power of the incident beam is transferred to the reflected one, and the reflection is described by Snell's law. Given the assumption that only $n_e$, $T_e$ close to the EC resonance affects the EC absorption calculation (absorption is localised within few mm from the EC resonance and the exact shape of the null and of the kinetic profiles is not relevant), the shape of the plasma in the poloidal field null region has been simplified to a circular poloidal cross-section where the major radius and the minor radius are Rsim=1.65 m and asim=0.5 m, respectively. This region is crossed both by the injected and by the reflected beam, in conditions where $B_{pol}/B_{tor}$<1%.

## 5 Conclusion and outlook

In the case of ITER First Plasma, burn-through modelling suggests that, at the second harmonic, power absorption is limited even in multi-pass configuration. A series of experiments were designed and performed on ASDEX Upgrade, aimed at answering the clear research question related to the EC absorption during this initial phase of the discharge, and confirm prediction. We chose to separate the EC heating effect on pre-ionization phase, starting with a robust ohmic breakdown and a controlled Ne impurity injection in the prefill phase, and started the heating soon after the current rise. We found that key parameters that strongly affect early impurity burn-through are onset and duration of EC pulse, that needs to be optimised. Furthermore, variations of more than 10% of the cold resonance position need be explored by modification of poloidal field coils currents, optimising the shape of the magnetic configuration. We found the ECH power thresholds as a function of Ne concentration, and the simulations reproduced quite successfully experimental outputs experimental results. Furthermore, sniffer probe signals show that absorption is below 30% for the first 20-25 ms of the EC pulse, and the same behaviour was reproduced by BKD0 and GRAY simulations.
Direct comparison with previous experiments carried out on TCV is not possible due to the different conditions explored: low loop voltage, which means the need to also assist pre-ionization phase by EC, and ECH power actively controlled to keep constant the derivative of the plasma current.
However, the model was applied to TCV (R/a: 0.88/0.25 m) to test scalability. Preliminary simulations show that 0.7 MW of power (82 GHz, X2) is capable of sustaining burn-through of half the concentration of impurities, but must be verified with further dedicated experiments. Open questions that play an important role and will have implication for ITER are related with pre-fill pressure, level of impurities released in the plasma in case of high stray radiation level and wall recycling.
**Acknowledgements** This work has been carried out within the framework of the EUROfusion Consortium, funded by the European Union via the Euratom Research and Training Programme (Grant Agreement No 101052200 — EUROfusion). Views and opinions expressed are however those of the author(s) only and do not necessarily reflect those of the European Union or the European Commission. Neither the European Union nor the European Commission can be held responsible for them.

## References

[1] G. Granucci, *et al.*, Nucl. Fusion **55** 093025 (2015)
[2] D. Ricci *et al.*, Proc. of the 46rd EPS Conf. on Plasma Physics O5-103 (2019)
[3] J. Stober *et al.*, Nucl. Fusion **51** 083031 (2011)
[4] D. Ricci *et al.* Proc. of the 43rd EPS conference on Plasma Physics O5.130 (2016)
[5] J. Stober *et al.*, Plasma Phys. Control. Fusion 62 024012 (2020)
[6] ITER Research Plan, ITER Technical Report ITR-18-003
[7] R. Dux,STRAHL user Manual, Preprint IPP (2006)
[8] https://open.adas.ac.uk
[9] D. Farina, Fusion Sci. Technol.**52**, 154 (2007)

# Electron cyclotron heating assisted start-up experiments in J-TEXT

*Junli* Zhang[1], *Peter C de* Vries[2], *Kazunobu* Nagasaki[3], *Donghui* Xia[1,*], *Wei* Jiang[4], *Zhoujun* Yang[1], *Zhifeng* Cheng[1], *Xin* Xu[1], *Zhijiang* Wang[1], *Nengchao* Wang[1], *Li* Gao[1], *Yonghua* Ding[1], *Zhipeng* Chen[1], *Zhongyong* Chen[1], and *Yuan* Pan[1] and J-TEXT team[**].

[1]International Joint Research Laboratory of Magnetic confinement Fusion and Plasma Physics, Huazhong University of Science and Technology, Wuhan 430074, People's Republic of China
[2]ITER Organization, Route de Vinon sur Verdon, CS 90 046, 13067 St Paul Lez Durance, France
[3]Institute of Advanced Energy, Kyoto University, Uji, Japan
[4]School of Physics, Huazhong University of Science and Technology

**Abstract.** Electron cyclotron heating (ECH) breakdown and burn-through assist has been adopted to make the ITER start-up, which uses a low toroidal electric field of about 0.3 V/m, more robust. Related studies have been carried out on many fusion machines such as DIII-D, JT-60U, Tore Supra and FTU. However, the required ECH power to ensure an effective breakdown assist at ITER is not yet clear. Experiment was also performed in J-TEXT to determine the minimum ECH power requirements to assist breakdown and develop a better physics description of the process. The breakdown loop voltage for a successful shot was reduced from 34 to 3.7 V (corresponding to 0.56 V/m) by 300 kW X2-mode ECH power. The critical ECH power to assist the successful formation of a tokamak discharge is about 200kW, and below this value, ionization caused by ECH power is observed, however, the tokamak discharge start-up does not succeed. The effect of different loop voltages and ECH pulse-width on start-up was also studied. As injecting ECH power at a proper time, low ECH power can achieve a similar pre-ionization result as the case with high power. Extremely low breakdown voltage leads to a higher toroidal field later when putting a capacity to continue discharging in J-TEXT. The earlier shutdown of ECH power caused a failed discharge.

## 1 Introduction

Start-up is the first stage of the tokamak operation. However, until now, there are still a lot of questions urged to be answered. For example, there are concerns about whether a low toroidal electric field of 0.3 V/m is enough to support robust breakdown in ITER first plasma. Many tokamaks such as DIII-D[1][2], JT-60U[3], Tore Supra[4] and FTU[5] carry out related experiments to verify the feasibility of low toroidal electric field start-up with electron cyclotron heating (ECH) assisted. These experiments provided a scientific basis for ECH assisted ITER start-up. However, the required ECH power to ensure an effective breakdown assist in ITER is not yet clear [6].

J-TEXT has conducted the ECH assisted start-up experiments since 2019. These experiments aimed to determine the minimum ECH requirements to assist in the breakdown and develop a better physics description of the process. This paper is organized as follows: A brief introduction to the experimental setup is given in the next section. Section 3 shows the experimental results and some analysis. Finally, this paper concludes with a summary.

*e-mail: xiadh@hust.edu.cn
**See the author list of "Y. Liang et al. 2019 Overview of the Recent Experimental Research on the J-TEXT Tokamak, Nucl. Fusion 59 112016"

## 2 Experimental setup

J-TEXT is a conventional middle-size iron-core tokamak with a major radius of 1.05 m. It operated at a minor radius of 0.25-0.29 m with a movable titanium-carbide-coated graphite limiter. The typical J-TEXT discharge in the limiter configuration is done with a toroidal field $B_t$ of 2.0 T, a plasma current $I_p$ of 200 kA, a pulse length of 800 ms, plasma densities $n_e$ of $1-7\times10^{19}m^{-3}$, and an electron temperature $T_e$ of 1 keV.[7]

The Ohmic field begins acting with a reversed current first, corresponding to a reversed peak in loop voltage at the beginning. Then the pre-magnetic capacitor is triggered at t=0.0 ms, which is an artificially defined time series for discharge. The capacitor bank, including the ionization capacitor, capacitor one (C1) and capacitor two (C2), works successively to discharge to the Ohmic field coils. The ionization capacitor with a high voltage can achieve a successful breakdown. C1 and C2 can provide volt-seconds to the discharge after the breakdown. Figure 1 shows the top view of related diagnostics and the ECH system. Key diagnostics includes polarimeter-interferometer (named 'polaris')[8], photo-diode array (PDA), vacuum gauge. Polaris provides information on line-integrated plasma density at the core. PDA measures the hydrogen-alpha radiation, which is related to the ionization condition of working gas, namely hydrogen gas. The vacuum gauge measures the pressure in the vessel. Gas puffing was employed to inject hydrogen gas by giving an electric pulse before the application of a toroidal electric field. The pulse is constant for all shots shown in this paper. A 105 GHz/500 kW/1 s ECH system was developed in 2019[9]. Two polarizers on the miter bend can achieve arbitrary polarization, allowing X2-mode ECH assisted start-up to be applied. Its launcher has the capacity of injecting ECH power with a toroidal injection angle of ±20° and a poloidal injection angle of ±20°, respectively. The ECH power is injected with a poloidal injection angle of +3° and a toroidal injection angle of 0° in this paper.

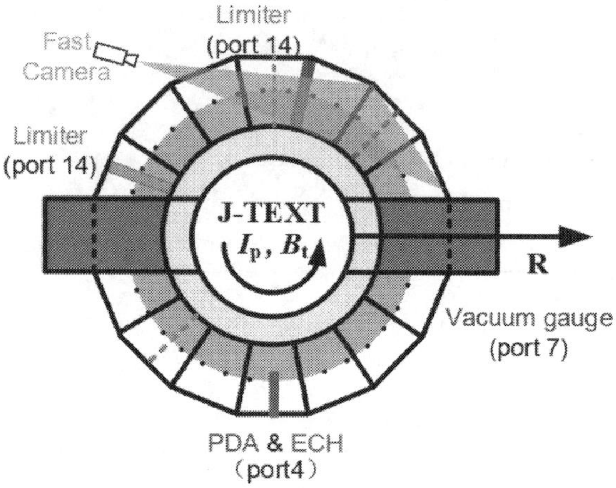


**Figure 1.** Distribution of related diagnostics and the ECH system on J-TEXT from a top view.

## 3 Experimental results and analysis

This is the first ECH assisted start-up experiment on J-TEXT. The error field was presented first to show the magnetic field configuration. A parameter scan of the voltage of the ionization capacitor was performed to explore the lower limit of breakdown voltage. The result of a typical ECH assisted start-up was given. Scanning of the ECH power and pulse-width showed its effect on start-up. The following parts give a more detailed description.

### 3.1 Error field

Prior to the plasma initiation, the magnetic field configuration is optimized to reduce the stray magnetic field and to obtain a favorable configuration for the tokamak start-up. It is expected to obtain a large enough field null structure, which will increase the effective connection length and achieve a robust start-up. Figure 2 shows the distribution of magnetic field configuration in cross section when t=0.0 s, which is calculated by EFUND code. The ECH resonance layer is around the center of the vessel (R=1.05 m) for a toroidal magnetic field of 1.87 T when 105 GHz ECH power is applied. In these experiments, the poloidal magnetic field is about 3 Gauss around the ECH resonance layer. The estimated connection length is 1590 m, which allows to achieve a successful breakdown in a large pre-fill pressure range [10].

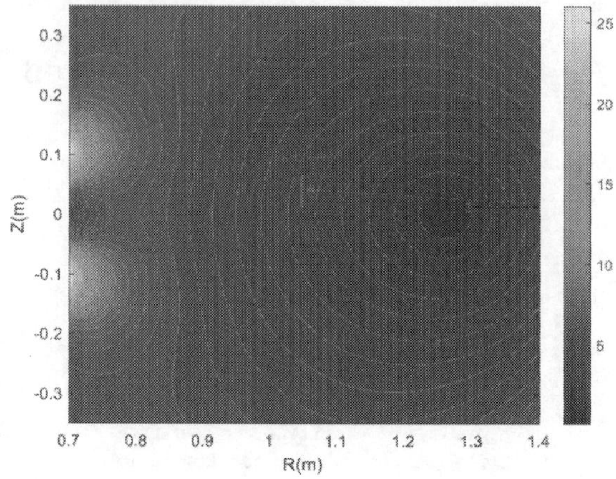


**Figure 2.** The magnetic field configuration (unit in Gauss) in cross-section When t=0 s. ECH resonance layer is around the center of the vacuum vessel (R=1.05 m, red square), where the stray-field is about 3 Gauss. The red dotted line is the schematic diagram of the EC propagation path

### 3.2 Typical ECH assisted start-up

For a reliable and robust start-up, J-TEXT usually discharges with a pre-magnetization and ionization capacitor of 1600 V. It will lead to an extremely high breakdown (loop) voltage of about 34 V (E=5.2 V/m). Scanning of the ionized capacitor with EC assisted was carried out to search for the limitation of breakdown voltage. Figure 3 illustrates the statistical results of the shots. It should be noted that the experiments were not carried out in one day. The pre-fill pressure is about 1.7 mPa. All the shots shown in figure 3 are assisted with 300 kW X2-mode ECH power. It can be easily observed that ECH power makes low loop voltage discharge possible, meaning expanding the operational space.

More details of typical temporal evolution are shown in Figure 4. The minimum breakdown voltage for a successful pure Ohmic heating start-up is about 14 V (E=2.1 V/m). Figure 4 presents the temporal evolution of a shot (#1065724) with this minimum breakdown voltage. However, the breakdown voltage can be as low as 3.7 V (E=0.56 V/m) with 150 kW X2-mode ECH assisted. The temporal evolution of corresponding shot #1068924 is shown in Figure 4. At J-TEXT, the iron-core can cause variations in the loop voltage during breakdown. Besides that, there are no special wall conditioning but Taylor-type discharge cleaning (TDC) and baking at 150 °C, which make it possible to gather more impurities in the vessel. Although this toroidal field is higher than 0.3 V/m, it demonstrates the capability of ECH power in assisted start-up.

When ECH power is applied, the electron density and hydrogen-alpha increase, 15 ms after the application of ECH power (but before the loop voltage). Pre-plasma density makes breakdown easier by generating more seed particles before the application of loop voltage. Less volt-seconds will be consumed for a lower breakdown voltage. However, low loop voltage also makes plasma current ramp up slowly, meaning more volt-seconds saved. As time goes by, volt-seconds tend to reach the same level. Note that when an extremely low breakdown voltage is applied, a higher loop voltage may be required later to achieve successful discharge, as shown in figure 4 at 0.012 s.

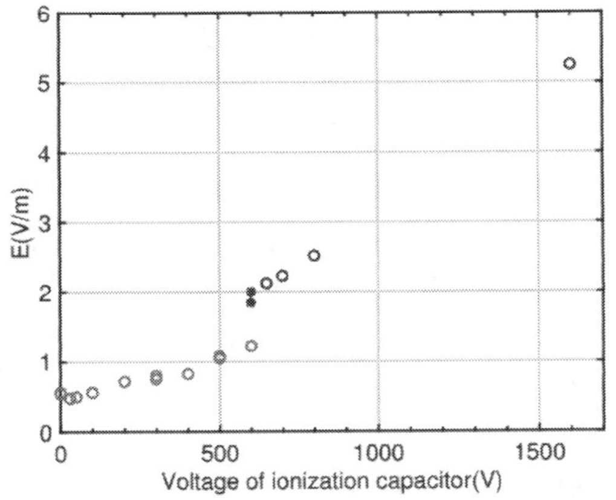


**Figure 3.** Operational space for pure Ohmic heating start-up and ECH assisted start-up on J-TEXT. A circle means the discharges were successful while an asterisk indicates the shot failed. The blue represents that it is a pure Ohmic heating start-up while the red represents EC assisted start-up

### 3.3 ECH power

Experiments with different ECH power on start-up were also conducted. It aims to search for the critical ECH power not just to achieve pre-ionization but for this ionization to be sufficient to effectively support the entire plasma initiation process.

Figure 5 shows the evolution of different signals during start-up with different ECH power assisted. It can be easily observed that high ECH power leads to quick and high-intensity ionization during pre-ionization phase, as shown in electron density and hydrogen-alpha radiation signals. It is not difficult to understand this effect. For an electron cyclotron beam with Gaussian-like profile, it results: $E \propto \sqrt{P}$, where $P$ and $E$ are the incident electron cyclotron power and the wave electric field amplitude, respectively. It indicates that high ECH power leads to a high electric field. When electrons travel through the wave electric field area, they will be accelerated in a shorter time to gain enough energy to achieve ionization or to higher energy, which will promote ionization in turn.

A more accurate expression, proposed by Farina, shows the maximum energy gain of an electron at second harmonic as $W_{max,n=2} \propto \sqrt{P}$ in non-linear interaction[11]. However, several factors, for example, Coulomb collisions, play a role in order for this local ionization to support a high-quality pre-plasma[12][13].

ECH power is also important after the application of loop voltage. It can be easily observed from figure 5 that for shots #1073819 and #1073820, some parameters, such as plasma current, line-integrated density at the core and the loop voltage, are nearly the same before t= 7 ms. Plasma current of shot #1073820 ramps up smoothly toward the pre-set value of 180 kA. However, the plasma current of shots #1073819 drops down as it nearly reaches about 10 kA. These two shots indicates the critical power for assisted start-up is between 200 kW and 250 kW. As the plasma current of shot #1073819 ramps to about 10 kA, it is possible to achieve a successful start-up if the injected ECH power is larger, which can form a better pre-plasma at higher density and temperature and a wider area, and provide more power for tokamak plasma. Hence, the critical power for the ECH assisted start-up is 200 kW. Strictly, more shots with different ECH power can provide a more accurate value. Limited by shots, we estimate this value from physical characterizations.

Tokamak start-up usually experiences breakdown and burn-through. Burn-through will require enough energy to achieve ionization of main species and impurities[6]. ECH power will help a lot to experience such a process, which can be verified from other shots #1073825 and #1073820. Here is a preliminary qualitative analysis. From the perspective of the energy balance equation, as shown in equation 1[10], enough Ohmic heating power density ($P_{\rm OH}$) and electron cyclotron absorbed power density ($P_{\rm ECH}$) is required to overcome the power loss density caused by equilibration ($P_{\rm equi}$), bremsstrahlung ($P_{\rm brem}$), impurity-related terms($\sum_I (P_{\rm ion} + P_{\rm line} + P_{\rm RRE} + P_{\rm DRE})$) and transport losses ($P^{\rm e}_{\rm con}$). The terms of $P_{\rm equi}$, $P_{\rm brem}$ and $\sum_I (P_{\rm ion} + P_{\rm line} + P_{\rm RRE} + P_{\rm DRE})$ can be ignored because of no closed flux surface and low electron energy. When input power, including Ohmic heating and ECH power, is not enough, the electron will have the possibility of cooling down and recombination, which will lead to higher plasma resistance. That's the major reason that loop voltage increases around t=8 ms.

$$\frac{3}{2}\frac{d}{dt}(n_e \kappa T_e) = P_{\rm OH} + P_{\rm ECH} - (P_{\rm Dion} + P_{\rm Drad}) - P_{\rm equi} - P_{\rm brem} - \sum_I (P_{\rm ion} + P_{\rm line} + P_{\rm RRE} + P_{\rm DRE}) - P^{\rm e}_{\rm con} \quad (1)$$

On the contrary, we can see that high power(#1073825 and #1073820) leads to an apparent higher electron density and a little higher plasma current. Enough power increases electron density and energy, which promotes the increase of plasma current and electron density in turn. However, when another capacitor (C1) comes into use around t=10 ms, there will be a peak, which can be easily explained by the discharge characteristics of capacitors. Volt-second consumption is nearly the same, while plasma density is visibly different. As other conditions are the same or simi-

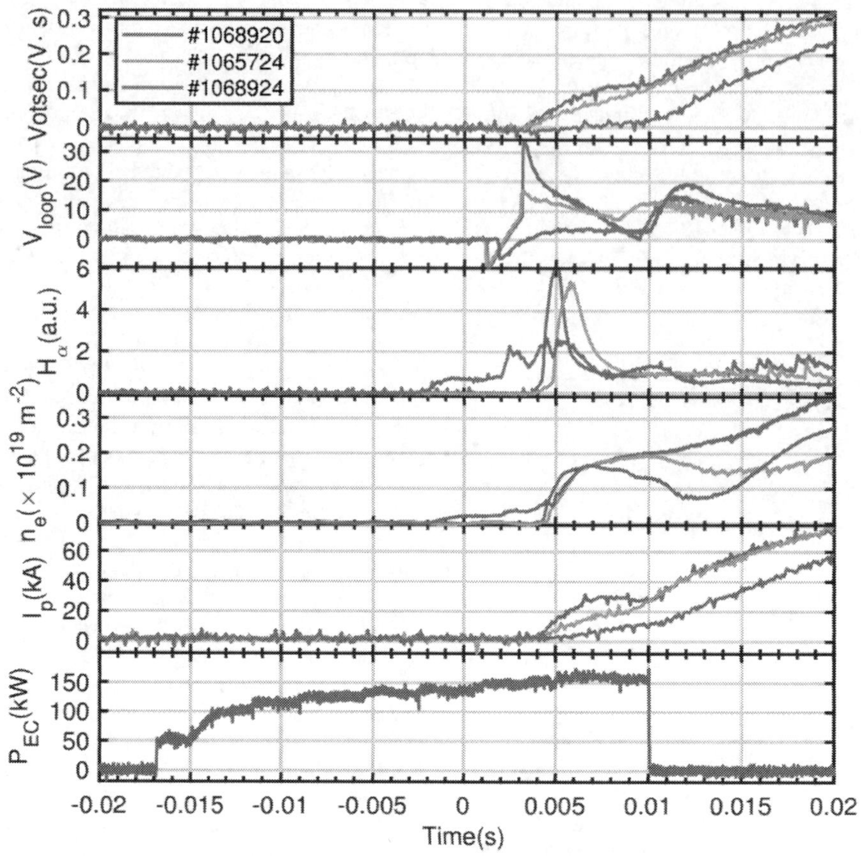


**Figure 4.** Evolution of different signals when discharging with ionization capacitor of 1600 V (red, #1068920), 800 V (orange, #1065724) and 1V (blue, #1068924). The last one is assisted with 300 kW ECH power while the two others are purely Ohmic heating. All the experimental densities presented in this paper are line-integrated densities.

lar, these differences should be the contribution of different ECH power.

### 3.4 ECH pulse-width

ECH pre-ionization provides more seed electrons, which makes it easier to achieve a successful start-up. Hence, a proper ECH pulse-width, which provides a high initial plasma density and energy at the time of applying loop voltage, may achieve a reliable and robust start-up. Related experiments were conducted to preliminarily investigate this question.

Figure 6 illustrates the evolution of key signals at different ECH pulses. For shot #1068928, ECH power is the highest, and the ECH pulse lasts for 36 ms. Its density and hydrogen-alpha appeared first. However, the intensity, especially hydrogen-alpha, is not the highest. By contrast, ECH power of shot #1068924 is applied at a proper time, succeeding in a start-up with a lower ECH power than all the other cases shown in Figure 6, which makes it the ideal shot for ECH assisted start-up. A long pulse is not the best because there is no closed flux surface. Hence, the particles are not in good confinement. After ECH power is applied for enough time, PWI may happen or the distribution of electrons is out of expectation. The latter requires further investigation. Compared with the other shots presented in Figure 6, shot #1068928 has the highest ECH power and rather long pulse-width (36 ms). The electrons ionized by this power can expand quickly along the radial direction. Finally, they can interact with the wall and possibly introduce impurities from the wall.

Usually, a higher ECH power can achieve quicker ionization[13], corresponding to a quick ramp-up of hydrogen alpha after ECH power is applied. However, we also find a couple of shots different from the others. A slightly higher ECH power of shot #1068923 was injected earlier than that of shot #1068929. However, the hydrogen-alpha and electron density of shot #1068923 have appeared lately. It is different from most of our experimental results. There isn't any convincing explanation until now.

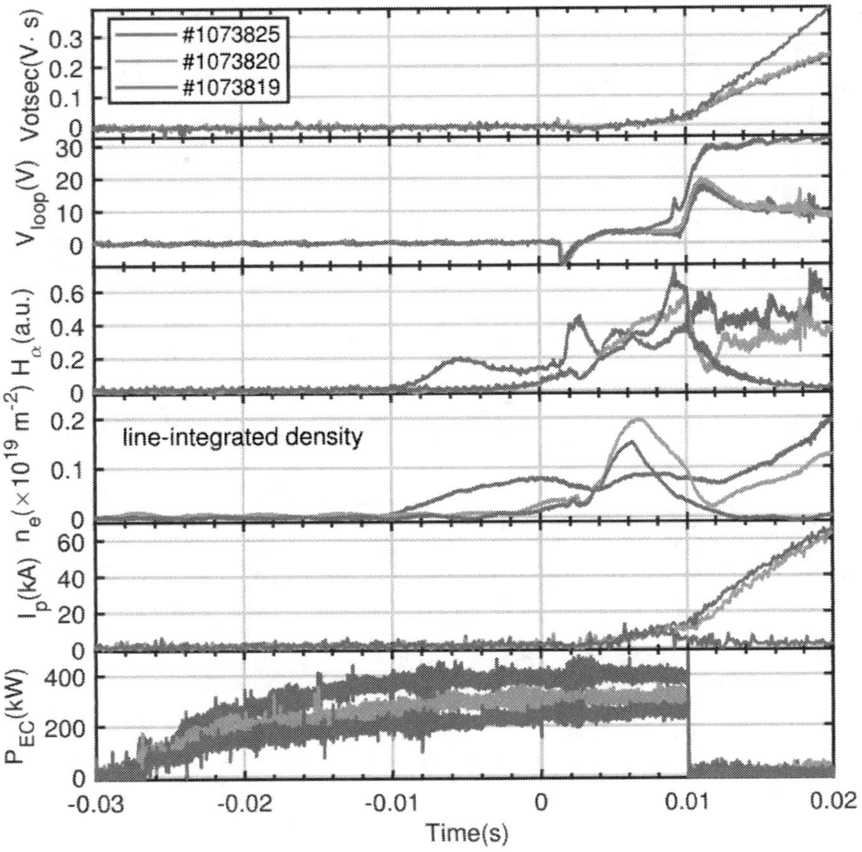


**Figure 5.** Evolution of key signals with different ECH power for assisted start-up.

After switching off the ECH power of shot #1068928, electron density and plasma current begin decreasing while loop voltage starts increasing. The reason is insufficient energy to support burn-through, which has been analyzed qualitatively.

## 4 Summary

This paper shows some ECH assisted start-up experimental results on J-TEXT. An estimated rather long connection length of 1590 m indicates a good magnetic field configuration. Scanning the ionization capacitor learned the operational space for pure Ohmic heating start-up and a much wider operation region for ECH assisted start-up. The lowest toroidal breakdown field is 0.56 V/m. Scanning ECH power demonstrates that critical averaged ECH power is about 200 kW. High power of 200 kW to 400 kW leads to faster and stronger ionization. Lower ECH power (below critical power) will fail during burn-through because of a lack of input power. These results are similar to recent DIII-D studies on ECH pre-ionization.[13] . Proper ECH timing is important. High ECH power and long ECH pulse-width lead to failed discharge, while low ECH power and short ECH pulse-width (shot #1068924) succeed.

## Acknowledgement

ITER is a Nuclear Facility INB-174. The views and opinions expressed herein do not necessarily reflect those of the ITER Organization. One of the authors Junli is very grateful for the help of the J-TEXT team. This work is supported by the National Magnetic Confinement Fusion Energy Research Project (2017YFE0301805) the National Natural Science Foundation of China (11905030, 11775090 and 51821005). Also, this work was partially supported by "PLADyS", JSPS Core-to-Core Program, A. Advanced Research Networks.

## References

[1] B. Lloyd, G. Jackson, T. Taylor, E. Lazarus, T. Luce, R. Prater, Nuclear Fusion **31**, 2031 (1991)

[2] G. Jackson, J. Degrassie, C. Moeller, R. Prater, Nuclear Fusion **47**, 257 (2007)

 https://doi.org/10.1051/epjconf/202327702002

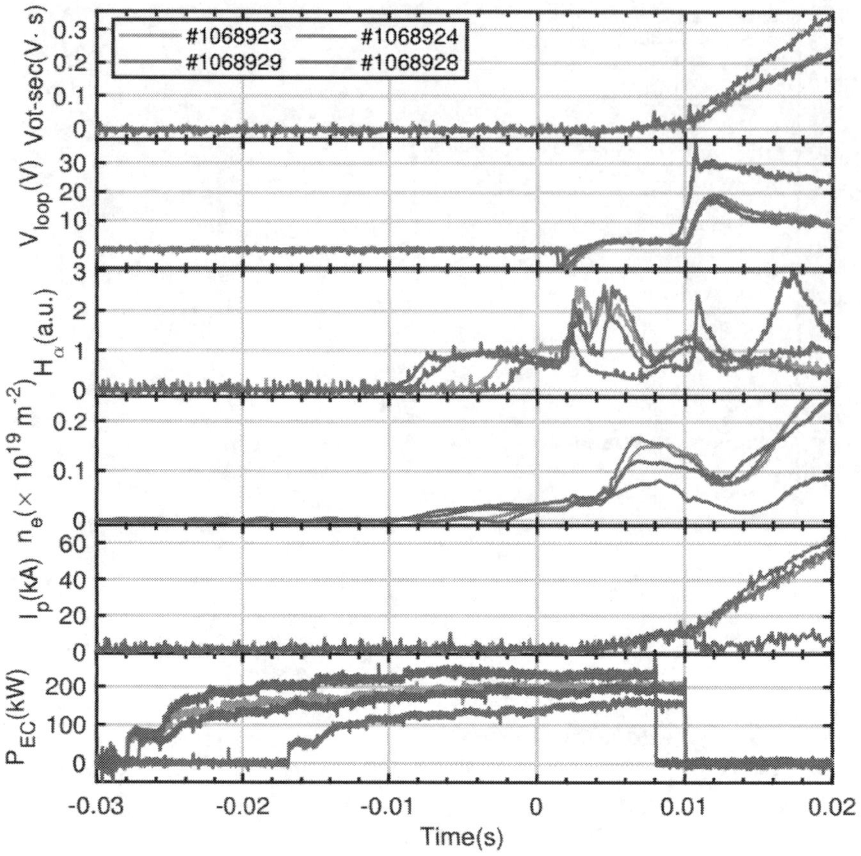


**Figure 6.** Evolution of key signals for different ECH pulse-width.

[3] K. Kajiwara, Y. Ikeda, M. Seki, S. Moriyama, T. Oikawa, T. Fujii et al., Nuclear Fusion **45**, 694 (2005)

[4] J. Bucalossi, P. Hertout, M. Lennholm, F. Saint-Laurent, F. Bouquey, C. Darbos, E. Traisnel, E. Trier, Nuclear Fusion **48**, 054005 (2008)

[5] G. Granucci, G. Ramponi, G. Calabrò, F. Crisanti, S. Nowak, G. Ramogida, O. Tudisco, W. Bin, A. Botrugno, P. Buratti et al., Nuclear Fusion **51**, 073042 (2011)

[6] P. De Vries, Y. Gribov, Nuclear Fusion **59**, 096043 (2019)

[7] Y. Liang, N. Wang, Y. Ding, Z. Chen, Z. Chen, Z. Yang, Q. Hu, Z. Cheng, L. Wang, Z. Jiang et al., Nuclear Fusion **59**, 112016 (2019)

[8] Y. Wang, L. Gao, P. Shi, X. Xu, Y. Zhou, Q. Yang, C. Yang, Q. Tao, C. Shen, Y. Wang et al., Plasma Science and Technology (2022)

[9] D. Xia, X. Chen, D. Xu, F.J. Gang, J. Zhang, N. Wang, Z. Yang, Z. Chen, Y. Ding, W. Zheng et al., Plasma Science and Technology **24**, 124010 ( 0) (2022)

[10] B. Lloyd, P. Carolan, C. Warrick, Plasma physics and controlled fusion **38**, 1627 (1996)

[11] D. Farina, Nuclear Fusion **58**, 066012 (2018)

[12] M. Carter, J. Callen, D. Batchelor, R. Goldfinger, The Physics of fluids **29**, 100 (1986)

[13] J. Sinha, P. de Vries, M. Walker, D. Battaglia, F. Turco, A. Hyatt, H. Kim, J. Stober, R. Yoneda, Y. Gribov et al., Nuclear Fusion **62**, 066013 (2022)

# Plasma heating and improvement of lower hybrid current drive efficiency by electron cyclotron waves on EAST

*Miaohui* Li[1], *Handong* Xu[1*], *Xiaojie* Wang[1], *Mao* Wang[1], *Bojiang* Ding[1], *Weiye* Xu[1], *Dajun* Wu[1], *Yunying* Tang[1], *Liyuan Zhang*[1], *Zege* Wu[1], *Jian* Wang[1], *Tao* Zhang[1], *Hanlin* Wang[1], *Qing* Zang[1], *Hailin* Zhao[1], *Haiqing* Liu[1], *Jinping* Qian[1], *Xianzu* Gong[1], *Fukun* Liu[1] *and Xiaolan* Zou[2]

[1]Institute of Plasma Physics, Chinese Academy of Sciences, Hefei 230031, P. R. China
[2]CEA, IRFM, Saint Paul-lez-Durance, France

**Abstract.** The electron cyclotron (EC) system on EAST consists of four gyrotrons with a frequency of 140 GHz (second harmonic of the extraordinary mode), each of which is expected to deliver a maximum power of 1.0 MW and be operated at 100-1000 s pulse length. Significant progress in long-pulse operation has been achieved recently, including the pulse duration up to 1056 s with EC power injected into plasma of 0.55 MW and the pulse duration of 310 s with EC power of 1.6 MW (output by 3 gyrotrons). High electron temperature ($T_e$ >12 keV) plasma measured by Thomson scattering was produced with the combination of EC and lower hybrid (LH) waves. It is found that the plasma heating effect depends on the EC power location greatly. By adjusting the EC power location, the plasma current profile can be modified. As a consequence of the increment of electron temperature by electron cyclotron resonance heating (ECRH), the lower hybrid current drive (LHCD) efficiency is improved, benefiting for the long-pulse operation. In addition, a synergy effect between EC and LH current drive was observed in steady-state operation on EAST.

## 1 Introduction

Electron cyclotron resonance heating (ECRH), as a highly efficient and controllable heating tool, has been widely used in tokamak magnetic confinement devices [1]. Usually, the EC power can be injected as narrow Gaussian beams, which gives rise to highly localized power deposition. Consequently, ECRH and electron cyclotron current drive (ECCD) are considered to be an effective tool to locally control the electron temperature and current profile in plasmas. An electron temperature ($T_e$) of up to 26 keV was achieved in JT-60U by injecting EC power of 2.9 MW into the centre of a reversed shear plasma produced by the lower hybrid (LH) waves [2]. EAST has four units of EC systems (No. 1– No. 4), each of which has one gyrotron with a frequency of 140 GHz (see Fig. 1) [3 - 5]. Additional two gyrotrons system is under development. However, the No. 1 gyrotron is currently undergoing repair (air leakage issue). EC waves are injected from the lower field side of the torus as an X-mode. In line with the long pulse operational capability of EAST, it is designed for a pulse length up to 1000 s. There are two high power lower hybrid current drive (LHCD) systems on EAST with different operating frequencies (2.45 GHz /4 MW, referred as LH1 [6] and 4.6 GHz /6 MW, referred as LH2 [7]). Both of the EC and LH systems are the main electron heating and current drive source for the long-pulse operation on EAST.

**Fig. 1.** Layout of ECRH system

## 2 Experimental results and analysis

### 2.1 progress in long-pulse and high $T_e$ operation

The first EC system was installed on EAST in 2015 with one gyrotron and the EC wave injected into plasma was ~ 0.5 MW. Thanks to the continuous increase in the number of EC systems and the improvement of the system, significant progress in high-power and long-pulse operation has been achieved recently. As shown in Fig. 2, the pulse duration has been extended to ~ 1056 s with $P_{EC}$ ~ 0.55 MW, corresponding to the EC energy injected into the plasma ~ 0.58 GJ. During this discharge, the EC power was shut down at 550 s, but it restarted automatically to work after 10 s (see Fig. 3). High power as 1.6 MW (output by 3 gyrotrons) injected into H-mode

* Corresponding author: xhd@ipp.ac.cn

plasmas with pulse duration ~ 310 s has been achieved. The increase in available input power made it possible to produce plasmas with high $T_e$ of up to 12 keV, which is measured by Thomson scattering diagnostic (see Fig. 4). The peak electron temperature was increased from 6.5 keV to 12 keV when 1.4 MW EC adding to the plasmas sustained by 2.3 MW LH power.

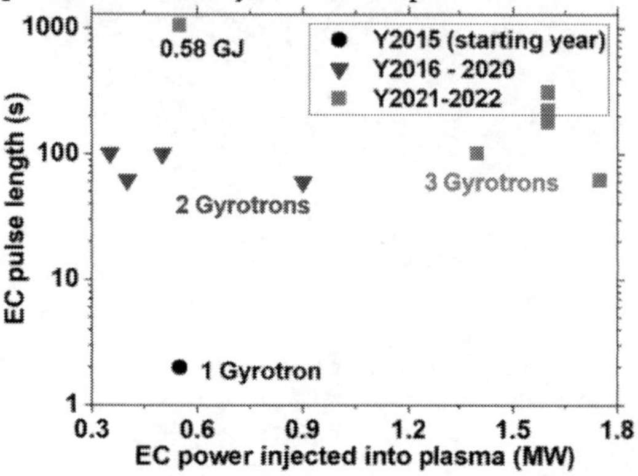


**Fig. 2.** Advances in the EC power injected into plasma and in their pulse duration on EAST.

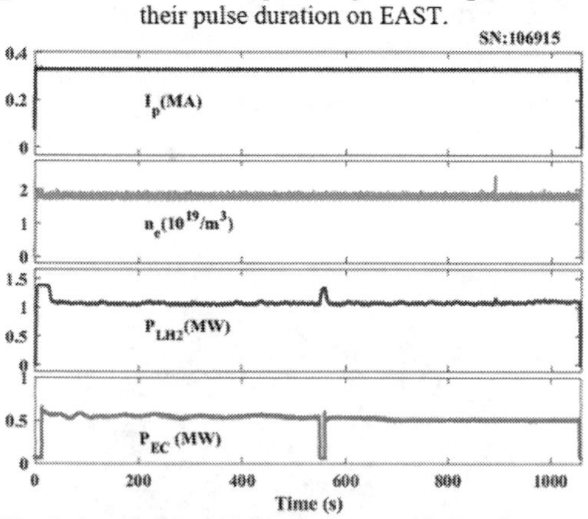


**Fig. 3.** A typical long-pulse discharge with duration up to 1056 s sustained by EC and LH waves.

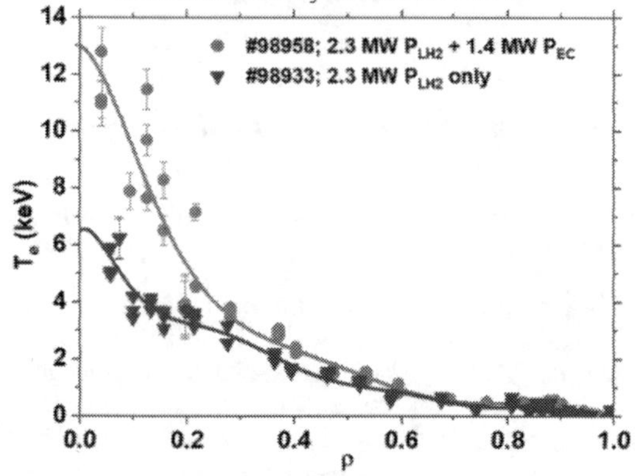


**Fig. 4.** Comparison of Electron temperature profiles measured by Thomson scattering with ECRH and without ECRH.

## 2.2 Effects of EC power location on plasma heating and current density profile

Each of the EC antennas consists of two mirrors located at the same port, one fixed focusing mirror and one movable plane mirror. The launching angle can be continuously varied with range of over 30 degrees in poloidal direction and ± 25 degrees in toroidal. Fig. 5 shows the increment of the plasma stored energy ($W_{MHD}$) and the central electron temperature when EC power injected into ohmic plasmas with different power deposition location realized by changing the poloidal angle. It is seen that an on – axis power deposition is critical for effective heating.

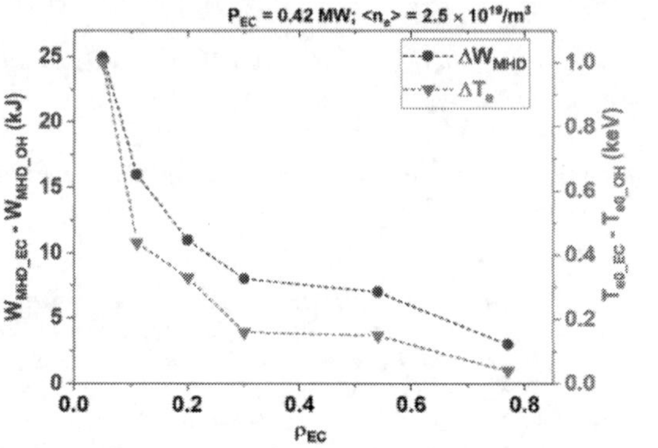


**Fig. 5.** The increment of plasma stored energy and central electron temperature after EC power injection versus EC power deposition location.

Shaping of the current density profiles by adjusting the EC power deposition is demonstrated. As shown in Fig. 6, when the EC power shifts from $\rho \sim 0.03$ to 0.3, the internal inductance ($l_i$) is decreased by ~ 0.3, indicating an off-axis ECRH is helpful to broaden the current profile. The power density profiles calculated by TORAY code are plotted in Fig. 7. This conclusion is further verified by the polarimeter /interferometer (POINT) measurement [8, 9] as illustrated in Fig. 8. The Faraday rotation angle is propotional to $\int n_e(z)\, B_{||}(z) dz$, where $B_{||}$ is the magnetic field component along the laser beam and dz is the plasma path length. Since

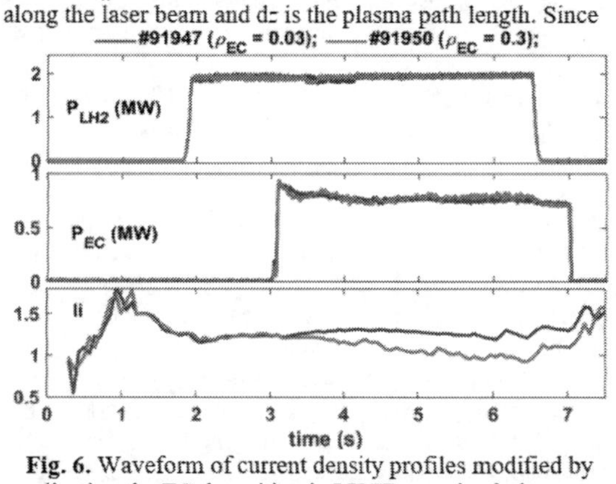


**Fig. 6.** Waveform of current density profiles modified by adjusting the EC deposition in LHCD sustained plasmas.

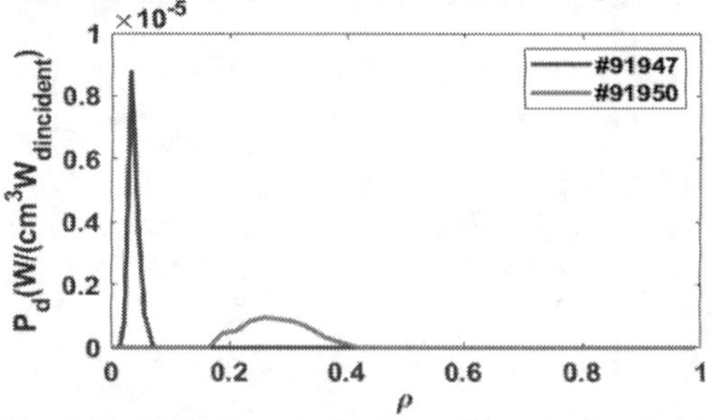


**Fig. 7.** Normalized power density profiles calculated with TORAY code

the density profiles are almost the same for these two discharges, the difference in Faraday rotation angle should be as a result of the different plasma current density profiles.

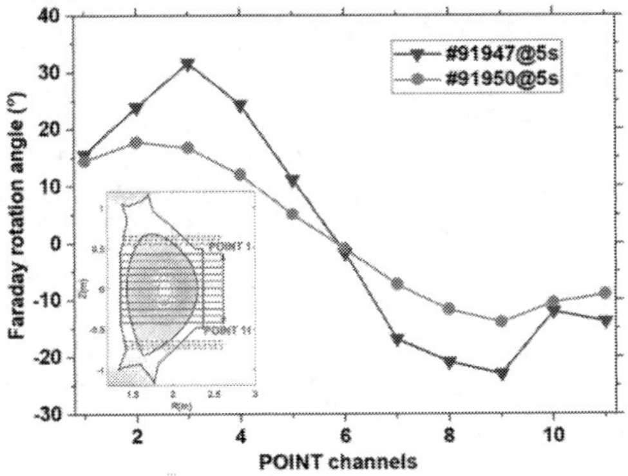


**Fig. 8.** Line-integrated Faraday rotation angle versus channels measured by POINT diagnostic. The optical layout of the 11 channels is also shown.

### 2.3 LHCD efficiency improved by ECRH

LHCD efficiency defined by

$$\eta = \frac{I_{\mathrm{LH}} n_e R}{P_{\mathrm{LH}}} (A/W/m^2) \qquad (1),$$

is recognized as a key factor for dimensioning an LH system, where $I_{\mathrm{LH}}$ is the plasma current driven by LH wave, and $R$ is the major radius of the plasma. According to the CD theory [10], LHCD efficiency increases with the square of the parallel resonant velocity, which is about 3.5 times of the thermal velocity $v_{\mathrm{th}} = \sqrt{kT_e/m_e}$. The experimental results also demonstrated that the LHCD efficiency is a significantly increasing function of $\langle T_e \rangle$ [11, 12].

Fig. 9 shows a typical discharge of LHCD efficiency improved by ECRH. With the EC power of 0.3 MW heating in the plasma core, the central electron

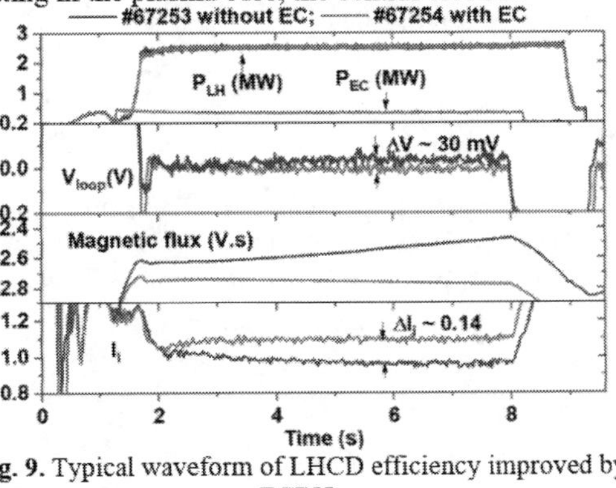


**Fig. 9.** Typical waveform of LHCD efficiency improved by ECRH.

temperature is increased by ~ 1.6 keV (see Fig. 10), leading to a higher CD efficiency indicated by a lower loop voltage (~ 30 mV). Moreover, a higher internal inductance ($l_i$) indicates that the LH current shifts to the inner region for the case with ECRH. Fig. 11 shows the LHCD efficiency with and without ECRH versus plasma density. Quasi-steady-state ($V_{\mathrm{loop}}$ < 100 mV) discharges are selected and the LH driven current is calculated by $I_{\mathrm{LH}} = I_{\mathrm{P}} - I_{\mathrm{EC}} - I_{\mathrm{OH}} - I_{\mathrm{BS}}$. It is seen that for similar density the LHCD efficiency in the plasmas with ECRH is higher by ~ 20% than that without ECRH. Besides, good CD effect ($\eta$ ~ 0.9 × $10^{19}$A/W/m$^2$) is obtained with the combination of LHCD and ECRH even when density increases up to 5.4 × $10^{19}$/m3.

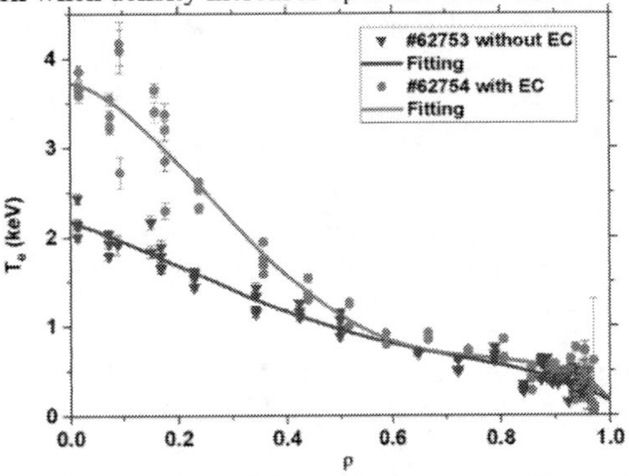


**Fig. 10.** Comparison of electron temperature profiles with and without ECRH.

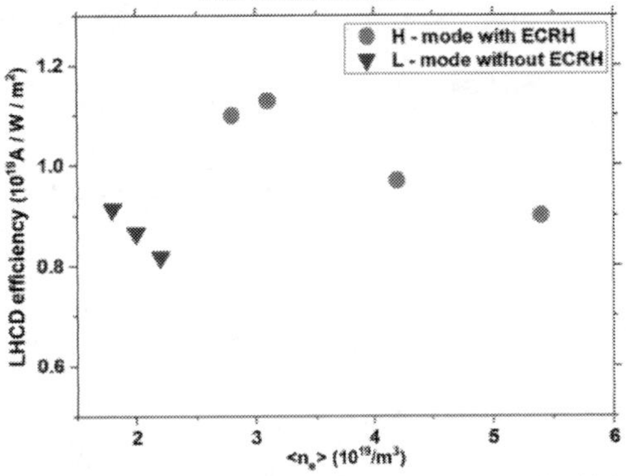


**Fig. 11.** Comparison of LHCD efficiency with and without ECRH.

It is well known that the interaction mechanism for EC waves with electrons is cyclotron damping, which accelerate electrons mainly in perpendicular direction; while for LH waves, it is dominant by Landau damping, which accelerates electrons mainly in parallel direction. When these two waves are injected simultaneously, if the particles accelerated by the EC waves reach the lowest limit velocity of the LH waves, or the EC power absorbed by the fast electron tail driven by the LH waves, a synergy effect between LHCD and ECCD will be created [13, 14]. This synergy effect was quantitatively demonstrated in steady-state experiments on Tore-Supra for the first time [15]. On EAST, the synergy effect between ECCD and LHCD is also observed in steady-state operation. The results obtained are well illustrated by the time history of discharge #106904, shown in Fig. 12. During this experiment, the LH power is feedback controlled by the magnetic flux consumption with constant plasma current ($I_p$ = 320kA) and constant density ($n_e$ = 1.8 × $10^{19}$/m$^3$), which is similar to the previous experiment on Tore Supra [15]. Due to the faulty algorithm, the magnetic flux consumption is not controlled constantly during the LH alone phase. However, from phase 1 and phase 2, we can infer that for fully non-inductive LHCD ($V_{\mathrm{loop}}$ = 0), it needs LH power of ($P_{\mathrm{LH}}$) 1.28 MW approximately with parallel refractive index $N_{\|}$ ~ 2.04. During the application of 0.55 MW ECCD with $N_{\|}$ ~ 0.34, the LH power drops ($\Delta P_{\mathrm{LH}}$) by approximately 0.3 MW (from 1.28 MW to 0.98 MW).

The bootstrap current ($I_{BS}$) calculated by Sauter model [16] is about 71 kA and the EC driven current calculated by TORAY code is in the range of 26 - 29 kA based on the measured electron and density profiles. Since the plasma current is kept constant and the loop voltage is zero, the additional current $\Delta I$, driven by the EC waves in the presence of LH waves can be obtained by the formula

$$\Delta I = (I_p - I_{BS})\frac{\Delta P_{LH}}{P_{LH}} \quad (2).$$

For discharge #106904, the additional current $\Delta I$ is ~ 59 kA. Consequently, the synergy factor defined as $F_{syn} = \Delta I / I_{EC}$ is estimate to be ~ 2.1 ± 0.1. Although the temperature shows a small difference during ECCD and before ECCD phases, this difference is not quantitatively large enough to explain the increase of the efficiency. As shown in Fig. 13, the ECE electron temperature during ECCD phase is only higher by ~ 5% than that before ECCD, while the LHCD efficiency is increased by 16% with the assumption that this part current $\Delta I$ - $I_{EC}$ is driven by LH wave.

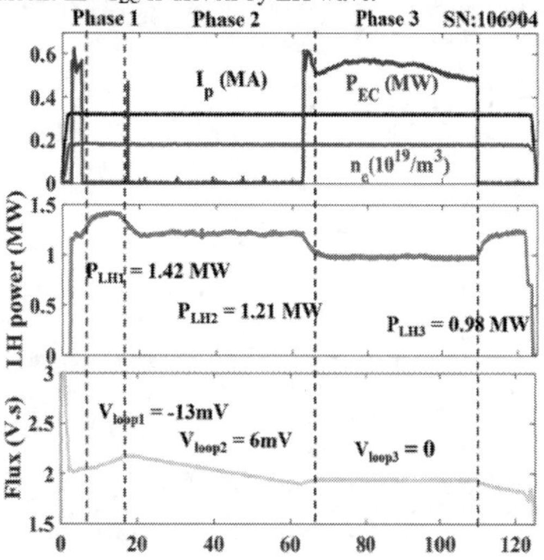


**Fig. 12.** The time history of a discharge during which the synergy effect between LHCD and ECCD is observed.

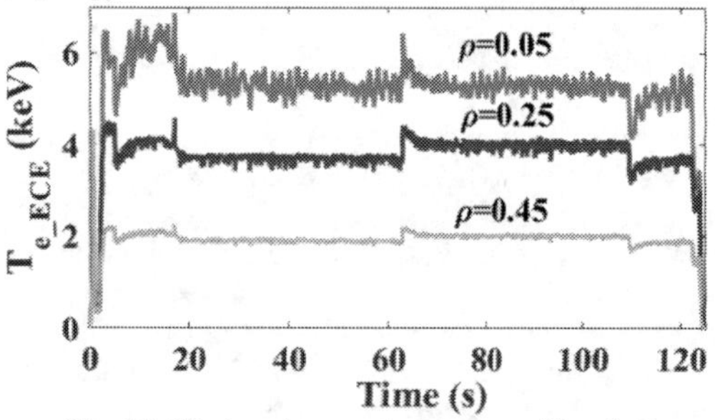


**Fig. 13.** Electron temperature measured by electron cyclotron emission (ECE) at various positions.

## 3 Summary and next work

With optimization of the operation conditions and improvement of the system, significant progress in long-pulse and high-power operation has been achieved with EC system, including pulse duration extended to ~ 1056 s with $P_{EC}$ ~ 0.55 MW, and pulse duration ~ 310 s with $P_{EC}$ up to 1.6 MW. High $T_{e0}$ ~ 12 keV plasmas was maintained for 100 s by the combination of 1.4 MW EC and 2.3 MW LH power. Thanks to a higher electron temperature in the plasma core with ECRH, the LH power deposition shifts to the inner region, thus giving a higher LHCD efficiency. A synergy effect between ECCD and LHCD was observed in steady-state plasmas. TORAY modelling predicts that the ECCD current can be > 80 kA at low density as shown in Fig. 14. Next work will investigate the ECCD efficiency and the synergy effects with different $N_{\parallel}$ of LH waves.

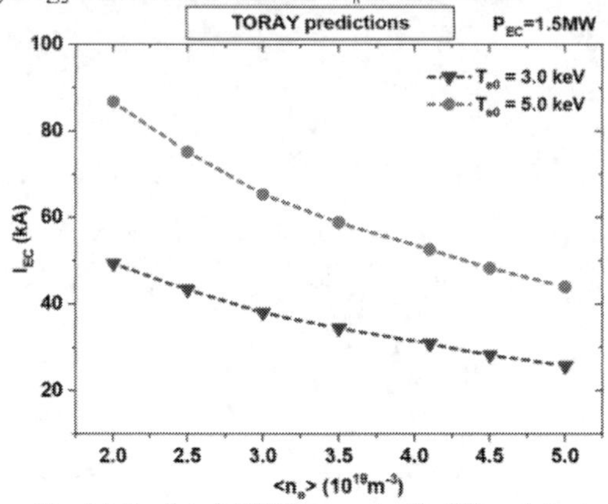


**Fig. 14.** Predicted ECCD current with different electron temperature and density.

## Acknowledgement

This work was supported by the Performance Improvement Project of EAST, Hefei Comprehensive National Science Centre, the National Key R&D Program of China (Grant No. 2017YFE0300401) and the National Natural Science Foundation of China (Nos. 11305211 and 11775259). The authors would like to thank the experts from GYCOM and CPI companies for their great efforts in the development of gyrotrons for EAST.

## References

1. ITER Physics Expert Group on Energetic Particles, Chapter 6: Plasma auxiliary heating and current drive, Nucl. Fusion **39**, 2495 (1999).
2. T. Suzuki et al., Nucl. Fusion **44**, 699 (2004).
3. H.D. Xu et al., Plasma Sci. Technol. **18**, 442 (2016).
4. X.J. Wang et al., Fusion Eng. Des **96-97**, 181 (2015).
5. H.D. Xu et al., Fusion Eng. Des.**164**, 112222 (2021).
6. J.F. Shan et al, *A New 4 MW LHCD System for EAST*, in Proceedings of the 23rd IAEA Fusion Energy Conference, Daejon, Korea (2010).
7. F.K. Liu et al., Fusion Engineering and Design **113**, 131 (2016).
8. H.Q. Liu et al., Rev Sci Instrum **87**, 11D903 (2016).

9. J.P. Qian et al., Nuclear Fusion **57**, 036008 (2017).
10. N.J. Fisch, Phys. Rev. Lett. **41**, 873 (1978).
11. K. Ushigusa et al., Nucl. Fusion **29**, 1052 (1989).
12. V. Pericoli Ridolfini et al., Phys. Rev. Lett. **82**, 93 (1999).
13. I. Fidone et al., Nucl. Fusion **27**, 579 (1987).
14. R. Dumont et al., Phys. Plasmas **7**, 4972 (2000).
15. G. Giruzzi, et al., Phys. Rev. Lett. **93**, 255002 (2004).
16. O. Sauter et al., Phys. Plasmas **6**, 2834 (1999)

# The development of 1 MW ECRH system on J-TEXT

*Xixuan* Chen[1], *Donghui* Xia[1,*], *Junli* Zhang[1], *Zhijiang* Wang[1], *Yuan* Pan[1] and J-TEXT team[1]

[1]International Joint Research Laboratory of Magnetic Confinement Fusion and Plasma Physics, Huazhong University of Science and Technology, 430074 Wuhan, China

**Abstract.** A 1 MW electron cyclotron resonance heating system has been designed and developed for J-TEXT tokamak not only to enhance the total heating power but also to provide another possible way to control the plasma parameters. The system is composed of two gyrotrons, both with an output power of 500 kW electron cyclotron wave power. The development progress of the system is elaborated and discussed. The commissioning tests for the gyrotrons were implemented, and each gyrotron realized its standard output successfully. With the assistance of this system, a variety of experiments have been carried out on J-TEXT, mainly related to plasma heating, plasma disruption, plasma start-up, current drive and so on.

## 1 Introduction

With advantages like highly localized power deposition and high coupling efficiency, electron cyclotron resonance heating (ECRH) has been widely applied on many magnetic confinement fusion devices (e.g., DIII-D, W7-X, HL-2A) [1-3] and is considered to be used for future fusion demonstration reactors. In order to improve the plasma parameters and broaden the operation range of Joint-TEXT (J-TEXT) tokamak, the development of the ECRH system has been initiated in 2017. After the integration design and setup of subsystems, a gyrotron which is at a frequency of 105 GHz with a power of 500 kW for 1 s was tested and put into physical experiments in 2019. Another gyrotron of same parameters has just finished its commissioning tests in 2022. Therefore, the J-TEXT ECRH system could realize an output power of 1 MW in total. The layout of the ECRH system is shown in Fig. 1 [4].

This paper is organized as follows. The introduction of the 1 MW ECRH system on J-TEXT is presented in section 2. The gyrotron commissioning tests are presented and discussed in section 3. The related physical experiments are presented in section 4.

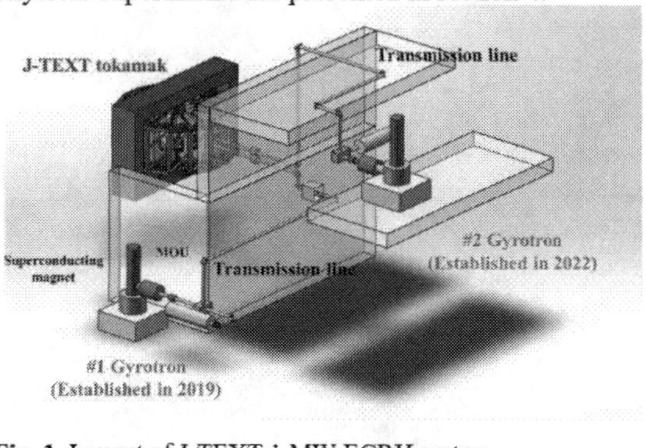


**Fig. 1.** Layout of J-TEXT 1 MW ECRH system

## 2 J-TEXT ECRH system

The J-TEXT tokamak is a conventional iron core tokamak reconstructed from the Texas Experimental Tokamak Upgrade (TEXT-U) [5-7]. The basic operation parameters of J-TEXT are listed in Table 1. As a long-term research program, the J-TEXT is aimed at developing fundamental physics and control mechanisms of high-temperature tokamak plasma confinement and stability.

**Table 1.** Basic operation parameters of J-TEXT.

| Parameters | Value |
|---|---|
| Major radius ($R_0$) | 1.05 m |
| Minor radius ($a$) | 25-29 cm |
| Toroidal field ($B_t$) | <2.4 T |
| Plasma current ($I_p$) | <240 kA |
| Plasma densities ($n_e$) | (1-7)×$10^{19}$ $m^{-3}$ |
| Electron temperatur ($T_e$) (With ohmic heating) | ~1 keV |

The ECRH system has been developed for J-TEXT to increase the core electron temperature and also to study the non-inductive current drive. This system is designed to work at the X2 mode for plasma heating and non-inductive current drive, considering its practical application.

* Corresponding author: xiadh@hust.edu.cn

As the core component of ECRH system, the gyrotron generates high power microwave of special frequency during a long pulse. The J-TEXT ECRH system contains two gyrotrons of 500 kW power each which are both manufactured by GYCOM Ltd., so in total the ECRH system could realize an output of 1 MW. The main characteristic parameters are given in Table 2. The first gyrotron finished its commissioning tests and was put into formal operation in J-TEXT physical experiments since 2019. The commissioning tests of the second gyrotron were just finished in 2022.

**Table 2.** Main characteristic parameters of the gyrotrons for J-TEXT ECRH system.

| Parameters | Value (#1 gyrotron) | Value (#2 gyrotron) |
|---|---|---|
| Frequency | 105 GHz | 105 GHz |
| Maximum output power | 500 kW | 500 kW |
| Maximum pulse duration | 1 s | 1 s |
| $TEM_{00}$ mode purity (after MOU) | 97% | 98% |
| Output efficiency | 45% | 45% |

Each gyrotron is equipped with a set of subsystems to realize its safe and stable operation, Fig. 2 shows the corresponding diagram.

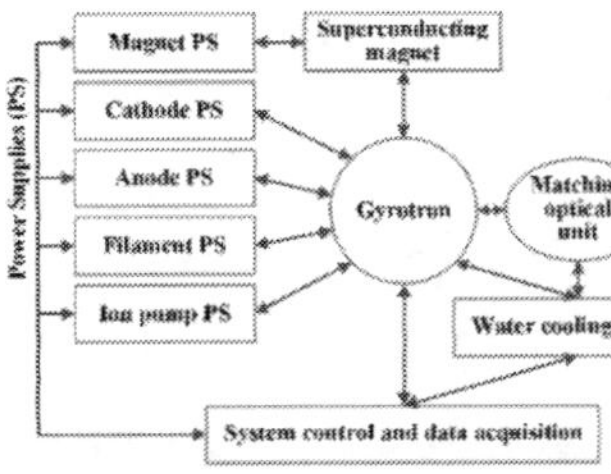


**Fig. 2.** Diagram of gyrotron operation platform.

There are various power supplies (PS) indispensable for gyrotron operation. Two high voltage power supplies which are the cathode power supply and anode power supply provide high voltage for establishing an accelerating electric field inside the gyrotron. Besides, to make the electrons form a helical motion, the superconducting magnet is needed to build a steady-state magnetic field, thus the magnet power supply is used to charge the superconducting magnet. What is worth mentioning here is that the superconducting magnet for the first gyrotron is cooled by liquid helium and nitrogen while the magnet for the second is cryogen-free. The filament power supply is used to heat the filament inside the gyrotron in order to generate active electrons. The ion pump power supply could maintain the vacuum of the gyrotron, which is very important for its operation. The electrical parameters of the power supplies are listed in Table 3.

**Table 3.** Electrical parameters of power supplies for the gyrotron.

| Power supply | Parameters (#1 gyrotron) | Parameters (#2 gyrotron) |
|---|---|---|
| Cathode power supply | -44 kV/25 A | -45 kV/24 A |
| Anode power supply | 30 kV/150 mA | 32 kV/200 mA |
| Magnet power supply | 100 A | 100 A |
| Ion pump power supply | 5 kV/10 mA | 5kV/10 mA |
| Filament power supply | 1500 VA | 2500 VA |

In addition to the power supplies, there is a specially designed control system for the gyrotron. The control system realizes functions including time sequence trigger, data acquisition, communication and most noteworthy, protection [8]. Besides, the ECRH system should be equipped with a cooling system to cool down the crucial components and meanwhile measure the output microwave power.

The subsystems mentioned above work together with the gyrotron in order to generate microwave as desired. The output microwave is then corrected by the mirrors in matching optical unit (MOU) which is connected with the output window of the gyrotron. To realize localized and efficient plasma heating, a specialized transmission line is constructed to guide the microwave into the plasma. The transmission line is mainly composed of corrugated waveguide, mitre bend, DC break and launcher [9], which are designed combining the characteristics of the output microwave. The transmission loss is mainly caused at mitre bends. There is either flat mirror or polarizer on each bend. The amount of these components and the possible transmission loss caused by them is listed in Table 4.

**Table 4.** The number of flat mirrors and polarizers and their possible transmission loss

| Component | Number | Possible transmission loss |
|---|---|---|
| Flat mirror | 3 | 1%~2% |
| Polarizer | 2 | 2%~3% |
| Total | | 7%~12% |

## 3 Commissioning tests of gyrotron

Before carrying out related physics experiments on J-TEXT tokamak, commissioning tests of the two gyrotrons are ought to be finished to check their status and to explore their operation characteristics.

The commissioning of a gyrotron usually contains the following stages. The first stage is cooling down of magnet in order to bring it to the superconducting state. The second stage is the alignment of the magnet, because the axis of the gyrotron should coincide that of the magnet. The third stage is gyrotron operation to find its best operation condition. Along with the gyrotron operation, MOU installation and alignment should be carried out in order to focus and direct the microwave beam correctly after it leaves the gyrotron. The final stage of the commissioning is also the most crucial part, that is to adjust all control parameters to increase gyrotron output power to the maximum value.

### 3.1 Cooling down of the magnet

As for magnet cooling, the temperature and magnet resistance change of the whole process is recorded, shown in Fig. 3. Fig. 3 (*a*) shows the resistance change and Fig. 3 (*b*) shows the temperature change. At the end of the cooling down process, the magnet resistance is 0.05 Ω and its temperature reached 3 K.

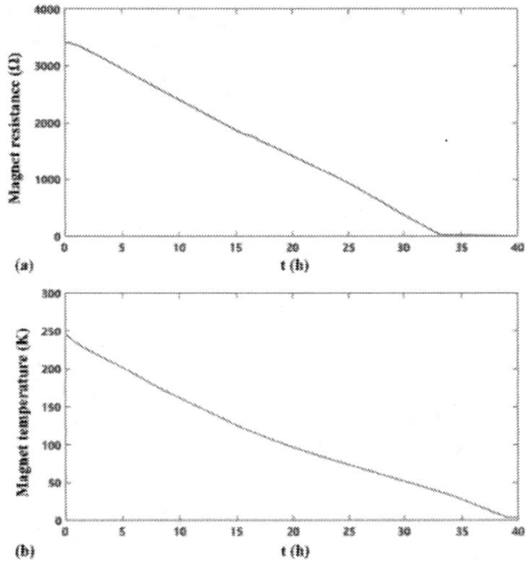


**Fig. 3.** Magnet resistance change (*a*) and magnet temperature change (*b*) during the cooling down process.

### 3.2 Magnet alignment

The magnet alignment is also very crucial and needs to be implemented precisely [10]. Without good alignment, there is a great chance that the gyrotron couldn't reach its best working condition, especially when a high-power output is required. The target of this step is to make the axis of the magnet coincide with that of the gyrotron as much as they could. That is to say, in this step, the higher the degree of two axes coincide with each other, the better alignment we have obtained. Figure. 4 shows the diagram of poor magnet alignment and good magnet alignment. A dummy tube which has a window to observe the fluorescent ring formed by electrons is used to replace the gyrotron in this step and its size is exactly the same as the gyrotron. When a fluorescent ring looks like orbit 1 in Fig. 4, it means good alignment has been obtained. While a fluorescent ring like orbit 2 takes place, more adjustment work should be done.

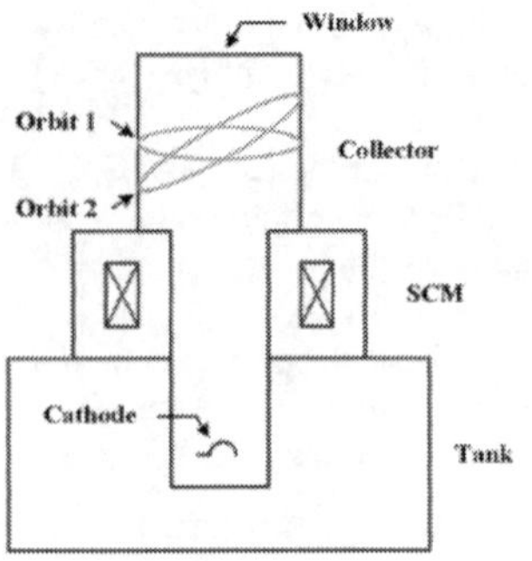


**Fig. 4.** Diagram of poor and good magnet alignment.

### 3.3 Gyrotron operation

The gyrotron could be operated without MOU at first to realize an output of low power and relatively low mode purity. Through this we could find a combination of suitable parameters for the gyrotron, which is also beneficial for MOU alignment.

Generally, the MOU contains a pair of phase correcting mirrors, the internal structure of which is shown in Figure. 5. To design an MOU for a gyrotron, the initial output wave of the gyrotron should be measured. Then the MOU is designed according to the actual output of the gyrotron.

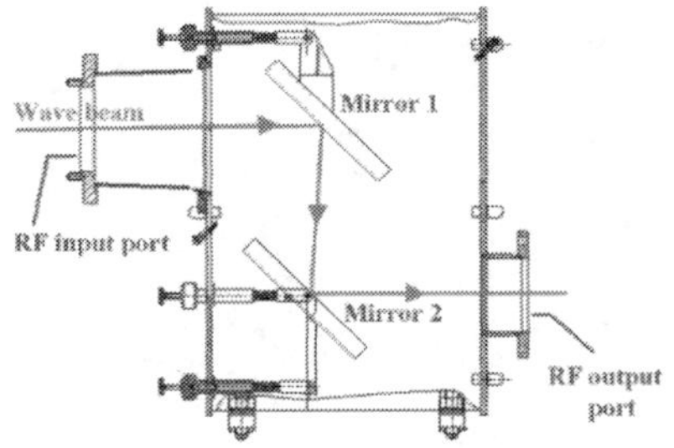


**Fig. 5.** Internal structure of MOU.

The most crucial goal of the commissioning is to find the most appropriate operation status for the gyrotron [11]. There are a few parameters that could have an influence on the gyrotron operation status, of them the most important are the magnet field intensity, the filament power, the cathode voltage and anode voltage. Besides understanding how those parameters could affect the gyrotron operation status, it is also very important to think out how to evaluate its operation status. Three ways are used to realize this target. The first is to build up a platform to measure the microwave signal, which is also called radio frequency (RF) signal and is obtained by directional coupler method [12]. The second is to use a spectrometer to measure the frequency of the output wave. The third is to measure its power with a dummy load, this method is called calorimetric method [13]. If the measured RF signal, frequency spectrum signal and output power are basically the same as results in the factory tests, then the gyrotron operation status is quite standard and good. Figure. 6 shows the measured RF signal and Fig. 7 gives the temperature rise curve which is used to calculate the microwave power. The area of the temperature rise curve represents the energy absorbed by cooling water. To calculate the microwave power, we first used a power supply with known power to heat the water, with formula (1) we could then calculate the microwave power because the pulse width of microwave is also clear.

$$P_{rf} = P_k \frac{D_k S_{rf}}{D_{rf} S_k} \tag{1}$$

here, $P_k$ is the power of the power supply, $D_k$ is the pulse width of the electric power, $S_{rf}$ is the area of the measured temperature rise curve when the microwave is absorbed by the water, $D_{rf}$ is the pulse width of the microwave power, and $S_k$ is the area of the measured temperature rise curve of the electric power.

As for the calibration of the RF-signal, the gyrotron should be operated under same parameters for 2 shots, and the microwave power are measured by calorimetric method and directional coupler method separately. The amplitude of the output signal from directional coupler method varied with different microwave power. The calculation results of calorimetric method could be used to calculate the RF-signal measured by directional coupler method.

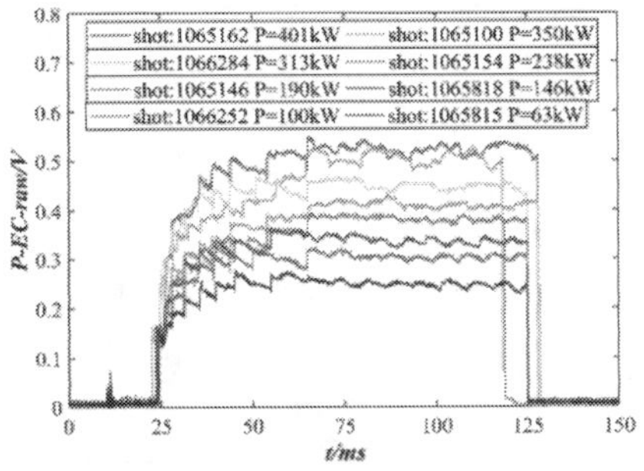


**Fig. 6.** Measured RF signal during gyrotron operation. Different shot has different power, therefore the RF signal has different value.

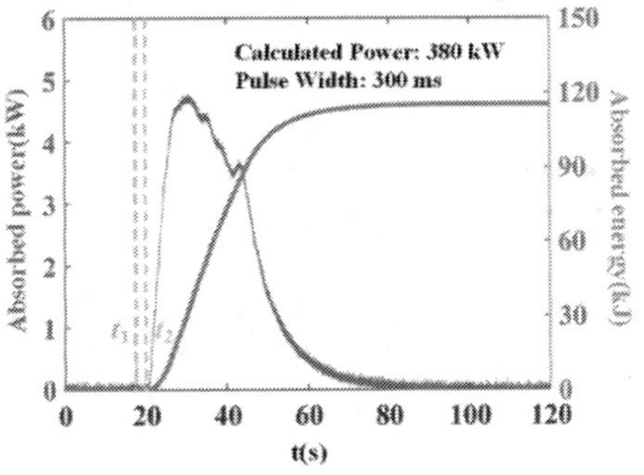


**Fig. 7.** A typical temperature rise curve during gyrotron operation, the left axis represents the power absorbed by water and the right axis represents the absorbed energy. Calculated power is 380 kW and the pulse width is 300 ms. $t_1$ is the microwave injection time while $t_2$ is the temperature rising time.

Through exploring the working characteristics of the gyrotron, its control parameters could be optimized continuously. Here a typical shot of one pulse is given in Figure. 8. In this shot, the cathode voltage is -40 kV, the anode voltage is 22 kV and the cathode current is 22 A, which corresponds to an output of 350 kW microwave power. Both gyrotrons have realized their maximum output power, however, considering the operation stability and safety, they are usually operated no more than 80% of the maximum output power.

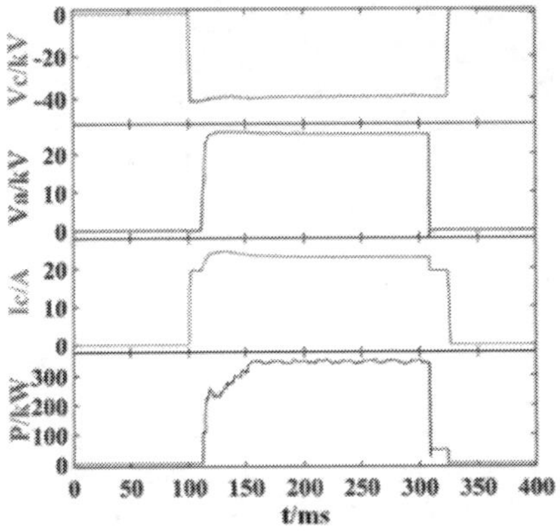


**Fig. 8.** Waveforms of a typical shot.

## 4 The experimental results with ECRH

The first gyrotron has been utilized in J-TEXT physical experiments right after the commissioning tests were finished. The microwave generated by the gyrotron was successfully injected into J-TEXT in 2019. The second gyrotron is not in use formally now because the commissioning tests were just finished recently. Until now, various kinds of related experiments have been carried out on J-TEXT, related to not only plasma heating but also plasma disruption, sawtooth control, plasma start-up, current drive and so on.

### 4.1 Electron thermal transport and temperature fluctuations of ECRH

Dedicated experiments have been carried out on the J-TEXT for investigating the electron thermal transport and plasma fluctuations before and during electron cyclotron resonance heating.

Fig. 9 [13] shows a typical discharge with ECRH for shot #1069645. The main plasma parameters are shown in Table 5.

**Table 5.** Main plasma parameters of shot #1069645

| Parameters | Value |
|---|---|
| $B_t$ | 1.875 T |
| $I_p$ | 175 kA |
| $n_e$ | (2-2.5)×10$^{19}$ m$^{-3}$ |

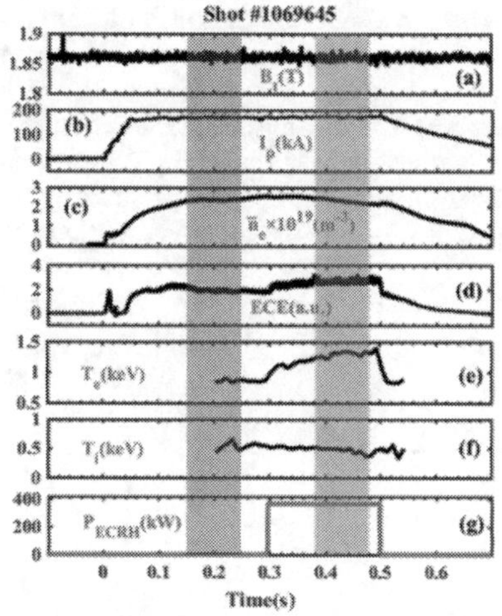


**Fig. 9.** Time evolution of plasma parameters for shot 1069645: (*a*) toroidal field $B_t$; (*b*) plasma current $I_p$; (*c*) line-averaged electron density $\bar{n}_e$; (*d*) electron temperature at *r*/*a* = 0.72 by ECE (no calibration); (*e*)-(*f*) the core of plasma electron and ion temperature obtained by XICS; (*g*) the ECRH power [14].

In Fig. 9, both the electron temperature at *r*/*a* = 0.72 measured by the electron cyclotron emission (ECE) diagnostic (Fig. 9(*d*)) and the core of electron temperature from the x-ray imaging crystal spectrometer (XICS) (Fig. 9(*e*)) increase significantly during ECRH. The core ion temperature (Fig. 9(*f*)) does not increase as the electron temperature but decreases with ECRH. The decrease of ion temperature in the plasma core indicates that ion heat transport may be enhanced.

### 4.2 Fast electron with ECCD

Fast electrons could be generated by electron cyclotron wave through wave-particle interaction when the resonant condition is fulfilled. The fast electron bremsstrahlung diagnostic on J-TEXT enables detailed studies of the generation and transport of fast electrons.

When the toroidal field is 1.875 T, the electron cyclotron wave absorption area is on-axis according to the calculation. A typical on-axis electron cyclotron current drive (ECCD) experiment with a number of fast electrons is shown in Fig. 10 [14]. In this shot, the plasma current is 150 kA, central electron density is kept at about $1 \times 10^{19}$ $m^{-3}$. With the EC wave injection at 250 ms, the line integrated intensity signal from the central chord of FEB IFEB begins to rise (Fig. 10(c)). The ECE signals at 80.5 GHz which the resonance layer is outside the plasma (+31.96 cm) are dominated by relativistically downshifted emission by the high-energy tail of the electron distribution function and can be employed to diagnose the population of fast electrons, as shown by the black line in Fig. 10(e).

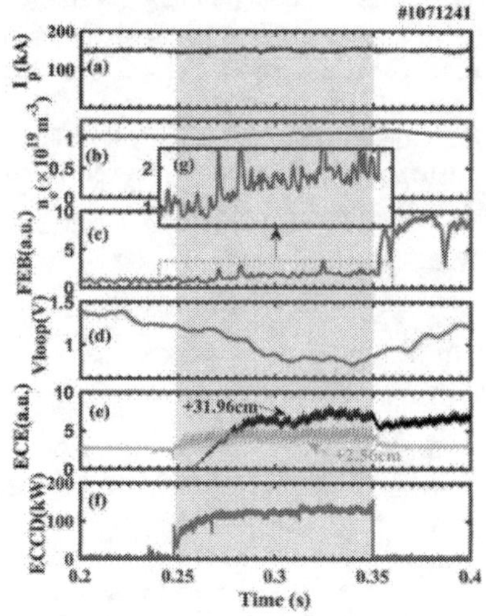


**Fig. 10.** Temporal evolution of typical on-axis ECCD experiment which generated a large number of fast electrons [15]. From the top to bottom, the waveforms are: (a) the plasma current, (b) the central line averaged electron density, (c) FEB signal at the radial position of 0 cm (from demodulator), (d) the loop voltage, (e) ECE signals at the radial position of +2.56 cm and +31.96 cm (out of plasma), (f) ECCD power. About 150 kW EC wave power is injected into plasma from 250 ms to 350 ms and (g) the enlarged FEB signal of dashed box.

## 5 Summary

J-TEXT 1 MW ECRH system contains 2 gyrotrons of 105 GHz/500 kW/1 s which are manufactured by GYCOM. The first gyrotron was tested and put into operation in 2019. The commissioning tests of the second gyrotron is just finished in 2022. During the commissioning process of the gyrotron, the cooling down and alignment of the magnet, the MOU alignment and the gyrotron operation were finished. The high power and long pulse output has been realized during the commissioning tests. Since 2019, the first gyrotron has been utilized in J-TEXT physical experiments, which not only expanded the parameter range of J-TEXT, but also promoted the ECRH and ECCD related research.

## ACKNOWLEDGMENTS

The authors want to appreciate J-TEXT team for valuable discussion and suggestion about electron cyclotron resonance heating. This work was supported by the National Key Research and Development Program of China (Grant 2017YFE0300200 and Grant 2017YFE0300204), the Key Research and Development Program of Hubei province (Grant 2021BAA167) and the National Natural Science Foundation of China (Grant 51821005).

## References


1. M. Cengher, J. Lohr, Y. A. Gorelov, R. Ellis, Egemen Kolemen, D. Ponce, S. Noraky, C. P. Moeller, IEEE Trans. Plasma Sci. **42**, 7 (2014)
2. W. Kasparek, P. Brand, H. Braune, G. Dammertz, V. Erckmann, G. Gantenbein, F. Hollmann, M. Grünert, H. Kumric, L. Jonitz, H.P. Laqua, W. Leonhardt, G. Michel, F. Noke, B. Plaum, F. Purps, M. Schmid, T. Schulz, K. Schwörer, M. Thumm, M. Weissgerber, Fusion Eng. Des. **74**, 243 (2005)
3. M. Huang, J. Rao, B. Li, J. Zhou, Z.H. Kang, H. Wang, B. Lu, D.H. Xia, C. Wang, K. Feng, M.W. Wang, G.Y. Chen, Y.N. Pu, Z.H. Lu, J.Q. Wang, X.R. Duan, Y. Liu, in Proceedings of the 17th Joint Workshop on Electron Cyclotron Emission and Electron Cyclotron Resonance Heating, 7-10 May 2012, Deurne, Netherlands (2012)
4. J.L. Zhang, D.H. Xia, C.H. Liu, Z.J. Wang, Y. Pan, IEEE Trans. Plasma Sci. **48**, 6 (2020)
5. Y. Liang, N.C. Wang, Y.H. Ding, Z.Y. Chen, Z.P. Chen, Z.J. Yang, Q.M. Hu, Z.F. Cheng, L. Wang, Z.H. Jiang, B. Rao, Z. Huang, Y. Li, W. Yan, D. Li, H. Liu, L. Zeng, Y. Huang, D.W. Huang, Z.F. Lin, W. Zheng, F.R. Hu, K.J. Zhao, M. Jiang, Y.J. Shi, H. Zhou, S.T. Peng, W.X. Guo, L. Gao, Z.J. Wang, M. Zhang, K.X. Yu, X.W. Hu, Q. Yu, G. Zhuang, K.W. Gentle, Y. Pan, the J-TEXT Team, Nucl. Fusion **59**, 112016 (2019)
6. G. Zhuang, Y. Pan1, X.W. Hu, Z.J. Wang, Y.H. Ding, M. Zhang, L. Gao, X.Q. Zhang, Z.J. Yang, K.X. Yu, K.W. Gentle, H. Huang, the J-TEXT Team, Nucl. Fusion **51**, 094020 (2011)
7. Y.H. Ding, Z.Y. Chen, Z.P. Chen, Z.J. Yang, N.C. Wang, Q.M. Hu, B. Rao, J. Chen, Z.F. Cheng, L. Gao, Z.H. Jiang, L. Wang, Z.J. Wang, X.Q. Zhang, W. Zheng, M. Zhang, G. Zhuang, Q.Q. Yu, Y.F. Liang, K.X. Yu, X.W. Hu, Y. Pan, K.W. Gentle, the J-TEXT team, Plasma Sci. Technol. **20**, 125101 (2018)
8. Z. Wang, G. Fan, D. Xia, X. Chen, W. Weng, IEEE Trans. Plasma Sci. **49**, 258 (2021)
9. J.L. Zhang, D.H. Xia, C.H. Liu, Z.J. Wang, Y. Pan, IEEE Trans. Plasma Sci. **48**, 1560 (2020)
10. D.H. Xia, M. Huang, J. Zhou, X.Y. Bai, T.L. Zheng, J. Rao, G. Zhuang, Plasma Sci. Technol. **16**, 410 (2014)
11. H. Wang, J. Rao, M. Huang, J. Zhou, C. Wang, Z.H. Kang, M.W. Wang, K. Feng, G.Y. Chen, B. Lu, J.Q. Wang, Fusion Eng. Des. **101**, 61 (2015)
12. T. Notake, H. Idei, S. Kubo, T. Shimozuma, Y. Yoshimura, S. Kobayashi, Y. Mizuno, S. Ito, Y. Takita, K. Ohkubo, W. Kasparek, T. Watari, R. Kumazawa, Rev. Sci. Instrum. **76**, 023504 (2005)
13. G. Michel, H. Laqua, W. Kasparek, Fusion Eng. Des. **56-57**, 655 (2001)
14. Z.J. Yang, Y.P. Zhang, X.H. Ren, F. Li, X. Xu, W. Yan, X.Y. Zhang, D.H. Xia, Z.C. Zhang, Y. Gao, X.Q. Zha, Q. Luo, Z.Y. Chen, Z.F. Cheng, Z.P. Chen, L. Gao, Y.H. Ding, Nucl. Fusion **61**, 086005 (2021)
15. X.B. Zhang, W. Yan, Z.Y. Chen, J.G. Fang, J.L. Zhang, Y. Li, X.X. Chen, Y.N. Wei, R.H. Tong, Z.F. Lin, Y. Zhong, L.K. Mou, F. Li, Z.H. Jiang, Z.J. Yang, N.C. Wang, Y. Pan, the J-TEXT team, Plasma Sci. Technol. **24**, 064007 (2022)

# ECRH experiments on Tokamaks SST-1 & Aditya-U and ECRH upgradation plan for SST-1

Braj Kishore Shukla, Jatin Patel, Harshida Patel, Dharmesh Purohit, Hardik Mistry, K.G. Parmar, SST-1 Team and ADITYA-U Team

Institute for Plasma Research, Gandhinagar-382428, India

**Abstract.** A 42GHz-500kW ECRH system [1-6] is used to carry out various experiments related to plasma breakdown and ECR heating on tokamaks SST-1 and Aditya-U. The system has been upgraded with new anode modulator power supply to launch two ECRH pulses to carry out breakdown and heating simultaneously. In SST-1, ECRH system is used routinely for plasma breakdown at fundamental harmonic, approximately 150kW power is launched for 70ms to 150ms duration and consistent plasma start-up is achieved in SST-1. In the recent experiments, second EC pulse is also launched at the plasma flat-top to heat the plasma, some heating signatures are seen but more experiments will be carried out to confirm the plasma heating with ECRH. In Aditya-U tokamak, simultaneous plasma breakdown and heating experiments are carried out successfully [2]. In the first pulse around 100kW power in fundamental O-mode is launched for 70ms duration for the breakdown at low-loop voltage and around 150kW ECRH power for 50ms duration is launched in second EC pulse to heat the plasma. In case of Aditya-U, plasma heating is observed clearly as soft X-ray signal increases sharply with ECRH. In Aditya-U tokamak, deuterium plasma experiments have been carried out and ECRH launched at the flat-top of deuterium plasma current. In deuterium plasma also ECR heating is observed as soft X-ray signal increases with ECH power. For SST-1, ECRH system is being upgraded with another 82.6GHz system, this system would be used to carry out plasma heating and start-up at second harmonic. The 82.6GHz system is already connected with the SST-1 tokamak, the old 82.6GHz-200kW Gyrotron will be upgraded to 400kW system to carry out effective heating experiments on SST-1 at higher ECRH power. The paper discusses the recent results of ECRH experiments carried out on tokamaks SST-1 & Aditya-U and presents the upgradation plan of EC system for SST-1.

## 1 Introduction

The 42GHz ECRH system has shown many interesting results towards low-loop voltage start-up and ECR heating in SST-1 and Aditya-U. Till now we have carried out experiments with single EC pulse and experiments were limited for either breakdown or heating. It is mainly due to limitation in power supply system. Recently the power supply system of 42GHz Gyrotron is upgraded and a new advance anode power supply is commissioned with the system. The rise and fall time of this power supply is ~1ms, which allow to operate the Gyrotron for more than one pulse. The data acquisition software is modified accordingly and interfaced with power supply as per the demand of the experiment. The pulse duration and ECRH power can be varied as per the experimental requirement. For the breakdown at fundamental harmonic, less ECH power is needed for breakdown while for heating higher power is recommended. The ECH power in both the pulses can be adjusted as per the need of experiment by selecting the suitable parameters of anode power supply. The ECRH two pulse experiments (breakdown and heating simultaneously) have been carried out on both the tokamaks

## 2 ECRH experiments in SST-1

In the SST-1, various experiments have been carried out related to low-loop voltage start-up and two EC pulse experiments. The recent ECRH results in SST-1 are shown in figure 1. Two pulse experiments are carried out for plasma breakdown and heating. The toroidal magnetic field of SST-1 tokamak is 1.5T and ECRH power in fundamental O-mode is launched from low field side. The first pulse of 150kW power for 50ms duration is launched for start-up and after a gap of 25ms another pulse is launched for 75ms.

The breakdown is achieved successfully with first pulse as H_alpha appears with ECRH. The appearance of $H\alpha$ with ECRH confirms the breakdown but variation of $H\alpha$ throughout the plasma shot indicates that burn-through may not be completed fully. It could be due to presence of impurity in the plasma as carbon increases during the plasma shot. The other reason of $H\alpha$ could be the emission from the edge. In the (Figure 1), there is increase in soft X-ray signal with the second pulse of ECRH but it is not conclusive for plasma heating with ECRH and more experiments are scheduled in next plasma campaigns. In SST-1, ECRH is mandatory for plasma start-up, so first pulse is needed for every plasma shots and second pulse would be applied when heating experiments would be carried out, these experiments would be carried out systematically in forthcoming campaign of SST-1.

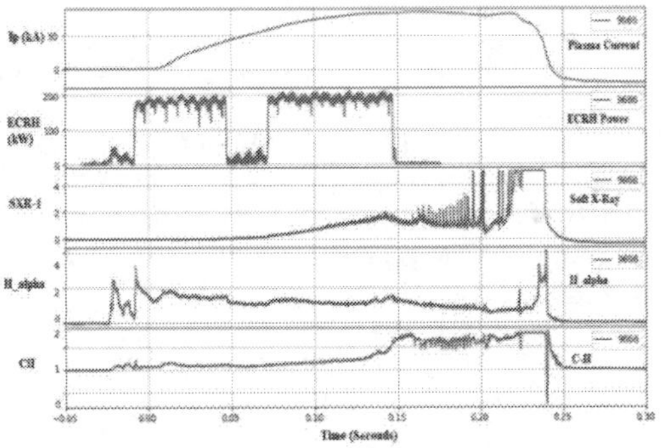

**Fig. 1.** SST-1 plasma shot with two ECRH pulses

## 3 ECRH experiments in Aditya-U

In Aditya-U tokamak, EC-assisted breakdown and heating experiments are carried out simultaneously. Two ECRH pulses launched in Aditya-U plasma, first pulse of ECRH is launched before the loop voltage and after a gap of 30ms another pulse is launched for plasma heating. The power in first pulse is 100kW and duration is 50ms while in second pulse 150kW power launched is launched for 100ms.

In Aditya-U tokamak, the operating toroidal magnetic field is 1.35T, so ECR layer lies inboard side the plasma. The standard Aditya-U plasma discharge is shown in figure 2. The loop voltage is around 13V and successful discharge of more than 200ms is achieved with plasma current close to 150kA.

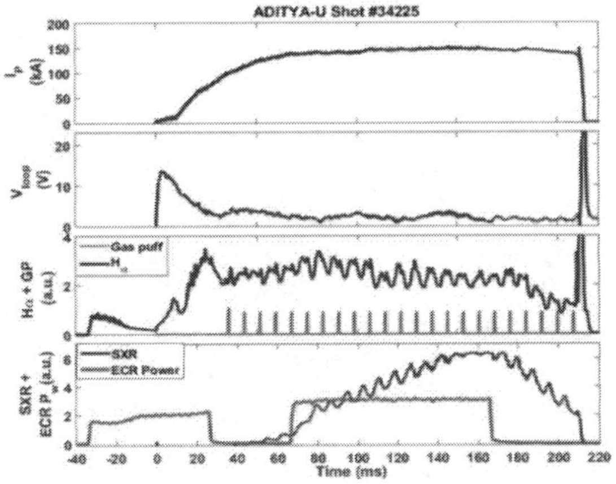


**Fig. 2.** ECRH two pulse in Aditya-U

In Aditya-U, the effect of plasma heating can be observed as soft X-ray (SXR) signal follow the ECH power (Figure 3). Although the soft X-ray signal starts increasing before the ECRH pulse so ECR heating effect may not be conclusive but reduction in soft X-ray signal immediate after ECRH pulse indicates the effect of ECRH. The soft X-ray (SXR) radiation is generated mainly by bremsstrahlung, which depends on both electron density $n_e$ and temperature $T_e$:

$$P_{brem} \sim Z_{eff}\, n_e^2 \sqrt{Te}$$

The contribution of impurity can be considered as constant in the plasma for both the cases (with and without ECRH) as tokamak conditions are maintained, so the increase in SXR flux at the detector due to ECRH power could be either due to increase in density or due to increase in electron temperature. Considering the Aditya-U plasma is fully ionized, the increase in soft X-ray could be due to increase in electron temperature. These ECR heating experiments are not yet completed on tokamak Aditya-U, it will be carried out again systematically with supporting diagnostics and systematic variation in the parameters like ECRH power and durations etc.

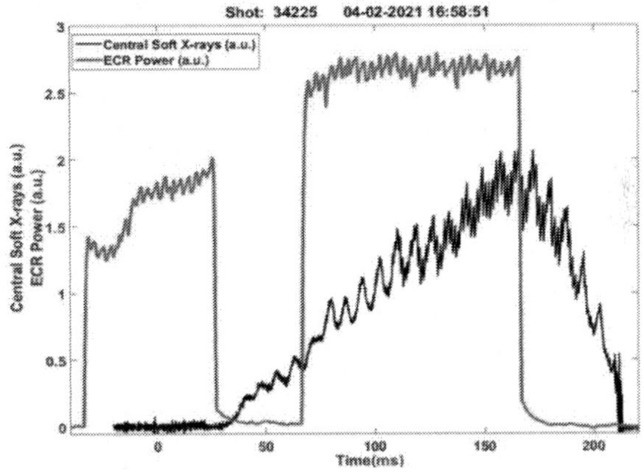


**Fig. 3.** Soft X-ray signal and ECRH in Aditya-U Plasma

## 4 ECR heating in Deuterium Plasma in Aditya-U

In Aditya-U, deuterium plasma experiments are carried and during this plasma campaign, ECR heating is also tried in deuterium plasma. The ECH power is launched at the flat top of plasma current. The ECRH power in fundamental O-mode is launched from the radial port (low field side of tokamak). The 200kW ECRH power for 50 ms duration is launched in deuterium plasma from 160ms to 210ms. The total plasma discharge length in this shot is around 245ms and plasma current close to 150kA (Figure 4). The effect of plasma heating with ECRH power is also seen with the increasing trend of soft X-ray with ECRH power, which indicates the EC wave coupling with deuterium plasma. The exact ECR heating cannot be explained as other diagnostics like ECE were not available during this shot, but decrease in soft X-ray after ECRH pulse indicates the ECR heating effect of deuterium plasma.

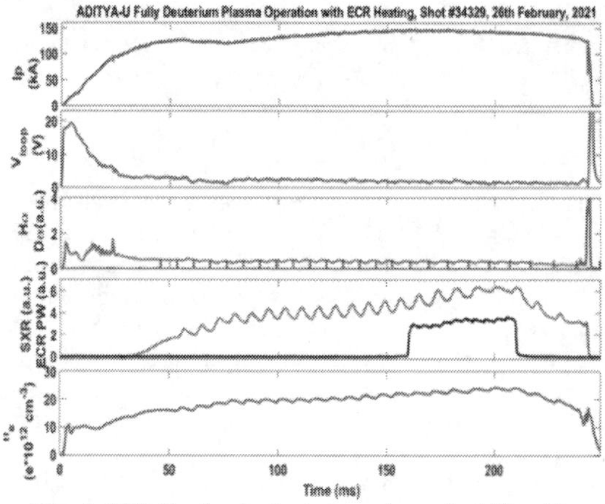

**Fig. 4.** ECR Heating in deuterium plasma in Aditya-U

## 5 Upgradation plan of ECRH system on SST-1

In the SST-1, 42GHz-500kW ECRH system is used to carry out breakdown and heating at fundamental harmonic (1.5T operating toroidal magnetic field). In 2010, another 82.6GHz-200kW ECRH system commissioned on SST-1 to carry out breakdown and heating at 3.0T operation (fundamental harmonic) and 1.5T operation (second harmonic). This system can deliver maximum 200kW ECRH power and considering 20% transmission loss, only 160kW ECRH will be available for ECRH experiments and this power is not sufficient for plasma start-up at second harmonic (1.5T) and ECR heating also demands higher power. This has been experimentally observed in SST-1 with 42GHz ECRH system [7]. At fundamental harmonic (1.5T operation) 150kW ECRH power is sufficient for plasma breakdown while for second harmonic breakdown (0.75T operation), more than 250kW ECRH power was required for reliable breakdown and plasma start-up.

In order to carry out plasma experiments at higher ECRH power, this old 82.6GHz is being upgraded in terms of power. The 82.6GHz ECRH system is planned to achieve double power (400kW) for 500 ms duration. This 82.6GHz ECRH system is already connected with the SST-1 tokamak, transmission line is laid and launcher mirrors are installed inside the tokamak. The Gyrotron will be repaired and upgraded in terms of power. With this upgradation, the entire ECRH system will be upgraded for SST-1 and total ECRH power (42GHz and 82.6GHz system) in SST-1 will be increased to 900kW which will facilitate to carry out several experiments related to plasma start-up and heating at both the harmonics.

## 6 Conclusion:

The 42GHz ECRH system for SST-1 and Aditya-U has been upgraded for multi ECRH pulse operation. The experiments related to ECRH assisted breakdown and ECR heating experiments can be carried out simultaneously in both the tokamaks. Depending on the experimental demand, the ECRH power and duration can be varied in both the ECRH pulses.

Initial experiments with two ECRH pulses have been carried on both the tokamaks SST-1 and Aditya-U both. The breakdown is consistent in both the tokamaks while ECR heating needs more experiments for its confirmation.

The 82.6GHz ECRH system which would be useful to carry out second harmonic ECR heating experiments on SST-1 at 1.5T magnetic field will be upgraded to double the power. The Gyrotron is under process of upgradation while transmission line is already connected with Gyrotron. This system will facilitate to carry out several ECRH experiments in SST-1 at higher EC power.

## References

1. Braj Kishore Shukla et.al. "EC-wave experiments on SST-1", , 31st ITPA meeting of TG SOL and Divertor Physics (18.01.–26.01.2022), ITER Online Presentation
2. R.L. Tanna, Tanmay Macwan, J. Ghosh, ….., B.K. Shukla, , …Y.C. Saxena, A. Sen, R. Pal and S. Chaturvedi, "Overview of Recent Experimental Results from the ADITYA-U Tokamak", 28th IAEA Fusion Energy Conference (FEC-2020), 10th – 15th May, 2021
3. "Fundamental O-mode ECRH assisted Low-loop voltage plasma start-up in tokamak ADITYA-U", B.K. Shukla, et.al., 46thEPS Conference on Plasma Physics, Milan (Italy), 8-12 July, 2019
4. BK.Shukla, et.al. "Gyrotron control system for plasma start-up and heating experiments in Aditya-U tokamak" Jatin kumar Patel, ICPSA-2019, Univ. of Lucknow, Nov 11-14, 2019
5. Braj K. Shukla et. al. "42GHz ECRH system on Aditya Upgrade" 30th Symposium on Fusion Technology, 2018 at Sicily-Italy September 16-21, 2018
6. Braj K shukla, et.al. Recent results on ECR assisted plasma startup, current drive and discharge wall conditioning in tokamak SST1: 29th symposium on Fusion Technology SOFT - 2016 Prague
7. B K Shukla et. al., "42GHz ECRH assisted Plasma Breakdown in tokamak SST-1", European Physical Journal (EPJ) Web of Conferences, Vol. 87, March 2015

https://doi.org/10.1051/epjconf/202327702006

# Operational domain for the new 3MW/1000s ECRH System on WEST

T. Fonghetti, R. Dumont, G. Giruzzi, J.-F. Artaud, J.-M. Bernard, F. Bouquey, C. Bourdelle, L. Delpech, J. Hillairet, P. Maget, P. Manas, P. Mollard, J. Morales, V. Ostuni, B. Robinet, and the WEST team

CEA, IRFM, F-13108 Saint-Paul-lez-Durance, France.

**Abstract.** The ECRH system formerly used in Tore Supra is being upgraded to start on WEST in 2023, at a power level of 1MW and frequency of 105 GHz. Its ultimate 3MW/1000s capability is expected to enlarge the WEST operational domain by increasing margins with respect to H-mode access, and by providing additional flexibility in terms of achievable scenarios using impurity and/or MHD control. This flexibility is made possible using an antenna based on three steerable mirrors for controlled power injection. In order to determine an appropriate range of EC wave injection angles for WEST scenarios, the fast and reliable ray-tracing code REMA has been interfaced with the WEST IMAS database. This allows the EC power damping rate to be quickly assessed, as well as deposition profiles to be predicted in realistic plasma conditions. Based on a typical WEST discharge at central magnetic field $B_0 \sim 3.6$ T, central line-averaged electron density $n_l \sim 4 \times 10^{19}$ m$^{-3}$ and central electron temperature $T_{e0} \sim 3$keV, ray-tracing calculations have been performed. Comprehensive poloidal and toroidal angle scans, as well as variations of $B_t$, $n_l$ and $T_{e0}$ with respect to the reference parameters have allowed an adequate range of injection angles to be determined for efficient use of ECRH and/or ECCD in typical WEST scenarios, and compared with the mechanical limits set by the antenna mechanical characteristics. In order to further characterize the effect of this new power source in WEST scenarios, EC wave deposition and current profiles from ray-tracing calculations have been included in integrated simulation codes. It has been shown that this additional power source could allow central electron heating to be achieved, potentially alleviating the issue of radiative collapse caused by impurities observed in some situations.

## 1 Introduction

Radiofrequency heating systems are widely used in magnetically confined fusion devices. Three of them have been used simultaneously in Tore Supra [1]: Lower-Hybrid (LH), Ion Cyclotron Resonance Heating (ICRH) and Electron Cyclotron Resonance Heating (ECRH). In the upgrade to the WEST configuration [2], the 118 GHz ECRH system has been discontinued because its characteristics were not consistent with the new design of the machine. As a result, the former gyrotrons will be replaced by 105GHz/1MW/1000s gyrotrons to suit the new WEST geometry and power requirements. This modification also includes an upgrade of most of the EC components as a result.

The new ECRH project has been started in 2021. The EC power is expected to be available in WEST starting in 2023 with a limited power (1MW) to reach its full 3MW/1000s capability in a staged approach [3]. From a plasma scenario standpoint, this additional power source is predicted to improve access to H-Mode. In addition, it is expected that the flexibility and localization of the EC wave absorption in the plasma will be useful in the context of impurity and MHD control.

Various numerical code developments have been carried out in order to provide support to the use of ECRH power on WEST and to prepare its implementation. Firstly, the ray-tracing REMA code [4] has been interfaced with the IMAS WEST database, in order to allow simulations using plasma parameters from existing shots to be readily performed. The operational domain in terms of injection angles for the EC antenna has been established as a result. Following this development, further simulations have been made possible. The potential role of the EC power to prevent radiative collapse due to core tungsten impurity accumulation, investigated, e.g., in ASDEX Upgrade, where ECRH has been demonstrated to be an efficient method to flatten tungsten density profiles [5], has been analysed in the context of integrated modelling. Experimental data from shot 55025, featuring a radiative collapse, has been used in simulations using the RAPTOR code [6]. By comparing the LH and EC power depositions, the expected advantages provided by the new ECRH system have been assessed.

## 2 Improvements of the REMA ray-tracing code

The ray-tracing code REMA [4] allows electron cyclotron waves propagation and absorption in a plasma to be simulated in an accurate and numerically efficient fashion. It has been extensively employed, among other studies, to simulate Tore Supra experiments [7]. In order to facilitate its use as a tool for the design and analysis of future experiments in WEST, some further developments have been recently carried out, and are summarized in Figure 1.

The REMA code requires the magnetic equilibrium, plasma parameters and the EC antenna configuration as inputs. Based on general wave and ray-tracing theory, the code provides the toroidal and poloidal projections of the ray trajectories, as well as the power and driven current density profiles and the total fraction of absorbed power. To facilitate its use, REMA has been directly interfaced with the WEST IMAS database. This has allowed integrated modelling simulations of existing pulses including an additional source of EC power to be performed. It is worth mentioning that the results of these simulations are also written in the IMAS format in preparation of potential further simulations.

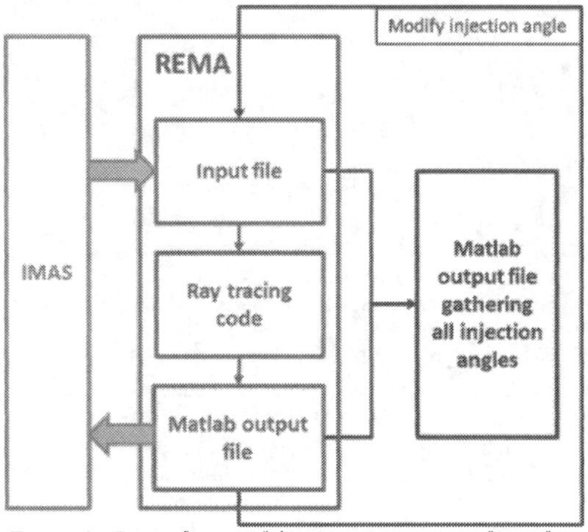


*Figure 1: General view of the improvements performed on the REMA code.*

The main objective of these developments is the determination of the operational domain for the EC antenna in terms of wave toroidal and poloidal injection angles. For this aim, a dedicated routine has been developed to sweep all mechanically reachable injection angles, and to build a synthetic view of the obtained results. As an illustration, Figure 2 displays the location of the EC absorption (in normalized radius) as a colour map over the whole toroidal/poloidal injection angle grid in the case of WEST shot 55799 [8]. Also shown are contours indicating the total absorbed power fraction. As an example, the outermost contour shows the 70% absorbed fraction in the first pass through the plasma.

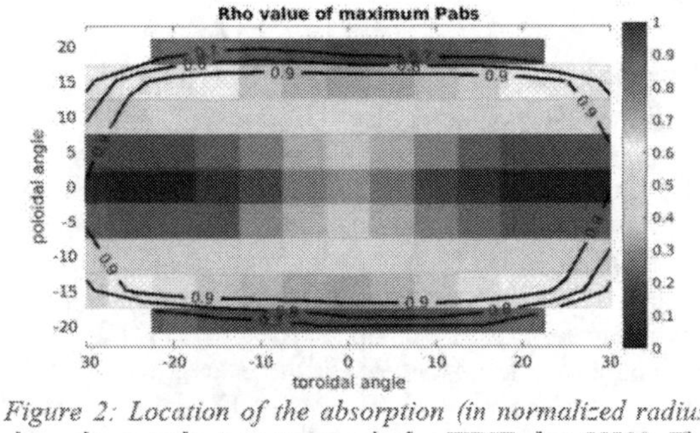


*Figure 2: Location of the absorption (in normalized radius depending on the injection angle for WEST shot 55799. The superimposed labeled black contours show the fraction of total absorbed power.*

## 3 Determination of the limits for the new ECRH system

As a preparation for the new ECRH system implementation on WEST, it is necessary to define its operational domain. It characterizes the injection angle limits, beyond which the waves are insufficiently absorbed by the plasma independently of its parameters. This is necessary for device safety, as poor absorption can result in levels of radiofrequency power in the vacuum vessel, potentially dangerous for the various systems installed in WEST.

### 3.1 Plasma profiles dependence for the wave absorption and propagation

In order to define those limits, the set of plasma conditions for which the operational domain is the widest needs to be identified. Any changes of the plasma parameter with respect to these conditions will be included in the maximum limits. This is why the propagation and absorption behaviour as a function of the plasma parameters needs to be precisely characterized.

As is well known, the electron density has a direct influence on the wave propagation properties. Indeed, larger densities result in larger wave refraction in the plasma, until a cut-off is attained and the wave is completely reflected. In WEST, because the magnetic field is high, the cut-off is expected at a local electron density of $n_{e,limit}$~$14\times10^{19}$m$^{-3}$. Considering a typical line-averaged electron density $n_l$~$6\times10^{19}$ m$^{-3}$, cut-off is rarely reached as $n_{e,max}$ is usually lower than $n_{e,limit}$. In order to establish the widest operational domain, appropriate injection angles should be found, that compensate for wave refraction in the case of high-density plasmas.

In Figure 3, it is observed that similar absorption locations (in the vicinity of the resonance layer, shown with a black line) are obtained for different injection angles when the line-averaged density is varied in the range expected in WEST scenarios, i.e. from $n_l$~$2\times10^{19}$m$^{-3}$ to $9\times10^{19}$m$^{-3}$.

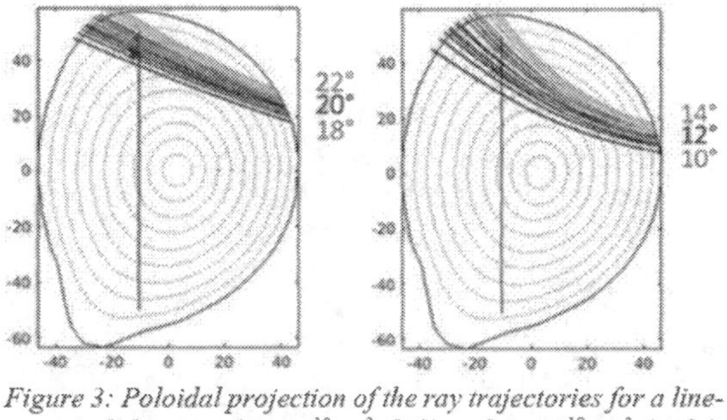


*Figure 3: Poloidal projection of the ray trajectories for a line-averaged density of $2\times10^{19}$ m$^{-3}$ (left) and $9\times10^{19}$ m$^{-3}$ (right). The vertical black line shows the EC resonance layer. The corresponding poloidal injection angles are indicated in the figure.*

The modification of the magnetic field has a negligible influence on the wave propagation properties but has a strong influence on the resonance location. Indeed, if the 105 GHz frequency has been chosen in order to get central absorption for the nominal WEST magnetic field and oblique propagation (for current drive applications), decreasing the magnetic field moves the resonance towards the high field side.

Modifying the electron temperature has little influence on the wave propagation. Larger temperatures are known to increase the absorbed power fraction. On the other hand, close to the plasma edge, waves are generally poorly absorbed by the plasma. Increasing the temperature therefore allows absorption locations closer to the plasma edge to be obtained with a sufficient absorbed fraction, thereby enlarging the operational domain.

As a conclusion, the optimal conditions for which the injection limits are maximised are low density, high temperature and nominal magnetic field. This corresponds in WEST to a line-averaged electron density of around $2\times10^{19}m^{-3}$, an electron temperature around 6keV and a central magnetic field around 3.7T.

### 3.2 Operational limits determination

With the global plasma conditions now being defined, it is useful to characterize the limits within which EC waves can be absorbed from the core to the edge, for all three steerable mirrors. For the middle mirror, it is possible to restrict the domain to the upper or lower part of the plasma, as they are approximately symmetric, except for the divertor area (see Figure 4).

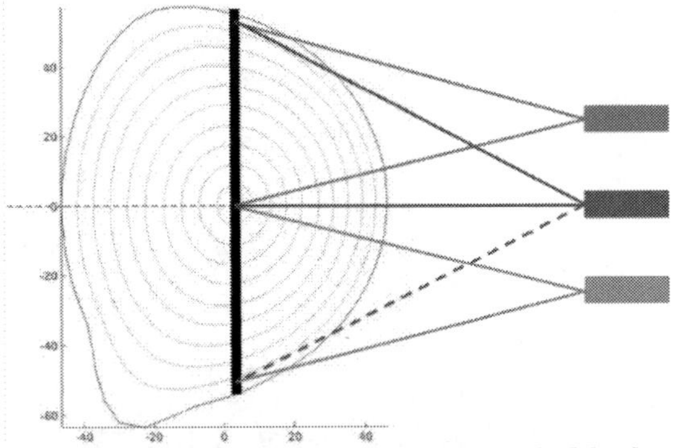

*Figure 4: Operational limits determination method for the three steerable mirrors. Only the upper part of the plasma is considered for the middle mirror.*

From the REMA simulations consisting of detailed injection angles scans described previously (see Figure 2), a global view of the system capabilities is obtained. Figure 5 shows results for the lower mirror, inside the limits defined by the antenna mirror movements.

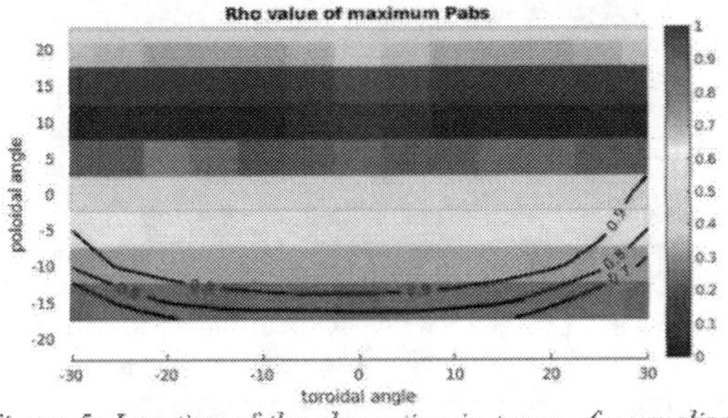


*Figure 5: Location of the absorption in terms of normalized radius (colormap) depending on the injection angles for the lower mirror in optimal conditions. Contours define a few values of the absorbed fraction (0.7, 0.8, 0.9).*

Repeating the process for the upper and middle mirrors, and refining the limits by running additional ray-tracing simulations, the operational domain, for every mirror has been obtained, as shown in Figure 6.

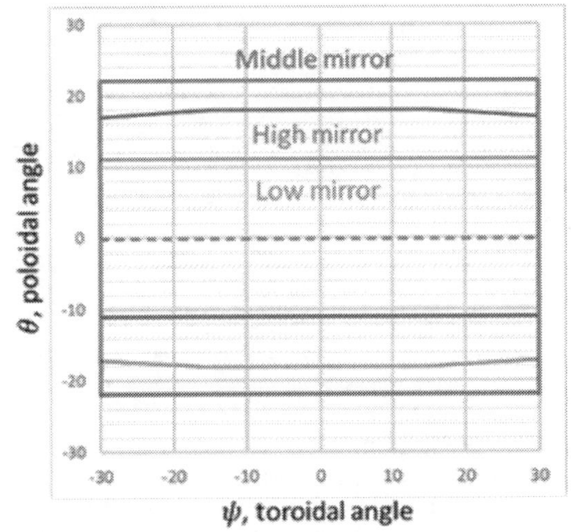


*Figure 6: Operational domain for the three steerable mirrors.*

### 3.3 Restricted absorption domain for different plasma conditions

Whereas the previous global studies establish that outside of the determined limit, waves are poorly absorbed, more detailed analyses are needed to better characterize the acceptable injection angles depending on the plasma conditions. A study was started to have a general view over the characterization of satisfying injection angles with respect to absorption properties for the first harmonic. For this purpose, three different plasmas are considered: low-temperature (~1keV at the centre), moderate-temperature with additional heating (~3keV) and high-temperature (~6keV). For these temperatures, the objective is to estimate the magnetic field domain compatible with the ECRH system, i.e. with an acceptable domain of injection angles and a sufficiently wide range of possible absorption locations. Then, simulations have been run at the magnetic field limit for a low-density ($n_l \sim 2\times10^{19}$ $m^{-3}$) and a high-density plasma ($n_l \sim 9\times10^{19}$ $m^{-3}$). The final domain with respect to wave absorption for the three different temperatures has been determined by combining the domains of low and high-density results. Figure 7 allows us to determine some general tendencies.

First observable result that is highly expectable is that higher is the temperature and wider is the domain with acceptable absorption. That way, off-axis injection is restricted as the local temperature is lower than the core one. To maximise the absorption, it seems that a centred poloidal and toroidal injection is preferable.

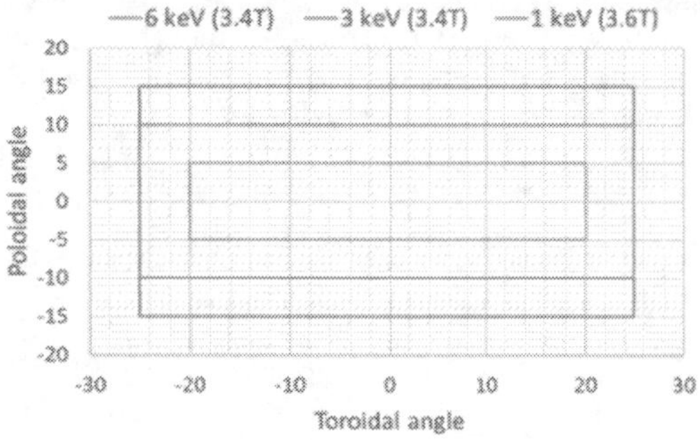


*Figure 7: Restricted domain for the middle mirror, depending on electron temperature and central magnetic field.*

## 4 Integrated modelling of WEST scenarios with EC power

The objective of the next study is to evaluate the potential positive impact of EC power in WEST scenarios, and to evaluate its influence compared to the other, already operating, radiofrequency heating systems (LH, in the present case).

### 4.1 Radiative collapse in WEST

As comprehensively presented in Ref. [9], WEST plasmas with LH power occasionally exhibit temperature collapses due to the power radiated by tungsten impurities in the plasma core exceeding the electron heat source. Shot 55025 displays an example of such a collapse (Figure 8). Around 9.5s, a fast decrease of the electron temperature limited to the central region (ρ<0.4) is observed. This coincides with a sudden increase of the tungsten density estimated from bolometry measurements, and a simultaneous decrease of the HXR (Hard X-ray) signal indicating a modification of the LH power deposition [9].

### 4.2 Impact of the ECRH system

The first study concerns the power deposition profiles. On one hand, the LH deposition profile has been simulated with a reduced model included in the METIS code [10]. The plasma kinetic profiles used have been fitted from experimental measurements. On the other hand, REMA has been run with adapted injection angles in order to obtain a centrally localized absorption in the conditions of WEST shot 55025. Figure 9 illustrates the differences between the computed LH and EC power deposition profiles.

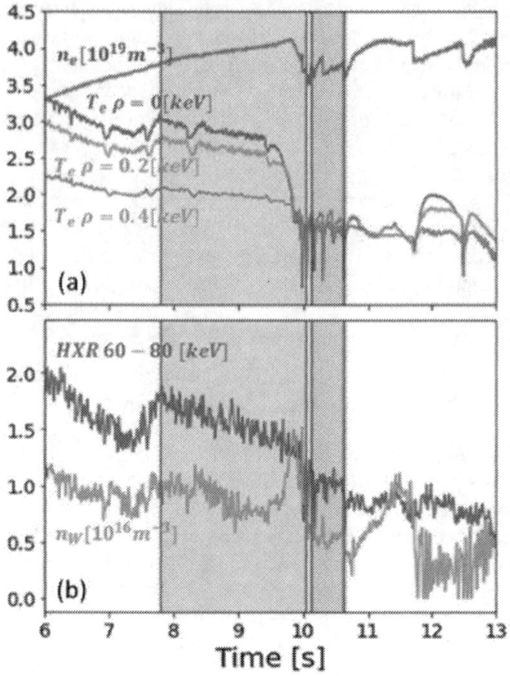


*Figure 8: WEST shot 55025. (a) Electron temperature at different radial locations, line-averaged electron density; (b) HXR emission in the energy band 60-80 keV and estimated central tungsten density (adapted from Ref. [9]).*

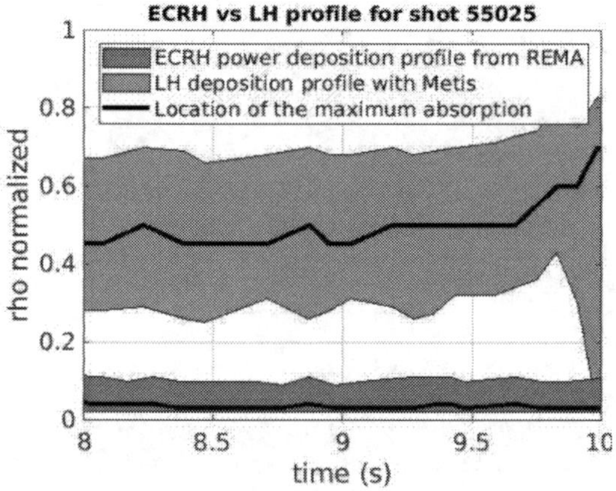


*Figure 9: Evolution of LH and EC power deposition profiles in shot 55025 (colored areas: full width at half maximum of the deposition; black line: location of deposition maximum)*

Firstly, typical LH deposition profiles, in the multi-pass regime typical of WEST, are radially wider radially than ECRH profiles. In addition, whereas ECRH allows power to be deposited in the core, the maximum of LH deposition is more off-axis. The second observation that makes ECRH interesting is the invariance of the deposition profiles with respect to the electron temperature decrease. When the collapse, characterized by a fast temperature decrease at t~9.5, occurs, the LH deposition moves more and more off-axis and the maximum location reaches $\rho\sim0.7$ at t~10s, thereby reducing the heat source on electrons. In comparison, the EC deposition profile is not affected. Its width and maximum remain constant, which allows efficient electron heating in the core to be maintained even in the presence of a radiated power increase caused by impurities.

Integrated simulations using the RAPTOR transport code [6] have been performed in order to further assess the benefits of an EC power source in WEST scenarios. This code allows the evolution of the electron temperature depending on turbulent transport and heat source to be simulated. For the present simulations, the Bohm-Gyrobohm [11] transport model has been used. In order to compare the electron temperature evolution with LH and EC power, shot 55025 has been considered, at time t=8s, i.e. during the flat-top phase of the discharge. If most of the data was taken from experiment and diagnostics, METIS interpretative shot allowed to fill the missing data. First simulations have been run using the METIS output IMAS datafile and varying the W concentration from 0 to $5\times10^{-4}$. Secondly, LH power (2.8MW) has been replaced by a central source of EC power, varied between 0 and 3MW and W concentration has also been varied. Resulting central electron temperature has been determined and is therefore displayed as a colour map in Figure 10.

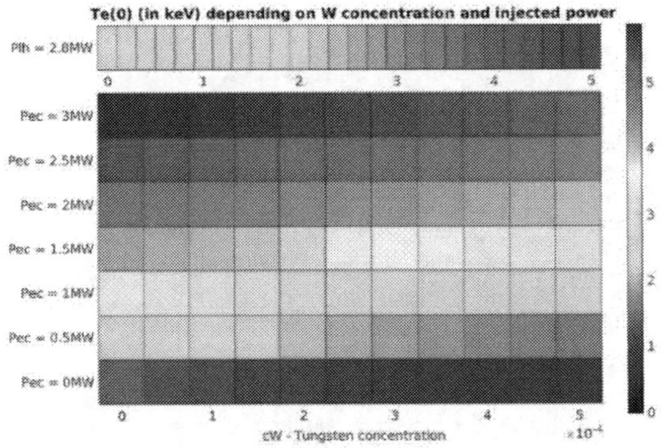


*Figure 10: Central electron temperature as a colour map depending on the injected power and the tungsten concentration.*

Firstly, for approximately equivalent injected power levels (2.8 MW for LH and 3MW for ECRH), it is observed that the electron temperature in the core gets twice higher with ECRH for low tungsten concentrations. At higher tungsten concentrations, it becomes around five times larger. The second quantitative observation is that the same electron temperature is obtained for EC power lever four times lower than the LH power level when the tungsten concentration is low. The difference is even larger for higher tungsten concentration. In terms of core electron heating, ECRH is therefore more efficient due to its more central deposition profile. This observation is not surprising, as the main role of LH waves in WEST is efficient current drive rather than central electron heating. ECRH, in principle, should prevent a radiative collapse at the levels of tungsten concentrations observed in some situations from occurring. These first studies therefore demonstrate the major benefit that can be anticipated from the future EC system in WEST.

## 5 Prospects

Additional integrated modelling simulations are now ongoing using RAPTOR, to better characterize the impact of the central heating and current drive from EC waves. The next objectives are to run refined simulations using the neural network-based transport model associated to the quasilinear gyrokinetic code QuaLiKiz (QLKNN) for turbulent transport [12] and LUKE to have a first principle based modelling of the LH deposition profiles [13]. Then, time-evolving integrated modelling of discharges exhibiting increasing of the core tungsten density in the presence of EC power will be performed. These simulations will be applied to a wide range of shots from the WEST database. Finally, in the future, the capabilities of the new ECRH/ECCD system to control MHD effects during the ramp-up and plateau phases in WEST plasmas will be studied.

## Acknowledgements

This work is the result of a collaboration with many workers at IRFM, most notably the ECRH team, and I would like to thank them all for their help and support.

## References

1. R.J. Dumont et al., Plasma Phys. Control. Fusion **56,** 075020 (2014).
2. J. Bucalossi et al., Nucl. Fusion **62**, 042007 (2022).
3. L. Delpech et al., Fusion Engineering and Design **186**, 113360 (2023).
4. V. Krivenski et al., Nucl. Fusion **25**, 127 (1985).
5. C. Angioni et al., Nucl. Fusion **57** 056015 (2017).
6. F. Felici and O. Sauter, Plasma Phys. Control. Fusion **54** 025002 (2012).
7. M. Goniche et al., Proc. 47th EPS Conference on Plasma Physics (2021).
8. M. Lennholm et al., Nucl. Fusion **43** 1458 (2003).
9. V. Ostuni et al., Nucl. Fusion **62** 106034 (2022).
10. J.F. Artaud et al., Nucl. Fusion **58**, 105001 (2018).
11. M. Erba, et al., Plasma Phys. Control. Fusion **39**, 261 (1997).
12. K.L. van de Plassche et al., Phys. Plasmas **27**, 022310 (2020).
13. Y. Peysson and J. Decker, Fusion Science and Technology **65** 22 (2014).

# Quantification of X3 absorption for ITER L-mode parameters in ASDEX Upgrade

*Jörg* Stober[1,*], *Martin* Schubert[1], *Mireille* Schneider[2], *Severin S.* Denk[3], *Rainer* Fischer[1], *Emanuele* Poli[1], *Dirk* Stieglitz[1], *Matthias* Willensdorfer[1], *Alberto* Loarte[2], *Daria* Ricci[4], ASDEX Upgrade Team[5], and Eurofusion MST-1 Team[6]

[1]Max-Planck-Institut für Plasmaphysik, Garching, Germany
[2]ITER Organization, St. Paul lez Durance, France
[3]Plasma Science and Fusion Center, Massachusetts Institute of Technology, Cambridge, USA
[4]Istituto per la Scienza e tecnologia dei Plasmi, Consiglio Nazionale delle Ricerche, Milano, Italy
[5]author list of *H. Meyer et al.* 2019 Nucl. Fusion **59** 112014
[6]author list of *B. Labit et al.* 2019 Nucl. Fusion **59** 086020

**Abstract.** For an early H-mode access in hydrogen, ITER considers operating at 1/3 of the full field using 170 GHz X-Mode for heating at the $3^{rd}$ harmonic. The optical thickness for such a heating scheme depends on $T_e^2$. It is rather low in the ohmic phase (with $T_e$ about 1-2 keV), but reaches high single pass absorption for the strongly EC heated plasma with $T_e$ exceeding 10 keV. Launching ECRH into an ohmic plasma may trigger a boot-strap process on $T_e$ if the additional power absorption due to increasing $T_e$ exceeds the additional power losses due to increased transport (which often tends to increase with input power). In this contribution we present measurements of the X3 absorption for the parameter range relevant for ITER, i.e. $n_e \approx 2\cdot 10^{19}\,m^{-3}$, $T_e \geq 2\,keV$ in order to back up theoretical estimates used for the modeling so far. In ASDEX Upgrade (AUG) such low densities cannot be reached in H-mode such that dominant heating with NBI is not an option. For moderate $T_e$, it is also not an option to use X3 heating as main heating, due to the excessive stray radiation threatening in-vessel components. This dilemma is solved with the 2-frequency EC system of AUG. The main central heating is done with the lower frequency of 105 GHz at the $2^{nd}$ harmonic and full single pass absorption. Up to 3.5 MW of ECRH are used at that frequency to vary $T_e$. Two other gyrotrons are used at 140 GHz to probe the X3 interaction close to the plasma center with a sequence of short blips. The expected values of single pass absorption are calculated with TORBEAM and vary from 7% to 70%. Below 40% single pass absorption the non-absorbed power triggers an arc in the tile gaps of the inner heat shield which screens the thermo-couples from the incoming beam such that they cannot be used. Between 40% and 80% single pass absorption, the predictions and measurements agree within the uncertainty of the measurement, unless we have clear evidence for non-linear interactions, which are not described by TORBEAM and which are not expected in ITER, but are due to some specific experimental choices for an isolated subset of our results.

## 1 X3-heating in ITER

The motivation for this work is given by the interest of ITER to operate early in its operational phase at 1/3 of the nominal magnetic field $B$ in order to verify predictions on the L/H power threshold and to allow an early start of ELM studies and ELM control [2]. During the first phase of plasma operation called PFPO-1, only ECRH will be available for plasma heating. In order to avoid time consuming and costly changes to the EC system, operation at the third harmonic of the cyclotron frequency $\omega_{ce}$ is desirable ($B = 1.8\,T$). At full field ($B = 5.3\,T$), EC operates at the fundamental $\omega_{ce}$. Since $\omega_{ce}$ drops linearly with $B$ both scenarios can be run with the same EC frequency ($170\,GHz$). Only the polarization of the beam has to be adapted; X-mode at the third harmonic (X3) vs. O-mode for the fundamental (O1), but this is already available since X-mode heating at half field is also foreseen (X2).

The optical depth for X3 absorption is not necessarily much larger than unity [1]. Thus, incomplete single pass absorption and related machine damage have to be taken into account. X3 absorption increases strongly with $T_e$, giving potentially rise to a bootstrap process such that the initially only partially absorbed EC-heating generates a fully absorbing target mainly by rising $T_e$.

First modeling results for such scenarios for ITER are published in [3]. Figure 1 reproduces two figures of this publication. The top figure shows the rise of the central $T_e$ after switching on $10-30\,MW$ of EC power at $275\,s$. The bottom part shows the corresponding decrease of the power not absorbed during the first path. Initially only 25% of the power is absorbed, increasing to close to 100% after a fraction of a second for $\geq 20\,MW$. The energy not absorbed during the first $300\,ms$ is about $150\,kJ$ per beam ($1\,MW$, 50% absorption, $0.3\,s$). In ASDEX Upgrade we have good experience with a limit of $80\,kJ$ non-absorbed power in short pulses (except for hitting windows

*e-mail: joerg.stober@ipp.mpg.de

EPJ Web of Conferences **277**, 02007 (2023) https://doi.org/10.1051/epjconf/202327702007
*EC21*

directly), but the power density in the beam footprint in ITER is probably more than a factor of 2 lower, such that $150\,kJ$ is a reasonable number. Still this exercise shows that the bootstrap process must not last significantly longer and it should reduce the non-absorbed power to only a few percent. The $10\,MW$ case in fig 1 would not be acceptable for machine safety.

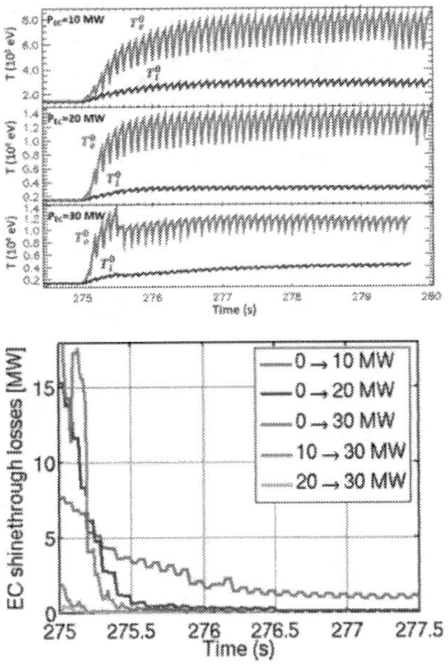


**Figure 1.** Modeling results for X3 heating in ITER from [3], figs. 6, 9. Top: evolution of central $T_e$, $T_i$ after switching on 10 – 30 $MW$ of ECRH. Bottom: evolution of the power not absorbed during the first path through the plasma.

Being close to the margin for machine safety raises concern on the accuracy of the model, which essentially has to balance the power densities absorbed from the microwave-beams (+ ohmic heating) for given $n_e$, $T_e$-profiles with transport effects describing how the kinetic profiles react on changes of the power profile. ITER has identified (1) the transport model and (2) the (linear)[1] absorption model as crucial elements and put this forward to the ITPA. For (1) the Transport and Confinement group was contacted. Results based significantly on EC heated discharges in AUG are published in [4]. For (2) the Integrated Operational Scenarios group was contacted and the work reported in this paper is a direct consequence.

## 2 X3 heating in ASDEX Upgrade

In AUG, X3 heating is used since more than a decade [5]. In contrast to the plans in ITER, the radial position of the resonance is chosen slightly on the low field side (LFS). This allows to have the X2 resonance still in the plasma on the high field side (HFS). Especially in H-modes this allows full absorption of the power not absorbed by the central X3 resonance. In fact, using EC in an ohmic plasma under these conditions, leads to an L/H-transition due to the power initially absorbed at the edge ([6], and references therein for more details on X3 usage). In cases when the X2 beam-dump is insufficient, holographic reflectors on the HFS may be used also for X3 (they were originally designed for O2 [5]). For 140 GHz, they maintain polarization and reflect and refocus the beam close to the plasma center. Since these reflectors are equipped with fast thermo-couples, the latter can be used to detect transmitted power as described during this workshop by *M. Schubert* [7]. Of course one has to lower $B$ such that the X2 resonance is no longer absorbing power from the beam after it passed through the X3 resonance.

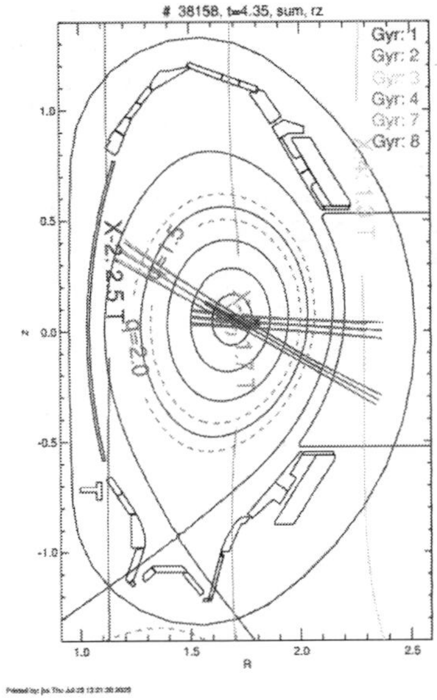


**Figure 2.** Initial setup for the EC system (poloidal projection): Beams 1-4 (horizontal) were tuned to 105 GHz, while beams 7,8, tuned to 140 GHz, target the thermo-couples on the upper HFS reflector tiles. $B_t = -1.7\,T$. The solidly colored areas along the beam path indicate where the majority of the power is deposited as calculated by TORBEAM. The cold resonances shown as blue and green lines correspond to 140 GHz. Note that in the later discharges (#39747-39751), beam 7 was replaced by beam 6.

To analyze scenarios with low single pass absorption, the X3 resonance itself is not suited for heating. This would be rather inefficient and dangerous for the machine. On the other hand, significant neutral beam injection drives the plasma into H-mode, such that the targeted ITER parameter range $10^{19}\,m^{-3} \leq n_e \leq 2 \cdot 10^{19}\,m^{-3}$ and $1.5\,keV \leq T_e \leq 6\,keV$ cannot be reached. A way out is to use a lower frequency EC beam to heat the plasma at the respective X2 resonance. Ideally this frequency would be 2/3 of the X3 probing frequency, Then the cold resonances would overlay. On AUG the gyrotrons are capable of 2 frequencies with the ratio 3/4 i.e. $105\,GHz$ / $140\,GHz$ (for details on the AUG EC system see [6]). In order to heat the plasma rather centrally with $105\,GHz$ X2, and the $140\,GHz$ X2 resonance being in the scrape-off layer on the

[1] See discussion in section 2.3.

HFS, one has to make use of the Doppler-term in the resonance condition and inject the 105 $GHz$ beams with significant toroidal angle ($\approx 20°$), i.e. with a large $N_{\parallel}$. These conditions can be realized for $B$ between 1.6 T and 1.7 T. Figure 2 shows the situation for 1.7 T.

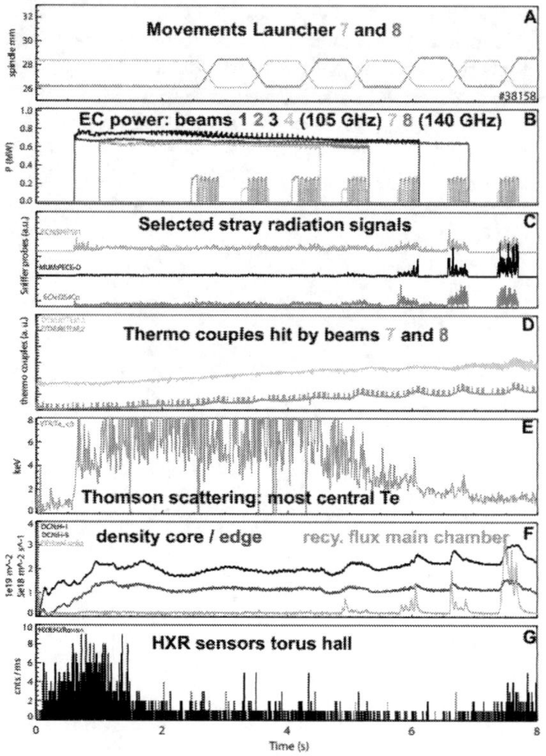


**Figure 3.** AUG #38158, $I_p$: 800 kA, $B_t$: -1.7 T. See text for detailed discussion.

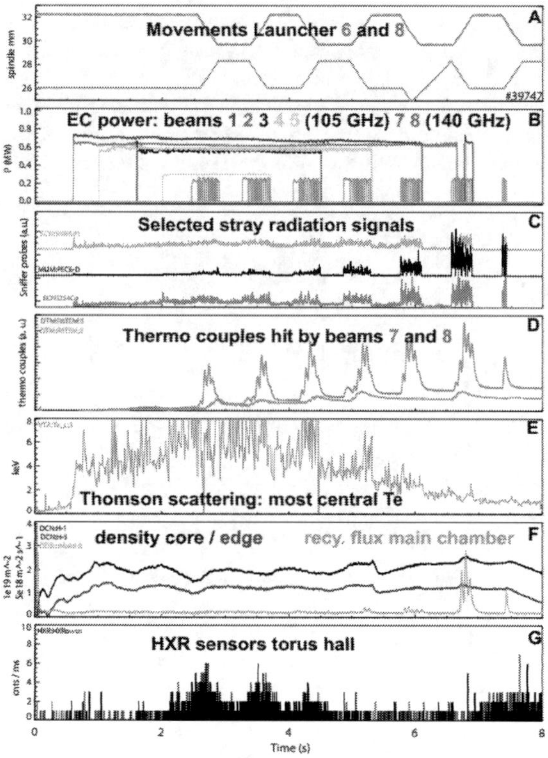


**Figure 4.** AUG #39747, $I_p$: 800 kA, $B_t$: -1.6 T. To ensure plasma initiation and to avoid locked-modes in the ramp-up phase the plasma was started with $B_t$: -1.7 T. The field was changed between 1.4-2.6 s. Otherwise identical control parameters as figure 3, except that beam 6 had to be used with 140 GHz since beam 7 was not available. The 105 GHz heating uses up to 5 beams (4 in fig. 3). As can be seen from box A, there was a mistake in programming the fifth sweep of launcher 8 in this specific discharge. Consequently, there is no rise of the TC on the respective reflector in box D.

## 2.1 X3 absorption experiments

The experiments were conducted as sketched above, resulting in 6 successful discharges of which two are shown in figures 3,4 performed on two days in July 2020 (#38158) and July 2021 (#39747-51). The experiments were conducted in hydrogen plasmas to avoid L/H transitions. On the days before the experiments the 140 GHz blips were injected into the empty vessel to fine-tune and fix the toroidal launching angles and to get the no-absorption level from the thermo-couples, using the same timing of launcher movements and 140 GHz blips as in the plasma experiments in the morning of the next day (figure3, boxes A,B). Five blips of 30 ms are applied while the beam is swept poloidally over the thermo couple. This allows to cope with slight variations of the refraction (essentially due to variations of the density peaking), in contrast to a static set-up. The toroidal variation of refraction is negligible according to TORBEAM [8]. For details of the analysis of the thermo-couple data see [7]. For higher expected absorption a longer pre-blip was applied in order to reach steady conditions during the phase of mirror movement. Also shown in box B are the traces for the 105 GHz beams. For the first 3 sweeps the power was constant (4 beams), then it decreased successively (3, 2, 1, 0 beams). Boxes C,D show the reaction of some stray-radiation-diodes and the thermo-couple (TC) data. The stray radiation increases significantly towards the end of the discharge, whereas the correlated rise of the TC-data seems to saturate. Box E shows that the variation of the 105 GHz power led to the expected variation of $T_e$, while $n_e$ (Box F) was kept at the envisaged low values. The $T_e$ data are determined from Thomson-Scattering (TS), since the stray radiation may damage the ECE. After determining the stray radiation levels, the sweeps with higher absorption were repeated with ECE, while those with lower absorption were repeated without the 140 GHz blips, but with ECE. Except for the cases with non-thermal electrons as discussed in section 2.2, ECE and TS agreed sufficiently well. Box G shows hard X-ray (HXR) data related to such relativistic electrons, and in figure 3 those are only found as a very small run-away population in the first sec. of the discharge. Also shown in box F is a trace of the recycling flux in the main chamber measured with an ionization gauge. It is strikingly in phase with the latter sweeps and the amplitude of the sniffer signals. The increase of the recycling flux must be related to a particle source driven by a sufficiently high level of stray radiation. This is also obvious from the reaction of the line-integrated density signals on these fluxes. As indicated in figure 2 the non-absorbed power crosses the X2 resonance in the scrape-off layer on the HFS. One may speculate that due to the relatively short connection length to the wall an arc can be fed by the wave at that location, which would probably lead to increased hydrogen desorption and at the same time shield the TCs from a part of the non-absorbed power. The effect would likely increase with the non-absorbed power and would also need a minimum power level to ignite the arc.

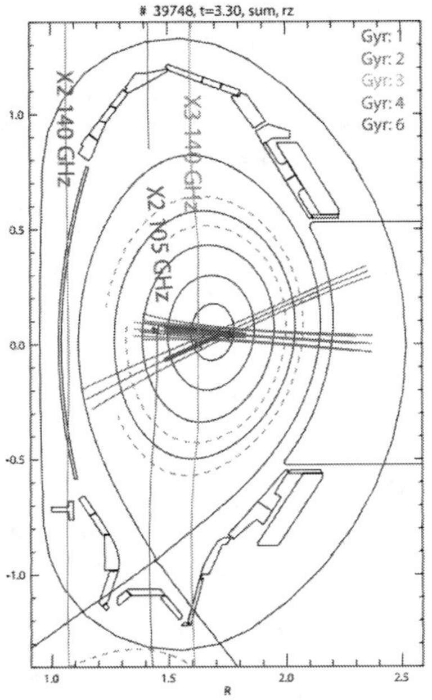


**Figure 5.** Beam propagation, cold resonances and regions of EC power deposition as calculated by TORBEAM for #39748 at 3.3 s (i.e. beam 8 off), which is the pre-heating phase of the second sweep. The phase is similar as in #39747 except that beam 8 is off and ECE data are available. The $T_e$ profiles used for the TORBEAM calculation are based on the LFS ECE channels (see discussion in section 2.2.

In the discharges from 2021 it was therefore attempted to lower $B$ further to shift the X2 resonance into the reflector tile. Since gyrotron 7 was under repair, beam 6 and the respective reflector tile were used instead. As seen in box D of figure 4 the response of this TC was significantly stronger, probably due to poorer thermal connection during the mounting (see [7] for details). Again one finds that the sniffer signals show the expected increase in transmitted power as $T_e$ drops, whereas the TCs do not show a significant change for the latter sweeps, for example sweep 4 and 6, i.e. the change of $B_t$ did not suppress this effect. The reflector tile for beam 6 can be seen with the video diagnostic. For #39747 (fig. 4) light emission is observed close to the position of the TC for sweeps 3-7[2]. It is very week for sweep 3 and strongest for sweep 6. Sweep 7 was incomplete due to a soft stop related to the OH coil limit. This is a direct confirmation of a discharge glowing in the vicinity of the TC. For sweeps 4-7 fig. 4 F shows a clear reaction of the recycling flux, but only for the pulses with beam 8. In any case we expect that the power measured with the TCs will be reduced under these conditions. We note from figure 5 that the cold X2 resonance for the 140 GHz blips is for beam 6 still just in front of the reflector. For beam 8 ($z_{heat\,shield} \approx 0.5\,m$ as in fig. 2, beam not shown in fig. 5), it is in the tile, respectively in the gap. Apparently also a resonance in a gap can lead to a significant particle source. Maybe the solid surface next to the resonance leads to an efficient fueling (production of secondary electrons) for the arc.

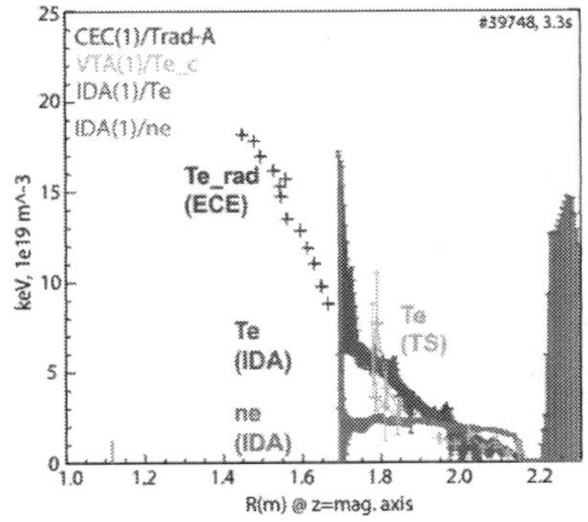


**Figure 6.** $n_e$ and $T_e$ profiles for AUG #39748 closest to 3.3s, no time averaging. $T_e$ data from Thomson scattering (TS) and ECE-emission ($T_{e,rad}$). The integrated data analysis (IDA) used the low field side (LFS) ECE data for $T_e$ and the interferometry and Lithium-beam data for $n_e$ [9].

## 2.2 Non-thermal (non-linear) effects

As mentioned above, the discharges were repeated with the ECE diagnostic active (and no 140 GHz blips where high sniffer levels had been observed). In some cases the radiation temperature was found to rise strongly for channels on the HFS (figure 6). This was only observed for $B_t = 1.6\,T$ and $P_{105GHz} \geq 2.5\,MW$. It has been checked with the radiation transport code ECRad [10] in combination with integrated data analysis [9] if this could be explained by shine-through effects due to limited optical thickness, as it was found to explain large radiation temperatures at the LFS edge of some plasmas. Still, no sensible $T_e$-profiles were identified which would explain the data under these assumptions. Thus, one has to take non-thermal effects into consideration. Since the $2^{nd}$ harmonic ECE emission of fast particles is shifted to lower frequencies due to the relativistic mass increase of the electrons, this radiation should be reabsorbed by slower electrons at larger major radius $R$ as the radiation travels to the LFS

[2]camera id 06Bul1, with low pass filter > 900 nm, same view with narrow $H_\alpha$ filter (id 06Bul) shows no obvious signal

where the detector is located. This means that the source of radiation must be located at larger $R$ than the corresponding cold X2 resonance, such that the radiation can escape. $3^{rd}$ harmonic emission from electrons with a relativistic $\gamma$-factor of a bit lower than 3/2 in the center of the plasma would fulfill this condition. $\gamma \leq 3/2$ corresponds to a kinetic energy below half of the electron rest energy, i.e. $\leq$ 256 keV (faster ones would not be seen, again due to the X2 resonance.). Electrons in the energy range should be visible on the HXR data. shown in box G of figure 4. For the phases with 4 and 5 EC-beams with 105 GHz, the signal level is clearly elevated. We also see that the HXR signals increase further during X3 blips (140 GHz), indicating that fast electrons are further accelerated, in other words the presence of the non-thermal electrons influences the absorption of the X3-blip. In such cases the latter needs to be estimated taking the distortion of the electron distribution into account.

The fast electron population is generated by a specific combination of the magnetic field and the large $N_{\parallel}$ of the 105 GHz injection (in spite of the net current being almost balanced). Figure 5 shows that the absorption region is largely Doppler-shifted since the cold resonance is located on the HFS behind the peak of the (thermal) electron temperature. The large shift means that the EC accelerates electrons with initial energies well above $kT$. If $B$ is larger, the cold resonance is closer to the plasma center and only electrons with a smaller parallel velocity are resonant in the plasma center, i.e. the heating goes more to the bulk of the electron distribution. This can be modeled with a code describing the collisional relaxation process, but as discussed in section 2.3 this is not the focus of this paper.

### 2.3 Comparison with linear theory

The non-thermal effects as described in the previous section are not expected to be relevant in ITER. The larger the machine, the better is the assumption, that an electron leaving an EC beam will thermalize before it re-enters another EC-beam. For the absorption calculation this allows to always assume a Maxwellian velocity-distribution. The absorption profile scales under those circumstances linearly with the input power, since no distortion builds up which would introduce a non-linearity by its dependence on input power. It is this kind of linear theory which was used in the calculations for fig. 1 and for which ITER is interested to test the validity. While in [3] the code GRAY was used, on AUG the code TORBEAM[8] is used. Both codes differ in the description of the Gaussian beam, but have been successfully bench marked for ITER [11]. In particular we set up TORBEAM to use the same absorption routine by *D.Farina*. Figure 7 compares the single pass absorption measured with the TCs to the calculation by TORBEAM. The fat blue crosses on the right correspond to the situations for which we expect the agreement to be good and indeed deviations are of the order of a few percent only. Red crosses indicate cases with non-thermal electrons, for which the TCs show higher single pass absorption than calculated with the linear code. This may be

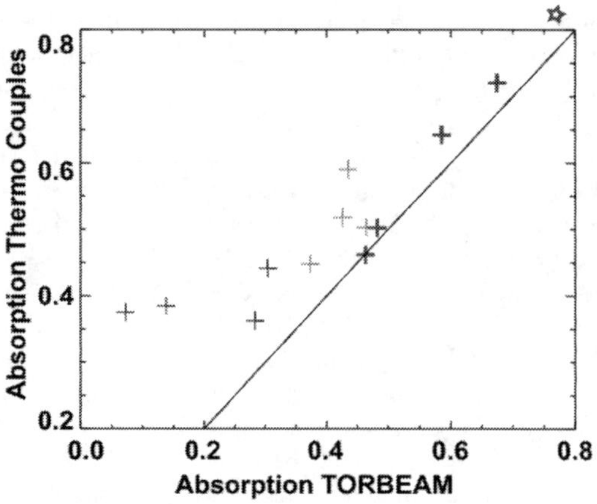


**Figure 7.** Comparison of the single-path absorption calculated by TORBEAM based on the $n_e$, $T_e$ profiles determined with IDA to the single path absorption determined with the TCs. Grey crosses correspond to cases with increased main chamber recycling fluxes or light on the video from the reflector. Red crosses correspond to cases with increased HXR levels (here IDA used TS data for $T_e$). The blue crosses are the remaining undisturbed data. All data except the blue star are from beam 6, which has the most structured TC data, reducing the error in the power estimation. The blue star corresponds to an early analysis of the $2^{nd}$ and $3^{rd}$ sweep with beam 8 in fig. 3 (TS data still to be reanalyzed with all the modeling established meanwhile [7]). The point is added here, since it corresponds to the highest single pass absorption achieved in the described experiments.

expected since X3 absorption depends strongly on the Larmor radius, such that velocity distributions with excessive fast electrons may absorb more than calculated assuming a Maxwellian. Finally the gray crosses correspond to cases with additional recycling fluxes or light in front of the TC as described above. Here the TCs miss part of the transmitted power because the plasma in front does absorb it, thus the apparent single-pass absorption is higher. Apart from those caveats the linear theory describes well the observations, though for a more limited range of single pass absorption (or $T_e$) than originally envisaged.

## 3 Conclusions and Outlook

Within the experimental limitations the linear theory (here represented by TORBEAM) describes well the measured X3 single-pass absorption. Unfortunately, single-pass absorption below 40% could not be accessed, due to effects of the X2-resonance close or inside the gaps of the inner heat-shield leading to de-gassing and arc formation distorting the power measurement in the tile. All sniffer probes signals increase significantly as the single-pass absorption calculated by TORBEAM drops below 40% (figure 8, but also figures 3,4). Still a quantitative analysis is difficult. The sketched parabola corresponds to a direct reflection, i.e. if only one path would be followed in both directions through the plasma. Of course multiple reflections are also important especially at low absorption and more distant probes.

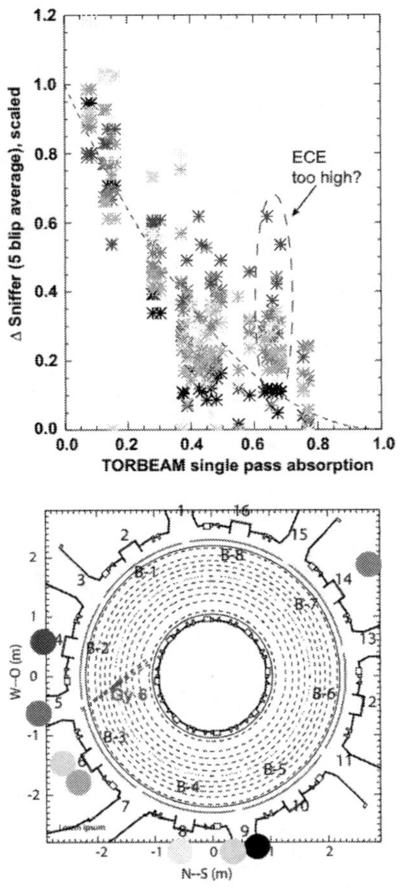


**Figure 8.** Top: Average increase of the sniffer signals for the five subsequent blips during the sweeps of beam 6 plotted against the single pass absorption calculated with TORBEAM. Some data appear twice: in case IDA $T_e$ was significantly different with ECE and TS, two TORBEAM runs were made using the different $T_e$ profiles. Most points in the vertically elongated dashed ellipse correspond to non-thermal electron distributions using ECE. The corresponding ones using TS lie at lower single.pas absorption (lower $T_e$, fig. 6). The sniffer increase was scaled with reference to the empty vessel reference.[a] The data is underlined with a parabola, see text. Bottom: Toroidal location of sniffer-probes and beam 6. Absolute values of more distant probes are lower, but this is hidden by the normalization as explained above.

[a]Additional scaling factors between 0.5 and 2.5 had to be introduced for the different probes, minimizing the scatter at the lowest and largest absorption and aiming at a value of unity for zero TORBEAM absorption. These factors were adjusted by eye. An explanation may be the lack of mode mixing by plasma fluctuations in the empty vessel, giving rise to less variable highly localized interference structures, no longer smeared out by time averages over the mode sensitive sniffers signals.

The appearance of the arcs on the HFS during these experiments may be avoided by reducing the beam power. Here the onset was observed when less than half the power was absorbed, suggesting a reduction of the gyrotron power by a factor of two will solve the issue, which may be technically still possible. Especially for ITER the lowest absorption levels may be relevant, since it is rather the $T_e$ value than the absorption, which should be compared . The latter scales favorably for ITER with the ratio $(B/\nabla B)/\lambda_{EC}$, which is more than a factor of 4 larger in ITER than in AUG. The absorption of 25% mentioned above for the ohmic ITER plasma corresponds to 6% for a plasma with similar $n_e$, $T_e$ in AUG, close to the lowest value in fig. 7 (7% for an ohmic plasma in AUG).

With respect to machine safety in ITER, the arcs observed in AUG on the HFS may trigger some analysis. Alternatively or as an initial test, ITER may envisage a regime with a HFS-X2-dump inside the plasma as referred to above for AUG. $B_t$ of 2.0-2.1 T seems feasible to achieve that. The disadvantage would be a higher L/H-threshold and a limited use of the upper launcher. The latter would only be an issue if more than 24 gyrotrons were available, i.e. if the EC-system would be extended. Since X3-heating does essentially come without current drive, issues like MHD-(de)stabilization can anyway not be tested. Finally we note, that the non-linear cases observed in AUG are nicely diagnosed and a good test-case for Fokker-Planck codes.

## Acknowledgment

The views and opinions expressed herein do not necessarily reflect those of the ITER Organization.
This work has been carried out within the framework of the EUROfusion Consortium, funded by the European Union via the Euratom Research and Training Programme (Grant Agreement No 101052200 — EUROfusion). Views and opinions expressed are however those of the author(s) only and do not necessarily reflect those of the European Union or the European Commission. Neither the European Union nor the European Commission can be held responsible for them.
*S.S. Denk* acknowledges support of his work by US DOE under DE-SC0016154.

## References

[1] V. Erckmann and U. Gasparino, Plasma Phys. Control. Fusion **36**, 1869 (1994)
[2] ITER Organization, Tech. Rep. ITR-18-003 (2018), Appendix G, `https://www.iter.org/doc/www/content/com/Lists/ITER Technical Reports/Attachments/9/ITER_Research_Plan_within_the_Staged_Approach_levIII_provversion.pdf`
[3] M. Schneider et al., Nucl. Fusion **59**, 126014 (2019)
[4] C. Kiefer et al., Nucl. Fusion **61**, 066035 (2021)
[5] H. Höhnle et al., Nucl. Fusion **51**, 083013 (2011)
[6] J. Stober et al., Plasma Phys. Control. Fusion **62**, 024012 (2020)
[7] M. Schubert et al., *Experiments with reduced single pass absorption at ASDEX Upgrade – instrumentation and applications* (2022), these conference proceedings
[8] E. Poli et al., Computer Phys. Comm. **225**, 36 (2018)
[9] R. Fischer et al., Fusion Sci. Technol. **76**, 879 (2020)
[10] S. Denk et al., Computer Phys. Comm. **253**, 107175 (2020)
[11] M. Schneider et al., Nucl. Fusion **61**, 126058 (2021)

# Experiments with reduced single pass absorption at ASDEX Upgrade – instrumentation and applications

*Martin* Schubert[1*], *Joerg* Stober[1], *Albrecht* Herrmann[1], *Eduard* Grigore[2], *Walter* Kasparek[3], *Carsten* Lechte[3], *Francesco* Monaco[1], *Bernhard* Petzold[1], *Burkhard* Plaum[3], *Emanuele* Poli[1], *Cristian* Ruset[2], *Stefan* Vorbrugg[1], *Dietmar* Wagner[1], and ASDEX Upgrade Team[4]

[1]Max-Planck-Institut f. Plasmaphysik, Boltzmannstr. 2, D-85748 Garching b. München, Germany
[2]Institute for Laser, Plasma and Radiation Physics, Atomistilor Street 409, RO-077125 Magurele-Bucharest, Romania
[3]Institut f. Grenzflächenverfahrenstechnik und Plasmatechnologie, Pfaffenwaldring 31, D-70569 Stuttgart, Germany
[4]see author list of *Ulrich* Stroth, *et al.*, Nucl. Fusion **62**, 042006 (2022)

**Abstract.** Reflecting gratings have been installed in the vacuum vessel of ASDEX Upgrade for all beamlines of the electron cyclotron resonance heating system. Potentially unabsorbed millimetre wave power after the first pass through the plasma is redirected towards the plasma centre. This increases the efficiency of heating schemes with reduced single pass absorption like O-2 or X-3. In order to monitor beam position and power, thermocouples were installed into the gratings. A numerical model was developed to evaluate the beam intensity during short pulses from the thermocouple measurement in a non-stationary environment. An experiment was carried out, where only the X-3 resonance is present in the plasma, and the millimetre wave beam shine-through was measured successfully as a function of the central plasma electron temperature. This allows to deduce the X-3 absorption experimentally. Scanning the launching angles, it seems possible to measure the 2D beam cross section after the first pass through the plasma.

## 1 Introduction

The electron cyclotron resonance heating system (ECRH) at ASDEX Upgrade (AUG) has been extended to eight Gyrotrons in total. Each Gyrotron operates at 105 and 140 GHz and is designed for up to 1 MW millimetre wave output power. Experimental topics at AUG include variations of the main magnetic field strength in the plasma and conditions with high plasma densities, potentially above the X-2 cut-off at 140 GHz, e.g. [1]. New ECRH schemes have been proposed [2] which can cope with the demand, at the cost of reduced single pass absorption, however. It is necessary to design reflecting gratings [3] and install them into the heat shield, i.e. the inner column inside AUG's vacuum vessel. The gratings are designed for 140 GHz, they are refocussing and redirecting each beam towards the plasma centre, thereby increasing the second-pass absorption and the total heating efficiency.

Part of the heat shield inside AUG consists of the steel W1.4901. Four of the gratings were manufactured out of this steel and a thin tungsten coating was applied, which is a newly developed technique. In order to monitor the position of each ECRH beam on the corresponding grating, thermocouples were installed. The fast response of the thermocouples opens the possibility to measure the local beam intensity after the first pass through the plasma with short pulses. Since experimental data on X-3 absorption at 140 GHz are rare, it was decided to try this measurement. The experimental setup uses the two available frequencies: (a) 105 GHz at different power levels and with nearly perfect absorption at the X-2 resonance in order to vary the central plasma electron temperature. (b) 140 GHz for the probing beam, where only the X-3 resonance is present in the plasma and the shine-through can be measured at the heat shield, using the thermocouples. The results on X-3 absorption are presented in a separate contribution to this conference.

In a more general approach to study the beam properties after the first pass through the plasma, we started to measure the local beam intensity as a function of the launching direction, in a set of reproducible and similar plasma discharges using the O-2 mode polarisation. Putting the measurements of several AUG experiments together, a kind of 2D image of the beam cross-section can be generated.

## 2 Two-pass scheme at ASDEX Upgrade

The design criteria were already written down [3] and the scheme has now been completed for all ECRH launchers (figure 1). We would like to point out that the oblique geometry is necessary in order to maximize the O-2 absorption [2]. For each launcher (figure 2) there is a corresponding reflecting grating (figure 3) and it is necessary that the unabsorbed beam power after the first pass is located on the corresponding grating, as long as the absorption is incomplete. Consequently, the heating

* Corresponding author: Martin.Schubert@ipp.mpg.de

geometry is fixed and no large variations of the launching direction are allowed.

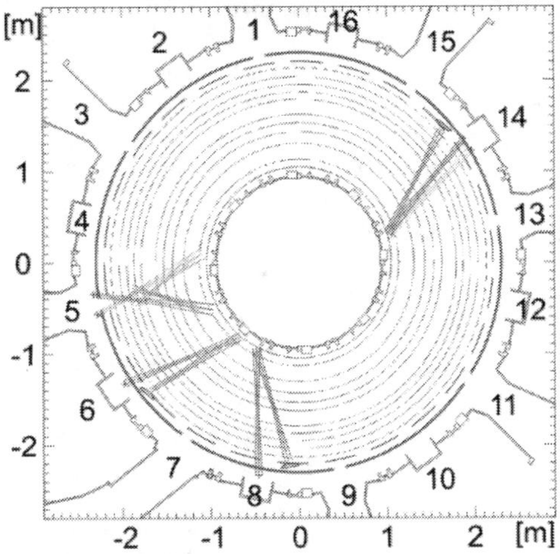


**Fig. 1.** Schematic top view on the plasma midplane cross section of ASDEX Upgrade, two-pass ECRH heating beams are shown in colour. The vacuum vessel consists of 16 segments. ECRH launchers are located in segments 5,6,8 and 14. Since the four launchers in segment 5 would shadow each other in this view, only two of them are shown for simplicity. The beam trajectories in segment 5 do not start or end in the midplane, they have a significant vertical component and appear shortened in the projection of this view.

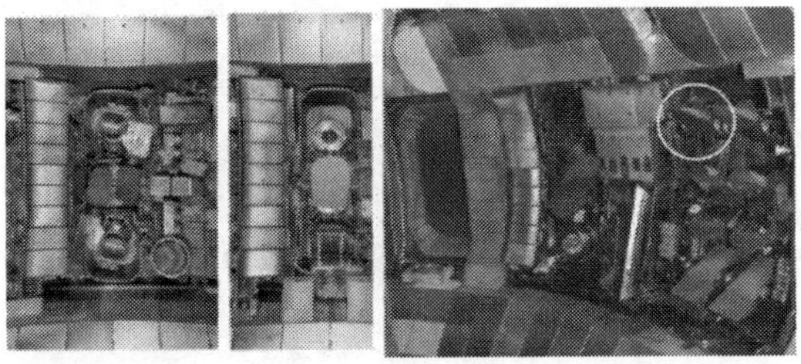

**Fig. 2.** ECRH launchers inside the AUG vacuum vessel, the view is radially outward. Left: Double launcher in segment 14. Middle: Single launcher in segment 8. Right: Segments 7,6,5 (from left to right) with the large port of the neutral beam input in segment 7, the single ECRH launcher in segment 6, and the four launchers in segment 5. The robust surfaces of the auxiliary limiters left of the launcher mirrors in segments 14, 8, and 6 are the design end points after the second pass. The launcher highlighted with a white circle in segment 5 (right photo) was used for the experiments described in section 4.

The inner column of AUG's vacuum vessel consists of tiles (figure 3). Two tiles at the corresponding position were merged into a double tile, to make the reflecting gratings sufficiently large. The heat shield around the midplane consists of W1.4901 steel, which concerns the gratings in segments 7 and 13 (figure 3). The other ones were machined out of graphite.

With tungsten coating, mm-wave power absorption of the grating can be reduced to approx. 0.5% [4]. For graphite, this coating is established at AUG for a long time, while for W1.4901 steel this had to be developed. Figure 4 shows the depth profile measured by glow discharge optical emission spectroscopy (GDOES) of a successful coating. A thin coating of up to 10 µm can be applied to this steel using a molybdenum interlayer.

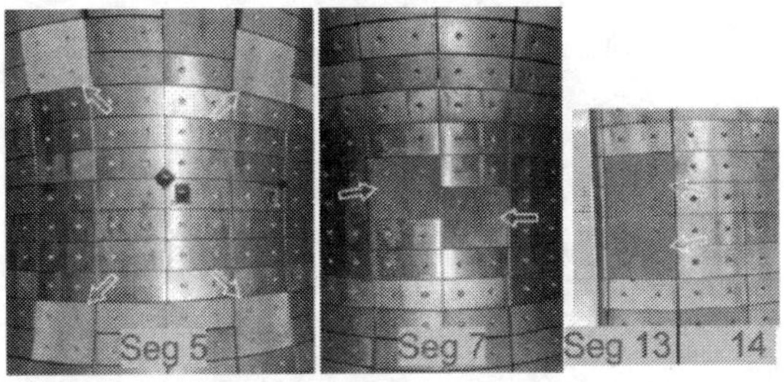


**Fig. 3.** AUG vacuum vessel heat shield, which consists of tiles. The view is radially inward towards the central column of the machine. The visible double tiles are the reflecting gratings that correspond to the launchers from figure 2. Left: Four gratings for the launchers of segment 5. Note, that the gratings are not in the midplane. Middle: Two gratings in segment 7 for the two single launchers of segments 6 and 8. Right: Two gratings in segments 13 for the double launcher of segment 14.

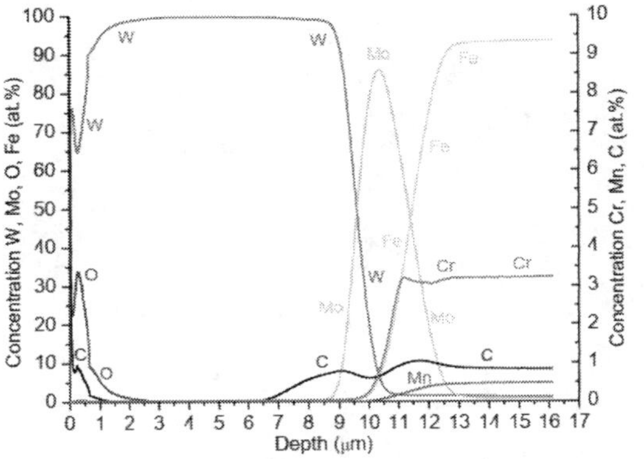


**Fig. 4.** Depth profiles of W, Mo, Fe, Cr, Mn, C and O measured by Glow Discharge Optical Emission Spectroscopy (GDOES) on W-coated sample made of W1.4901 steel. Near the surface of the sample (0…8 µm), we have almost pure tungsten (W). The molybdenum (Mo) interlayer is found in a depth of approx. 10 µm. For depths > 12 µm we find the expected composition of W1.4901 steel (Fe, Cr, C, Mn etc). GDOES is sputtering the sample and is, thus, destructive.

## 3 Thermocouple measurement

Fast thermocouples can be used to monitor ECRH beam positions in scenarios with potential shine-through [2], and this technique has been developed continuously [5]. We chose a NiCr-Ni (type K) thermocouple with an outer diameter of 0.25 mm [6]. Sensitivity 40.96 µV/K and maximum specified temperature approx. 1200 °C. The jacket of the thermocouple is made of Inconel 600 and it is connected to a flexible cable within a cylindrical joint with outer diameter 4 mm and a length of 25 mm. We have tested different mount options and found that the best compromise between longevity and signal amplitude is achieved in a flush mount (figure 5).

### 3.1 Flush mount

Several tools have been developed in order to facilitate the flush mount, cf. figure 6. We have put emphasis on sufficient protection and shielding of the cables in order to prevent damage during operation with mm-waves.

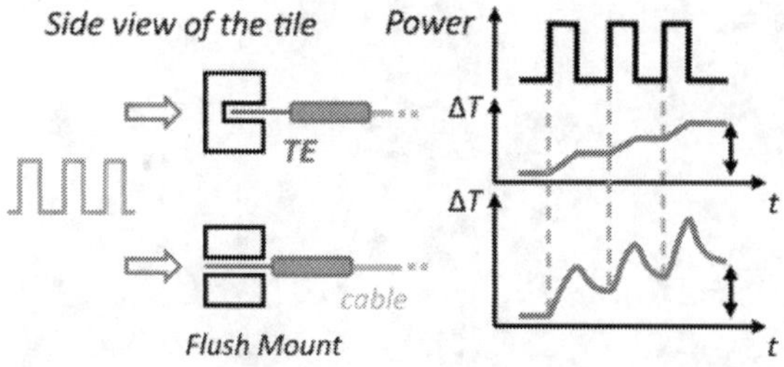


**Fig. 5.** Schematic comparison of thermocouple (TE) mount options and signal response on an incoming pulse train of mm-wave power. Top: Mounted from the rear inside the bulk material. Bottom: Flush mount, the borehole goes through the surface and the TE is partly exposed to the mm-wave. The On/Off time of the power is of the same order as the relaxation time of the thermocouple. If the total heat capacity of the absorbing material is comparable, then the total temperature rise after the last pulse plus relaxation time is also comparable (double sided arrows on the right).

**Fig. 6.** Left: Transparent CAD view of two double tiles with built-in channels for the cables. Right: Tools which ensure a precise flush mount. During the attachment procedure, the opening of the borehole at the plasma exposed surface is temporarily blocked (upper right photo). This opening is normally located on the bottom of a groove of the grating. The view onto the rear side (lower right) shows the mechanics used to fixate cables and cylindrical joints.

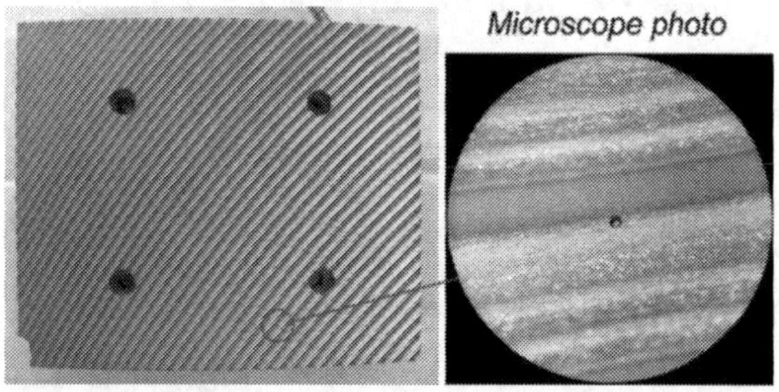


**Fig. 7.** Complete assembly of a reflecting grating, tungsten coated and equipped with thermocouples. Left photo: plasma facing side, width and height approx. 20 cm and 18 cm. The four large openings are for attaching the tile to the heat shield. Right: magnified area around a 0.5 mm borehole, containing a flush mounted thermocouple with a diameter of 0.25 mm.

In the manufacturing process the thermocouples are attached to the tile with a graphite glue, which is a mixture of graphite powder and phenolic resin. The tile including thermocouples and cables is baked, until the graphite glue is sufficiently stable. After complete assembly, tile and grating are inspected (figure 7). If the flush mount was successful, then within each borehole the tip of the thermocouple is clearly visible.

During the recent AUG experimental campaigns we did not encounter thermocouple sensor failures specific to the flush mount. In the AUG case, the specified temperature ranges of the thermocouple sensor and jacket [6] are not exceeded. This should be regularly checked, depending on the planned application.

### 3.2 Evaluation model

During experiments with applied mm-wave power, Gyrotron On/Off times are recorded with a precision better than 0.1 milliseconds (ms) by directional couplers in the transmission line. The thermocouple signals are filtered with a 300 Hz Bessel low-pass and recorded with 1 ms time resolution. Taking into account the relevant transit times and the transfer function of the filter, we observe a characteristic time delay of typically 10 ms between the mm-wave power arriving at the borehole and the start of the signal rise in the recorded thermocouple signal. The relaxation time of the signal, when the mm-power is switched off, is of the same order. This indicates that, when evaluating short pulses, the model should be comprehensive enough to account for the effect of these time constants.

Thermocouples, which are not hit by the mm-wave, show a continuous, uniform, and slow rise of the temperature due to the background heat load from the plasma. The rise time is larger than the relaxation time of the thermocouple. We therefore model this background heat load $p_{bg}$ homogeneously across the tile, based on the direct measurement from those thermocouples.

When evaluating pulsed mm-wave power similar to figure 5 (bottom), we use a simplified 2D model around each thermocouple in order to account for the dynamics of the heat exchange between the absorbing surface and the thermocouple sensor, which is located several 10 µm below the surface [6], see also figure 8.

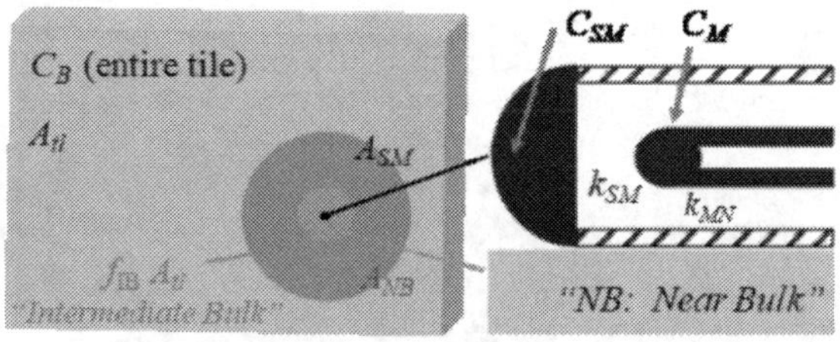


**Fig. 8.** Schematic visualisation of model temperature zones around and inside a thermocouple. Left: The entire tile with total heat capacity $C_B$ and exposed surface area $A_{tl}$. Cylindrical zones around the thermocouple, which expand into the depth: near bulk $A_{NB}$ (grey) and intermediate bulk $f_{IB}\,A_{tl}$ (fraction of $A_{tl}$ dark green). Right: side view of the thermocouple cross section [6]. Absorbing surface area $A_{SM}$ with heat capacity $C_{SM}$. The metal-metal joint with the heat capacity $C_M$ is the position of the temperature measurement $T_M$ and it is located several 10 µm below the surface [6]. The heat flux has to cross insulating material with model heat conductances $k_{SM}$ and $k_{MN}$.

The model has three grid points in radial direction parallel to the tile surface, and one additional grid point in the depth, where the measurement $T_M$ takes place (figure 8). The three radial grid points are: (i) absorbing tip of the thermocouple with surface area $A_{SM}$, heat capacity $C_{SM}$, temperature $T_{SM}$, (ii) the near bulk with surface area $A_{NB} = f_{NB}\, A_{tl}$, temperature $T_{NB}$, and (iii) the intermediate bulk with surface area $A_{IB} = f_{IB}\, A_{tl}$, temperature $T_{IB}$.

Both near and intermediate bulk are characterized as fractions $f_{NB}$ and $f_{IB}$ of the entire tile, i.e. with heat capacities $C_{NB}=f_{NB}\, C_B$, and $C_{IB}=f_{IB}\, C_B$, respectively.

We introduce heat conductances: $k_{SM}$ between the thermocouple tip and the measurement sensor ($C_M$, $T_M$), $k_{MN}$ between the sensor and the near bulk, and $k_{NB}$ between near and intermediate bulk. For simplicity, we allow heat transfer from $T_{NB}$ to $T_{IB}$ (via $k_{NB}$) as a loss term for $T_{NB}$, but we set $T_{IB}$ as a boundary condition, which changes its value only due to the heat flux from the surface, ignoring the incoming radial transfer from the near bulk. In doing so, we assume that at this position, the incoming heat flow from the near bulk is small compared to the flow from the surface, and we end up with a system of three coupled differential equations for $T_{SM}$, $T_M$, $T_{NB}$, and a separate equation for $T_{IB}$:

$$C_{SM}\dot{T}_{SM} = A_{SM}\,(p_{bg} + p_A) - k_{SM}\,(T_{SM} - T_M) \qquad (1)$$

$$C_M\,\dot{T}_M = k_{SM}\,(T_{SM} - T_M) - k_{MN}\,(T_M - T_{NB}) \qquad (2)$$

$$f_{NB}\,C_B\,\dot{T}_{NB} = f_{NB}\,A_{tl}\,(p_{bg} + p_A) + k_{MN}\,(T_M - T_{NB}) - k_{NB}\,(T_{NB} - T_{IB}) \qquad (3)$$

$$f_{IB}\,C_B\,\dot{T}_{IB} = f_{IB}\,A_{tl}\,(p_{bg} + r_{IB}\cdot p_A) \qquad (4)$$

In this system $p_A$ is the absorbed mm-wave intensity, $p_{bg}$ is the (homogeneous) background intensity, measured by the distant thermocouples, and the term $r_{IB}{\cdot}p_A$ is to model the heat flux to the intermediate bulk as a fraction of the mm-wave intensity. The fraction $r_{IB}$ had to be introduced in order to correct the baseline of the signal, whenever the incoming mm-wave intensity is zero for a duration longer than the relaxation time of the thermocouple. It is an indication, that the beam width is indeed smaller than the dimensions of the tile.

A cooling term can be included in each equation, which is due to the fact, that the tile is mounted on a cooling channel, see figure 6, CAD drawing on the left: the vertically elongated, blue structure is in fact a tube with cooling water. The cooling time constant is of the order of several minutes and was measured, using long acquisition times. Within the short pulse analysis of our experiments, it can be neglected.

The system of equations (1-4) can be solved in time, using a Runge-Kutta fourth order scheme, depending on the parameter values. The Runge-Kutta solver is a subroutine to an optimizer [7], which does a least square fit of the model $T_M$ to the measured signal, using as free parameters: the absorbed intensity $p_A$, most of the material constants (except $C_B$, $A_{SM}$, $A_{tl}$, which can be measured) and the fractions $f_{NB}$ and $r_{IB}$ ($f_{IB}$ is indeed irrelevant due to our choice to ignore the incoming radial transfer from the near bulk).

## 4 Experimental results

In this paper we restricted the evaluation to a short pulse analysis, assuming the absorbed mm-wave intensity $p_A$ constant during each pulse. It is necessary to calibrate the system in vacuum, measuring the thermocouple response with respect to a well-defined incoming beam power. We neglect the influence of the incoming wave polarisation, because (a) the gratings are optimized to be independent on the polarisation [2] and (b) the reflection geometry and, thus, the angles of incidence are nearly constant. The possible dependency on the polarisation is, therefore, expected to be small.

### 4.1 Launcher sweep and vacuum calibration

It is possible that the beam will be deflected due to beam refraction in the plasma, and therefore we continuously scan the beam across the thermocouple. For that purpose, we use one of the fast steerable launchers of AUG in segment 5 (cf. figure 2). During one sweep of the launcher mirror, we apply 5 pulses of Gyrotron power, which end up on different positions across the thermocouple (figure 9).

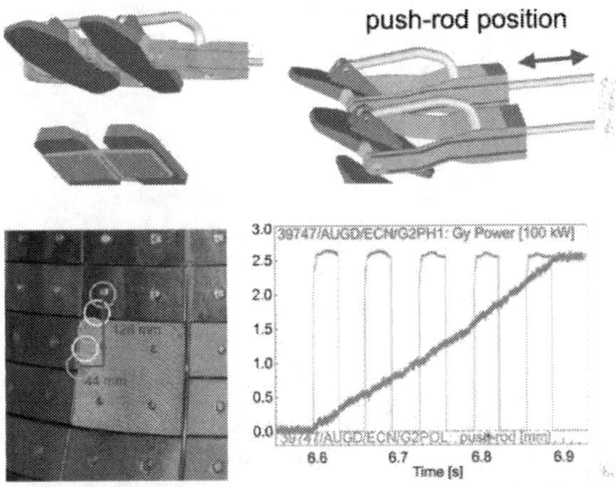


**Fig. 9.** Exemplary visualisation of the applied mm-wave power and timing of the launcher sweep. The inclination of the launcher mirror is controlled by the push-rod position (top). Five pulses are applied during one sweep of the launcher. The pulses end up on different positions on the heat shield and around one selected thermocouple of the corresponding reflecting grating (photo bottom left). The timing of the push-rod position and of the applied mm-wave power are shown.

This sweep is large enough to clearly identify the maximum response of the thermocouple to the applied mm-wave power, in both cases with and without plasma (e.g. figure 10). According to the TORBEAM simulation [8], the vertical beam refraction is more important than the horizontal one. A fast horizontal sweep is technically not possible with this launcher.

The signal quality in the vacuum case (figure 10) is very good, such that the free parameters (section 3.2) can be fitted together with the pulse amplitudes. The parameter values of the fit result seem reasonably consistent: e.g. $C_{SM}/k_{SM}$ = 30 ms, corresponding to the main time constant of the thermocouple measurement.

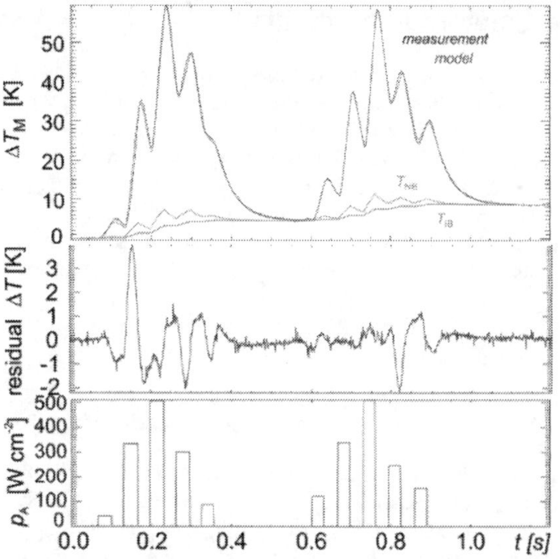


**Fig. 10.** Measured signal response to pulses of mm-wave power without plasma, compared with the modelling. Pulse amplitudes (bottom row) are the fit results. The main part of the residual (middle row) can be understood from the fact, that the model intensity amplitude is assumed to be constant within one pulse. In reality, during the fast launcher sweep for example in the second pulse (at approx. 0.15 sec) the absorbed mm-wave intensity increases during the measurement. The model fits an average value and, therefore, the residual, which is model minus measurement, is positive during the first part of the pulse. The observer should keep in mind, that there is a time lag between the mm-wave power and the measurement. The data was recorded as AUG experiment #39744.

The fitted quantity $k_{SM}/A_{SM}$=15000 W/(K m$^2$) can be approximated by $\lambda/d$ where $\lambda$ is the specific heat conductivity and $d$ the distance between the absorbing surface and the measurement position ($T_M$). In the case of a typical insulator with $\lambda$ approx. 1 W/(K m) this corresponds to $d$ of the order $10^{-4}$ m, which is consistent with the specification of the thermocouple [6].

An additional fitted signal delay of the order of 1 ms of the thermocouple measurement with respect to the power improves the fit quality. It should also be mentioned, that the fit of parameters, which are related to material constants, is part of the calibration. When evaluating a measurement, only beam related quantities should be fitted again, i.e. pulse amplitudes $p_A$ and the fraction to the intermediate bulk $r_{IB}$.

## 4.2 Absorption measurement X-3

Experiments were carried out to measure the plasma absorption at the X-3 resonance as a function of the plasma electron temperature. We show an exemplary overview of experimental data in figure 11. Heating and X-3 probing geometry are visualized in figure 12. For further evaluation see the contribution by J. Stober [9]. The signal evaluation as described in section 4.1 yields the results of figure 13. The material constants from the vacuum calibration are used. Fit results are: $p_A$ (one value for each pulse) and $r_{IB}$ (only one value for the entire time interval). If the absorbed power during one pulse is not constant, then deviations between model and measurement can be observed in the residual, e.g. during the longer pulse at $t$ approx. 4.9 sec, where the background plasma (cf. figure 11) seems to be transient.

In order to evaluate the X-3 absorption, the fitted $p_A$ with plasma (figure 13) are normalised by the launched Gyrotron power of Gyrotron 6 (figure 11) and the five pulses during one launcher sweep are summed up. The same procedure is done in the vacuum case, where in particular the launched Gyrotron power was different. The shine through is evaluated by the ratio of the plasma measurement versus the vacuum measurement.

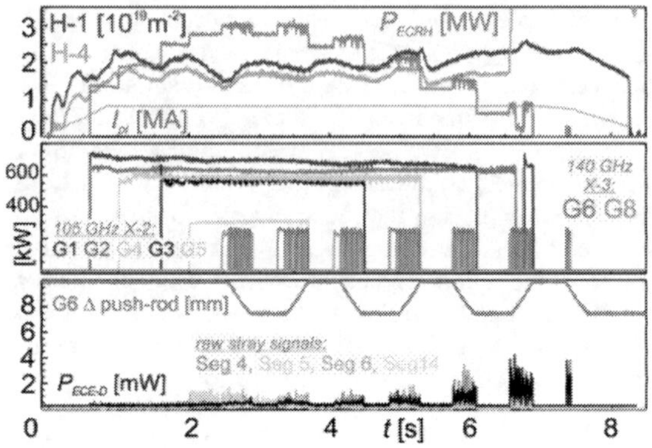


**Fig. 11.** Comprehensive overview of AUG discharge #39747 with total ECRH power data, evolution of plasma density and plasma current on top, individual Gyrotron power signals in the middle row, and Launcher 6 push-rod scans, as well as stray radiation signals in the bottom row. The blue density signals on top are line-integrated interferometry data. Middle row: the purpose of the group of 105 GHz Gyrotrons is to vary the plasma heating such, that different levels of plasma electron temperature exist. Bottom row: Stray radiation, in particular which is measured by the ECE, is significant only during the 140 GHz X-3 pulses and increases with decreasing plasma electron temperature, as expected.

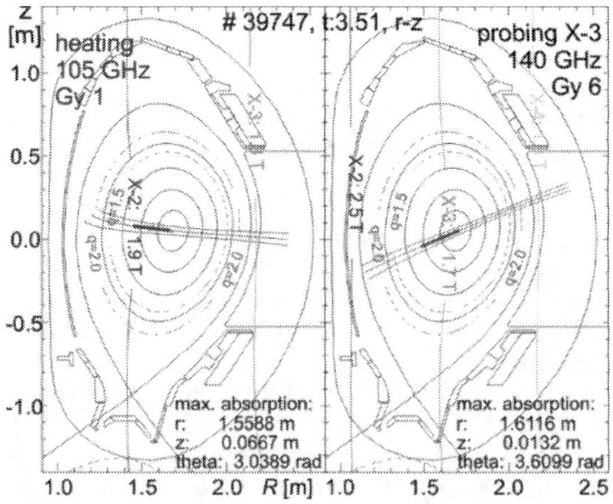


**Fig. 12.** Cross section, AUG #39747 showing the geometry [8] of the 105 GHz X-2 heating (left) and the 140 GHz probing (right). The main magnetic field strength was chosen such, that for 140 GHz only the X-3 resonance is present in the plasma. The heating beams (left) were launched with a toroidal angle, so that the Doppler shifted resonance is closer to the centre.

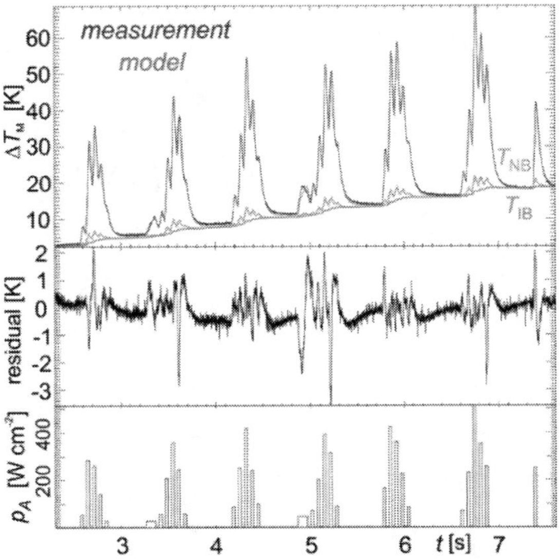


**Fig. 13.** Evaluation of thermocouple measurement, AUG #39747 showing in the top row: measurement (black) and the temperature at the model grid points $T_M$ (red), $T_{NB}$ (grey), $T_{IB}$ (green). The amplitudes of the model temperature $T_{SM}$ are much larger and not shown for simplicity. Second row: residual = model temperature minus measurement. The residual is subject to minimization in the fit procedure using the $p_A$ as fit parameters. The resulting $p_A$ are shown in the bottom row. The parameter $r_{IB}$ was also fitted, but assumed constant during the experiment.

### 4.3 Shine-through power in O-2 polarisation

The evaluation model described in section 3.2 was developed very recently. Earlier, the amplitude of the pulse response of one thermocouple to the pulsed mm-wave power in O-Mode was recorded as a function of the launching direction (figure 14).

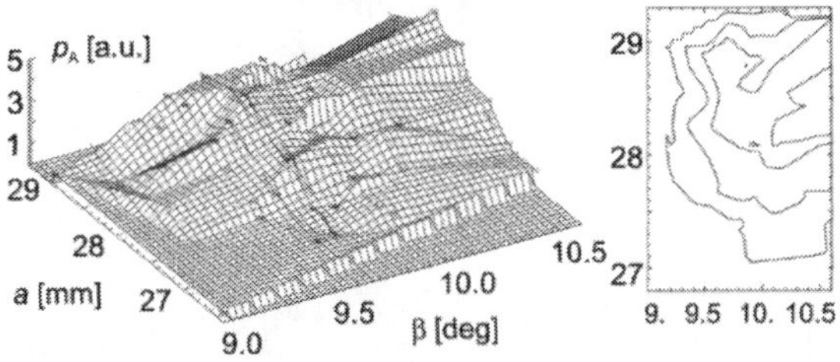


**Fig. 14.** Pulse amplitude response of one thermocouple on the reflecting grating of Gyrotron #8 to TORBEAM [8] evaluated O-2 mode shine through power as a function of the push-rod position $a$ [mm] and the rotation of the mirror support around the push-rod β [deg]. The plot on the left was generated using Delauney triangulation and corresponding visualization [7] for irregularly gridded data. Individual measurements are plotted in orange (2017 campaign) and blue (2019 campaign). Points connected with a dashed line are from a launcher sweep as described in section 4.1. The inset on the right is a contour plot of the same data, with contour levels 0.9, 1.8, 2.7 and 3.6. Within the plot range of $a$ [mm], the centre of the beam moves approx. 10 cm on the heat shield, and approx. 7 cm within the plot range of β. The coordinates are only approximatively corresponding to the vertical and horizontal directions, because the geometry is oblique (cf. figure 9).

This was repeated in a particular plasma experiment called "Standard H-mode", which is carried out frequently and considered reproducible. Short pulses (approx. 30 ms) of Gyrotron no. 8 (Elisey 3M) were used for these experiments. Vacuum calibration data were not measured. Figure 14 shows data from the 2017 and 2019 experimental campaigns. With the thermocouple position being fixed, it should be possible to transform the variation of the launching direction into a cross section of the shine-through beam. This needs beam tracing and reliable plasma kinetic profiles in order to evaluate the shine-through power. We propose to use the simplified analysis as shown in figure 14 in order to verify the launcher setting.

## 5 Summary

The thermocouple mount and the corresponding evaluation model were developed further in order to provide a measuring tool to study mm-wave deposited power on the heat shield of ASDEX Upgrade with high time resolution. A local monitoring of shine-through intensity was demonstrated exemplarily. A systematic comparison of the beam geometry and size with beam tracing predictions seems possible.

Careful mechanical work by H. Schütz is gratefully acknowledged. Photos in figures 6 and 7 as well as the launcher CAD model (figure 9) are also courtesy of H. Schütz. This work has been carried out within the framework of the EUROfusion Consortium, funded by the European Union via the Euratom Research and Training Programme (Grant Agreement No 101052200 — EUROfusion). Views and opinions expressed are however those of the author(s) only and do not necessarily reflect those of the European Union or the European Commission. Neither the European Union nor the European Commission can be held responsible for them.

## References

1. P. Lang, *et al.*, Nucl. Fusion **59**, 026003 (2019)
2. H. Höhnle, *et al.*, Nucl. Fusion **51**, 083013 (2011)
3. M. Schubert, *et al.*, EPJ Web Conf **203**, 02009 (2019)
4. W. Kasparek, *et al.*, Int. J. Infrared Millim. Waves **22**, 1695 (2001)
5. M. Schubert, *et al.*, 43rd Conf. Plasma Phys, ECA **40A**, P1.026 (2016)
6. RÖSSEL-Messtechnik GmbH, *Mantelthermo-elemente*, Produktinformation 071 (12/2020)
7. ITT Visual Information Solutions, *IDL Version 7.1 Reference Guide* (2009)
8. E. Poli, A. Peeters, G. Pereverzev, Comput. Phys. Commun. **136**, 90 (2001)
9. J. Stober, et al., *Quantification of X3 absorption for ITER L-mode paramteres in ASDEX Upgrade*, these conference proceedings (2022)

# Simulation and reflectivity measurements of the ITER first plasma beam dump material

*Andreas* Hentrich[1*]

[1]University of Stuttgart, Institute of Interfacial Process Engineering and Plasma Technology, Germany

**Abstract.** When constructing a beam dump for a fusion device, one faces several challenges and constrictions: It must be vacuum compatible, resistant to temperatures and radiation, should neither gas out nor deteriorate, should take up as less space as possible and obviously also should absorb microwave radiation well. As possible candidates, the group of ceramic metal oxide layers have been suggested, from which three different material systems have been produced by atmospheric plasma spraying and investigated: Titanium oxide, chromium oxide and an aluminum-titanium oxide compound material. Both pure titanium oxide and chromium oxide don't show the absorption necessary, but $Al_2O_3/TiO_2$ (60/40) is very promising and will be presented in the following.

## 1 Sample, measurement & model description

The samples were produced and characterized regarding thickness and surface roughness by Venancio Garcia at the IFKB stuttgart, the production process and a cross section in shown in figure 1. The thickness of the samples could be reliably determined upto ±5µm by the production process. In total, 11 samples with thicknesses between 30 and 180µm have been produced and investigated.

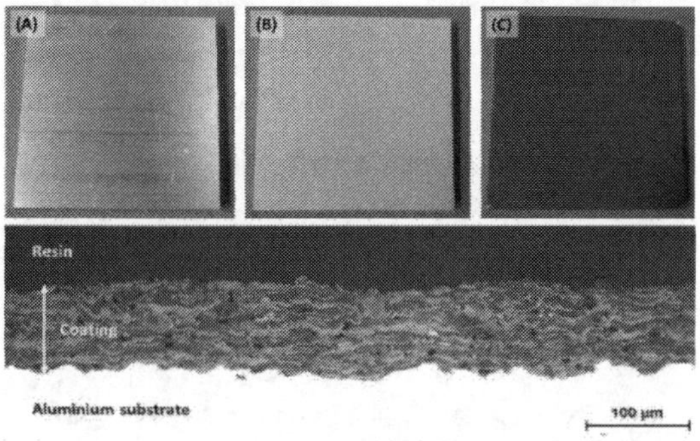


**Fig. 1.** (A) substrate, (B) after grit blasting, (C) after plasma spraying, (D) cross-section

Afterwards, the reflectivity of the samples was measured. For this, a network analyzer was used to generate the microwave and the local oscillator. To get the intended measurement frequency of 170GHz, a harmonic generator was used and with the help of an harmonic mixer and the LO, the signal was modulated back down to a measurable frequency. For emitting the microwave onto the sample and receiving the reflection of it, two optimized horn antennas were used.

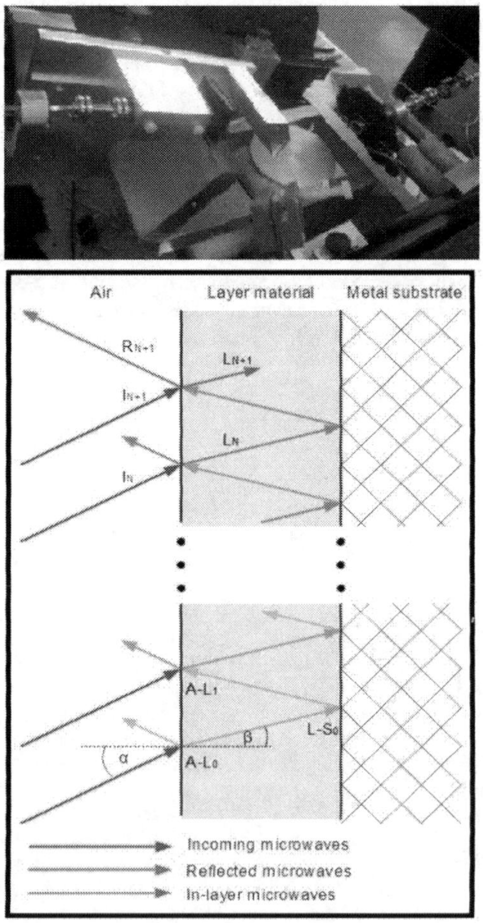


**Fig. 2.** The measurement setup and a sketch of the model

* Corresponding author: andreas.hentrich@igvp.uni-stuttgart.de

As reference an aluminum mirror was used. The dependency of the reflectivity on layer thickness, polarization and incidence angle was measured.

In parallel, a plane wave model was developed to describe the situation at hand from a theoretical point of view. It neither uses the approximation of low nor high absorbing materials and considers the existence of multiple reflections, but is limited to a single frequency and assumes smooth surfaces everywhere. The substrate is simply treated as perfect electric conductor compared to the layer material. The relevant parameters of the model are: frequency and polarization of the wave, incidence angle and layer thickness for the geometry and refractive index and absorption coefficient as material parameters. Both the wave and the geometry parameters are known in advance either by setup and/or previous measurement, just the material parameters are undetermined and were thereby aquired by fitting, for further details of the model or to the fitting refer to [1].

## 2 Experimental findings

Measurement and simulation have been found to be in reasonable agreement, the refractive index was determined to be 4.734 and the absorption coefficient $A_{Layer} = \omega * \varepsilon'' + \sigma$ to be 93.3 S/m (equal to $s^3A^2/(m^3kg)$), which is comparable to literature values for similar materials at 140GHz [2]. The peak of absorption lies at 90µm and depending on incidence angle and polarization, between 50 and more than 90% of the incoming microwaves were absorbed.

Through the model, the reason of the advantageous absorption properties compared to the other two materials could be determined: Surprisingly, the compound material does not absorb better, because it has a higher absorption coefficient than the other materials, but exactly the opposite of it. This leads to an increased resonator effect, whereby the microwave travels for a much longer time inside the layer, by far offsetting the decreased absorption per length unit.

The second important effect measured is, that the resonant thickness is almost independent of the incidence angle. Normally one would expect by geometry, that a higher incidence angle leads to a bigger phase shift, when passing through the layer, compared to a lower incidence angle for the same layer thickness. Thereby it is reasonable to assume, that higher incidence angles have smaller resonant thicknesses. This effect could be firmly rejected for the investigated material and again the reasons could be concluded from the model: At first, the refractive index of the material is comparably high, thereby decreasing the difference in angle for the diffracted wave inside the medium of an high incidence angle wave compared to a low incidence angle one. Secondly, for higher incidence, the wave not just travels longer inside the material, but also longer inside the air before interfering with itself, whereby the two phase shifts partially cancel out. This has positive implications considering the possible use to absorb stray radiation or when the geometry of the situation changes, because the optimal layer thickness will stay (almost) constant.

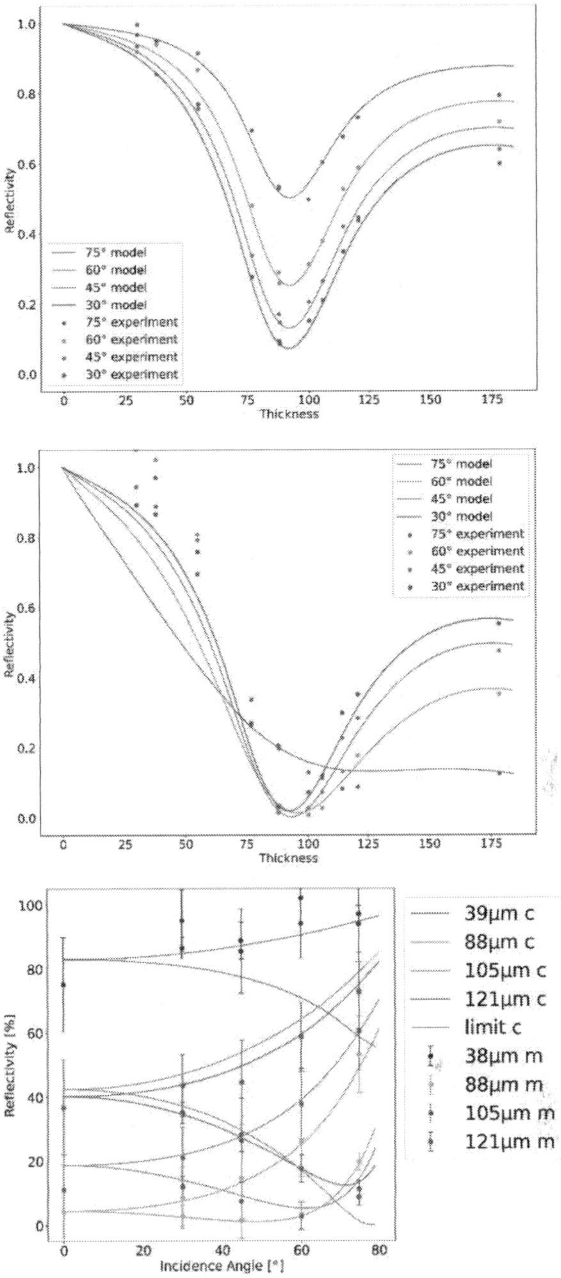


**Fig. 3.** Thickness dependency of the reflectivity for orthogonal and parallel polarization respectively, additionally the incidence angle dependency. The points are the measured values, the lines the respective calculated ones.

## 3 Simulation results

The model further allows to enhance the understanding of resonant absorbers. In figure 4 the reflectivity as function of thickness at increasing absorption is shown.

As expected, when the absorption coefficient is very low, the microwave won't be disturbed by the layer and is simply refllected at the metal substrate. The exception is the resonant thickness at around 90µm, at

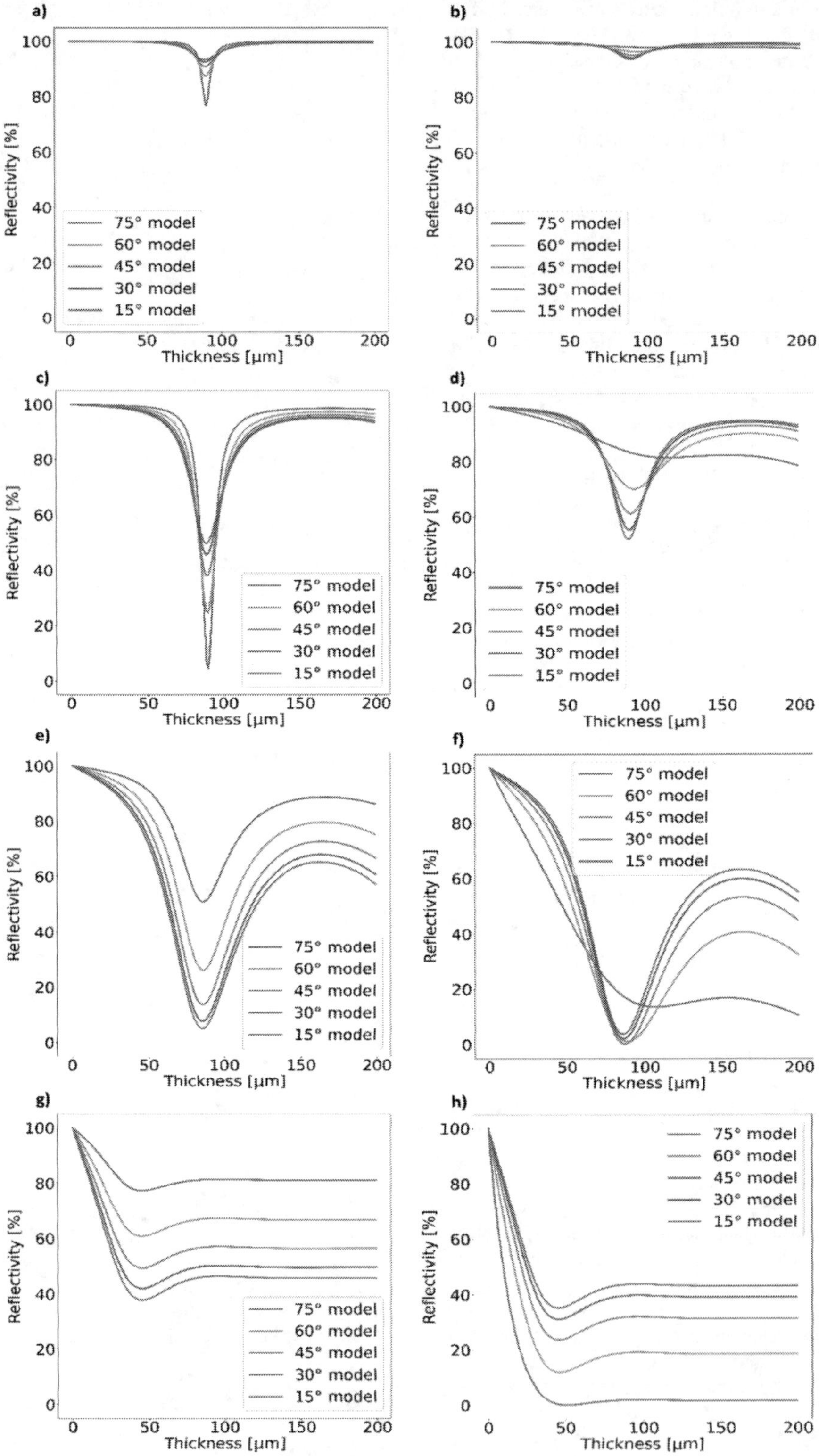


**Fig. 4.** Modeled data for the result of a change in absorption coefficient on the thickness dependency with a refractive index of 5. Left column is orthogonal, right one parallel polarization, the absorption coefficient from top to bottom is 1, 10, 100, 1000.

 https://doi.org/10.1051/epjconf/202327702009

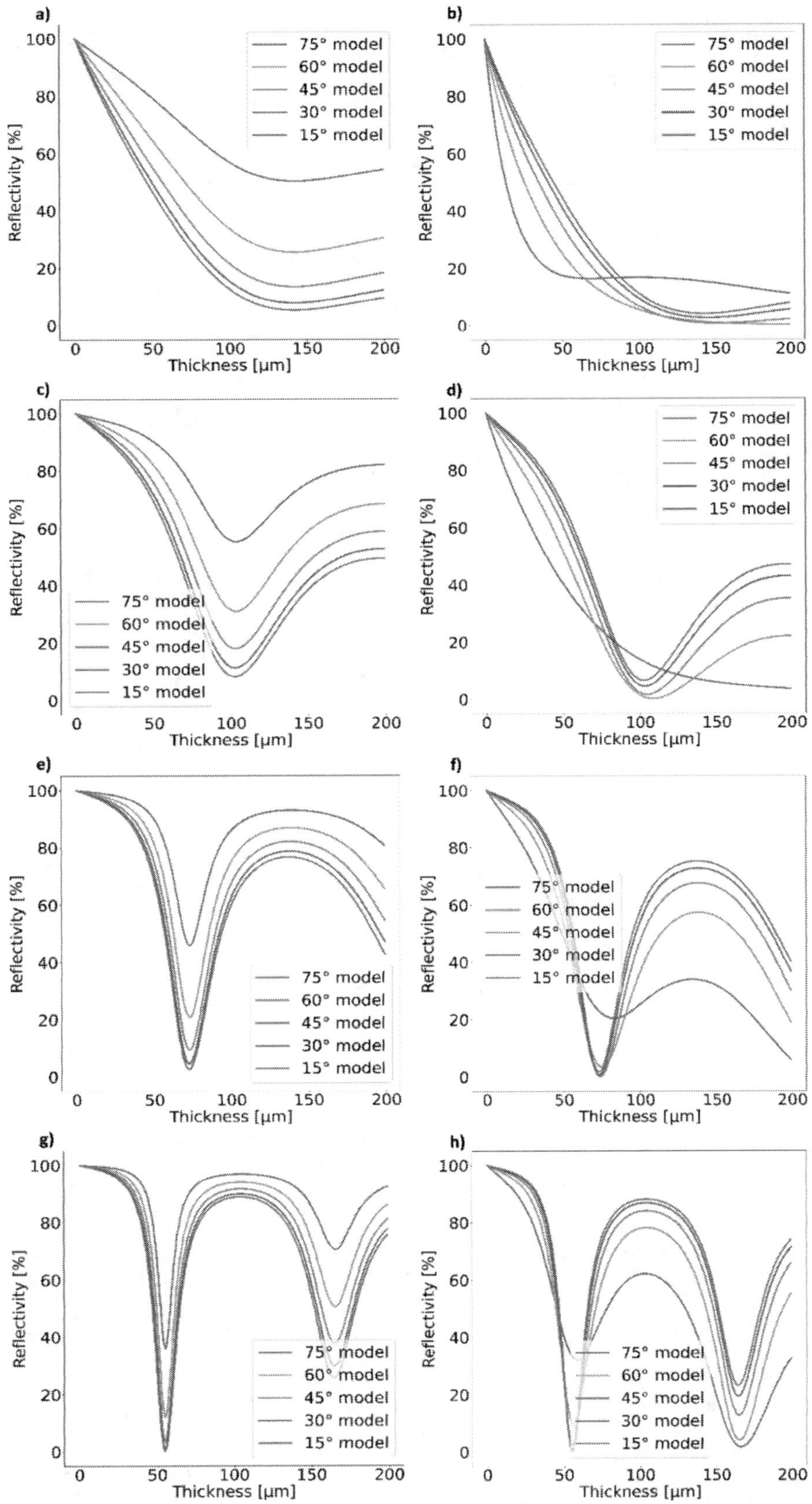


**Fig. 5.** Modeled data for the result of a change in refractive index on the thickness dependency with an absorption coefficient of 100. Left column is orthogonal, right one parallel polarization, the absorption coefficient from top to bottom is 2,4,6,8.

which the phase shift induced by the passage through the layer exactly leads to destructive interference at the layer to air interface. Thereby the microwave travels for much longer distances inside the layer, whereby even with small absorption per distance an effect is visible.

For a higher absorption coefficient (c and d), the absorption at the resonant thickness will increase dramatically. At the same time, the absorption for a microwave, which travels just once through the layer, is still negligible. At this point the resonantor properties are absolutely dominant.

Increasing it even further (e and f), to a level comparable to the measured material, the effects become more complicated, because now two contributions are overlapping for the reflectivity: At first, the resonator effect is still very relevant, the resonant thickness is clearly visible. In contrast to before, even at not resonant thicknesses relevant absorption occurs. This means that even after a single passing a relevant amount of energy is absorbed, whereby the destructive interference and thereby the resonator is perturbed. This increases the effect of the first time the wave hits the air-layer interface.

For very high absorption coefficients (g and h), the dependency simplifies a lot: Almost the complete energy is absorbed by passing once through the layer. This means no destructive interference can occur, the resonator effect vanishes, just the reflection coefficient at the first air-layer interface is relevant. Thereby the thickness becomes mostly irrelevant, the whole energy traveling into the layer will be absorbed either way. This complete surpression of the resonator effect leads to the curious effect described before, that a material with such a high absorption coefficient is actually a worse absorber than a resonant layer with a lower one.

Another effect, which can be seen, is that the increase of absorption coefficient will shift the resonant thickness. This is due to the fact, that for materials with significant absorption the effect of the absorption on the wavelength can not be neglected anymore [3], whereby the wave length decreases and logically the phase shift per length traveled in the material increases, leading to a lowering of the resonant thickness.

Additionally one can estimate the effects of the refractive index on the absorption properties (figure 5). When starting with low refractive indices (a and b), the absorption peak around the resonant frequency is excessively broad and it is generally speaking a very good absorber. One of the reasons is, that the small difference in refractive index at the air-layer interface leads to a very small initial reflection. The second reason is, that a smaller refractive index for the same absorption coefficient leads to a larger attenuation constant [1]. The effect described before, that the resonant thickness is mostly independent from the incident angle, is definitely not the case for these kind of materials, which is also expected, considering the already discussed surpression mechanism.

Increasing the refractive index (c and d) will lead to a decrease of the effective absorption and more reflection at the initial air-layer interface, thereby decreasing the absorption drastically at most thicknesses. The obvious exception is the resonant thickness, which is now shifted to lower values, but interestingly has very similar absorption properties compared to before. In this parameter region the effects of a decreasing effective absorption coefficient and an increase of passages through the layer cancel out almost perfectly.

With the next increase (e and f), the absorption under non-resonant conditions becomes very small. Additionally the absorption peak becomes increasingly sharp, while the absorption values are still similar to before. With these sudden shifts in reflectivity it also becomes increasingly hard to produce a layer at the resonant thickness, because within the expected thickness error of the plasma spraying of $\pm 10\mu m$ the reflectivity already changes significantly.

Finally, with a refractive index of 8 (g and h), a second absorption peak shows up in the displayed thickness region. The trends from before simply continue: Sharper resonance peak, less absorption outside the resonance, but almost identical to the one before directly at the resonant thickness. A small, but interesting effect is also visible when comparing the two polarizations in (g) and (h): While in the orthogonal case, the 75° incident angle is the worst absorbing case for both visible absorption thicknesses, it's actually the best absorbing one for the second peak of the parallel polarization. This is due to the decrease of the resonator contribution for higher thicknesses.

## 4 Summary and future research

The current state of work has been presented. This included a short presentation on the production process, the measurement and the theory side of oxide layer materials used for microwave absorption. Because these have been described excessively in previous work [1], a focus was on the simulation. These were done using a plane wave model, which takes into consideration the not purely dielectric nature of the materials and multiple reflexions. It has been shown, that measurement and simulation are in reasonable agreement, so the model was now used to predict some of the behaviour of these kind of materials. The main results were: A resonance occurs when the thickness of the layer results in a phase change in a single passage equal to $\pi/2$, which can be used for efficient absorption, this thickness is for 170GHz in the order of magnitude of 100μm. There exists an optimal absorption coefficient, both below and above this value the effective absorption is reduced, for a frequency of 170GHz the value is usually between 50 and 500S/m, depending on the refractive index. Speaking of refractive index, one between 4 and 6 seems to be favourable. Above 6, the absorption peak is not broad enough for the expected thickness error of the production process. Below 4, the resonant thickness is strongly incidence angle dependent.

The next step is to find a way to change the material parameters in the production process by altering the percentage of titanium oxide in the compound material,

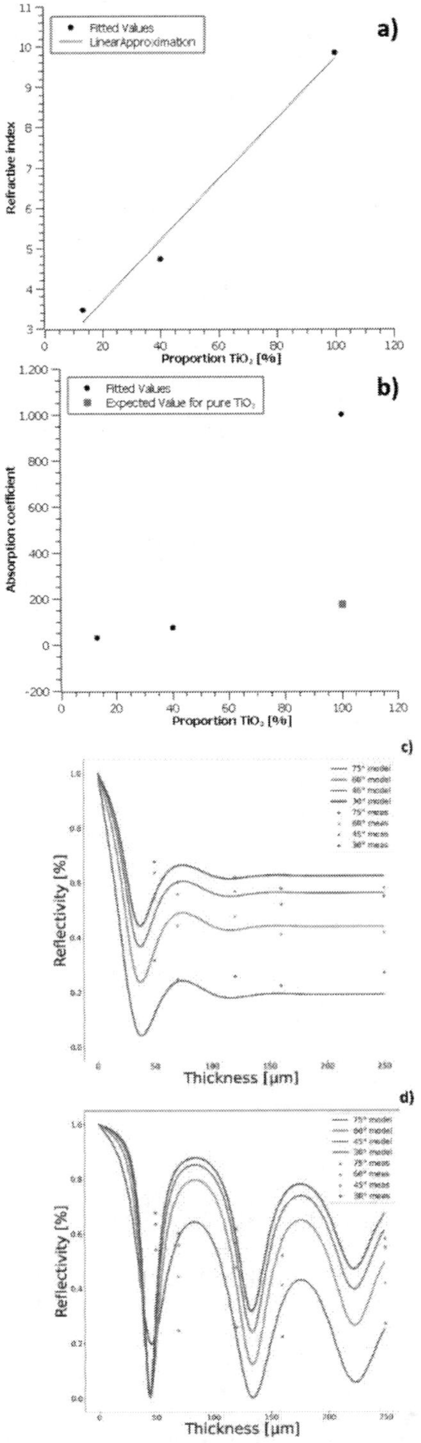


**Fig. 6.** The effect of an increase in $TiO_2$ proportion on the refractive index (a) and absorption coefficient (b). In red the linear fit and the expected value from a linear fit respectively is shown. In (c) and (d), the resulting thickness dependency for pure $TiO_2$ is shown, when considering an absorption coefficient of 1000S/m (c) and one of 175S/m (d).

which would allow the production of beam dumps by design. This process is already in progress, till now three material systems have been investigated: pure titanium oxide, the presented 40% one and a 13% one, with the rest being aluminum oxide. From microscopic images (compare figure 1) it is known, that the two oxides mostly stay in separate phases, a large change in atomic structure is therefore not expected. Also we can assume, that compared to the 1.76mm wavelength of the microwave, the materials are almost homogeneous, thereby a linear change of the material parameters is at least a reasonable first expectation.

When investigating the refractive index dependency (figure 6a), a linear dependency seems to be confirmed. In contrast, when looking at the absorption coefficient (figure 6b), the expected value from a linear extrapolation for pure $TiO_2$ is just a fifth of the value estimated by the optimized parameters, which come from fitting the model to the measurements. This leads to two possible conclusions: The linear dependency is simply not correct for the absorption coefficient or something with the optimization went wrong. A closer look into the thickness dependency of pure TiO2 makes the second option more likely: In 6c the fit is displayed, when using the optimized absorption coefficient of around 1000 S/m. There is a reasonable agreement with the measurement points. The problem is, when using the 175S/m instead coming from the linear extrapolation, it is also not completely unreasonable. This showcases a fundamental problem: With the high expected refractive index of $TiO_2$, three absorption peaks can be expected in this thickness range. The five measured thicknesses are simply a too low resolution to describe the three absorption peaks sufficiently, leading to inconclusive results for this material. The obvious conclusion is: more research is necessary, especially more thicknesses for titanium oxide to determine the material parameters accurately and in general more material compositions, to evaluate if the linear dependency hold true.

The biggest shortcoming of this study so far is, that all measurements and experiments were performed at room temperature and not at the upto 500°C, which is expected in the later device for the beam dump. This is far from ideal, thereby making temperature dependency plus stability of the material the obvious next focus of research. Additionally in the scope of the approval process of the material for ITER, the vacuum compatibility needs to be proven.

## References

1. A.Hentrich et al., *Resonant Atmospheric Plasma-Sprayed Ceramic Layers Effectively absorb Microwaves at 170 GHz*, J Infrared Millim Terahertz Waves (2022)
2. R. Wacker, F. Leuterer, D. Wagner, H. Hailer, W. Kasparek, *Characterization of absorber materials for high-power millimetre waves*, in Proceedings of the Conference: Infrared and Millimeter Waves (2002)
3. Y.Duan and H.Guan, *Microwave absorbing materials*, Chapter 1, Pan Stanford Publishing Pte. Ltd., Singapore (2016)

# Database study of turbulent electron temperature fluctuation measurements at ASDEX Upgrade

*Christian* Yoo[1*], *Rachel* Bielajew[1], *Garrard D.* Conway[2], *Pedro A.* Molina Cabrera[1,2], *Pablo* Rodriguez-Fernandez[1], *Branka* Vanovac[1], *Anne E.* White[1], the EUROfusion MST1 Team[3], and the ASDEX Upgrade Team[4]

[1]Massachusetts Institute of Technology, Cambridge, MA 02139, USA
[2]Max Planck Institute for Plasma Physics, 85748 Garching, Germany
[3]See author list of B. Labit et al. 2019 Nucl. Fusion **59** 086020
[4]See author list of U. Stroth et al. 2022 Nucl. Fusion **62** 042006

**Abstract.** In this work, an automated method for the analysis of data from the correlation electron cyclotron emission (CECE) diagnostic is applied to discharges in the ASDEX Upgrade (AUG) tokamak. This recently developed, automated method provides an efficient means of accurately analysing large quantities of experimental turbulence data, enabling the development of the largest database of CECE measurements of tokamak plasmas to-date. The turbulence database provides the opportunity to search for large-scale trends in experimental data to improve our understanding of transport-relevant plasma turbulence. The results of physics-based investigations utilizing this turbulence database will be reported on separately from this work.

## 1 Introduction

Turbulence is known to play a major role in the transport of energy and particles out of tokamak fusion reactors [1], lowering the associated energy and particle confinement times and reducing the ability of tokamaks to produce net energy. Accurate models of turbulence and transport are thus vital to the design of fusion reactors that will be used for energy production [2].

The correlation electron cyclotron emission (CECE) diagnostic installed on the ASDEX Upgrade (AUG) tokamak detects second harmonic, X-mode, electron cyclotron emission in order to measure broadband, long-wavelength ($k_\theta\rho_s \lesssim 0.3$) turbulent electron temperature fluctuations perpendicular to the mean magnetic field ($\delta T_{e\perp}/T_e$) [3], yielding insight into turbulence-driven transport. These temperature fluctuations are among the key parameters driving the turbulent electrostatic electron heat flux in tokamak plasmas [4]. Since the original implementation of the CECE diagnostic on W7-AS [5], follow-on studies on devices around the world have reported on important, transport-relevant experimental trends in turbulent electron temperature fluctuations using CECE data (see, e.g., ref's [6-9]). Additionally, data from the CECE diagnostic has and continues to serve as a key component in the validation of advanced gyrokinetic and gyrofluid models of plasma turbulence [10-11].

Up to this point, however, experimental turbulence studies based on CECE data have generally focused on traditional shot-to-shot comparisons using discharges with similar experimental conditions. Studies of CECE data containing multiple discharges have generally been limited to about a dozen or fewer discharges (see, e.g., a relatively large study in ref [10]). Indeed, to the best of the knowledge of the authors, a large-scale experimental turbulence database based on CECE measurements of tokamak plasmas has not been reported on prior to this work.

Experimental databases provide the opportunity to search for large-scale trends not necessarily accessible in traditional shot-by-shot comparisons. Studies based on large-scale databases as detailed in, for example, ref.'s [12] and [13] have led to the development of important experimental scalings that have guided the design and operation of tokamaks. In the study of turbulence in tokamak plasmas specifically, probes have been used to build a multi-machine database of measurements in the edge of tokamak plasmas [14] while reflectometry signals taken during ten years' worth of operation of the Tore Supra tokamak were analysed in a large database study of core frequency spectra [15]. The current work reports on the development of a large-scale turbulence database using the CECE diagnostic installed on AUG.

The development of this turbulence database has benefited from the combination of multiple factors. An ongoing, active collaboration between scientists at the MIT Plasma Science and Fusion Center and the Max Planck Institute for Plasma Physics has enabled measurements using the CECE diagnostic at AUG for multiple years. As a passive, non-perturbative system, the CECE diagnostic at AUG can operate on a nearly continuous basis. Interpretation of CECE data as turbulent electron temperature fluctuations is relatively straightforward under certain conditions that include

* Corresponding author: cyoo@mit.edu

long steady-state periods, high optical depth, measurement localizations far from cut-off, and a nearly perpendicular line of sight between the plasma and the diagnostic optics. Since its installation in April 2021, an additional 24-channel CECE system has operated in a separate toroidal sector in tandem with the 24-channel CECE system originally detailed in ref. [3], doubling the measurement capacity of the diagnostic. Access to years' worth of data from a large suite of diagnostics at AUG and advanced analysis systems such as the Integrated Data Analysis (IDA) multi-diagnostic analysis system [16] provides valuable information that is paired with the CECE data in the turbulence database. Altogether, these contributing factors have enabled the development of a large-scale experimental turbulence database that is being applied towards the search for large-scale trends to help improve our understanding of plasma turbulence and transport.

## 2 Automated method for the analysis of CECE measurements

In order to create a database of experimental CECE measurements, an automated method was developed for the analysis of the diagnostic's data. For each discharge analysed, the automated method: 1. determines the optimal time windows for CECE analysis 2. verifies that the plasma conditions during the selected time intervals are well-suited for the interpretation of CECE measurements as turbulent electron temperature fluctuation amplitudes perpendicular to the mean magnetic field ($\delta T_{e\perp}/T_e$) 3. verifies the quality of the raw voltage data measured by the diagnostic 4. filters the auto-power, cross-power, and coherence spectra calculated from the raw data 5. calculates $\delta T_{e\perp}/T_e$ based on the coherence spectra 6. checks for the presence of coherent modes that can amplify the magnitude of the coherence spectra beyond the level attributed solely to drift-wave-type turbulence and 7. processes the $\delta T_{e\perp}/T_e$ values from the CECE diagnostic into the turbulence database along with local and global plasma and engineering parameter values from a host of additional diagnostics available in the AUG shot-file system.

### 2.1 Determination of analysis time windows

The determination of time windows for the analysis of CECE measurements by the automated method is based on two often competing requirements. On the one hand, the maximization of these time intervals leads to better statistics, as CECE data is time-averaged to distinguish turbulence from thermal noise [5]. On the other hand, the goal of the database study is to assess trends in $\delta T_{e\perp}/T_e$ across a wide range of parameter space, the accuracy of which are improved when the database parameters are in steady-state as this minimizes the statistical uncertainty in the parameter values. The long duration, steady-state conditions well-suited for CECE measurements and database studies cannot be taken for granted across experiments at AUG, as many discharges feature experimental conditions that evolve continuously throughout the discharges.

Figure 1 shows the time series of the three parameters used in the steady-state analysis algorithm: the plasma current ($I_p$) [MA], electron temperature ($T_e$) [keV], and electron density ($n_e$) [$10^{19}$ m$^{-3}$], with $T_e$ and $n_e$ profile data provided by IDA and defined as the local values of these parameters at each CECE measurement location. These parameters are particularly important to analyse in steady-state as they serve as the basis for multiple important physical parameters including the localization of the CECE measurements, optical depth, cutoff frequencies, collisionality, and gradients in $T_e$ and $n_e$. Here steady-state is defined as the state in which the mean values of $T_e$ and $n_e$ remain within designated limits, as a percentage of their initial values during the particular time window, at all CECE measurement locations during the plasma current flat-top period. While one could account for additional parameters (such as the gradients that drive turbulence) in the steady-state algorithm, doing so generally decreases the probability that the algorithm will detect steady-state intervals during a given discharge. Additionally, while $I_p$, $T_e$, and $n_e$ data are nearly always available in AUG shot-files for use in the steady-state algorithm for a given discharge, certain diagnostic data (such as $T_i$ data) are not.

The mean value of each of the parameters is tracked in 10-msec time-averaged increments. A minimum time interval constraint of 0.5 seconds is implemented to achieve CECE sensitivity limits on the order of 0.2-0.3% (as per equation 7 in ref [3]). The steady-state algorithm utilizes an iteratively increasing steady-state threshold, from a starting point of 2% up to a pre-defined maximum of 10%. This prioritizes finding the most steady-state intervals when they are available while still enabling the algorithm to obtain relatively steady-state periods during even highly dynamic discharges, as shown by the three detected steady-state intervals highlighted in blue in Figure 1.

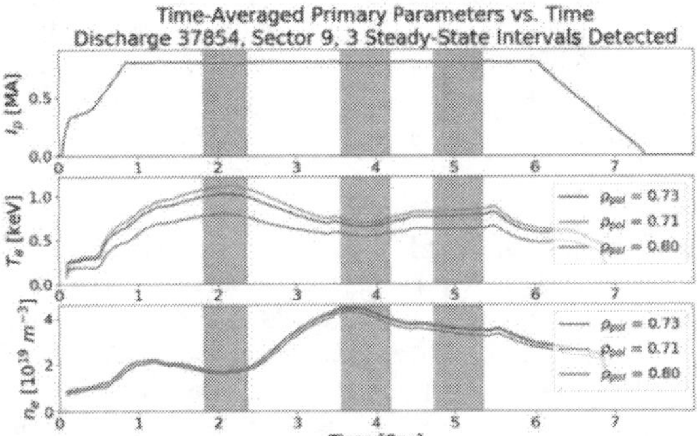


**Fig. 1.** Time intervals for the analysis of CECE data are selected based on the degree to which the local electron temperature and density are in steady-state at each CECE measurement location within the plasma current flat-top period. The curves shown for the electron temperature and density are evaluated at three radial positions: the radially innermost, middle, and outmost CECE measurement locations for this discharge.

## 2.2 Verifying plasma conditions

Sufficient optical depth is important for ensuring that fluctuations in the radiation temperature measured by the CECE diagnostic can be interpreted as electron temperature fluctuations and not fluctuations in electron density [5]. As shown in Figure 2, the analysis process calculates the time-averaged optical depth over the range of CECE measurements (shown in the highlighted region) during the selected time windows. Optical depth values of 2 (red) and 4 (green) mark typically cited lower limits as to what is considered sufficient optical depth for CECE measurements. [17]

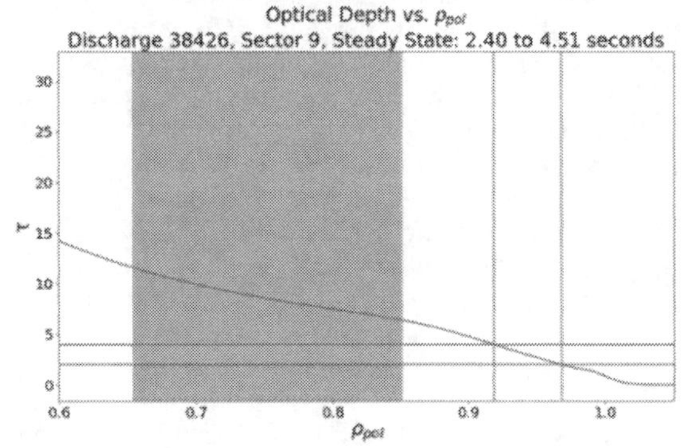


**Fig. 2.** The optical depth (τ) at each location in the plasma measured by the CECE diagnostic (highlighted region) is tracked by the analysis method. Higher optical depth minimizes the impact of density fluctuations on CECE measurements [5, 17]. The red and green lines denote optical depth levels of 2 and 4, respectively.

The analysis also identifies CECE channels that are measuring in regions of the plasma that are cutoff (i.e., regions where the right-hand cutoff frequency is greater than the second harmonic electron cyclotron frequency) [18], in which case the ECE signals at the intended measurement locations do not reach the diagnostic's focusing optics and detectors. Additionally, the analysis identifies CECE channels that have corresponding second harmonic electron cyclotron frequencies that are less than 105.5% of the respective right-hand cutoff frequencies. This 105.5% frequency threshold is more conservative as a limit than the right-hand cutoff frequency and indicates the level below which refraction has a non-negligible effect on ECE measurements. This limit corresponds to an equivalent threshold, determined based on ray tracing studies, which requires that the local electron density be less than 85% of the cutoff density in order for refraction to have a negligible effect on ECE measurements [19-21]. CECE channels whose second harmonic electron cyclotron frequencies are greater than this limit, as shown in Figure 3, are included for further analysis while those channels with lower frequencies are removed from the analysis process.

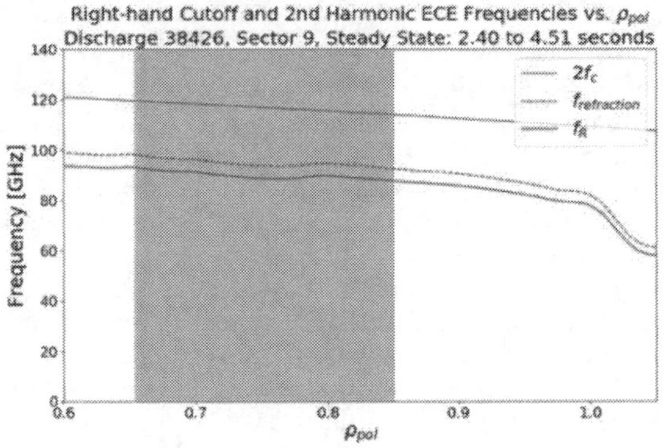


**Fig. 3.** CECE measurement locations (highlighted region) are not cutoff if the second harmonic electron cyclotron frequency (solid green curve) is above the right-hand cutoff frequency (solid red curve). Additionally, the effects of refraction on data processed by the automated analysis are minimized by identifying and removing from further analysis any CECE channels that have second harmonic electron cyclotron frequencies below a 105.5% multiple of the right-hand cutoff frequency (dashed red curve) [19-21].

## 2.3 Verifying the quality of the raw CECE data

The raw voltage signals detected by each CECE channel are checked for a finite response to the presence of the plasma as well as for saturation, the latter of which is illustrated by the red curve in Figure 4, as this prevents the accurate processing of the time-series voltage data into auto-power and cross-power spectral density and coherence frequency spectra using Fourier analysis. Any channels that do not pass these checks are removed from the analysis. Additionally, the mean voltage level of each CECE channel is calculated during each analysis time window in order to determine whether the voltages are satisfactorily within the linear response region of the CECE detectors.

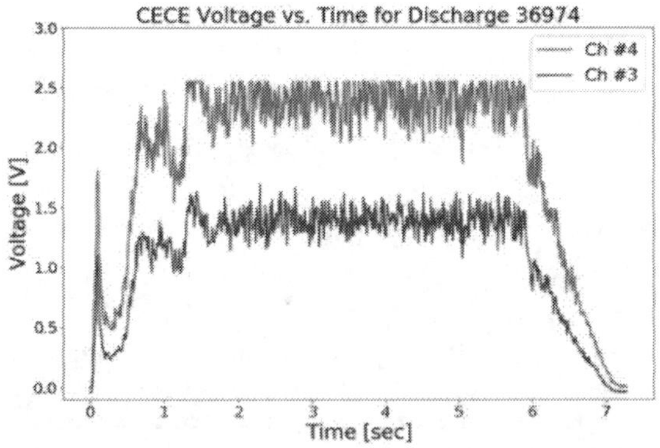


**Fig. 4.** The voltage time traces of each CECE channel are checked for a finite response to the presence of the plasma as well as for saturation. Channels not passing these checks are removed from further analysis for a given analysis interval. The red curve shows a channel whose signal has saturated at the maximum voltage level of the data acquisition system. The blue curve would pass the checks of the analysis.

## 2.4 Filtering CECE spectra

A frequency-based filter is applied to the auto-power, cross-power, and coherence frequency spectra in order to remove large, narrowband peaks. These peaks, shown in red in Figure 5 on top of the broadband spectrum attributed to drift-wave-type turbulence, have been identified as being due to electronics noise since they have been observed in the spectra even when no plasma is present. While the peaks appear at similar frequencies across CECE channels and experiments, there is some variation of the peak frequencies measured by any individual diagnostic channel during even the same plasma discharge.

As a result of the temporal variation in the frequencies of the peaks and in order to account for the possibility of narrowband peaks arising in the data at entirely different frequencies, a threshold-based, moving-average filter is applied to identify and remove the narrowband peaks for each channel during each analysis time window. Thresholds based on the mean and standard deviation of each moving frequency increment are used to identify the peaks from the underlying broadband spectra. To reduce the potential of over-filtered spectra introducing an additional source of error into the results of the CECE analysis, any spectra that have been filtered beyond a designated percentage of 10% of their signal bandwidth are removed from further analysis.

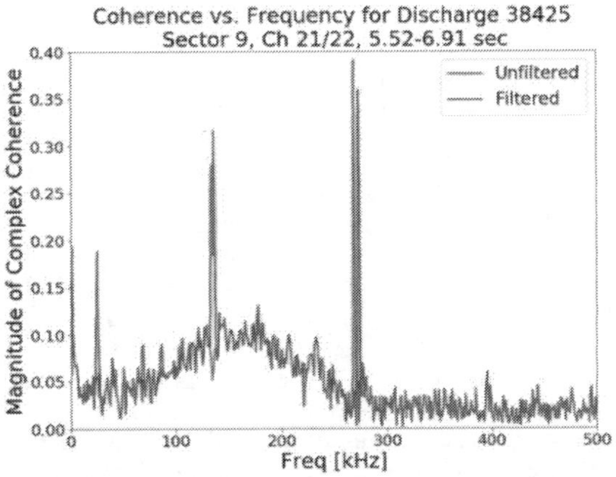


**Fig. 5.** Narrowband peaks (red) attributed to electronics noise are identified and removed from the CECE auto-power and cross-power spectral density and coherence frequency spectra using a threshold-based, moving average filter.

## 2.5 Checking for coherent modes

In order to identify coherent modes due, for example, to MHD activity that might be present within the coherence spectra, the analysis process calculates the spectrogram of data from a Mirnov coil measuring the radial magnetic field. As shown in Figure 6, the spectrograms can be used to identify possible contributions to the coherence spectra that are not due to drift-wave-type turbulence at any given time during a plasma discharge. Any coherence spectra whose signal bandwidths overlap with the frequencies of any such modes identified in the magnetics are flagged, although the spectra are not removed from the analysis process in order to enable the possibility of studying the effect of such modes on CECE measurements.

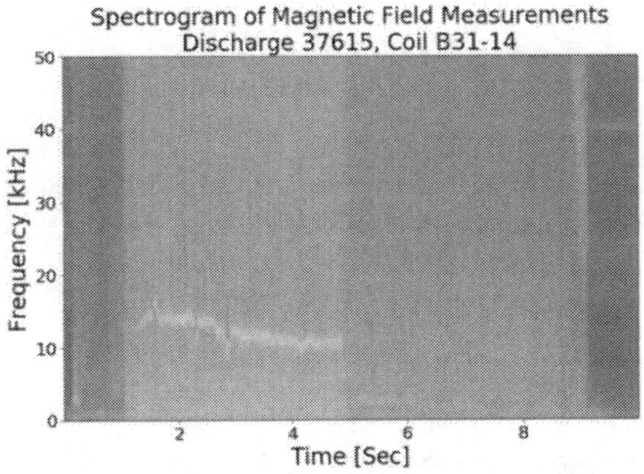


**Fig. 6.** A spectrogram of radial magnetic field measurements from a Mirnov coil is used to identify coherent modes (as shown by the narrow yellow trace) that can cause the CECE coherence spectra to be greater in magnitude than they would be solely as a result of drift-wave-type turbulence. Coherence spectra whose signal bandwidths overlap with the frequencies of the coherent modes identified by the magnetics are flagged.

## 2.6 Calculating $\delta T_{e\perp}/T_e$

As per equation 1 in ref. [3], calculating $\delta T_{e\perp}/T_e$ involves the integration of the coherence spectra over the signal bandwidth and the subtraction of a background coherence level. In order to determine the frequency limits for integration, a threshold is applied based on the standard deviation of moving frequency increments within the ECE-ECE phase angle spectra for each frequency-adjacent pair of CECE channels. As shown in Figure 7, the dip in the magnitude of the standard deviation of the ECE-ECE phase angle tracks well with the region of frequency space in which the magnitude of the complex coherence is large, enabling an accurate identification of the signal bandwidth. A fixed frequency width (i.e., 20 kHz) above the coherence integration region is used to select the background subtraction region.

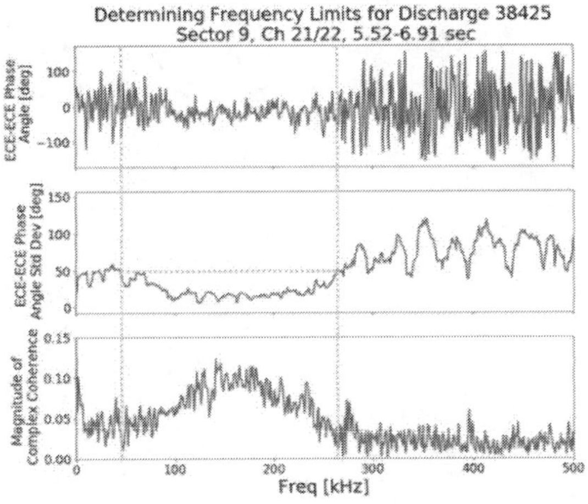

**Fig. 7.** The standard deviation of the ECE-ECE phase angle spectra between frequency-adjacent channels is used to determine the frequency limits (vertical, yellow dotted lines) over which the CECE channels should be integrated during the calculation of $\delta T_{e\perp}/T_e$.

### 2.7 Processing diagnostic data into database

Upon completing the calculation of $\delta T_{e\perp}/T_e$ values for a given analysis interval and discharge, the processed data is written to the turbulence database. In addition, the time-averaged values of numerous local plasma parameters, evaluated at the same radial positions as the $\delta T_{e\perp}/T_e$ measurements, and the time-averaged values of multiple global plasma and engineering parameters, all of which are accessed through the AUG shot-file system, are written to the turbulence database.

## 3 Turbulence database

### 3.1 Database parameters

At the present time, over 30 parameters, including $\delta T_{e\perp}/T_e$ and local and global plasma and engineering parameters, are written to the database along with their statistical uncertainties for each analysis interval. The inclusion of multiple tens of parameters in the turbulence database provides the opportunity to search for large-scale trends in $\delta T_{e\perp}/T_e$ across a wide-range of experimental operating conditions. As shown in Table 1, these parameters are relevant to: 1. turbulence drive and damping 2. ion and electron kinetics 3. the magnetic field, safety factor, and rotation 4. auxiliary heating power and radiated power 5. stored energy, current, and confinement 6. plasma geometry and 7. fuel composition. $\gamma_{peak}$ corresponds to the peak value of each coherence spectrum while $\gamma_{FWHM}$ refers to the full-width at half-maximum of a Gaussian fit to each coherence spectrum. $\nabla T_e$, $\nabla T_i$, $\nabla n_e$ represent the gradients in electron temperature, ion temperature, and electron density, respectively. $\upsilon_{eff}$ is the effective collisionality as defined in ref. [22]. $B_{tot}$ is the total magnetic field, $q_s$ is the safety factor, and $v_{rot}$ is the plasma rotation velocity. $P_{total}$ is the total auxiliary heating power, $P_{ohmic}$ is the ohmic heating power, $P_{NBI}$ is the neutral beam injection power, $P_{ECRH}$ is the Electron Cyclotron Resonance Heating power, $P_{ICRH}$ is the Ion Cyclotron Resonance Heating power, and $P_{rad}$ is the radiated power. $W_{mhd}$ is the MHD stored energy, $\beta_N$ is the normalized beta, $I_p$ is the plasma current, and $\tau_E$ is the energy confinement time. R is the major radius, a is the minor radius, $\varepsilon$ is the inverse aspect ratio, $\kappa$ is the elongation, $\delta_{upper}$ is the upper triangularity, and $\delta_{lower}$ is the lower triangularity. A is the fuel mass number and $Z_{eff}$ is the plasma effective atomic number.

| | | |
|---|---|---|
| | **Measured CECE Parameters** | |
| | $\delta T_{e\perp}/T_e$, $\gamma_{peak}$, $\gamma_{FWHM}$ | |
| | **Parameters associated with CECE Measurements** | |
| **Local** | Turbulence Drive and Damping | $\nabla T_e$, $\nabla T_i$, $\nabla n_e$, $v_{eff}$ |
| | Ion and Electron Kinetics | $T_e$, $T_i$, $n_e$ |
| | Field, Safety Factor, Rotation | $B_{tot}$, $q_s$, $v_{rot}$ |
| **Global** | Power | $P_{tot}$, $P_{ohmic}$, $P_{NBI}$, $P_{ECRH}$, $P_{ICRH}$, $P_{rad}$ |
| | Energy, Current, Confinement | $W_{mhd}$, $\beta_N$, $I_p$, $\tau_E$ |
| | Geometry | R, a, $\varepsilon$, $\kappa$, $\delta_{upper}$, $\delta_{lower}$ |
| | Fuel composition | A, $Z_{eff}$ |

**Table 1.** The turbulence database contains over 30 local and global plasma and engineering parameters along with their statistical uncertainties for each CECE analysis interval.

### 3.2 Overview of database contents

The first iteration of the database was developed by running the automated analysis method on 700 discharges measured by the CECE system installed on AUG Sector 9 from May 2020 to January 2021. Among the 700 discharges (including AUG discharge numbers 37,700 - 38,399), 166 of them contained data that passed the checks of the automated analysis. Through the analysis of the 166 discharges, 4,286 datapoints were written to the database. Of this number, 3,936 datapoints were located in the core (defined here as corresponding to $\rho_{tor} < 0.8$, where $\rho_{tor}$ is the square root of the normalized toroidal magnetic flux) of the AUG plasmas.

## 4 Conclusion

The current work reports on the development of a large-scale turbulence database using the CECE diagnostic installed on AUG. A recently developed, automated method provides an efficient means of accurately analysing large quantities of experimental turbulence data from the CECE diagnostic. The data processed by the automated method is stored along with local and global plasma and engineering parameters, evaluated during the same time intervals and at the same radial locations as the CECE measurements. The application of this method to data from discharges at AUG has enabled the development of the largest turbulence database of CECE measurements of a tokamak plasma to-date. The turbulence database provides the opportunity to search for large-scale trends in experimental data to help improve our understanding of transport-relevant plasma turbulence. Work is in progress using the database to study the effects of collisionality and gradients on the saturated amplitude of turbulence measured by the CECE diagnostic. The database could also be applied towards large-scale validation studies of advanced computational models of plasma turbulence. The results of physics-based investigations utilizing this turbulence database will be reported on separately from this work.

## Acknowledgements

This work is supported by the US Department of Energy under grants DE-SC0014264, DE-SC0006419, and DE-SC0017381.

This work has been carried out within the framework of the EUROfusion Consortium, funded by the European Union via the Euratom Research and Training Programme (Grant Agreement No 101052200 – EUROfusion). Views and opinions expressed are however those of the authors only and do not necessarily reflect those of the European Union or the European Commission. Neither the European Union nor the European Commission can be held responsible for them. This material is based upon work supported by the National Science Foundation Graduate Research Fellowship under Grant No. 1745302 and 2141064. Any opinion, findings, and conclusions or recommendations expressed in this material are those of the authors and do not necessarily reflect the views of the National Science Foundation.

## References

1. W. Horton, Reviews of Modern Physics **71**, 3 (1999)
2. P. Rodriguez-Fernandez, N.T. Howard, M.J. Greenwald, A.J. Creely, J.W. Hughes, J.C. Wright, C. Holland, Y. Lin, F. Sciortino, and the SPARC team, J. Plasma Physics **86**, 865860503 (2020)
3. A.J. Creely, S.J. Freethy, W.M. Burke, G.D. Conway, R. Leccacorvi, W.C. Parkin, D.R. Terry, A.E. White, Rev. Sci. Instrum. **89** 053503 (2018)
4. A.J. Wootton, B.A. Carreras, H. Matsumoto, K. McGuire, W.A. Peebles, Ch.P. Ritz, P.W. Terry, S.J. Zweben, Physics of Fluids B: Plasma Physics **2**, 2879 (1990)
5. S. Sattler, H.J. Hartfuss, W7-AS Team, Physical Review Letters **72**, 5 (1994)
6. R. Bielajew, G.D. Conway, M. Griener, T. Happel, K. Höfler, N.T. Howard, A.E. Hubbard, W. McCarthy, P.A. Molina Cabrera, T. Nishizawa, P. Rodriguez-Fernandez, D. Silvagni, B. Vanovac, D. Wendler, C. Yoo, A.E. White, ASDEX Upgrade Team, Phys. Plasmas **29**, 052504 (2022)
7. M. Fontana, L. Porte, S. Coda, O. Sauter, The TCV Team, Nucl. Fusion **58**, 024002 (2018)
8. C. Sung, A.E. White, N.T. Howard, C.Y. Oi, J.E. Rice, C. Gao, P. Ennever, M. Porkolab, F. Parra, D. Mikkelsen, D. Ernst, J. Walk, J.W. Hughes, J. Irby, C. Kasten, A.E. Hubbard, M.J. Greenwald, the Alcator C-Mod Team, Nucl. Fusion **53**, 083010 (2013)
9. A.E. White, P. Phillips, D.G. Whyte, A.E. Hubbard, C. Sung, J.W. Hughes, A. Dominguez, J. Terry, I. Cziegler, Nucl. Fusion **51**, 113005 (2011)
10. A.J. Creely, P. Rodriguez-Fernandez, G.D. Conway, S.J. Freethy, N.T. Howard, A.E. White, the ASDEX Upgrade Team, Plasma Phys. Control. Fusion **61**, 085022 (2019)
11. A.E. White, J. Plasma Physics, **85**, 925850102 (2019)
12. M. Greenwald, J.L. Terry, S.M. Wolfe, S. Ejima, M.G. Bell, S.M. Kaye, G.H. Neilson, Nucl. Fusion **28**, 2199 (1988)
13. R.J. Goldston, Plasma Phys. Control. Fusion **26**, 87 (1984)
14. P. Simon, M. Ramisch, A.A. Beletskii, A. Dinklage, M. Endler, S. Marsen, B. Nold, U. Stroth, P. Tamain, R. Wilcox, Plasma Phys. Control. Fusion **56**, 095015 (2014)
15. Y. Sun, R. Sabot, S. Heuraux, X. Garbet, S. Hacquin, G. Hornung, G. Verdoolaege, Phys. Plasmas **26**, 032307 (2019)
16. R. Fischer, C.J. Fuchs, B. Kurzan, W. Suttrop, E. Wolfrum, ASDEX Upgrade Team, Fusion Science and Technology **58**, 2 (2010)
17. T.D. Rempel, R.F. Gandy, A.J. Wootton, Rev. Sci. Instrum. **65**, 6 (1994)
18. M. Bornatici, R. Cano, O. De Barbieri, F. Engelmann, Nucl. Fusion **23**, 1153 (1983)
19. A.E. White, P. Phillips, D.G. Whyte, A.E. Hubbard, C. Sung, J.W. Hughes, A. Dominguez, J. Terry, I. Cziegler, Nucl. Fusion **51**, 113005 (2011)
20. A.E. White, L. Schmitz, W.A. Peebles, T.A. Carter, T.L. Rhodes. E.J. Doyle, P.A. Gourdain, J.C. Hillesheim, G. Wang, C. Holland, G.R. Tynan, M.E. Austin, G.R. McKee, M.W. Shafer, K.H. Burrell, J. Candy, J.C. DeBoo, R. Prater, G.M. Staebler, R.E. Waltz, M.A. Makowski, Rev. Sci. Instrum. **79**, 103505 (2008)
21. A.G. Lynn, P.E. Phillips, A. Hubbard, Rev. Sci. Instrum. **75**, 10 (2004)
22. R.M. McDermott, C. Angioni, G.D. Conway, R. Dux, E. Fable, R. Fischer, T. Pütterich, F. Ryter, E. Viezzer, the ASDEX Upgrade Team, Nucl. Fusion **54**, 043009 (2014)

EPJ Web of Conferences 277, 03002 (2023) https://doi.org/10.1051/epjconf/202327703002
EC21

# Electron temperature fluctuation measurements with Correlation Electron Cyclotron Emission in L-mode and I-mode plasmas at ASDEX Upgrade

*Rachel* Bielajew[1,*], *Garrard D* Conway[2], *Tim* Happel[2], *Klara* Höfler[3,2], *Pedro A* Molina Cabrera[4], *Ulrike* Plank[2], *Pablo* Rodriguez-Fernandez[1], *Davide* Silvagni[2], *Branka* Vanovac[1], *Christian* Yoo[1], *Anne* White[1], and The ASDEX Upgrade Team[5]

[1]Plasma Science and Fusion Center, Massachusetts Institute of Technology, Cambridge, MA 02139, USA
[2]Max Plank Institute for Plasma Physics, 85748 Garching, Germany
[3]Department E28, Technical University of Munich, 85748 Garching, Germany
[4]École Polytechnique Fédérale de Lausanne, Swiss Plasma Center, CH-1015 Lausanne, Switzerland
[5]See author list of U. Stroth et al. 2022 Nucl. Fusion **62** 042006

**Abstract.** The Correlation Electron Cyclotron Emission (CECE) diagnostic at ASDEX Upgrade (AUG) is used to investigate the features of outer core and pedestal ($\rho_{pol}$ = 0.85-1.0) turbulence across confinement regime transitions. The I-mode confinement regime is a promising operational scenario for future fusion reactors because it features high energy confinement without high particle confinement, but the nature of the edge and pedestal turbulence in I-mode plasmas is still under investigation. The edge Weakly Coherent Mode (WCM) appears in the I-mode pedestal and may play a role in transport. In this work we explore electron temperature ($T_e$) fluctuations in the plasma outer core and pedestal using a 24-channel high radial resolution CECE radiometer. CECE measurements provide turbulence information including the $T_e$ fluctuation amplitude, turbulent spectra, and radial localization of turbulent features. With CECE measurements we show that the WCM is localized in the pedestal region in both L-mode and I-mode and is measured in optically thick plasmas with a $T_e$ fluctuation amplitude of 2.3%. Broadband drift wave turbulence is measured in the outer core with a $T_e$ fluctuation amplitude of <1%. A second CECE system recently installed at AUG allowed for non-standard fluctuation measurements during L-mode and I-mode experiments. The second CECE system was toroidally separated from the primary system, allowing measurements of the long-range toroidal correlation of the WCM indicating its low toroidal mode number. A reflectometer sharing a line of sight with the second CECE system enabled density-temperature cross-phase ($\alpha_{n_eT_e}$) measurements. The WCM $\alpha_{n_eT_e}$ changes between L-mode and I-mode as the $T_e$ gradient steepens.

## 1 Introduction

Electron temperature ($T_e$) fluctuation measurements provide important information on plasma confinement scaling and transitions between L-mode and high confinement regimes, as well as the role of edge fluctuations in these regimes. Using the Correlation Electron Cyclotron Emission (CECE) diagnostic at ASDEX Upgrade (AUG), these measurements are obtained with high temporal (4 MHz) and spatial (2 mm) resolution to study turbulence behavior in the high confinement regime I-mode. I-mode is characterized by a heat transport barrier without a particle transport barrier, so it achieves high energy confinement while allowing impurity exhaust and staying below stability boundaries that lead to Edge-Localized Modes (ELMs) [1–3]. I-mode is typically achieved at AUG in the unfavorable $\nabla B$ drift configuration because higher heating power can be applied before a transition to ELMy H-mode as compared to the favorable drift configuration.

The nature of turbulence in I-mode and its role in sustaining the I-mode regime is still under investigation, and CECE measurements reveal important properties of the turbulence spanning the region from the outer core (normalized radius $\rho_{pol}$ < 0.9) through the pedestal to the last closed flux surface (LCFS) at $\rho_{pol}$ = 1.0. The Weakly Coherent Mode (WCM) is a broadband, mid-frequency (40-100 kHz) pedestal fluctuation with a larger density fluctuation amplitude than electron temperature fluctuation amplitude [4, 5]. The WCM has been hypothesized to be important in the transport properties of I-mode, similar to how edge coherent modes in other ELM-free high confinement regimes cause particle transport in place of ELMs (e.g. the Quasi-Coherent Mode in Enhanced D-alpha H-mode and the Edge Harmonic Oscillation in Quiescent H-mode [6, 7]). The WCM has been observed in the density and electron temperature fluctuation signals of AUG L-modes as well [5, 8]. A Low Frequency Edge Oscillation (LFEO) is also observed as a coherent and low-frequency mode in the pedestal of I-modes, and may be a Geodesic Acoustic Mode (GAM) [9]. The GAM is proposed to couple with the WCM and cause its broad structure in frequency space [10, 11].

*e-mail: bielajew@mit.edu

 https://doi.org/10.1051/epjconf/202327703002

The rest of this paper will present a study of L-modes and I-modes using CECE and coupled diagnostics to probe the nature of the outer core and edge turbulence. In Section 2 we present CECE measurements of $T_e$ fluctuations, which investigate the nature of the WCM and LFEO as well as outer core and pedestal top turbulence. A comparison of favorable and unfavorable drift configuration $T_e$ turbulence is also presented. In Section 3 we present measurements of the WCM using non-standard hardware configurations, including long-range correlation measurements using toroidally separated CECE systems and $n_eT_e$ cross-phase measurements using a coupled reflectometer.

## 2 CECE measurements

The first CECE diagnostic on AUG shares optics with the 1-D ECE system and has a set of interchangeable radio frequency (RF) sections for tuning measurements to different regions of the plasma. For the following measurements, an RF section with a tunable Gunn oscillator and receiving range of 109-124 GHz was used for edge and pedestal optimized measurements. The RF section sits on a sliding rail so that the antenna beam waist (radius ~7.6 mm) can be moved radially in the plasma. For these measurements the waist was optimized for pedestal measurements. The intermediate frequency (IF) section of the CECE diagnostic features a comb of 24 channels with fixed filters of 200 MHz bandwidth and center frequencies spaced 250 MHz apart. Neighboring CECE channels sit within a turbulence correlation length and measure the same $T_e$ fluctuations but do not share thermal noise. By correlating neighboring channels, coherency spectra ($\gamma_c$) of $T_e$ turbulence are obtained. By integrating in frequency space over the turbulent features in the spectra and subtracting the background coherent noise, the normalized fluctuation amplitude ($\delta T_e/T_e$) perpendicular to the magnetic field can be obtained [12].

CECE is used to study $T_e$ turbulence in the outer core through the pedestal region in L-modes and I-modes. Figure 1 shows the $T_e$ fluctuation amplitude over radius during an I-mode phase of a discharge. The fluctuation amplitude rises smoothly from ~ 0.5% in the outer core ($\rho_{pol}$ < 0.90) to ~ 1% at the pedestal top ($\rho_{pol}$ = 0.96). Then the fluctuation amplitude increases sharply between $\rho_{pol}$ = 0.97 and the LCFS due to fluctuations associated with the WCM. At its maximum amplitude at $\rho_{pol}$ = 0.99 the WCM fluctuation amplitude is 2.3%. The plasma becomes optically thin ($\tau$ < 2) near the LCFS, but the WCM peak amplitude occurs in an optically thick region of the plasma. The fluctuation amplitude over radius gives information about the radial range of the WCM. The WCM fluctuations cover a $\rho_{pol}$ range of ~ 0.02, corresponding to physical width of approximately 1.5 cm. This discharge also had L-mode phases featuring the WCM at the same radial location as in I-mode. There is a reduction in the averaged outer core ($\rho_{pol}$ < 0.95) $\delta T_e/T_e$ fluctuation level between L-mode and I-mode from 0.52% to 0.48% but this small change is within error bars.

A spectrogram of $T_e$ fluctuations from a single CECE channel at the WCM peak at $\rho_{pol}$ = 0.99 is shown in Fig-

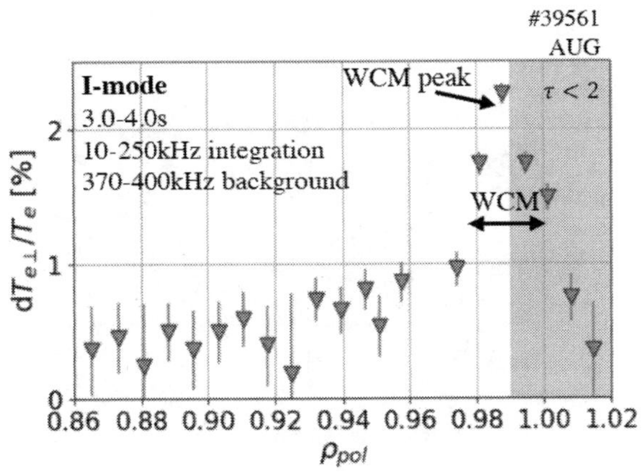


**Figure 1.** Electron temperature fluctuation level over radius for an I-mode phase of a discharge. The turbulence integration window was 10-250 kHz and the background subtraction window was 370-400 kHz in the calculation of fluctuation amplitude. The region of marginal optical depth $\tau$ < 2 are shaded in grey. Fluctuation levels rise from ~ 0.5% in the outer core to 2.3% at the WCM peak near $\rho_{pol}$ = 0.99.

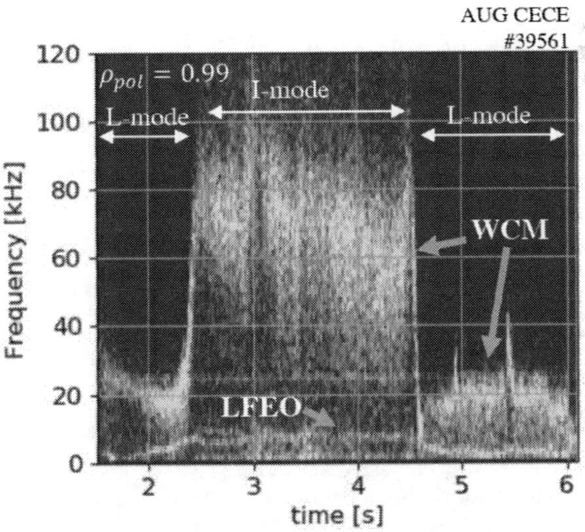


**Figure 2.** Spectrogram from CECE channel 19 near the WCM peak amplitude of $\rho_{pol}$ = 0.99. The discharge develops from L-mode (<2.4s) to I-mode (2.4-4.5s) to L-mode (>4.5s) and the WCM appears as the 20 kHz mode in L-mode which spins up to the 40-90 kHz in I-mode, and then back to 20 kHz in L-mode. LFEO is the narrowband mode at 8 kHz which appears in the I-mode phase only.

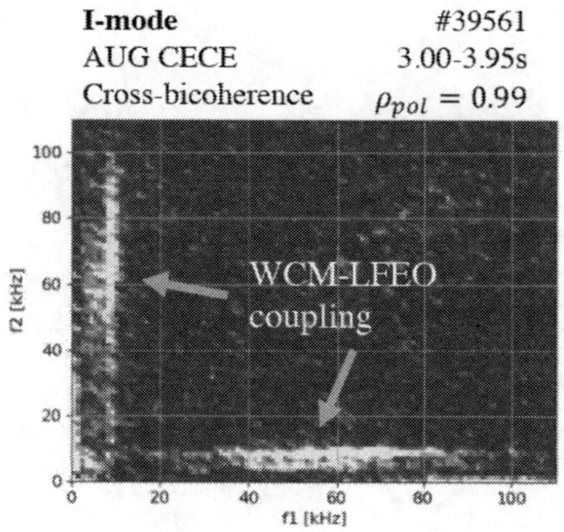


**Figure 3.** Cross-bicoherence of neighboring CECE channels near the WCM peak location of $\rho_{pol} = 0.99$. Coupling is seen between the WCM (40-80 kHz) and the LFEO (8 kHz). FFT bins of 4096 were used for the analysis with 1855 overlapping ensembles.

ure 2. This discharge develops from L-mode (<2.4s) to I-mode (2.4-4.5s) then back to L-mode (>4.5s). The WCM appears as the 20 kHz mode in L-mode which spins up to the 40-90 kHz in I-mode, and then back to 20 kHz in L-mode. The frequency shift between L-mode and I-mode may correspond with a change of radial electric field between L-mode and I-mode and the Doppler shift of fluctuations due to increased $E \times B$ velocity. The LFEO can also be seen during the I-mode phase of the discharge as the coherent mode at $\sim$ 8 kHz.

The WCM and LFEO are coupled as seen in Figure 3, which shows the cross-bicoherence between neighboring CECE channels near the WCM location of $\rho_{pol} = 0.99$. The analysis shows that the the narrow-band LFEO at 8 kHz is coupled with the broadband WCM at 40-90 kHz.

In addition to the L-mode/I-mode comparison, L-mode phases were compared in matched favorable and unfavorable $\nabla B$ drift configuration discharges. Figure 4 shows the CECE coherency spectra at the pedestal top ($\rho_{pol} = 0.96$) and the WCM location ($\rho_{pol} = 0.99$) during three ECRH power steps. Both the favorable and unfavorable configuration discharges are in L-mode for the first two power step (0.2 MW and 0.6 MW) and in these phases discharge parameters including temperature and density are matched. During the lowest power step, there is very little difference in the turbulence at the pedestal top between the favorable and unfavorable configuration discharges (Fig. 4a). During this low power step, the beginnings of the formation of the WCM can be seen at $\rho_{pol} = 0.99$ in both the favorable and unfavorable configurations as the broadband mode centered around 30 kHz (Fig. 4d). The WCM-like feature is stronger (a higher value of $|\gamma_c|$) in the unfavorable configuration as compared to the favorable configuration. During the second power step, at the pedestal top the fluctuations in the unfavorable configuration are slightly stronger than the fluctuations in the favorable configuration (Fig. 4b). At the WCM location, in unfavorable configuration the WCM feature appears more strongly than at lower power. In favorable configuration the WCM is no longer visible as a distinct broadband mode (Fig. 4e). At the highest power step, the unfavorable configuration discharge remains in L-mode and the fluctuations at the pedestal top are still visible in the coherency spectra (Fig. 4c) while the WCM is clearly visible in the spectra at $\rho_{pol} = 0.99$ (Fig. 4f). The favorable configuration discharge has transitioned to ELMy H-mode, so the plasma parameters no longer match with the unfavorable configuration discharge and the fluctuations are mostly beneath the sensitivity limit at both the pedestal top and $\rho_{pol} = 0.99$. Whether the disappearance of the WCM is a result of $E \times B$ shear remains to be investigated.

## 3 CECE diagnostic extension measurements

In addition to the standard CECE analysis presented using the 24-channel comb, nonstandard measurements were taken using a second CECE system and a coupled reflectometer to investigate additional physics of the WCM.

### 3.1 Long-range correlations

A second CECE system was installed at AUG at a second port which was toroidally separated by 60° from the primary CECE system. As with the primary system, the new CECE system also features a 24-channel comb with IF filters of 200 MHz bandwidth with centers spaced apart by 250 MHz. The RF front ends are interchangeable, and for these experiments the diagnostic was optimized for edge measurements with a receiving range of 105-113 GHz. This CECE diagnostic was installed on the same oversized waveguide as the Doppler Reflectometers and shares their antenna and focusing mirror.

The secondary CECE system was operated during the L-mode/I-mode discharge previously presented. The safety factor during this discharge at $\rho_{pol} = 0.99$ and the toroidal separation between the two CECE systems place the edge CECE channels in the two systems on approximately the same magnetic field line. The correlation between CECE channels near the WCM location ($\rho_{pol} = 0.99$) is shown in Figure 5, as the $\gamma_c$ spectra between these toroidally separated channels. The standard CECE $\gamma_c$ spectra of two neighboring channels in the same radial comb is overlaid with normalized amplitude for comparison. Both the WCM and LFEO are clearly visible in the long-range correlations. This toroidal correlation indicates that the both the WCM and LFEO have low toroidal mode number.

### 3.2 $n_eT_e$ cross-phase measurements

The phase between density and temperature fluctuations, $\alpha_{n_eT_e}$, has previously been used to quantify the mixture of ion temperature gradient (ITG) versus trapped electron

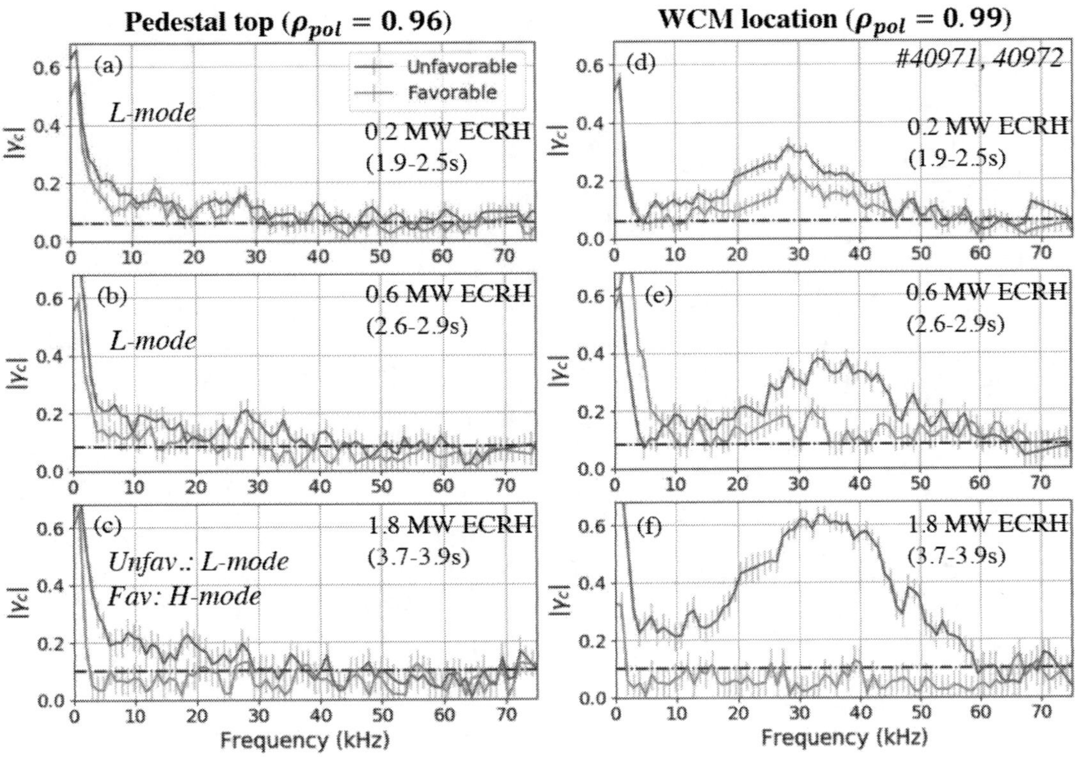


**Figure 4.** CECE coherency spectra in matched favorable (blue, 40972) and unfavorable (red, 40971) configuration discharge phases. Pedestal top (a-c) and the WCM location (d-f) are shown at three different ECRH power steps: 0.2 MW (a, d), 0.6 MW (b,e), and 1.8 MW (c,f). The unfavorable configuration plasma remains in L-mode for all power steps while the favorable configuration is in H-mode at the highest power step. WCM formation can be seen in all power steps in unfavorable configuration, and only the lowest power step in favorable configuration.

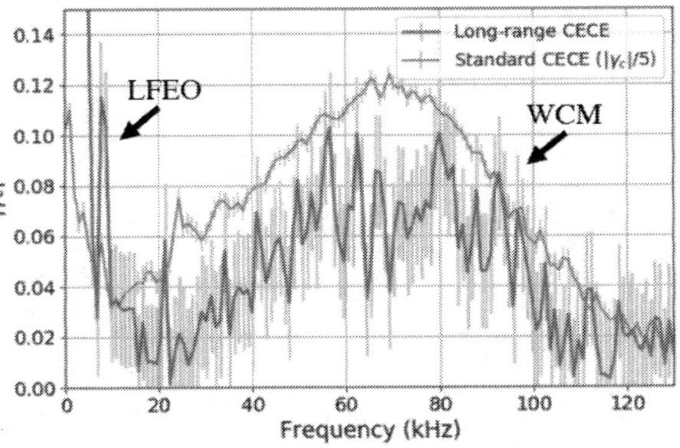


**Figure 5.** CECE coherency spectra obtained by correlating channels near the WCM location at $\rho_{pol}$ = 0.99 from CECE systems toroidally separated by ~60° (green). CECE coherency spectra from standard CECE analysis with neighboring channels in a single system is overlaid in grey, with a normalization factor for clear visualization. The WCM is visible in both the long-range and local correlations.

mode (TEM) turubulence in core plasmas, and as a metric for validating gyrokinetic codes [13–15]. This measurement provides valuable information about the nature of the plasma turbulence. The $n_eT_e$ phase diagnostic was operated during the L-mode/I-mode experiments by coupling the secondary CECE system and the Doppler Reflectometers in this sector of the tokamak using a 3dB wire-grid splitter and the same data acquisition system for both CECE and reflectometer signals. If radial alignment between the CECE channels and the reflectometer cutoff is achieved via appropriate choice of reflectometer frequency then $\alpha_{n_eT_e}$ may be obtained from the correlation between the reflectometer and CECE signals.

The $\alpha_{n_eT_e}$ analysis is shown in Figure 6 during the L-mode and I-mode phases of the discharge. There is coherency between the reflectometer and the CECE channel in the WCM range of frequencies in both L-mode and I-mode, as shown by Fig. 6a and c. From these correlations, $\alpha_{n_eT_e}$ is extracted and shown in degrees in Fig. 6b and d. Over the range of frequencies where there is coherency between the reflectometer and CECE signals, there is an evident, out-of-phase $n_eT_e$ cross phase in both L-mode and I-mode. The L-mode WCM cross phase is farther out-of-phase than the I-mode WCM, with $\alpha_{n_eT_e} = -171°$ at the peak CECE-reflectometer in L-mode as compared

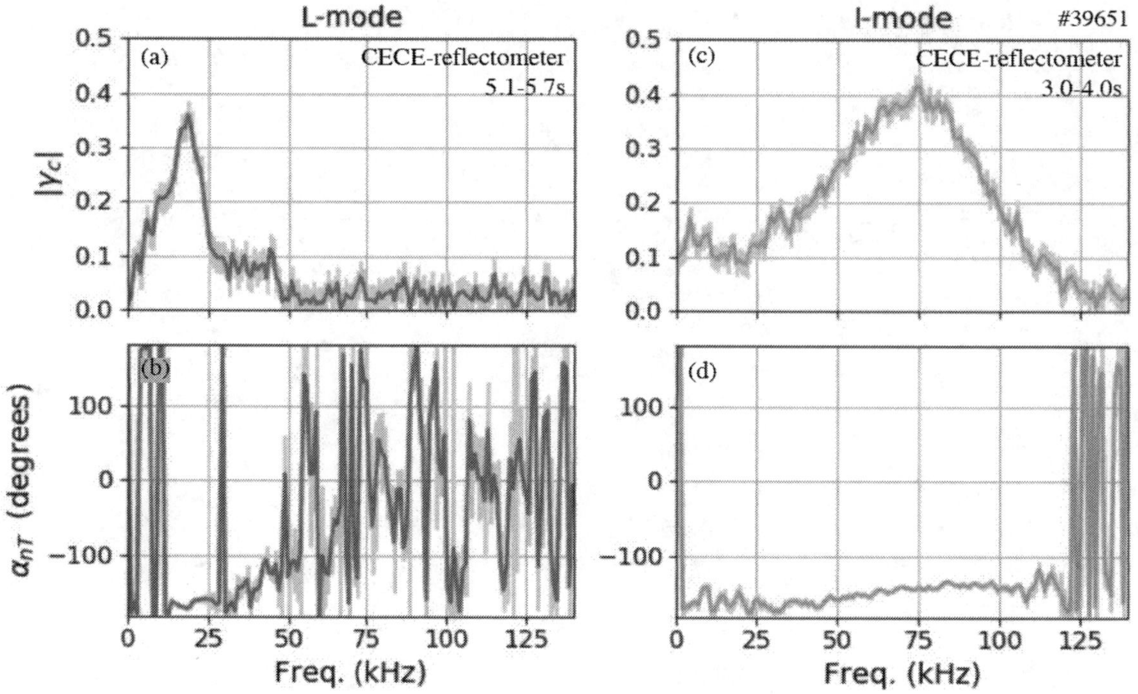


**Figure 6.** Coherency between a CECE channel and V-band reflectometer at $\rho_{pol}$ = 0.98 during (a) L-mode and (c) I-mode phases, showing the WCM at 20 kHz in L-mode and 40-100 kHz in I-mode. The $n_eT_e$ phase over frequency obtained from CECE-reflectometer correlations shown in (b) L-mode and (d) I-mode. The $n_eT_e$ phase at the spectral peak of the WCM is $-171°$ in L-mode and shifts to $-143°$ in I-mode.

to $-143°$ at the peak coherency in I-mode. This change from out-of-phase to less out-of-phase between L-mode and I-mode occurs as $\nabla T_e$ steepens between L-mode and I-mode, so is consistent with previous measurements and simulations that have shown $\alpha_{n_eT_e}$ to change with $\nabla T_e$ due to increased TEM drive [13, 15]. However, the WCM $\alpha_{n_eT_e}$ values observed here are farther out-of-phase than previous core turbulence $\alpha_{n_eT_e}$ measurements, indicating that the WCM may be of a different nature than core turbulence. The $n_eT_e$ cross-phase does not enter directly into cross-field heat and particle transport equations. Therefore, these measured $\alpha_{n_eT_e}$ values alone do not give information about the direction of heat versus particle transport for the WCM without further assumptions about the $n_e\phi$ cross-phase, the relationship between $n_e\phi$, $T_e\phi$, and $n_eT_e$ cross-phases, the $n_e$ to $T_e$ fluctuation amplitude ratio, as well as assuming the WCM to be electrostatic in nature.

The slope of $\alpha_{n_eT_e}$ over the frequency range of CECE-reflectometer coherency can provide information about the poloidal velocity of the WCM. Because of a small poloidal separation between the reflectometer and CECE, there is a finite time delay between the WCM's appearance across the reflectometer cutoff and the CECE resonance, which corresponds to a change in $\alpha_{n_eT_e}$ with frequency. The I-mode WCM $\alpha_{n_eT_e}$ has a change of approximately 25° over a frequency range of 60 kHz, which corresponds to a time delay of 1.2 $\mu$s. The CECE channel and reflectometer are separated by approximately 4 mm. This corresponds to a poloidal velocity of approximately 3 km/s, which is similar to previously reported values at the L-mode and I-mode WCM location [8]. This estimation is performed in the laboratory frame, and the background plasma velocity would be needed to calculate the WCM's phase velocity in the plasma frame. The L-mode WCM is too narrowband to determine if $\alpha_{n_eT_e}$ has a slope over its frequency range.

The out-of-phase $\alpha_{n_eT_e}$ measurements also have implications for quantifying the effect of $n_e$ fluctuations on the radiated temperature signals collected by CECE as the optical depth becomes marginal ($\tau < 2$) near the LCFS. In L-mode phases, the WCM peak often occurs in this marginal region with $1 < \tau < 2$. Following the analysis of Reference [16], the I-mode $\alpha_{n_eT_e}$ values which are neither in-phase nor completely out-of-phase suggests that the CECE measurements of the WCM are robustly $T_e$ fluctuations with little impact from $n_e$ fluctuations. The L-mode $\alpha_{n_eT_e}$ is close to completely out-of-phase, which indicates that $n_e$ fluctuations may cause the apparent radiated temperature fluctuation amplitude to be less than the actual $T_e$ fluctuation amplitude at the L-mode WCM location.

## 4 Conclusions

A 24-channel high radial and time resolution CECE system was used to study outer core and pedestal turbulence in L-modes and I-modes at AUG. The characteristic I-mode fluctuations of the WCM and the LFEO are both visible in CECE spectra and these two modes are shown to be coupled by bicoherence analysis. Radial profiles of the $T_e$ fluctuation amplitude shows that turbulence increases from $\sim 0.5\%$ in the outer core to $\sim 1\%$ at the pedestal top. Then, the $T_e$ fluctuation amplitude increases sharply due to the localized WCM fluctuations ($\rho_{pol} = 0.98 - 1.0$), with a peak around $\rho_{pol} = 0.99$ where $\delta T_e/T_e = 2.3\%$. The WCM is also present in L-mode as a more coherent mode at lower frequency than in I-mode. An investigation of turbulence was performed in matched favorable and unfavorable $\nabla B$ drift configuration L-mode plasmas. CECE coherency spectra show that the WCM is present in both favorable and unfavorable configurations at low heating power, but at slightly higher heating power in unfavorable configuration the WCM is strengthened while in favorable configuration the WCM is no longer visible. At low power, the outer core fluctuations are matched in favorable and unfavorable configurations, while at higher power the unfavorable configuration has stronger fluctuations than the favorable configuration.

Non-standard CECE diagnostic extensions allowed for a deeper investigation of the physics of the edge fluctuations in L-modes and I-modes. Long-range correlations performed across toroidally separated CECE systems showed that both the WCM and the LFEO are correlated over long toroidal distances and therefore have low toroidal mode numbers. The coupled reflectometer-CECE system allowed for $\alpha_{n_eT_e}$ measurements which provide valuable information about the nature of the WCM including how it changes between L-mode and I-mode with $\nabla T_e$ and its poloidal velocity. The value of $\alpha_{n_eT_e}$ alone cannot be used to determine the direction particle versus heat transport from the WCM. These cross-phase measurements in conjunction with optical depth calculations can also assist in quantifying the impact of $n_e$ fluctuations on the radiated temperature fluctuation signals.

## 5 Acknowledgements

This work has been carried out within the framework of the EUROfusion Consortium, and has received funding from the Euratom Research and Training Programme (Grant Agreement No 101052200 — EUROfusion). The views and opinions expressed herein do not necessarily reflect those of the European Commission. This work was also supported by the US Department of Energy under grants DE-SC0014264, DE-SC0006419, and DE-SC0017381.

## References

[1] D.G. Whyte, A.E. Hubbard, J.W. Hughes, B. Lipschultz, J.E. Rice, E.S. Marmar, M. Greenwald, I. Cziegler, A. Dominguez, T. Golfinopoulos et al., Nuclear Fusion **50**, 105005 (2010)

[2] T. Happel, P. Manz, F. Ryter, P. Hennequin, A. Hetzenecker, G.D. Conway, L. Guimarais, C. Honoré, U. Stroth, E. Viezzer, Nuclear Fusion **56** 064004 (2016)

[3] D. Silvagni, T. Eich, T. Happel, G.F. Harrer, M. Griener, M. Dunne, M. Cavedon, M. Faitsch, L. Gil, D. Nille et al., Nuclear Fusion **60** 126028 (2020)

[4] A.E. White, P. Phillips, D.G. Whyte, A.E. Hubbard, C. Sung, J.W. Hughes, A. Dominguez, J. Terry, I. Cziegler, Nuclear Fusion **51** 113005 (2011)

[5] T. Happel, M. Griener, D. Silvagni, S.J. Freethy, P. Hennequin, F. Janky, P. Manz, D. Prisiazhniuk, F. Ryter, M. Bernert et al., Nuclear Materials and Energy **18** 159-165 (2019)

[6] A.E. Hubbard, R.L. Boivin, R.S. Granetz, M.Greenwald, J.W. Hughes, I.H. Hutchinson, J. Irby, B. LaBombard, Y. Lin, E.S. Marmar et al., Physics of Plasmas **8** 2033 (2001)

[7] K.H. Burrell, W.P. West, E.J. Doyle, M.E. Austin, T.A. Casper, P. Gohil, C.M. Greenfield, R.J. Groebner, A.W. Hyatt, R.J. Jayakumar et al., Physics of Plasmas **12** 056121 (2005)

[8] R. Bielajew, G.D. Conway, M. Griener, T. Happel, K. Höfler, N.T. Howard, A.E. Hubbard, W. McCarthy, P.A. Molina Cabrera, T. Nishizawa et al., Physics of Plasmas **29** 052504 (2022)

[9] W. McCarthy Ph.D. Thesis, Massachusetts Institute of Technology Cambridge, MA (2022)

[10] I. Cziegler, P.H. Diamond, N. Fedorczak, P. Manz, G.R. Tynan, M. Xu, R.M. Churchill, A.E. Hubbard, B. Lipschultz, J.M. Sierchio et al., Physics of Plasmas **20** 055904 (2013)

[11] P. Manz, P. Lauber, V.E. Nikolaeva, T. Happel, F. Ryter, G. Birkenmeier, A. Bogomolov, G.D. Conway, M.E. Manso, M. Maraschek et al., Nuclear Fusion **55** 83004 (2015)

[12] A.J. Creely, S.J. Freethy, W.M. Burke, G.D. Conway, R. Leccacorvi, W.C. Parkin, D.R. Terry, A.E. White, Review of Scientific Instruments **89** 53503 (2018)

[13] J.C. Hillesheim, J.C. DeBoo, W.A. Peebles, T.A. Carter, G. Wang, T.L. Rhodes, L. Schmitz, G.R. McKee, Z. Yan, G.M. Staebler et al., Physics of Plasmas **20** 056115 (2013)

[14] A.E. White, W.A. Peebles, T.L. Rhodes, C.H. Holland, G. Wang, L. Schmitz, T.A. Carter, J.C. Hillesheim, E.J. Doyle, L. Zeng et al., Physics of Plasmas **17** 056103 (2010)

[15] S.J. Freethy, T. Görler, A.J. Creely, G.D. Conway, S.S. Denk, T. Happel, C. Koenen, P. Hennequin, A. E. White, Physics of Plasmas **25** 055903 (2018)

[16] T.D. Rempel, R.F. Gandy, A.J. Wootton, Review of Scientific Instruments, **65**, 2044-2048 (1994)

# Electron temperature fluctuation levels of the quasi-coherent mode across the plasma radius

*Branka* Vanovac[1,*], *Jörg* Stober[2], *Elisabeth* Wolfrum[2], *Matthias* Willensdorfer[2], *Luís* Gil[3], *Michael* Faitsch[2], *Rachel* Bielajew[1], *Christian* Yoo[1], *Garrard* Conway[2], *Severin* Denk[1], *Rachael* McDermott[2], *Anne* White[1], and *ASDEX* Upgrade Team[4]

[1]Massachusetts Institute of Technology, Plasma Science and Fusion Center, Cambridge, MA 02139, USA
[2]Max Plank Institute for Plasma Physics, 85748 Garching, Germany
[3]Instituto de Plasmas e Fusão Nuclear, Instituto Superior Técnico, Universidade de Lisboa, 1049-001 Lisboa, Portugal
[4]See author list of U. Stroth et al. 2022 Nucl. Fusion 62 042006

**Abstract.** EDA H-mode is an ELM-free regime in which the edge quasi-coherent mode (QCM) replaces the ELMs. The estimated location of the quasi-coherent mode is in a partly optically thin region of steep gradients localized between $\rho_{pol} = 0.96$ -1. Relative fluctuations of radiation temperature between 15 and 80 kHz are about 7% with significant density contribution. In the electron cyclotron emission (ECE) channels with resonances in the plasma core, a mode with the same frequency as the quasi-coherent mode is measured. The peak amplitude of both core and edge modes matches the strongest electron temperature gradient in the core and the edge, respectively. The ECE core and edge signals are out of phase. The radiation transport forward model (ECRad) shows that the refraction explains the phase relation between the edge and the core ECE channels. The phase correlates with the sign of the core $\nabla T_e$. The amplitude of the fluctuations in the core decreases with decreasing gradients, which is the trend seen in the experiment. The amplitude ratio of the core and edge fluctuation is a factor of five in the experiment; this ratio remains a factor of a hundred in the modeling.

## 1 Introduction

The EDA H-mode (Enhanced $D_\alpha$ high confinement mode) regime is a regime free of edge localized modes (ELMs) with good energy confinement properties [1]. Recently it has been a significant focus of the ASDEX Upgrade experimental program and its main features are reported in [2]. At ASDEX Upgrade, the dominant external heating is the electron cyclotron resonance heating (ECRH) and the rotation profiles are hollow, similar to observations in some L-mode plasmas [3]. The ELMs are replaced by a quasi-coherent mode thought to regulate the transport and keep the pedestal stable against large type-I ELMs. The EDA H-mode regime is routinely obtained at ASDEX upgrade and the neutral beam injection NBI power window is extended with Ar seeding [4]. As mentioned, the prominent signature of EDA H-mode is the quasi-coherent mode measured at the plasma edge with different diagnostics; thermal He-beam emission spectroscopy (THB), electron cyclotron emission (ECE), and reflectometers. However, in some core ECE channels, a fluctuation appears with the same frequency as the quasi-coherent mode and is absent in the signals of other core diagnostics i.e., soft X-rays. The origins of the core ECE emission have been debated, and the main question remains whether it is a consequence of the strong refraction on the edge density or actual core fluctuations.

*e-mail: vanovac@mit.edu

The quasi-coherent mode (QCM) in the plasma edge is a feature of the EDA ELM-free regime, but it can also coexist with ELMs in ELMy H-modes, most likely appearing between ELMs. However, in the context of this work, we will refer only to the quasi-coherent mode where ELMs are absent and describe it in terms of its frequency and associated temperature fluctuation amplitudes.

The relative temperature fluctuation levels across the pedestal are assessed via the Correlation Electron Cyclotron Emission (CECE) instrument [5] covering the region of $\rho_{pol} = [0.85 - 1]$ while the standard ECE diagnostics is used for core-edge amplitude comparison and comparison to the modeling. EC Radiation transport is assessed with ECRad [6], a radiation transport code suitable for treating electron cyclotron emission with specifics of CECE and ECE diagnostics such as launching angles and IF bandwidth. Refraction is included in ECRad, allowing studies on the refractive effects on the core emission.

## 2 Features of quasi-coherent mode

### 2.1 Frequency

The peak frequency of the quasi-coherent mode at ASDEX Upgrade during its stationary phase varies between 15 and 60 kHz. The calculated power spectral density from the ECE measurements at three different radial locations is displayed in figure 1. The peak frequency is determined

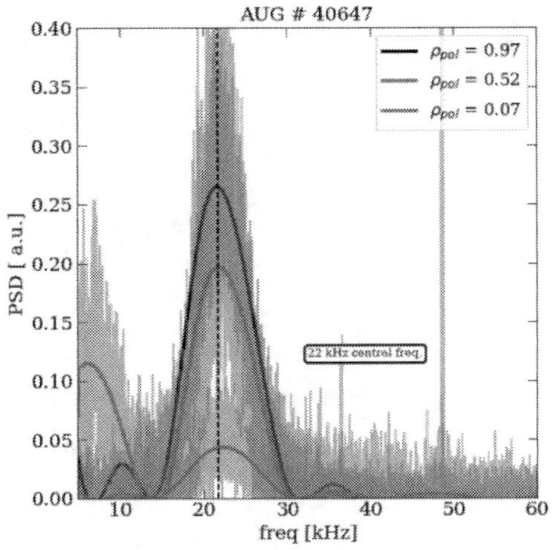


**Figure 1.** AUG #40647: Power spectral density of ECE measurements at three different locations as indicated by different colors $\rho_{pol} = 0.97$ in black, $\rho_{pol} = 0.52$ in purple and $\rho_{pol} = 0.07$ in red. The black dashed line marks the center frequency of about 22 kHz.

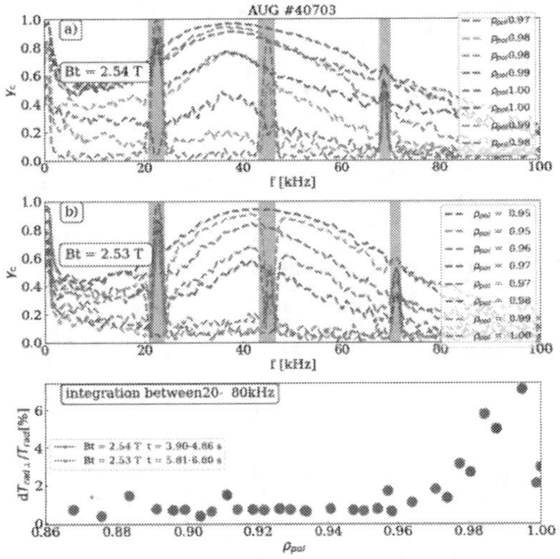


**Figure 2.** AUG #40703 a) Cross-coherence between subsequent CECE channels for magnetic field strength of 2.54 T. b) Cross-coherence between subsequent correlation ECE channels for magnetic field strength of 2.53 T. c) Relative radiation temperature fluctuations integrated between 20 and 80 kHz from the CECE measurements for $B_t = 2.54$ T in red and for $B_t = 2.53$ T in magenta, respectively. Grey regions are marking peaks of strong noise in the channels.

from the spline fit to the data shown in gray. The black curve shows the power spectrum measured at the edge at $\rho_{pol} = 0.97$, the purple one at $\rho_{pol} = 0.52$ at the mid-radius, while the red curve corresponds to the measurement in the plasma core at $\rho_{pol} = 0.07$. The peak frequency for both the edge and the core is identical at f = 22 kHz, indicating the same feature measured in both plasma regions. The power spectra broaden due to the changes in the frequency of the quasi-coherent mode during the integration time.

## 2.2 Localisation and amplitude

We use the CECE system for the localization and $\delta T_{rad}$ estimate. The CECE system at ASDEX Upgrade is a comb of 24 channels with fixed IF filters of 200 MHz bandwidth and 250 MHz spacing of the center frequencies, translating into approximately 2 mm of spatial resolution. Neighboring CECE channels measure correlated temperature fluctuations but uncorrelated thermal noise. By correlating neighbouring channels, coherency spectra $\gamma_c$ of turbulence are obtained. We obtain the fluctuation amplitude by integrating coherency in frequency space and subtracting the background noise. A dedicated experiment with two magnetic field strengths enables more accurate localization of the QCM. With changing magnetic field strength, the measurement position of the ECE & CECE channels changes while the mode location remains the same. Consequently, the mode amplitude moves from one channel to the next. In figure 2 we show the CECE measurements during discharge #40703, for both field strengths. Figure 2 a) and 2 b) show the coherency spectra $\gamma_c$ against normalized radius. Figure 2 c) displays the relative temperature fluctuation levels versus normalized plasma radius $\rho_{pol}$ for Bt = 2.54 T in red and for Bt = 2.53 T in magenta. The mode affects the radial range between $\rho_{pol} = 0.96$ and the separatrix in both cases. Due to the low optical depth outside of $\rho_{pol} = 0.97$, the fluctuation amplitudes of 7% can be a consequence of density perturbation in this region [7].
Profile ECE measurements integrated between 15 and 60 kHz are shown in figure 3 for the same discharge #40703. Corresponding temperature profiles are shown in 3 a) and profile gradients in 3 b). For both magnetic field strengths, the maximum amplitude distribution (shown in 3 c)) matches the peak of the electron temperature gradients in the edge and the core of the plasma.

A typical feature of EDA H-mode is the high electron density, however, in our experiment ECE radiation is always below the cutoff and the ECE profile measurements are routinely obtained. The electron temperature and density profiles for two cases comparing two discharges are shown in figure 4. The discharge #40647 had plasma current $I_p$ = 0.6 MA while the discharge #40417 had $I_p$ = 0.7 MA and consequently higher density. Both discharges had the same magnetic field strength and the ECRH heating scheme. The higher current discharge had NBI blips applied for charge exchange measurements.

The corresponding temperature fluctuations across the pedestal for both cases averaged over 2 s of the discharges are shown in figure 5. The behavior of the radiation temperature fluctuations for two different density profiles differs. The $T_{rad}$ fluctuations for the low pedestal density case are increasing from 0.97 and peak just before the separatrix. The $T_{rad}$ fluctuations for the high pedestal density case are increased across an entire measurement region. This major difference in the shape of the radial ampli-

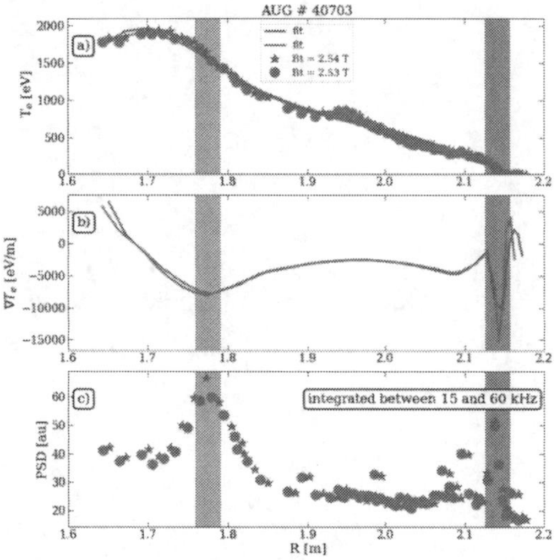


**Figure 3.** AUG #40703 with changing magnetic field; 2.53 T in magenta and 2.54 T in red a) The electron temperature profiles, b) the gradient of the electron temperature, c) amplitude distribution integrated between 15 and 60 kHz corresponding to the quasi-coherent mode

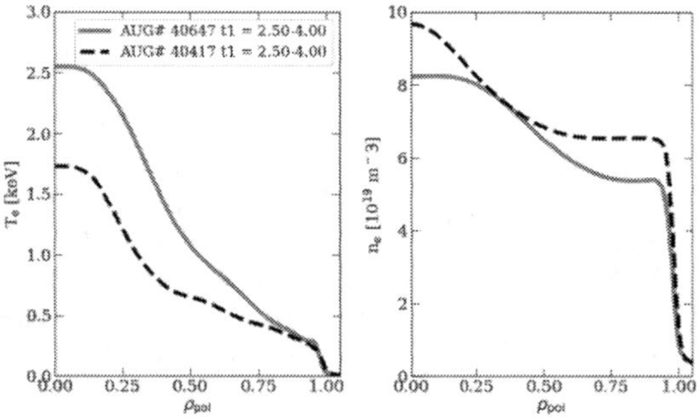


**Figure 4.** Profiles obtained with integrated data analysis (IDA) for low AUG #40647 and high AUG #40417 pedestal density discharges averaged during time interval t = 2.5 - 4 s. Low pedestal density is displayed in red while high pedestal density case is shown in black.

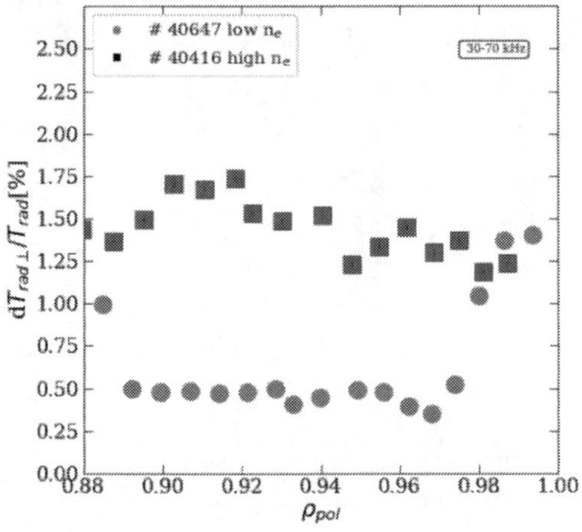


**Figure 5.** Radial profiles of the relative radiation temperature fluctuations measured by the correlation ECE instrument for two different density values; low pedestal density #40647 in red and high pedestal density #40417 in black.

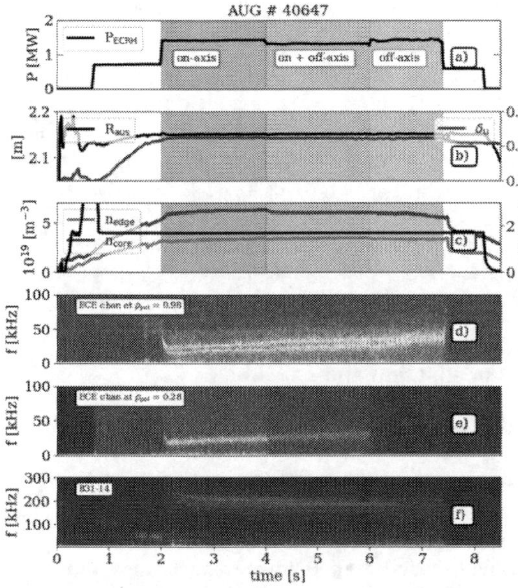


**Figure 6.** AUG #40647: Temporal evolution of plasma parameters during EDA H-mode. a) ECRH heating scheme. b) Outer plasma boundary $R_{aus}$ (black) and upper triangularity $\delta_u$ (blue). c) edge and core densities together with the deuterium fuelling. d) Spectrogram of the edge ECE channel. e) Spectrogram of the ECE core channel. f) Spectrogram of the magnetic coil B31-14 measuring changes in the radial magnetic field. Time intervals of different heating schemes are marked in colored surfaces: blue - on-axis heating; green - on and off-axis heating; orange - off-axis heating.

tude distribution of the temperature fluctuations suggests its density dependence.

### 2.3 Changing core $T_e$ gradients

Figure 6 displays an EDA H-mode designed to study the effect of core temperature gradient on the quasi-coherent activity measured in the core and edge ECE channels. The temperature gradient during the discharge was changed by the ECRH deposition shifting from central to off-axis heating. The discharge has a plasma current of $I_p$ = 600 kA and a magnetic field $B_T$ = 2.5 T. Figure 6 a) illustrates the ECRH heating scheme. Between 2-4 s two central gyrotrons heat the plasma. The second phase, between 4-6 s combines one central gyrotron and one mid-radius gyrotron. In the last phase, between 6-7.4 s, both gyrotrons deposit the power at mid-radius. The total power is about 1.4 MW throughout the discharge. The plasma had a fixed shape as represented by the upper triangularity $\delta_u$

and the outer plasma boundary $R_{aus}$ as displayed in figure 6 b). The core and edge densities as well as the fueling are shown in figure 6 c) and remained constant throughout the discharge. Spectrograms of the edge and core ECE channels, displayed in figure 6 d) and 6 e) show the quasi-coherent mode. Figure 6 f) displays the spectrogram of the magnetic coil measuring radial magnetic field fluctuations at the outer midplane. The strongest modes in the coil are around 200 kHz frequency, usually with toroidal mode numbers n = 5,6 and 7. The fundamental frequency of this harmonic structure has a frequency similar to the quasi-coherent mode in this example. The profile comparison between on and off-axis ECRH heating during the discharge #40647 is displayed in figure 7. The phase with on-axis heating is marked in blue, and the phase with off-axis heating is in orange. The average profiles measured with ECE are displayed as symbols, and corresponding fits as lines in figure 7 a). The estimated gradients of the measured and fitted profiles are shown in 7 b). The radial distribution of the ECE amplitude measurement integrated between 15 and 60 kHz is displayed in 7 c). The location of the peak amplitude and matching gradients is emphasized with a shaded region the core being in red and the edge marked in black.

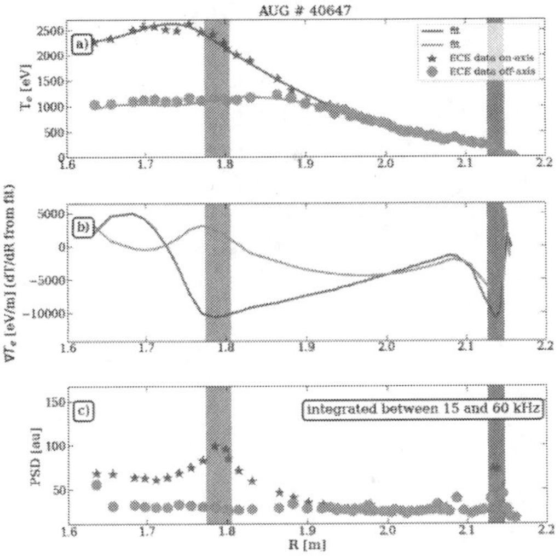


**Figure 7.** AUG #40647 profiles for the two different heating schemes. On-axis ECRH heating is shown in blue while off-axis is shown in orange. a) Profile of the electron temperature $T_e$ measured with ECE in symbols and the corresponding fits in solid lines. b) Gradients of the electron temperature fits. c) Amplitude distribution of Fourier filtered data between 15 and 60 kHz. The two shaded regions correspond to the maximum perturbation amplitude in the core (in red) and at the edge (in black).

## 3 Refraction effects

The edge quasi-coherent mode is the primary candidate for particle transport across the pedestal. However, the measurements of the quasi-coherent mode in the core ECE channels remain ambiguous. One hypothesis is that the mode measured in the core ECE channels is due to the refraction of the beams on a high-density perturbation at the edge. Another important ingredient is the steepness of the temperature profile.

Density fluctuations are measured with a thermal He beam and reflectometers. Another supporting argument is the absence of the core mode in other core diagnostics, i.e. soft X-ray measurements. To account for the possible refraction from the significant density perturbation at the edge, we employ the radiation transport forward model ECRad [6] where the geometry of the ECE system is included. We model the steep and the flat temperature profile scenarios of the discharge shown in figure 7. In the modeling we use a single ray, hence the observation volume is neglected. As input to the ECRad we use a two-dimensional $T_e$ and $n_e$ grid. Raytracing is then performed and the radiation transport equation is solved along the ray path for each time-point, resulting in time-resolved radiation temperature. The 2D data used for the modeling is the output of a recent JOREK simulation for a regime with small ELMs [8] regime, where a rotating mode at 20 kHz with n = 10 was observed. This set of data is, at present, suitable to treat the quasi-coherent mode as the perturbation structure of the small ELMs scenario has similar features as the quasi-coherent mode in terms of frequency, toroidal mode number and the extent across the pedestal. The JOREK simulation is performed for a different discharge than the one treated in this work. Therefore, JOREK data is further modified to be closer to the shape and magnitude of the background electron density and temperature profiles of the discharge #40647.

The final modified profiles are fed separately into ECRad for the two cases. The modeling is performed for fifteen-time points to obtain time-resolved radiation temperatures. We then calculate the relative radiation temperature fluctuations $(T_{rad} - \langle T_{rad} \rangle)/\langle T_{rad} \rangle$ for each time point where $\langle T_{rad} \rangle$ is an equilibrium temperature without perturbation. The result for the case of the steep electron temperature profile is displayed in figure 8. We compare two different radial locations. The core channel is at $\rho_{pol} = 0.21$ shown in red and the edge ECE channel at $\rho_{pol} = 0.98$ in black. The edge and the core fluctuations are in anti-

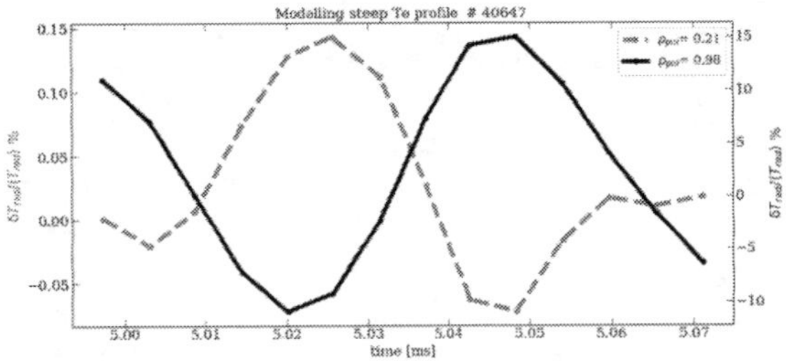


**Figure 8.** ECRad modeling: modelled time evolution of the radiation temperature fluctuations for the on-axis heating case. Black solid line shows the channel at the plasma edge while red dashed line represents the behaviour in the core.

https://doi.org/10.1051/epjconf/202327703003

phase. The maximum amplitude of the core channel is about 0.15 %, while the radiation temperature fluctuations at the edge are 15 %, making it a factor of 100 difference between the edge and the core, in this modeling exercise.

Now we compare this modeling result to the experimental measurements. The time-resolved ECE measurements (experiment) in the core and the edge of the plasma, for the on-axis heating scenario, are shown in figure 9. The channel at $\rho_{\rm pol}$ = 0.2 is displayed in red while the edge channel at $\rho_{\rm pol}$ = 0.98 is shown in black. The phase relation between the edge and the core changes over time. However, for the majority of the time, the two signals are close to anti-phase, similar to the modeling (see figure 8). Core and edge amplitude differ by a factor of 5.

A modeling is then performed for the case with an off-axis

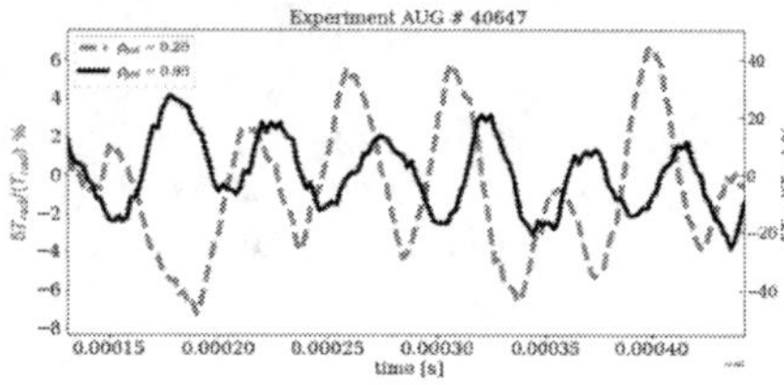


**Figure 9.** AUG #40647: time evolution of the $\delta T_e/\langle Te\rangle$ obtained from the raw ECE data in the core $\rho_{\rm pol}$ = 0.2 (red) and at the edge at $\rho_{\rm pol}$ = 0.98 (black).

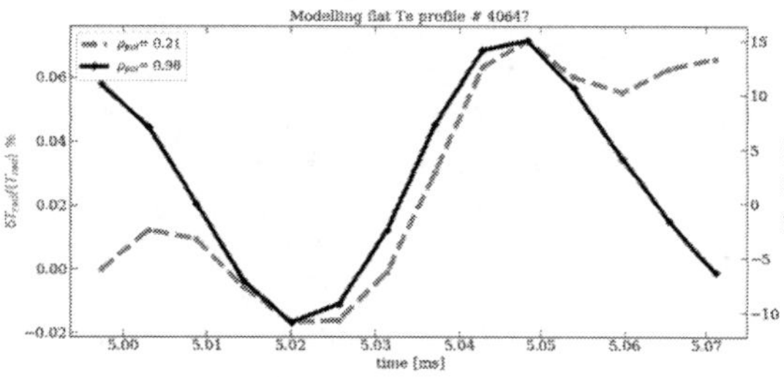


**Figure 10.** ECRad modeling: modelled time evolution of the radiation temperature fluctuations for the off-axis heating case. Black solid line shows the channel at the plasma edge while red dashed line represents the behaviour in the core.

ECRH heating scheme in which case the flat core temperature profile is flat. A comparison between relative temperature fluctuations of the two different radial locations for the off-axis scenario is shown in figure 10. The core channel is at $\rho_{\rm pol}$ = 0.21 in red and the edge ECE channel at $\rho_{\rm pol}$ = 0.98 in black. The amplitude of the core fluctuation is about 0.06 %, which is 2.5 times smaller than in the case of the steep gradient. The edge fluctuation amplitude remains unchanged, which is expected since the only change to the modeling input is the temperature profile in the plasma core. The core and the edge channels are in phase. This is a consequence of the sign of the temperature gradient. In the case of the flat $\nabla T_e$, the core temperature leans towards the hollow core profile, allowing for the temperature gradient of the opposite sign if compared to the steep gradient case. This comparison is best seen in figure 7b).

For better clarity, figure 11 shows the ray path of one of the core ECE frequencies. The equilibrium ray path without perturbation is shown in black, while the red corresponds to the ray path with edge perturbation included. Corresponding resonances are marked with squared symbols. The bending of the ray path happens at the edge, which is seen in the zoomed edge region. As a consequence, the resonances are vertically displaced in the core.

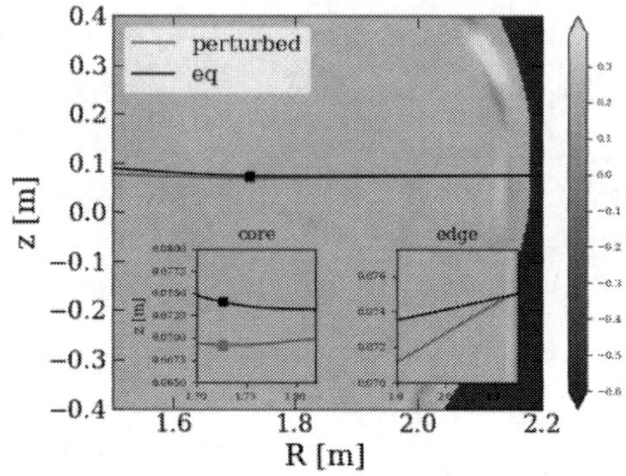


**Figure 11.** 2D density contour plot with density perturbation at the edge used in the forward model. The ECE ray is refracted at the edge which can be seen in the zoomed picture and the resonance (in red) is displaced with respect to the equilibrium (black) position.

## 4 Conclusion and Outlook

The correlation ECE instrument makes it possible to obtain the relative temperature fluctuation levels of the quasi-coherent mode at the plasma edge. Estimated fluctuation $T_{\rm rad}$ levels integrated between 20-80 kHz are around 7%. The location of the quasi-coherent mode is measured to be in the region of steep gradients between $\rho_{\rm pol}$ = 0.96 -1 of insufficient optical depth, where density contributes to the measured radiation temperature fluctuations. Two scenarios with different plasma currents show differences in the trends when comparing the relative temperature fluctuations estimated with the correlation ECE instrument. With higher density, the fluctuations amplitude pattern seems to be flat across the gradient region. For the lower density case, they have a pronounced peaking close to the separatrix. This is further to be investigated.

The profile ECE system at ASDEX Upgrade allows for comparing the fluctuation levels across the entire plasma radius. Channels in the core and edge measure oscillations at the frequency of the quasi-coherent mode. Maximum mode amplitudes from the core and edge ECE measurements match the location of the steepest $T_e$ gradient. The ECE instrument also enables phase measurements between the edge and core. With a steep gradient, the edge-core channels are anti-phase.

We challenge the refraction hypothesis by modeling the EC radiation transport with strong density and temperature perturbation at the edge, using 2D profiles. In the steep gradient phase, the core/edge behavior reproduces the experiment and the two channels are in anti-phase. The flat gradient case has in-phase behavior which correlates with the sign of the Te gradient in the core. The difference between the core and edge in the experiment is about a factor of five, while in the modeling, factor 100 is found.

Possible sources for the increased levels in the experiment are (i) The quasi-coherent mode potentially has a higher amplitude and broader radial extent at the edge than the JOREK small ELM case used as input. (ii) The ECE LOS for the modeled case was vertically closer to the axis, leading to a more perpendicular incidence angle and hence less temperature variation with the vertical refraction. (iii) The plasma potentially exhibits a rigid radial oscillation with the QCM frequency due to the strong variation of plasma pressure. So far, this work is showing that the refractive effects could play a role in the formation of the core ECE mode's signature. However, to further test this hypothesis, an extension of the work presented here is needed. Therefore, scans in the shape of the temperature profile and the amplitude of the density perturbation are planned for future work.

## Acknowledgment

The author thanks Andres Cathey for providing JOREK data for input to the forward modeling. This work is supported by the US DOE under Award DE-SC0014264. This work has been carried out within the framework of the EUROfusion Consortium, funded by the European Union via the Euratom Research and Training Programme (Grant Agreement No 101052200 - EUROfusion). Views and opinions expressed are however those of the author(s) only and do not necessarily reflect those of the European Union or the European Commission. Neither the European Union nor the European Commission can be held responsible for them.

### References

[1] M. Greenwald, R. Boivin, P. Bonoli, R. Budny, C. Fiore, J. Goetz, R. Granetz, A. Hubbard, I. Hutchinson, J. Irby et al., Physics of Plasmas **6**, 1943 (1999), `https://doi.org/10.1063/1.873451`

[2] L. Gil, C. Silva, T. Happel, G. Birkenmeier, G. Conway, L. Guimarais, A. Kallenbach, T. Pütterich, J. Santos, P. Schneider et al., Nuclear Fusion **60**, 054003 (2020)

[3] R. McDermott, C. Angioni, G. Conway, R. Dux, E. Fable, R. Fischer, T. Pütterich, F. Ryter, E.V. and, Nuclear Fusion **54**, 043009 (2014)

[4] A. Kallenbach, M. Bernert, P. David, M.G. Dunne, R. Dux, E. Fable, R. Fischer, L. Gil, T. Görler, F. Janky et al., Nuclear Fusion **61**, 016002 (2020)

[5] A.J. Creely, S.J. Freethy, W.M. Burke, G.D. Conway, R. Leccacorvi, W.C. Parkin, D.R. Terry, A.E. White, Review of Scientific Instruments **89**, 053503 (2018), `https://doi.org/10.1063/1.5005507`

[6] S. Denk, R. Fischer, E. Poli, O. Maj, S. Nielsen, J. Rasmussen, M. Stejner, M. Willensdorfer, Computer Physics Communications **253**, 107175 (2020)

[7] Vanovac, B., Denk, S.S., Wolfrum, E., Willensdorfer, M., Suttrop, W., Fischer, R., Luhmann, N.C., ASDEX Upgrade Team, EPJ Web Conf. **203**, 02011 (2019)

[8] A. Cathey, M. Hoelzl, G. Harrer, M.G. Dunne, G.T.A. Huijsmans, K. Lackner, S.J.P. Pamela, E. Wolfrum, S. Günter, Plasma Physics and Controlled Fusion **64**, 054011 (2022)

# Radial localization of electron temperature pedestal and ELM-like events using ECE measurements at Wendelstein 7-X

*N.* Chaudhary[1,*], *M.* Hirsch[1], *T.* Andreeva[1], *J.* Geiger[1], *U.* Hoefel[1], *K.* Rahbarnia[1], *G. A.* Wurden[2], *R. C.* Wolf[1], and the W7-X Team[**]

[1]Max Planck Institute for Plasma Physics, Wendelsteinstrasse-1, 17489, Greifswald
[2]Los Alamos National Laboratory, Los Alamos, NM, USA

**Abstract.** A magnetic configuration scan was performed at Wendelstein 7-X stellarator by varying the rotational transform to analyze the plasma confinement for magnetic configurations with different edge magnetic island locations and sizes. For the magnetic configurations, where the 5/5 island chain was moved inside the last closed flux surface, it was observed with electron cyclotron emission measurements that an electron temperature, $T_e$, pedestal develops in the plasma buildup phase and followed by the edge localized mode (ELM)-like crashes. From the mapping of the island to the plasma radius from HINT equilibrium, it was found that the $T_e$ pedestal is formed at the island location on the high field side of the plasma. The ELM-like crashes occur at the location of the pedestal and the transport barrier is broken typically with an energy loss of 3-4% during a single ELM-like event. The frequency and the amplitude of the ELM-like crashes were observed to be changing with island size, plasma heating power and density. Additionally during the plasma decay, after the heating was switched-off, a transition to degraded plasma confinement state was observed with changed $T_e$ profile gradients, faster decay rate of diamagnetic energy, and increased H-alpha levels.

## 1 Introduction

Wendelstein 7-X (W7-X) is a high iota and low shear optimized stellarator concerning magnetic field geometry with a scientific objective to attain 30 min of detached steady-state plasmas [1–3]. Due to the low magnetic shear, it is possible to shape the separatrix by large islands constituting an island divertor [4]. During the operational campaign OP 1.2 of W7-X, a magnetic configuration scan was performed by varying the rotational transform with magnetic configurations between iota = 5/4 and iota = 5/5 at the boundary [5, 6]. During the configuration scan the location of the 5/5 magnetic island chain was moved inside the last closed flux surface, and the island size was varied by changing the control coil current. For these magnetic configuration scan experiments, it has already been reported that the plasma energy increased for the configurations where the 5/5 island chain is close to the plasma boundary but still inside the last close flux surface [5]. However, the reason behind the improved confinement for certain configuration from the scan is not fully understood yet.

The electron cyclotron emission (ECE) at W7-X is measured by a heterodyne radiometer, in the spectral range of the second harmonic extraordinary mode (X2-mode) from 126 to 162 GHz with a collection optics located at the low field side (LFS) observing a bean shaped plasma cross section [7, 8]. The 32 channel radiometer has a temporal resolution of the order of μs which enables the measurement of the rapid fluctuations in $T_e$, for example, the ELM-like crashes possibly caused by the formation of the $T_e$ pedestal, and moreover the diagnostic provides a measurement of the spatial location of the $T_e$ fluctuations in the plasma. For the magnetic configuration scan experiments where the 5/5 island chain was moved inwards, it was observed with the ECE diagnostic that a $T_e$ pedestal develops in the plasma indicating the formation of the transport barrier. Furthermore, it was observed that the transport barrier is broken by a fast (2-3 ms) crash which has characteristic like a tokamak ELM [9], and hence for this work these crashes are referred to as ELM-like crashes. Additionally, these crashes are also reported to be observed with multiple diagnostics at W7-X and are referred to as island localized mode (ILM)s [6, 10, 11].

The paper reports on the experimental observation of the $T_e$ pedestal and the ELM-like crashes and provides the spatial location of the $T_e$ pedestal and the ELM-like crashes with respect to the location of the magnetic island in the plasma. The ECE measurements were used to localize the $T_e$ pedestal and ELM-like crashes in the plasma and an equilibrium calculation done with HINT code [12, 13] was used to determine the location of the magnetic island. Moreover, this work focuses on investigating the changes in the $T_e$ profile gradients for different magnetic configurations to understand the effect of island location and size on the transport in the plasma and hence overall confinement.

*e-mail: neha.chaudhary@ipp.mpg.de
**See *Thomas Sunn Pedersen et al 2022 Nucl. Fusion 62 042022* for the W7-X Team

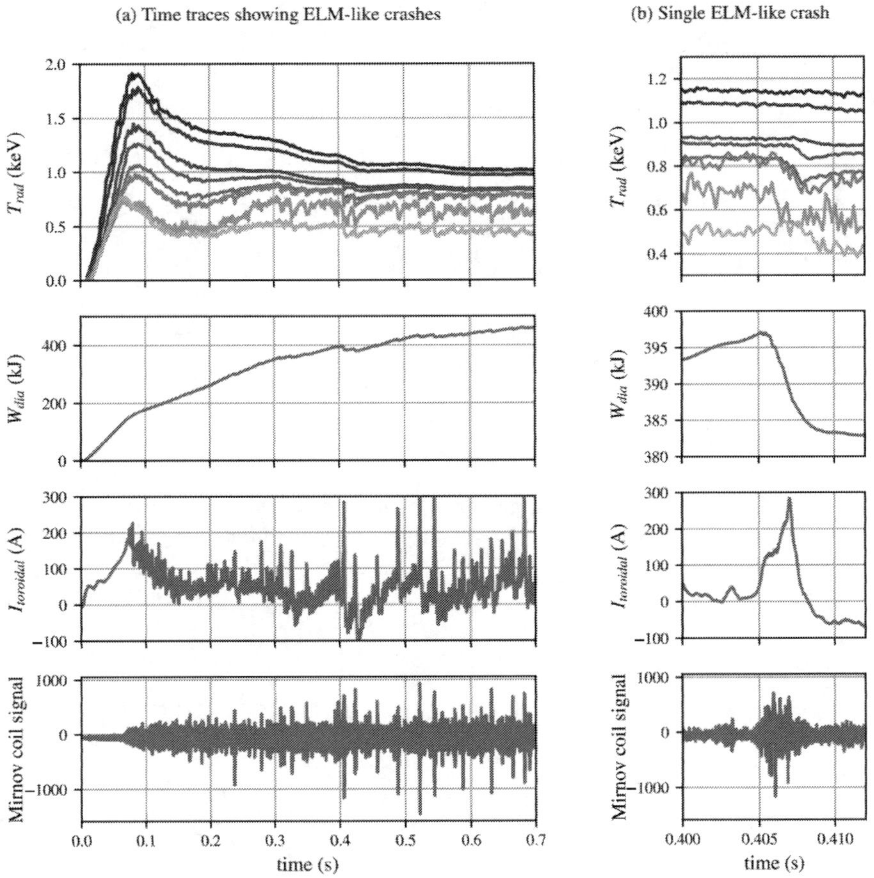


Figure 1: (a) Time traces of the ECE radiation temperature, $T_{rad}$, at high field side (HFS) frequencies, the diamagnetic energy, $W_{dia}$, the toroidal current, $I_{toroidal}$, and the magnetic fluctuations measured by Mirnov coils are shown for the plasma discharge #20180927.026 where the 5/5 island chain was moved inside the last closed flux surface. (b) Zoomed in time traces of (a) at 0.4 s showing the temporal evolution of different parameter during a single ELM-like crash.

The experimental observations are discussed in details in the following sections.

## 2 ELM-like crash observation in multiple plasma diagnostics

The ELM-like crashes were observed in multiple plasma diagnostic measurements during the magnetic configuration scan experiments for the plasma discharges where the magnetic island chain was moved inside the last closed flux surface. Figure 1a shows the time trace of ECE radiation temperature, $T_{rad}$, at the HFS frequencies (144.86-153.56 GHz). The line of sight of the ECE diagnostics crosses the X-point of the magnetic island at HFS and the ELM-like crashes were observed in the $T_{rad}$ for these HFS channels. However a more thorough mapping is presented in the next section. The ELM-like crashes were more pronounced in the high resolution temporal measurements of toroidal current, $I_{toroidal}$, and the magnetic fluctuations.

Figure 1b shows the zoomed in time traces during a single ELM-like crash lasting typically 2-3 ms. From the ECE time traces, it can be observed that the crash occurs at $T_{rad} \approx 0.6$ keV. The occurrence of crash in temperature lead to propagation of a heat wave and the amplitude of this heat wave decrease in the direction away from the location of crash. The $W_{dia}$ measurements show that a single ELM-like crash was accompanied by an energy loss of on average 3.5%. In addition, $I_{toroidal}$ also increased during the ELM-like crash with simultaneously increase in magnetic fluctuations activity observed with Mirnov coils [14].

## 3 Electron temperature pedestal

At the start of the magnetic configuration scan experiments specially for the configurations where the magnetic island chain was moved inside the plasma [5], it was observed from the ECE diagnostic that an electron pedestal develops during the plasma buildup phase. Figure 2 shows the ECE time traces for the plasma startup phase, and around 0.27 s

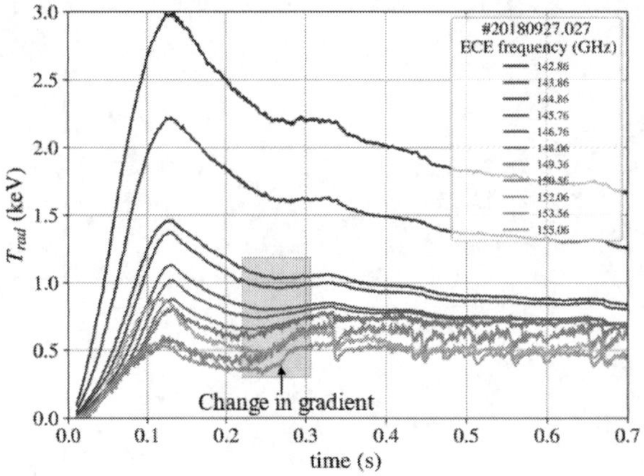


Figure 2: Time traces of ECE channels measuring plasma core to edge (142.86-155.06 GHz) region are shown for #20180927.027. At the start of the plasma discharge around 0.27 s, the temperature gradient changes (the increasing $T_{rad}$ at HFS channels) leading to development of an edge pedestal which is broken by an ELM-like crash at approximately 0.34 s.

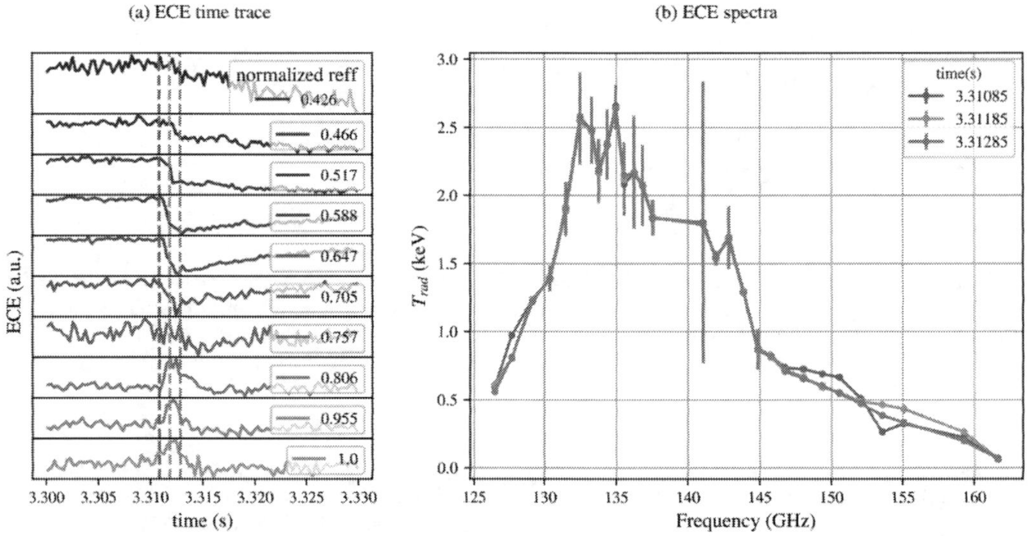


Figure 3: (a) A single crash in $T_e$ shown using ECE data for #20180927.026. Three time points from this crash have been chosen to see the ECE spectra during the crash. The channel around which the neighbouring channels are out of phase is the location of the crash and correspond to inversion radius ($r_{eff}$ = 0.757) of the crash in the plasma. (b) The corresponding ECE spectra before (blue curve) and during (orange and green curve) the crash are shown. It can be clearly observed that the location of crash coincides with the location of the pedestal.

in the discharge (marked in figure), it can be seen that the edge channels (lower than 1 keV) start to show change in electron temperature gradient and a transport barrier develops leading to the delayed increase in the core temperature. The pedestal develops at the start of the plasma by flattening of the radiation temperature near the plasma edge and this is followed by an ELM-like crash in the electron temperature at the location of the pedestal. The frequency and amplitude of ELM-like crashes vary depending on the plasma pressure.

### 3.1 Pedestal breakage and heat wave propagation

Figure 3a shows from ECE measurements, a single ELM-like crash with inversion radius at normalized effective plasma radius, $r_{eff}$ = 0.757 (from cold resonance mapping with HINT). The corresponding heat wave propaga-

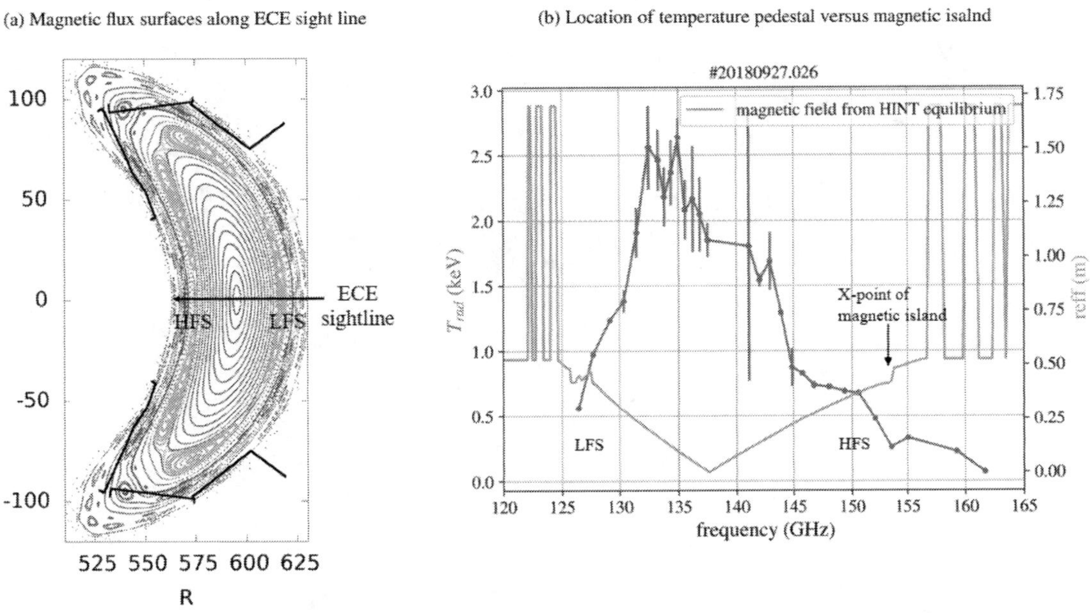


Figure 4: (a) Poincaré plot showing the magnetic flux surface for finite pressure with magnetic island chain (orange color) moved inside the last-closed-flux-surface. The sight line of ECE diagnostic crosses the island O-point on LFS passing the plasma center, and X-point on the HFS. (b) An ECE spectrum is shown where the pedestal can be seen around the channels from 150 to 155 GHz and the magnetic field calculated from the HINT equilibrium is shown as a function of effective plasma radius. The location of the magnetic island is marked in the figure.

tion through ECE channels can be observed with inverted phase of signals at neighbouring channels to the location of crash and the heat wave moves outwards with decaying amplitude in both directions. Three time points are selected (marked in blue, orange and green) before and during the crash to show the temporal evolution of ECE spectra. In figure 3b, a pedestal in ECE spectrum is observed before the crash (blue color) around the channels 150-155 GHz, and the pedestal breaks during ELM-like crash (orange and green color). The typical duration of the crash is 2 – 3 ms with longer recovery time and the frequency of the crash changes with different plasma parameters.

### 3.2 Pedestal location with respect to magnetic island

Figure 4a shows the Poincaré plot of the magnetic flux surfaces with the magnetic island chain shifted in the edge confinement region. The ECE diagnostic measures perpendicular to the magnetic field direction and the radiation collection optics is located at the LFS. And hence crosses the O-point of island from LFS channels and X-point of island from HFS channels. Figure 4b shows the measured ECE spectrum and calculated magnetic field along the ECE line of sight with HINT code which is a magnetic equilibrium code that also includes magnetic islands in the calculation. The magnetic island location is marked in the figure and represents the discontinuity in the magnetic field values. It can be seen that the pedestal develops approximately at the location of the magnetic island in the plasma. This suggests that presence of island impacts the plasma confinement via development of a transport barrier, however the exact mechanism is yet to be understood.

## 4 H to L-mode back transition

Figure 5 shows a possible back transition from a high confinement state to a lower one from different diagnostic measurements. The top plot shows the ECE time traces during the plasma decay phase after the heating was stopped at 4.5 s. At around 4.8 s, a change in the slope of $T_{rad}$ decay rate can be observed in different HFS channels resulting in changed electron temperature gradients accompanied by an ELM-like crash. Simultaneously, an increase in the $W_{dia}$ decay rate was observed alongside increased H-alpha levels and magnetic fluctuations. Hence, the change in temperature gradient and $W_{dia}$ decay rate suggest that the plasma confinement has changed to a degraded state, and this could be a possible fingerprint of a H to L-mode transition. However, a clear forward transition from L to H-mode is so far not observed as the temperature pedestal develops already within the plasma buildup phase.

## 5 Discussion

For the magnetic configurations where the magnetic island chain was moved inside the edge confinement region, a

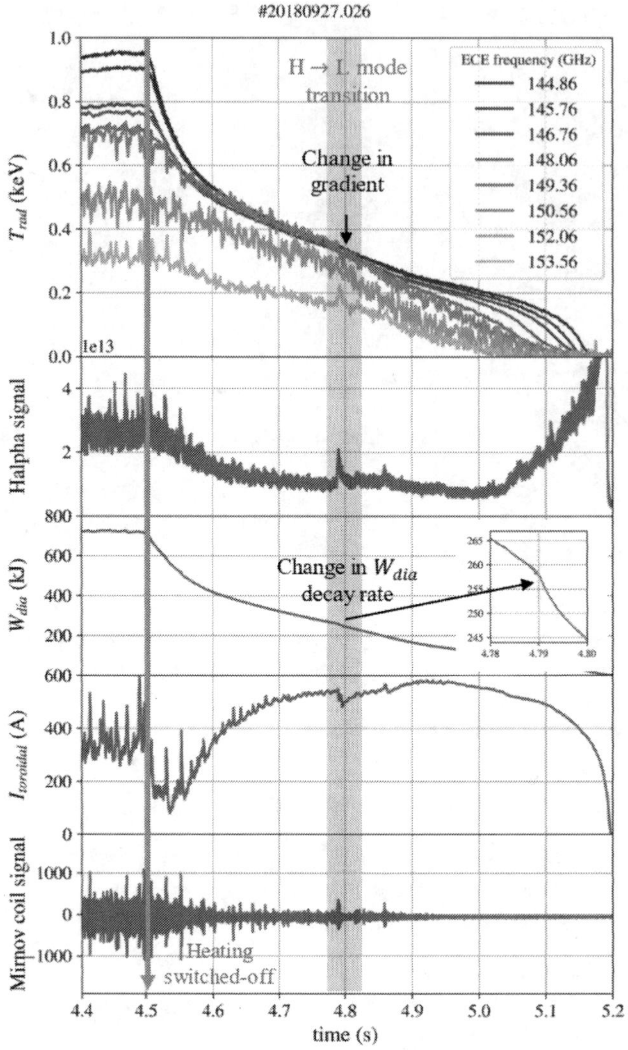


Figure 5: Time traces of ECE, H-alpha, $W_{dia}$, $I_{tor}$, and magnetic fluctuations are shown for the plasma discharge #20180927.026 where the magnetic island chain was moved inside the edge confinement region. Plasma heating was switched-off at 4.5 s, and at approximately 4.8 s, transition to a degraded confinement state can be observed through increased $W_{dia}$ decay rate and changed temperature gradient with occurrence of an ELM-like crash (top plot).

temperature pedestal, as a consequence of a transport barrier, develops at the rational island location suggesting that islands affect the transport in that regime. The transport barrier is broken by ELM-like crash with an inversion radius $r_{eff}$ = 0.757 at the location of pedestal. The ELM-like crash characteristics change with varying iota profile through changed island location and size, plasma density, heating, and hence vary with changes in the plasma pressure profile, and a typical ELM-like crash results in an energy loss of 3-5%. The pedestal builds up already during the plasma startup phase in W7-X stellarator, and hence no clear forward L to H-mode transition is seen for all the available configurations during iota scan experiments from OP1.2 of W7-X. However, a possible backward H to L-mode transition is observed during the plasma decay phase, after the heating is switched off, from different diagnostic measurements and the plasma parameters threshold values for back transition vary with different magnetic configurations. A detailed analysis of magnetic island impacting the transport and leading to the temperature pedestal developmental and an improved confinement state, alongside quantitative analysis for the H to L-mode back transition threshold values for different plasma parameter and magnetic configurations is yet to be followed.

## Acknowledgment

This work has been carried out within the framework of the EUROfusion Consortium, funded by the European Union via the Euratom Research and Training Programme (Grant Agreement No 101052200 — EUROfusion). Views and opinions expressed are however those of the author(s) only and do not necessarily reflect those of the European Union or the European Commission. Neither the European Union nor the European Commission can be held responsible for them.

## References

[1] C. Beidler et al., Nature **596**, 221 (2021)

[2] T. Klinger et al., Nuclear Fusion **59**, 112004 (2019)

[3] R. Wolf et al., Physics of Plasmas **26**, 082504 (2019)

[4] T.S. Pedersen et al., Nuclear Fusion **62**, 042022 (2022)

[5] T. Andreeva et al., Nuclear Fusion **62**, 026032 (2022)

[6] J. Geiger et al., *Confinement and equilibrium with internal Islands in a configuration scan with respect to iota in W7-X*, in *Proc. 28th IAEA Fusion Energy Conf.(FEC-2020, Virtual Event)* (2021)

[7] M. Hirsch et al., *ECE diagnostic for the initial operation of Wendelstein 7-X*, in *EPJ Web of Conferences* (EDP Sciences, 2019), Vol. 203, p. 03007

[8] U. Hoefel et al., Review of Scientific Instruments **90**, 043502 (2019)

[9] H. Zohm, Plasma Physics and Controlled Fusion **38**, 105 (1996)

[10] G. Wurden et al., *Structure of island localized modes in Wendelstein 7-X*, in *46th EPS Conference on Plasma Physics* (European Physical Society, 2019)

[11] G. Wurden et al., *Quasi-continuous low frequency edge fluctuations in the W7-X stellarator*, in *45th EPS Conference on Plasma Physics* (European Physical Society, 2018)

[12] K. Harafuji, T. Hayashi, T. Sato, Journal of computational physics **81**, 169 (1989)

[13] Y. Suzuki, N. Nakajima, K. Watanabe, Y. Nakamura, T. Hayashi, Nuclear Fusion **46**, L19 (2006)

[14] K. Rahbarnia et al., Plasma Physics and Controlled Fusion **63**, 015005 (2020)

# A model of non-Maxwellian electron distribution function for the analysis of ECE data in JET discharges

G. Giruzzi[1*], M. Fontana[2], F.P. Orsitto[3], E. de la Luna[4], R. Dumont[1], L. Figini[5], M. Maslov[2], S. Mazzi[6], S. Schmuck[5], L. Senni[3], C. Sozzi[5], C. Challis[2], D. Frigione[3], J. Garcia[1], L. Garzotti[2], J. Hobirk[7], A. Kappatou[7], D. Keeling[2], E. Lerche[2], C. Maggi[2], J. Mailloux[2], F. Rimini[2], D. Van Eester[8], and JET contributors[a]

[1] CEA, IRFM F-13108 Saint Paul-lez-Durance, France
[2] United Kingdom Atomic Energy Authority, Culham Centre for Fusion Energy, Culham Science Centre, Abingdon, Oxon, OX14 3DB, UK
[3] ENEA Department Fusion and Technology for Nuclear Safety, C R Frascati , 00044 Frascati, Italy
[4] National Fusion Laboratory, CIEMAT, Madrid, Spain
[5] Istituto per la Scienza e Tecnologia dei Plasmi, CNR, 20125 Milano, Italy
[6] EPFL, Swiss Plasma Center, CH-1015 Lausanne, Switzerland
[7] Max-Planck-Institut für Plasmaphysik, D-85748 Garching, Germany
[8] Laboratory for Plasma Physics, LPP-ERM/KMS, Brussels, Belgium
[a] See J Mailloux et al., Nucl. Fusion **62**, 042026 (2022)

**Abstract.** Recent experiments performed in JET at high level of plasma heating, in preparation of, and during the DT campaign have shown significant discrepancies between electron temperature measurements by Thomson Scattering (TS) and Electron Cyclotron Emission (ECE). In order to perform a systematic analysis of this phenomenon, a simple model of bipolar distortion of the electron distribution function has been developed, allowing analytic calculation of the EC emission and absorption coefficients. Extensive comparisons of the modelled ECE spectra (at both the 2nd and the 3rd harmonic extraordinary mode) with experimental measurements display good agreement when bulk electron distribution distortions around 1-2 times the electron thermal velocity are used and prove useful for a first level of analysis of this effect.

## 1 Introduction

Discrepancies between electron temperature measurements by Thomson Scattering (TS) and Electron Cyclotron Emission (ECE) have been often observed in high-temperature tokamak plasmas, in particular on TFTR [1], JET [2] and FTU [3]. Such observations, made on different machines, by different types of instruments, using different calibration methods, are too ubiquitous to be ascribed to instrumental effects; they rather call for explanations based on physics phenomena. The hypothesis that the discrepancy could be associated to non-Maxwellian bulk electron distributions has been put forward in the past [4,5] and appears as a plausible explanation in the case of a plasma heated by EC waves, as FTU [3]. For TFTR and JET, electron heating rather takes place because of the interaction of the electron distribution either with a fast ion tail driven by Neutral Beam Injection (NBI) and/or Ion Cyclotron Resonance Heating (ICRH), or by energetic alpha particles produced by fusion reactions in DT (Deuterium-Tritium) plasmas. Two mechanisms are known to produce small bipolar distortions of the electron distribution in the presence of energetic ions: collisional relaxation [6] or Landau damping of kinetic Alfvén waves, as observed in the magnetosheath [7,8].

Recent experiments performed at JET at high level of plasma heating, in preparation of, and during the DT campaign have shown again TS-ECE discrepancies on an extensive database [9]. ECE is observed to be higher or lower than TS, depending on the plasma scenario. Moreover, ECE measured by a Martin-Puplett interferometer on a broad frequency range displays differences between 2nd and 3rd harmonics extraordinary (X) mode, which, at high temperatures (> 4 keV) and high densities are expected to yield the same radiation temperature. In order to perform a systematic analysis of this effect, a simple model of bipolar distortion of the electron distribution function has been developed, allowing analytic calculation of the EC emission and absorption coefficients. Bulk electron distribution distortions around 1-2 times the electron thermal velocity are considered for a first level of analysis of this effect. In this paper, the model is described in Sec. 2. Comparisons of the modelled ECE spectra (at both the 2nd and the 3rd harmonic) with experimental measurements are presented in Sec. 3. Conclusions and possible impact on ITER are presented in Sec. 4.

* Corresponding author: gerardo.giruzzi@cea.fr

 https://doi.org/10.1051/epjconf/202327703005

## 2 Electron distribution function model

A toy model of isotropic perturbation $f_1(p)$ of the electron distribution function $f(p)$, where $p$ is the modulus of the electron momentum, is developed as follows. We take the relativistic Maxwellian as the unperturbed distribution:

$f_M = Ae^{-\mu(\gamma-1)}$, where

$$A = \frac{\mu e^{-\mu}}{4\pi K_2(\mu)(mc)^3}\ ,\ \mu = \frac{mc^2}{T_e}\ ,\gamma\ = \sqrt{1+\frac{p^2}{(mc)^2}}$$

$m$ is the electron rest mass, $c$ the speed of light, $T_e$ the electron temperature and $K_2$ the modified Bessel function of the second kind. The perturbed distribution function is defined as:

$$f = A(e^{-\mu(\gamma-1)}+f_1)$$

and a suitable bipolar isotropic form of the perturbation $f_1$ is given as a function of three parameters $f_0$, $p_0$ and $\delta$ by:

$$f_{1u} = f_0 sin\left[\frac{\pi}{\delta}(p-p_0)\right] \text{ for } p_0-\delta < p < p_0+\delta$$

Various types of anisotropic forms can also be defined by multiplying $f_{1u}$ by functions of the pitch-angle $\theta$, e.g.,

$$f_{1s} = f_{1u} sin\theta\ ,\ f_{1c2} = f_{1u} cos^2\theta\ ,$$
$$f_{1s2} = f_{1u} sin^2(2\theta)\ , f_{1c} = f_{1u} cos\theta\ .$$

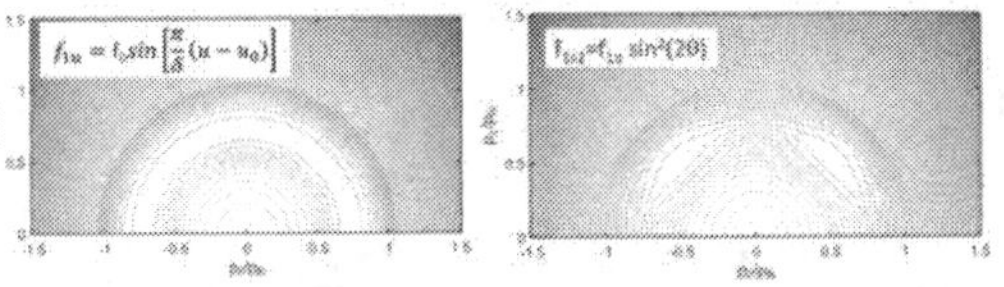

**Fig. 1.** Two examples of the model distribution function, perturbed isotropically (left) and anisotropically (right). Level curves in the parallel and perpendicular normalised momentum plane.

Two examples of isotropic and anisotropic perturbed distribution functions of this kind are shown in Fig. 1. All these functions allow analytical calculations of the electron cyclotron absorption ($\alpha$) and emission ($\beta$) coefficients for perpendicular propagation. The general expressions of the absorption and emission coefficients for an arbitrary electron distribution function, as momentum-space integrals, are well known and available in a number of papers. Starting, for instance, from Eqs. 10-13 of Ref. [10], for the extraordinary wave and $n > 1$, these expressions are given by:

$$(\alpha,\beta) = A_0 \sum_n(\alpha_n,\beta_n),\ A_0 = \frac{2\pi^2}{N_X}\frac{\omega}{c}\frac{\omega_p^2}{\omega^2}\left|1-\frac{i\varepsilon_{12}}{\varepsilon_{11}}\right|^2\frac{\mu e^{-\mu}}{4\pi K_2(\mu)}$$

where $n$ is the harmonic number, $\omega$ the wave frequency (times $2\pi$), $\omega_p$ and $\omega_c$ the plasma and the electron cyclotron frequencies, $N_x$ the cold refractive index and $\varepsilon$ the cold plasma dielectric tensor;

$$\alpha_n = A_n\left[\frac{\mu u_n}{n\omega_c/\omega}e^{-\mu\left(\frac{n\omega_c}{\omega}-1\right)} - Q_n h(u_n)\frac{\pi}{\delta/mc}f_0 cos\left(\frac{\pi}{\delta/mc}(u_n-u_0)\right)\right]$$

$$A_n = \mu^{n-1}\frac{n\omega_c}{\omega}u_n^{2n}\left(\frac{N_X\omega}{\sqrt{\mu}\omega_c}\right)^{2(n-1)}\frac{B(n+1,1/2)}{[2^n(n-1)!]^2},\quad u_n = \left[\left(\frac{n\omega_c}{\omega}\right)^2-1\right]^{1/2},$$

$$u_0 = \frac{p_0}{mc}\ ,\quad h(u_n) = H\left(u_n-u_0+\frac{\delta}{mc}\right)H\left(u_0+\frac{\delta}{mc}-u_n\right)$$

$B(x,y)$ is the Beta function and $H$ the Heaviside function.

$$\beta_n = A_n\frac{mc^2u_n}{n\omega_c/\omega}\left[e^{-\mu\left(\frac{n\omega_c}{\omega}-1\right)} + Q_n h(u_n) f_0 sin\left(\frac{\pi}{\delta/mc}(u_n-u_0)\right)\right].$$

The function $Q_n$ depends on the choice of the perturbed distribution function as follows:

$$f_1 = f_{1u} \rightarrow Q_n = 1\ ,\ f_1 = f_{1u} sin\theta \rightarrow Q_n = \frac{B(n+3/2,1/2)}{B(n+1,1/2)}$$

$$f_1 = f_{1u} cos^2\theta \rightarrow Q_n = \frac{B(n+1,3/2)}{B(n+1,1/2)}\ ,$$

$$f_1 = f_{1u} sin^2(2\theta) \rightarrow Q_n = 4\frac{B(n+2,3/2)}{B(n+1,1/2)}\ ,\quad f_1 = f_{1u} cos\theta \rightarrow Q_n = \frac{B(n+1,1)}{B(n+1,1/2)}$$

In the following, only the isotropic perturbation $f_{1u}$ is considered.

Using these expressions of the emission and absorption coefficients, the radiation temperature, measured along a line of sight in the equatorial plane (as it is approximately the case for JET), is given by:

$$T_{rad}(\omega) = \int_{R_0-a}^{R_0+a} dR\,\beta(R)\,exp(-\int_R^{R_0+a}\alpha(R')dR')\ ,$$

where $R$ is the major radius coordinate and $a$, $R_0$ are the minor and major radii. An example of the absorption and emission coefficients, as well as of the radiation function $F_{rad} = \beta exp(-\int\alpha dR)$, i.e., the integrand appearing in the radiation temperature expression, at various wave frequencies and for typical JET parameters is presented in Fig. 2. This figure shows that $\alpha$ and $\beta$ are broad functions of $R$, nevertheless, emission is well localised in space owing to the exponential re-absorption term that multiplies $\beta$ in the radiation function expression.

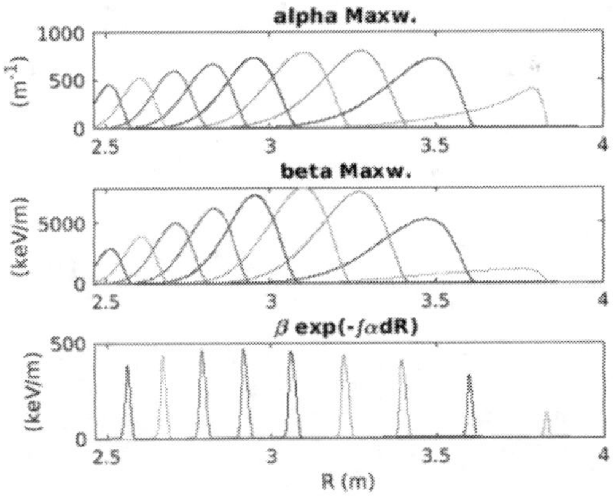


**Fig. 2.** 2nd harmonic EC absorption, emission coefficients and radiation function $\beta exp(-\int\alpha dR)$ for a Maxwellian plasma and JET parameters. $T_e$ = 10 keV.

Owing to the resonance condition $\gamma = n\omega_c(R)/\omega$, for a given wave frequency the localisation in $R$ corresponds to a localisation in electron kinetic energy $E = mc^2(\gamma-1) = mc^2(R_c/R-1)$, where $R_c$ is the cold resonance location, defined by $n\omega_c(R_c) = \omega$. The location of the maximum of the radiation function $F_{rad}$

in normalised momentum $p/p_{th}$ (where $p_{th}=mv_{th}=(mT_e)^{1/2}$) depends on the harmonic number and on the electron temperature, as shown in Fig. 3, where the maximum is plotted together with the widths at half height. This figure illustrates a fundamental property of the ECE diagnostic: different harmonics probe different parts of the electron distribution function, both momentum and width decreasing with the electron temperature. This means that ECE at 2nd and 3rd harmonics can be used to constrain the electron distribution function in the region $p_{th}<p<2p_{th}$, in a more and more precise way for higher and higher temperatures.

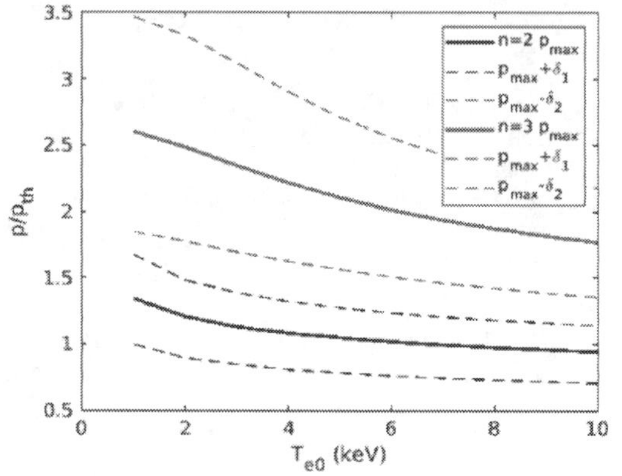


**Fig. 3.** Momentum value (normalised to thermal momentum) at the maximum of the radiation function $\beta exp(-\int \alpha dR)$ (solid lines) for the 2nd (blue) and the 3rd (red) harmonics. Dashed lines correspond to the width at half height of the radiation function.

Two examples of the previously defined isotropic model perturbation are shown in Fig. 4, together with the ranges seen by the 2nd and 3rd harmonics, for $T_e$=7 keV.

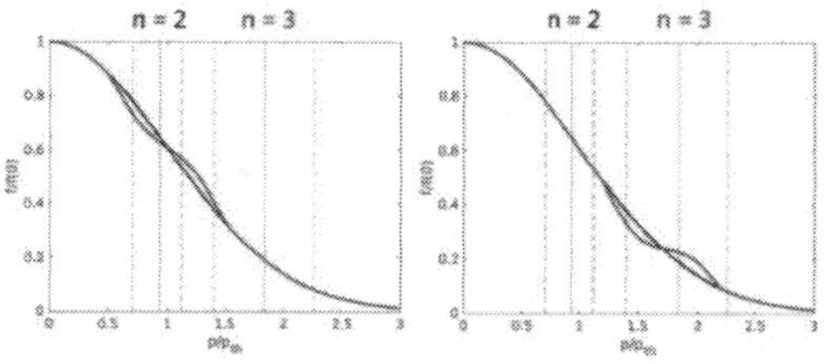


**Fig. 4.** Model electron momentum distribution function (red) compared to the Maxwellian (blue), both normalised to the value at p=0. The vertical lines define the ranges seen by ECE at the 2nd and 3rd harmonics, for $T_e$=7 keV. Examples of perturbations mainly affecting the 2nd (left) and the 3rd (right) harmonics.

It appears that the two harmonics can be differently affected by perturbations localised at different momenta, i.e., with a different $p_0$ parameter. In general, the smaller the absolute value of the distribution function derivative in a given harmonic range, the larger the corresponding radiation temperature. Moreover, since the locations and widths of the momentum regions probed by the two harmonics vary with temperature, the same perturbation will affect the $T_e$ profile measured by ECE more or less significantly, depending on the temperature. An example is shown in Fig. 5, where the 2nd harmonic ECE profiles are computed for typical JET parameters and two different values of the central temperature $T_{e0}$: 3 and 10 keV. For the same perturbation parameters ($f_0$ = 0.03, $p_0/p_{th}$ = 1, $\delta/p_{th}$ = 0.25), the radiation temperature is significantly affected at high temperature, but practically unaffected at low temperature, in agreement with experimental observations [1, 2].

The strong sensitivity of ECE to very small perturbations of the electron distribution (a few percent in the example of Fig. 5) is due to the presence of the exponential term in $F_{rad}$ and to the fact that the absorption coefficient is an integral in momentum space of perpendicular derivative of the distribution function [10]. The same perturbation would have very little effect on the Thomson scattering measurement, which is simply proportional to the distribution function (see, e.g., Eqs. 5.8, 5.9 of [11]). This is illustrated in Fig. 6, where $F_{rad}$ is plotted as a function of the normalised electron kinetic energy, together with the equivalent quantity for Thomson scattering (scattered radiation spectrum). Clearly, the impact is completely different. This gives the main key to understand why the two measurements of the electron temperature can give different results if the distribution function is not Maxwellian.

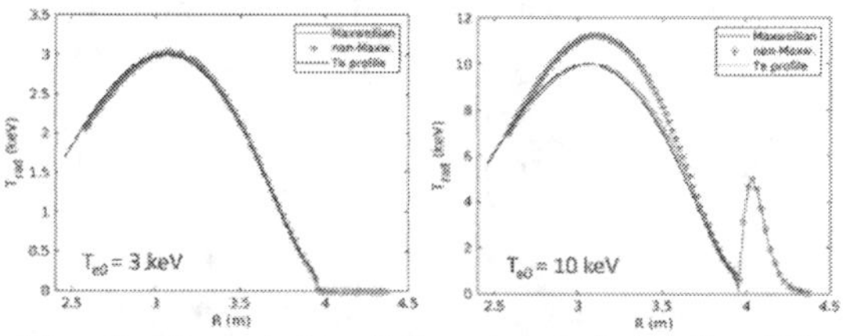


**Fig. 5.** Computed radiation temperatures for JET parameters and two different central temperatures. Assumed $T_e$ profile, Maxwellian distribution and perturbed distribution cases are shown. R is defined, for each frequency, as the cold resonance position. Perturbation parameters: $f_0$ = 0.03, $p_0/p_{th}$ = 1, $\delta/p_{th}$ = 0.25.

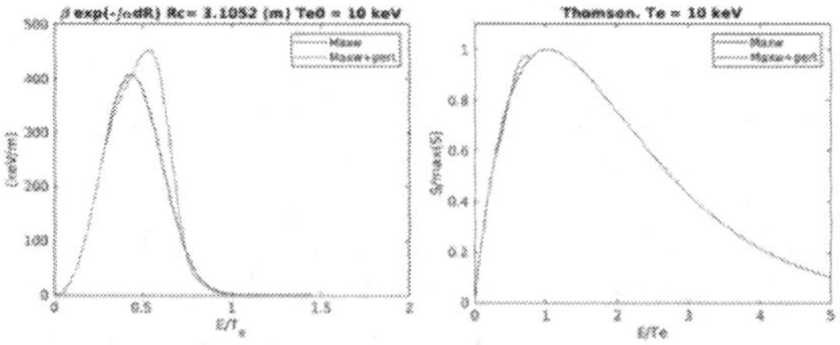


**Fig. 6.** Radiation function vs electron kinetic energy normalised to $T_e$ for 2nd harmonic ECE (left) and normalised TS scattered radiation spectrum (right). Maxwellian (blue) and perturbed (red) distributions. Perturbation parameters: $f_0$ = 0.03, $p_0/p_{th}$ = 1, $\delta/p_{th}$ = 0.25.

## 3 Comparison with ECE and TS data

The distribution function model can now be used as an analysis tool of the extensive JET database described in

Ref. [9]. Examples of various subsets of the database are shown, corresponding to specific experimental scenarios.

In Figs. 7-9, ECE central temperatures measured at both the 2nd and 3rd harmonics by means of the JET Martin-Puplett interferometer [12] are shown vs the corresponding temperature measured by Thomson scattering via the so-called LIDAR system [13] (left panels). These temperatures are averaged over a region covering 10% of the minor radius around the centre. In the right panels, the central temperatures measured at the 3rd harmonic are directly plotted vs the corresponding 2nd harmonic ones. In Fig. 7 only, error bars are also shown. On both panels, the corresponding quantities computed using the above described model are also shown, choosing (by trial and error) a set of distribution function parameters that optimise the agreement with data, identical for both harmonics. The perturbation set of parameters used is displayed at the top of the figures. In all cases, a wall reflection coefficient of 0.55 has been assumed, according to previous evaluations for the JET machine [12]; however, the results are very weakly dependent on this parameter. The three figures refer to three different subsets of data, corresponding to different experimental scenarios [9]:

- Fig. 7: DD baseline discharges with low gas, low Neon injection and pellets, characterised by small ELMs and partially detached divertor [14].
- Fig. 8: DT discharges in the baseline scenario.
- Fig. 9: DT discharges in the hybrid scenario.

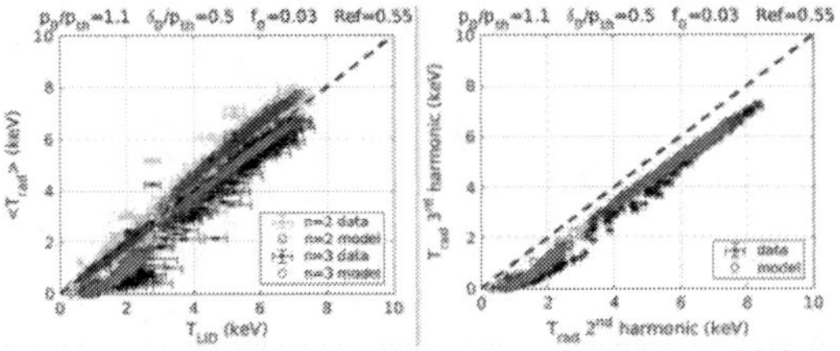


**Fig. 7.** Subset of JET Deuterium (DD) discharges with low gas, low Neon injection and pellets (pulse numbers: 96990, 96992, 96993, 96994, 96996, 96998, 96999). Comparison between data and model, for perturbation parameters: $f_0$ = 0.03, $p_0/p_{th}$ = 1.1, $\delta/p_{th}$ = 0.5. Left: central ECE radiation temperature vs LIDAR central temperature (both averaged over a region covering 10% of the minor radius around the centre). Right: 3rd harmonic vs 2nd harmonic central radiation temperatures. Error bars are also shown.

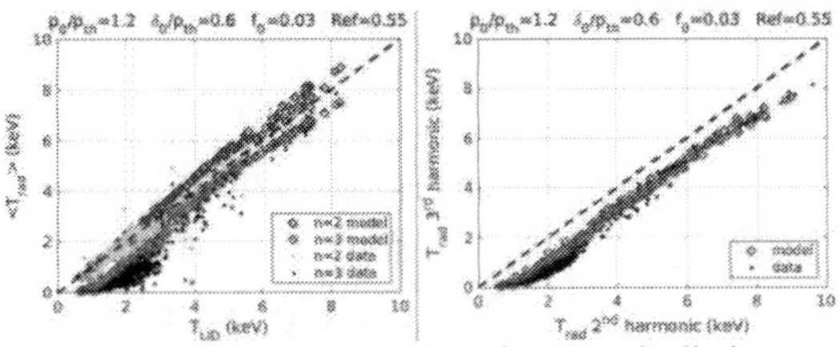


**Fig. 8.** As in Fig. 7, for the DT baseline scenario discharges data set (pulse numbers: 99520, 99795, 99796, 99797, 99799, 99805, 99861, 99862, 99863, 99878, 99943, 99944, 99948). Perturbation parameters: $f_0$ = 0.03, $p_0/p_{th}$ = 1.2, $\delta/p_{th}$ = 0.6.

In all these figures, the 3rd harmonic points are well below the corresponding TS measurements at low temperature ($T_{e0}$ < 4 keV, approximately), because of the low optical thickness.

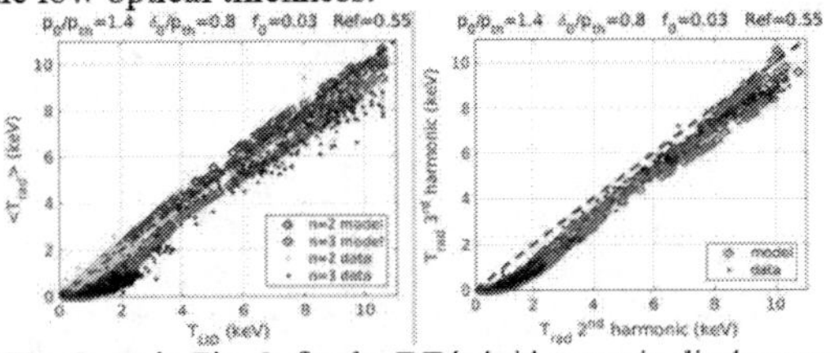


**Fig. 9.** As in Fig. 8, for the DT hybrid scenario discharges data set (pulse numbers: 99448, 99449, 99450, 99452, 99455, 99541, 99542, 99543, 99544, 99594, 99595, 99596, 99760, 99761, 99866, 99867, 99868, 99869, 99908, 99910, 99912, 99914, 99949, 99950, 99951, 99953). Perturbation parameters: $f_0$ = 0.03, $p_0/p_{th}$ = 1.4, $\delta/p_{th}$ = 0.8.

In Figs. 7 and 8, it appears that the temperature measured by the 3rd harmonic, in the optically thick range, is close to the LIDAR temperature, or slightly lower. Following the discussion of Figs. 3 and 4, this means that the distribution function is nearly unperturbed in the velocity range covered by the 3rd harmonic (~ $2v_{th}$). On the other hand, the 2nd harmonic temperature is significantly higher than the TS one at high $T_{e0}$, meaning that some flattening of the distribution function takes place around $v_{th}$. Indeed, the optimum fit is obtained for $p_0/p_{th}$ = *1.1 – 1.2*. This situation is very similar to that of past JET experiments [2].

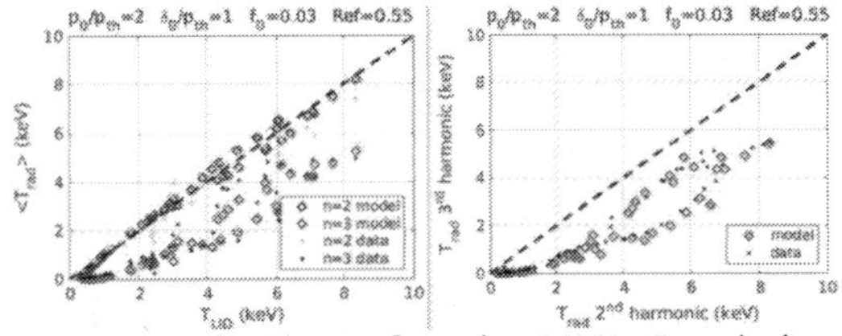


**Fig. 10.** As in Fig. 9, for pulse 96850. Perturbation parameters: $f_0$ = 0.03, $p_0/p_{th}$ = 2, $\delta/p_{th}$ = 1.

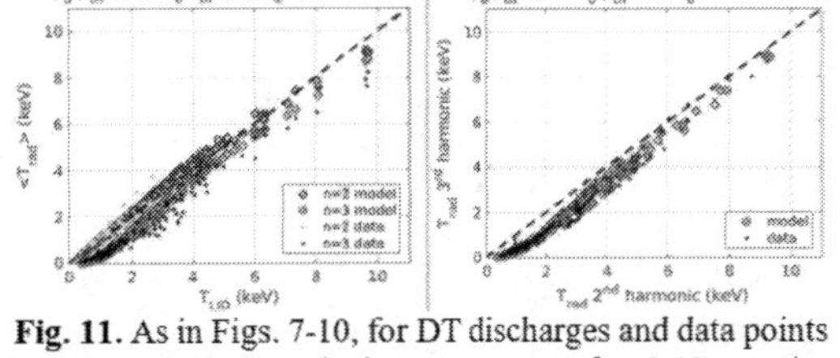


**Fig. 11.** As in Figs. 7-10, for DT discharges and data points with NBI only. Perturbation parameters: $f_0$ = 0.03, $p_0/p_{th}$ = 1.35, $\delta/p_{th}$ = 0.6.

In Fig. 9 (DT hybrid scenario), the 3rd harmonic tends to yield lower temperatures and the optimum fit is obtained for $p_0/p_{th}$ = *1.4*. There are also cases in which the 3rd harmonic is strongly affected by the perturbation, whereas the 2nd one is weakly affected. In general, phases with ICRH only have these characteristics. As an example, it is interesting to show how the model behaves for the simulation of an individual pulse (96850), characterised by low density and, in the high temperature phase, ICRH only. Figure 10 shows that in this case the 2nd harmonic is nearly unperturbed, whereas the 3rd harmonic has non-Maxwellian features that are well reproduced by a broad perturbation located at $p_0/p_{th} = 2$.

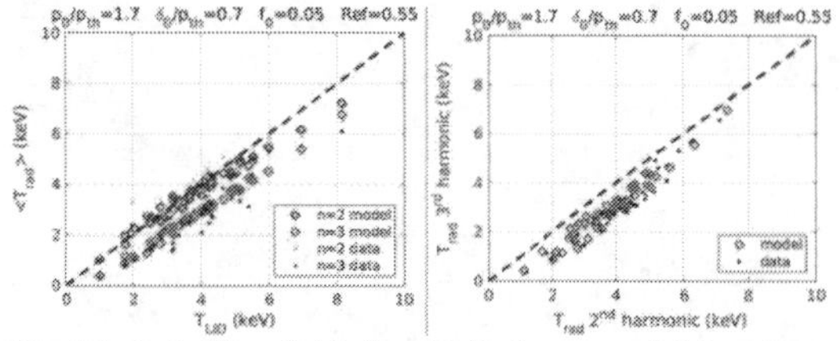


**Fig. 12.** As in Figs. 7-10, for DT discharges and data points with ICRH only. Perturbation parameters: $f_0 = 0.05$, $p_0/p_{th} = 1.7$, $\delta/p_{th} = 0.7$.

The fact that ICRH tends to produce perturbations at somehow higher momenta is a general trend, that can be illustrated by considering all the DT discharges (belonging to both baseline and hybrid scenarios) and selecting data points with NBI only and with ICRH only. There is a significant number of them. Results for these two data sets are shown in Figs. 11 and 12, respectively. ICRH only phases of the DT discharges are clearly perturbed at higher momenta than NBI only ones, i.e., $p_0/p_{th} = 1.7$ with respect to 1.35. The intensity of the perturbation is also higher ($f_0 = 0.05$ instead of 0.03). This is possibly related to the different types of ion tails that the two heating systems produce and/or to their different mechanisms and intensity of direct interaction with the electrons.

These examples have shown that the most sensitive parameter of the model is the perturbation location in momentum space, $p_0$. This suggests a possible use of the model as an analysis tool, in order to detect trends in the database with respect to various distinctive quantities. This is illustrated in Fig. 13. Using the full database, the $p_0$ value is determined by fitting the measured ECE central temperature (at both harmonics) for selected values of four quantities: central density, ratio of heating power and central density, Alfvén velocity normalised to electron thermal velocity and fast ion beta (this quantity is not a measured one, but is obtained from results of NBI and ICRH modelling codes that are available in the JET database). Slight adjustments of the other model parameters (perturbation width and intensity) are made in some cases. Figure 13 shows that a regular behaviour of $p_0$ is obtained in all these cases; the trends observed can be used to guide the search for an interpretation of the experimentally observed non-Maxwellian features.

## 4 Conclusions

The discrepancy observed in JET plasmas [2,4,9] between ECE and TS measurements has been analysed with the hypothesis that it could be associated with a non-Maxwellian bulk electron distribution function. This hypothesis has been already formulated in the past [4,5], however, it remained to be quantitatively assessed on an extensive database. In order to perform such a quantitative comparison, a toy model of non-Maxwellian electron distribution function has been developed, allowing analytical computation of the X-mode ECE spectra and massive comparison with the JET central temperature database [9], with more than 12000 data points.

Using this model, various points have been clarified:

- an electron distribution function perturbation of a few percent, localised around 1-2 thermal velocities, is sufficient to explain the level of discrepancy observed;
- such a perturbation would be practically invisible to TS diagnostics, at least those of JET and other existing machines; however, specific TS systems can be conceived to this end [15];
- even if such a perturbation is present at any temperature, it becomes visible and more and more significant at higher and higher temperatures, because the radiation function becomes narrower and shifts to lower velocities;
- at high temperature ($T_{e0}$ > 4-5 keV), X-mode measurements at both 2nd and 3rd harmonic (possibly also higher harmonics and/or 1st harmonic O-mode) are essential in order to properly constrain the distribution function in different velocity ranges.

An important open question is whether a perturbation of this kind could significantly affect measurements of the ECE profile in ITER. Of course, since the cause of the perturbation is at present still unknown, there is no reason to assume that the same

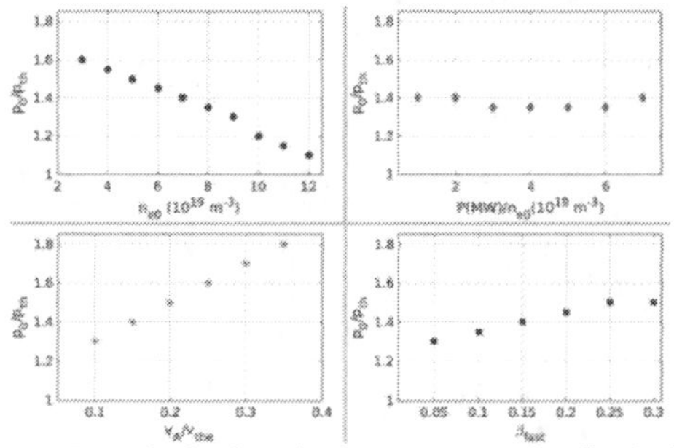

**Fig. 13.** Perturbation location in momentum that fits the full database for selected values of various quantities. Central density (top left), ratio of heating power and central density (top right), Alfvén velocity normalised to electron thermal velocity (bottom left) and fast ion beta (bottom right).

mechanisms acting in JET (and TFTR) could also be present and significant in an ITER plasma. The model only allows quantifying whether a given perturbation would affect the ECE temperature profile in a sizeable way. This is what is shown by the example of Fig. 14. Two non-Maxwellian distributions are considered, with perturbations of the same intensity as those observed in JET plasmas, similar width and two different momentum localisations: $p_0/p_{th} = 0.75$ and $p_0/p_{th} = 1$. The impact on the 2nd harmonic ECE profile is shown in the left panel (red and blue stars, respectively). It appears that at the high temperatures expected in ITER (25 keV in this example) the effect can be stronger and acting in two opposite ways, depending on the momentum localisation of the perturbation. Because of the large temperature variations on an ITER profile,

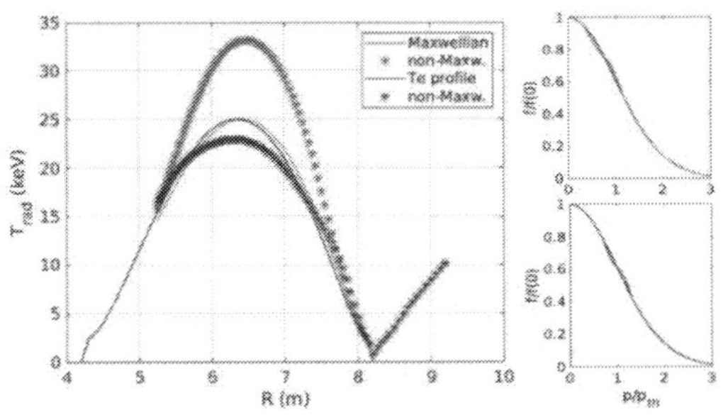


**Fig. 14.** 2[nd] harmonic ECE temperature profile computed for an ITER plasma with $T_{e0}$ = 25 keV (left panel). R is defined, for each frequency, as the cold resonance position. Input $T_e$ profile (black line). Maxwellian case (cyan line) and two examples obtained with perturbed distribution functions, with perturbation parameters: $f_0 = 0.03$, $\delta/p_{th} = 0.3$, $p_0/p_{th} = 0.75$ (top right panel, red) and $p_0/p_{th} = 1$ (bottom right panel, blue).

both effects can be observed on the same ECE profile: note that in the 10-15 keV range, the blue ECE profile is higher than the Maxwellian, whereas it becomes lower beyond 15 keV. Therefore, it would not be just a matter of central temperature value, but a distortion of the full profile. The model described here should be regarded as an analysis tool only. It is not linked to a theoretical explanation about the origin of the electron distribution function perturbation. At present, we could just formulate hypotheses. Because of the nature of the JET high performance discharges analysed [9], which is analogous to those of TFTR [1], likely explanations involve the fast ion population, which is ubiquitous in these discharges (driven by NBI, ICRH and the fusion reactions themselves for DT pulses). In addition, ICRH also directly interacts with the electrons. Fast ions might interact with the electrons via two mechanisms:

- collisional relaxation of the fast ion tail on the electron distribution;
- interaction of fast ion driven MHD modes with the electrons (e.g., via Landau damping)

About the first mechanism, it is generally assumed that the electron distribution function basically remains Maxwellian, because interaction with the fast ion tail takes place at low velocities (around thermal or sub-thermal), where collisions are very strong. However, we have demonstrated that a tiny perturbation (~a few percent) is sufficient to explain the observed effects. In this velocity range, asymptotic expansions of the collision operator cannot be used and, looking for small effects, even the usual linearization of the collision operator could be questionable. Therefore, a solution of the kinetic equation with the full integro-differential collision operator is likely to be needed, which is not generally available in the literature. In Ref. [6], this type of problem is solved for inertial fusion applications and the resulting perturbation has a clear bipolar structure, in the thermal velocity range. Work is now ongoing in order to solve the problem numerically in the JET parameter range and for various input fast ion distributions (NBI, ICRH driven tails or alpha particle distributions can be computed by means of Monte-Carlo or Fokker-Planck codes).

About the second mechanism, direct observation by probes in the magnetosheath have revealed the presence of bipolar distortions of the electron distribution function, around the electron thermal velocity [7]. Gyrokinetic simulations have provided a convincing interpretation of such observations in terms of Landau damping of kinetic Alfvén waves [8], which are also known to be present in tokamak plasmas. Nevertheless, magnetosheath plasmas have very different characteristics with respect to tokamak plasmas (for instance, they are nearly collisionless), therefore, these results cannot be easily generalised to the case of interest here. However, they have inspired analogous gyrokinetic calculations that are now in progress, for parameters close to those of the JET experiments. Note that, in principle, other MHD modes (not necessarily directly excited by fast ions) could also interact with the electrons and cause distribution function distortions.

In conclusion, as for the diagnosis of the electron temperature profile in high temperature plasmas, TS appears rather insensitive to small perturbations of the electron distribution function, therefore, in this respect, it is expected to provide a reliable measurement of the electron temperature. Conversely, ECE can definitely be affected by tiny perturbations of the electron distribution (a few percent) localised in the range 1-2 $v_{th}$, in a different way at different harmonics and in different temperature ranges. Therefore, the temperature measurement can be considered as less reliable than that of TS. However, this high sensitivity of ECE can be exploited to constrain the electron distribution function in order to extract information on its detailed shape and explore fundamental physics effects, such as, e.g., those related to fast ion physics.

**Acknowledgements:** This work has been carried out within the framework of the EUROfusion Consortium and has received funding from the Euratom research and training programme 2014-2018 and 2019-2020 under grant agreement No 633053. The views and opinions expressed herein do not necessarily reflect those of the European Commission.

## References

[1] G. Taylor, R. Harvey, Fus. Sci. Techn. **55**, 64 (2009)
[2] E. de la Luna et al., Rev. Sci. Instr. **74**, 1414 (2003)
[3] G. Pucella et al., Nucl. Fusion **62**, 042004 (2022)
[4] V. Krivenski et al., 29th EPS Conf. on Plasma Physics, ECA Vol. 26B, O-1.03 (2002)
[5] V. Krivenski, Fus. Eng. Des. **53**, 23 (2001)
[6] B. Appelbe et al., Phys. Plasmas **26**, 102704 (2019)
[7] C.H.K. Chen et al., Nat. Commun. **10**, 740 (2019)
[8] S.A. Horvath et al., Phys. Plasmas **27**, 102901 (2020)
[9] M. Fontana et al., this workshop.
[10] G. Giruzzi, Nucl. Fusion **28**, 1413 (1988)
[11] S.L. Prunty, Phys. Scr. **89**, 128001 (2014)
[12] S. Schmuck et al., Rev. Sci. Instr. **87**, 093506 (2016)
[13] M. Maslov et al., JINST **8**, C11009 (2013)
[14] J. Garcia et al., Phys. Plasmas **29**, 032505 (2022)
[15] K.V. Beausang et al., Rev. Sci. Instr. **82**, 033514 (2011)

# Investigation of $T_e$ measurements discrepancies between ECE and Thomson diagnostics in high-performance plasmas in JET

*M.* Fontana[1,*], *G.* Giruzzi[2], *F. P.* Orsitto[3], *E.* de la Luna[4], *R.* Dumont[2], *L.* Figini[5], *D.* Kos[1], *M.* Maslov[1], *S.* Schmuck[5], *C.* Sozzi[5], *C. D.* Challis[1], *D.* Frigione[3], *J.* Garcia[2], *L.* Garzotti[1], *J.* Hobirk[6], *A.* Kappatou[6], *D.* Keeling[1], *E.* Lerche[7], *C.* Maggi[1], *J.* Mailloux[1], *F.* Rimini[1], *D.* Van Eester[1], and *JET contributors***

[1]United Kingdom Atomic Energy Authority, Culham Centre for Fusion Energy, Culham Science Centre, Abingdon, Oxon, OX14 3DB, UK
[2]CEA, IRFM F-13108 Saint Paul-lez-Durance, France
[3]ENEA Department Fusion and Technology for Nuclear Safety, C R Frascati , 00044 Frascati, Italy
[4]National Fusion Laboratory, CIEMAT, Madrid, Spain
[5]Istituto per la Scienza e Tecnologia dei Plasmi, CNR, 20125 Milano, Italy
[6]Max-Planck-Institut für Plasmaphysik, D-85748 Garching, Germany
[7]Laboratory for Plasma Physics, LPP-ERM/KMS, Brussels, Belgium

**Abstract.** For high-temperature JET and TFTR discharges, electron cyclotron emission (ECE) measurements of central electron temperature were systematically found to be up to 20% higher than those taken with Thomson scattering. In recent high-performance JET discharges, central $T_e$ measurements, performed with LIDAR Thomson scattering and the X-mode ECE interferometer, have been studied in a large database, including deuterium (DD), and deuterium-tritium plasmas (DT). Discrepancies between $T_e$ measurements have been observed outside of the experimental uncertainties. ECE measurements, at high $T_e$, have been found to be higher or lower than those of LIDAR, depending on the specific plasma scenario. In addition, discrepancies between the peaks of the second and third harmonic ranges of the ECE spectrum have been interpreted as evidence for the presence of non-Maxwellian features in the electron distribution function. These comparisons seem to suggest that such features can be found in most of the high-performance scenarios selected in this JET database.

## 1 Introduction

Measurement of electron temperature ($T_e$) is one of the basic diagnostic requirements of any fusion experiment. Among the most common techniques used to evaluate $T_e$, are electron cyclotron emission (ECE) [1] and incoherent Thomson scattering (TS) [2], both of which have been widely employed in a variety of experiments for many decades. Despite the extensive experience operating these diagnostics and analysing the data they produce, it is known that values of $T_e$ measured using ECE and TS can diverge substantially in some specific conditions. This was first observed on TFTR [3, 4] and on JET [5]. In pulses heated with neutral beam injection (NBI) and ion cyclotron resonance heating (ICRH), at high temperature ($T_e$>4-5 keV), it was found that ECE measurements of core $T_e$ were systematically higher (up to 20%) than those taken using TS. In high $T_e$ pulses at JET in 2004, ECE spectra in mainly extraordinary (X) mode polarisation showed higher second harmonic (X2) peak temperatures than the ones in the third harmonic (X3) range. At the same time, however, X3 peak temperatures showed good agreement with TS measurements [6]. But due to the high optical thickness ($\tau$) of these plasmas in the third harmonic range and assuming a Maxwellian electron energy distribution function (EEDF), the two harmonics should match well. It has been proposed [7, 8] that the discrepancies between ECE and TS measurements, together with the anomalous ECE spectra, could be explained by a distortion of the EEDF in the bulk energy range, that is to say at energies close to the thermal energy of the core electrons. ECE and TS, in fact, operate based on different physical principles and respond differently to distortions of the EEDF. ECE measurements are particularly sensitive to the shape of the distribution function, with the absorption coefficient depending from the derivative of the EEDF [9]. To this point, no physical explanations for such EEDF distortions have been clearly identified yet. When a set of very similar experiments was repeated at JET in 2006, however, no discrepancy between ECE and TS, nor hints of non-maxwellian ECE spectra, were observed [6, 10]. A possible explanation was found in the higher hydrogen (H) concentration used in the 2006 pulses. In the intervening years, in fact, the lower limit of allowed hydrogen concentration at JET had been increased to reduce fast-ion losses that had been found to risk damage to the vessel

*e-mail: matteo.fontana@ukaea.uk
**See the author list of 'Overview of JET results for optimising ITER operation' by J. Mailloux et al. to be published in Nuclear Fusion Special issue: Overview and Summary Papers from the 28th Fusion Energy Conference (Nice, France, 10-15 May 2021)

walls. Higher H concentration affects the power distribution of ICRH between H and D ions and is believed to have reduced the temperature of the high-energy tail of the hydrogen ion distribution [6]. No more discrepancies between ECE and TS were observed in high-temperature ($T_e$ up to 8 keV) experiments at JET from 2006 to 2018. Similar results were also obtained in experiments at Alcator C-mod. In plasmas heated with various configurations of ICRH, no discrepancy between ECE and TS was reported even though $T_e$ up to 8 keV was reached [11]. From these experiments, it was concluded that these effects are not simply related to high electron temperatures, but must be influenced by other causes as well. Discrepancies between ECE and TS have instead been recently observed in very high-temperature ($T_e$ up to 14 keV), EC heated pulses on FTU [12]. There, TS temperatures were found to greatly exceed ECE measured ones (up to 50% higher) in phases where $T_e$>8 keV, while they agreed inside the experimental uncertainty for lower temperatures.

The discrepancy between ECE and TS is extremely relevant for next-generation experiments such as ITER and future reactor-like devices. Accurate and consistent measurements of electron temperature are necessary both for physics studies and for the operation of a commercial facility. The fact that discrepancies between TS and ECE, supposedly two of the best understood and established diagnostic techniques in plasma physics, appeared in some of the most reactor-relevant experiments realized until now, including in the few experiments that used duterium-tritium (DT) mixtures, makes the problem even more important. In the past years, JET has worked to develop high-performance plasma scenarios after the installation of the ITER-like wall [13]. These efforts have culminated in the first experimental campaign using DT fuel since almost thirty years, which has marked a new record in generated fusion energy. This occasion presented an excellent opportunity to investigate the issue of ECE-TS discrepancies. For this, the core temperature measurements on JET have been collected in a large database, that covers different plasma conditions and isotopic compositions including DT plasmas. Study of this database highlighted different behaviours for various plasma scenarios. For $T_e$>5 keV, cases with $T_{e,ECE} > T_{e,TS}$ and with $T_{e,ECE} < T_{e,TS}$ have both been observed. Related ECE spectra seem to display features that could be attributed to the presence of EEDFs with non-Maxwellian characteristics.

Section 2 of this paper contains a description of the main diagnostics used to measure the core electron temperature at JET. Section 3 describes the criteria used to build the discharges database and the main characteristics of the scenarios to which the selected pulses belong. In Section 4, the observations based on the temperature measurements collected in the database are discussed. Finally, Section 5 summarizes the results and anticipates some of the work currently in progress.

## 2 Core $T_e$ diagnostics at JET

Various choices are available for both ECE and Thomson diagnostics at JET. The standard ECE diagnostics available at JET are two Martin-Puplett interferometers [14] and one radiometer [15] observing the plasma through horizontal lines of sight, close to the vessel midplane (see Figure 1). The two interferometers collect radiation mainly of the ordinary (O) and extraordinary (X) wave mode, respectively. Both interferometers are absolutely calibrated, through the use of in-vessel sources over a wide spectral range (50-500 GHz). The temperature profiles (60 Hz acquisition frequency) can cover both low and high field side (LFS and HFS) of the tokamak (limited by harmonic overlapping for the X-mode) for the large range of the toroidal magnetic field available at JET (1.7-3.8 T). The spectral resolution is 3.66 GHz, corresponding to about 10 cm around the plasma axis. The 96-channels ECE radiometer is one of the main diagnostics used to study temperature fluctuations at JET, thanks to its high radial resolution (~2 cm) and acquisition frequency (from 5 to 200 kHz). The radiometer provides measurements on a range of frequencies that is changed pulse-by-pulse to follow to the plasma radial profile for X2 emission. Due to the different available configurations of mixers that can be selected to cover JET profiles at various magnetic field settings, the radiometer is cross-calibrated against the ECE interferometer on a pulse-by-pulse basis.

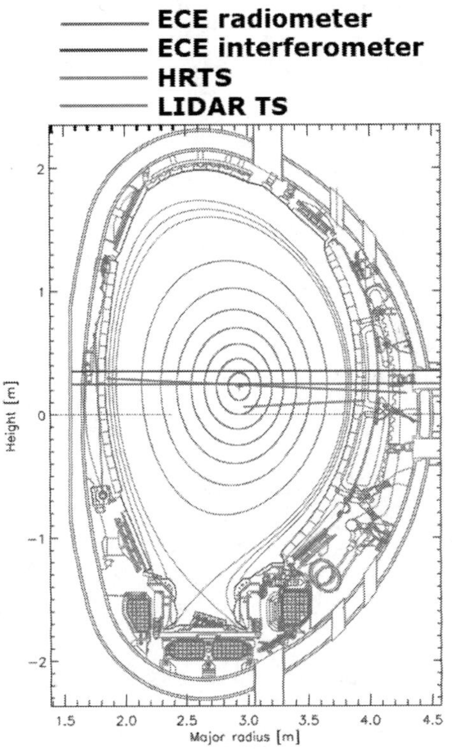


Figure 1: Lines of sight of the main electron temperature diagnostics on JET.

JET is also equipped with a high resolution Thomson scattering system (HRTS) [16, 17] and a LIDAR Thomson scattering diagnostic [18–20]. Both diagnostics are independently calibrated. LIDAR Thomson scattering pro-

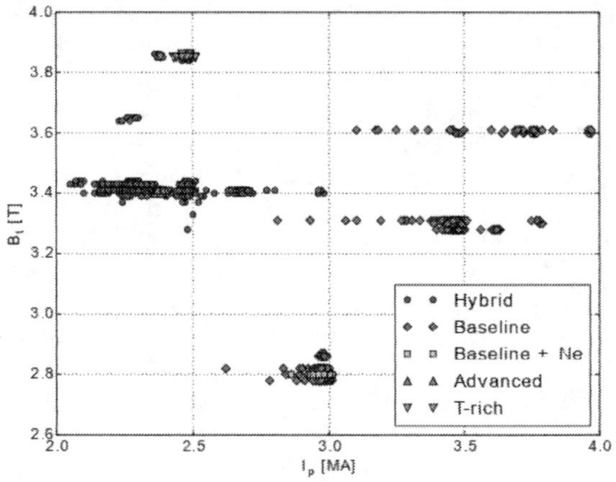


Figure 2: Plasma current ($I_p$) and toroidal field $B_t$ of database points corresponding to flat-top phases for each pulse.

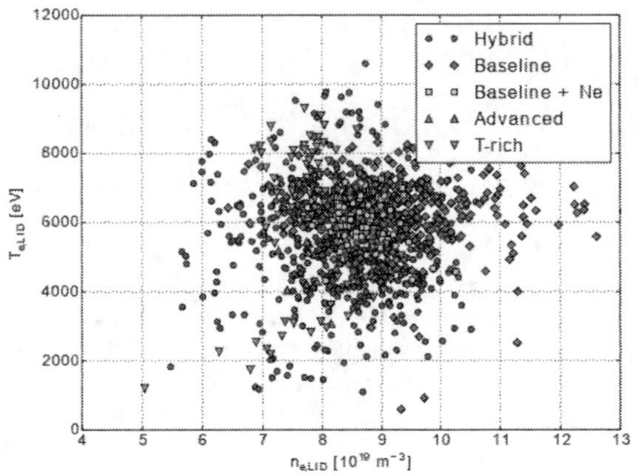


Figure 3: Electron density ($n_e$), and temperature ($T_e$) of database points corresponding to flat-top phases for each pulse.

vides density and temperature profiles covering the plasma LFS, core, and part of the HFS. The laser has a repetition frequency of 4 Hz, and every spatial point covers about 7 cm radially. HRTS provides better temporal (20 Hz) and spatial resolution (2 cm) compared with LIDAR. One of the main strenghts of HRTS is the excellent coverage of the plasma pedestal. However, as shown in figure 1, its line of sight is positioned slightly below the equatorial midplane of the JET vessel and is directly slightly downward. For this, in the majority of JET configurations, HRTS is limited to the LFS profile and does not manage to fully reach the plasma core, usually stopping at $\rho$>0.2 in normalized radial coordinates. Since the effects described in this work have been mostly observed in proximity to the plasma axis, HRTS was excluded from the analysis.

Also measurements from the ECE radiometer have not been included in this study. Due to the relatively low repetition rate of LIDAR, in fact, the high temporal resolution of the ECE radiometer is not necessary for comparing the two techniques. Furthermore, the radiometer does not allow comparison of second and third harmonics. In conclusion, the diagnostics that will be referred to in the rest of the paper for ECE and TS measurement are the X-mode ECE interferometer and the LIDAR Thomson scattering system respectively.

## 3 Pulse selection for the database

ECE and TS measurements have been compared for pulses from several JET campaigns from 2018 to 2022. These experiments include DD and DT plasmas covering a variety of high-performance plasma scenarios, designed to demonstrate high fusion power for about five seconds with the ITER-like wall. The two main ones are the so-called baseline and hybrid scenarios. The baseline scenario employs high field (2.8-3.5 T), high current (3-4 MA) and high density (7-11 $\times 10^{19} m^{-3}$ in the core), using pellet injection to control edge localised modes. The goal is to reach stable, high-performance conditions similar as to what is foreseen for the standard ITER scenario [21]. Some DD pulses in this category included various levels of Ne seeding to improve confinement through controlled edge radiation. Hybrid pulses work at high field (3-3.9 T) but lower current (2-2.8 MA) compared to baseline ones, aiming for high $\beta$ operation. The particular current profile developed for these pulses ($q_{95}$ > 3, $q_0 \geq$ 1) strongly reduces the occurrence of deleterious MHD modes in the plasma flat-top phase [22]. During the DT experiments, some of these pulses were designed to optimize non-thermal fusion reactions, with T-rich plasmas heated by D-only NBI. Finally, the database also includes pulses belonging to experiments exploring advanced scenarios, studying internal transport barriers, energetic particle modes and afterglow [23].

For each pulse, LIDAR times were selected where the laser energy was above 0.4 J and $T_e$>1 keV. For each of these points, the closest ECE time, with a maximum difference of 10 ms, was identified. ECE spectra were converted into radial profiles using a magnetic reconstruction based only on magnetic measurements with no further constraints. For each ECE and LIDAR $T_e$ radial profile, a representative value of the centre is determined by averaging the measurement points included between 2.85-3.15 m. This averaging window helps to smooth out the effects of the different lines of sight for the two diagnostics (see Figure 1) and the uncertainties in the localization of the plasma axis position. The values used to compare the second and third harmonics of the ECE spectra were calculated averaging $T_e$ in the interval +/- 5% around the frequency corresponding to the plasma centre according to the equilibrium reconstruction.

The points that make up the database cover the evolution of the pulses from ramp-up to ramp-down or disruption. During the flat top phase for these pulses, the combinations of toroidal magnetic field ($B_t$), plasma current ($I_p$) and electron density ($n_e$) and temperature ($T_e$) covered by the different scenarios are shown in Figures 2 and 3 re-

https://doi.org/10.1051/epjconf/202327703006

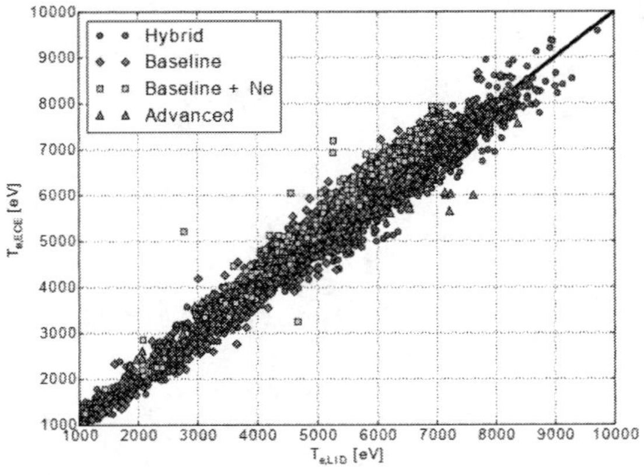


Figure 4: Comparison of central ECE and LIDAR temperatures in DD pulses.

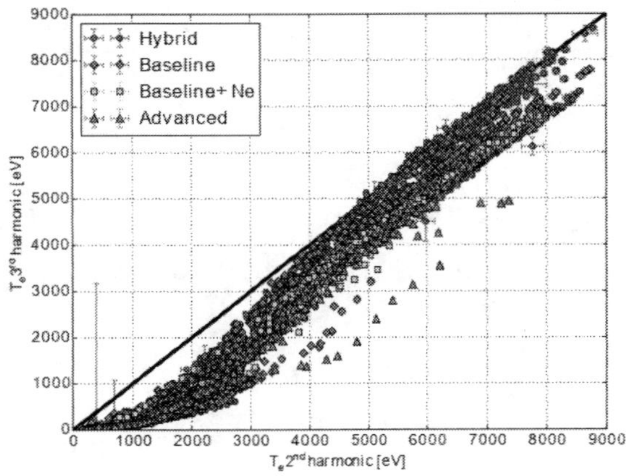


Figure 6: Comparison of peak temperatures for second and third harmonics X-mode in DD pulses.

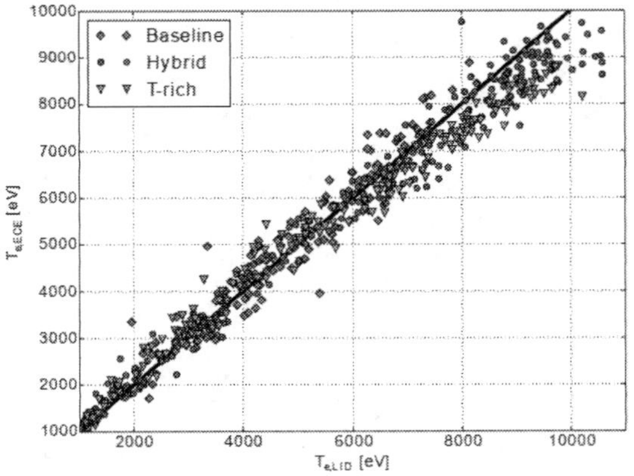


Figure 5: Comparison of central ECE and LIDAR temperatures for DT pulses.

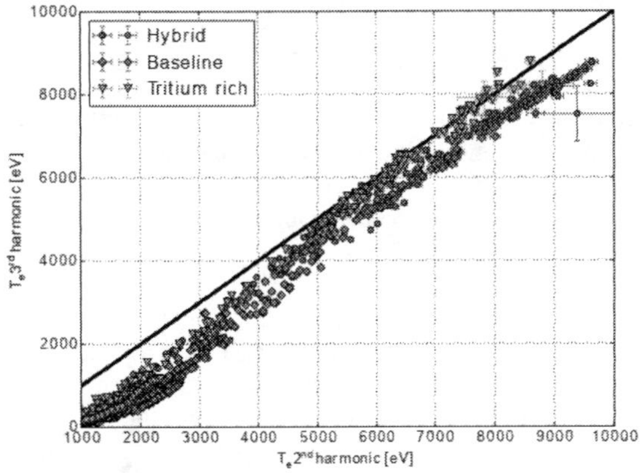


Figure 7: Comparison of peak temperatures for second and third harmonics X-mode in DD pulses.

spectively. Each point corresponds to a LIDAR time included in the database. These pulses were heated via NBI and/or ICRH. The maximum heating power reached was 38.5 MW, in some of the hybrid pulses. In the following, only pulses with $T_e > 5$ keV in the core will be discussed.

## 4 Discussion

The central averages of $T_e$ for LIDAR and ECE, calculated as detailed in Section 3, have been compared, separating DD pulses from DT pulses. For the DD pulses, in Figure 4, $T_{e,ECE}$ and $T_{e,LID}$ are quite similar for most pulses. However, it is possible to distinguish significant differences comparing the different scenarios. Baseline pulses with Ne seeding, in particular, appear to have $T_{e,ECE} > T_{e,LID}$, similar to what was observed in past experiments at JET and TFTR [4, 5], even though to a lower extent. A clear difference is visible between these baseline pulses and those with no Ne seeding, and even more when they are compared with the hybrid pulses. A similar situation can be observed for DT pulses in figure 5. Here the largest discrepancies ($T_{e,LID} > T_{e,ECE}$) are observed for hybrid pulses, and especially T-rich experiments, while baseline pulses tend to remain closer to $T_{e,LID} = T_{e,ECE}$.

The discrepancies between $T_{e,LID}$ and $T_{e,ECE}$ have been suggested to be related to the presence of non-Maxwellian features in the EEDF [7]. This explanation is consistent with the higher X2 peak temperature, compared with X3 and Thomson scattering diagnostics, observed for some of these pulses. In figure 6, we can observe that, as expected, for low $T_e$, X3 peak temperature is consistently lower compared with X2. This is due to the low $\tau_{X3}$ at these points. With increasing temperature, however, $\tau_{X3}$ increases and, in presence of a Maxwellian EEDF, $T_{e,X3}$ should approach $T_{e,X2}$. In this case this seems to happen only for a fraction of the baseline and hybrid pulses while, in the rest of the points, $T_{e,X3}$ remains consistently below $T_{e,X2}$ even at very high $T_e$. Also for the DT pulses, only some pulses display $T_{e,X3} \sim T_{e,X2}$, as shown in Figure 7. In this case however, the different scenarios remain rather well separated. In particular the T-rich pulses display the best agreement between second and third harmonic at high $T_e$. This increased uniformity in each scenario is due to

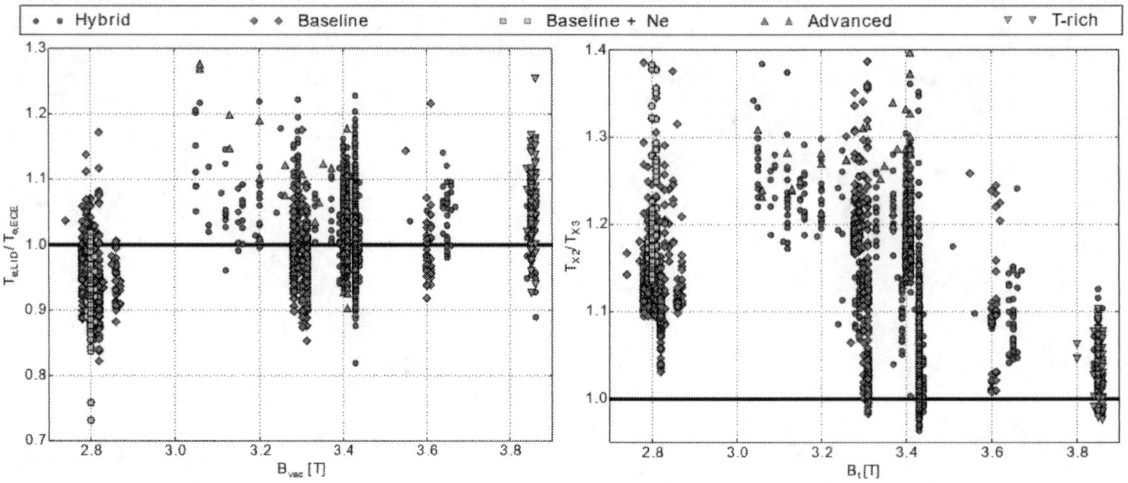


Figure 8: Ratio of ECE and LIDAR measurements (a) and ratio between the peaks of X2 and X3 emission (b) as a function of the vacuum magnetic field $B_{vac}$.

the fact that almost no time was dedicated to scenario development in the DT campaign, in order to minimize the consumption of nuclear fuel and neutron budget. As a consequence, the parameters of each discharge changed little for a given scenario.

A possible explanation for $T_{e,X3} < T_{e,X2}$ that does not imply the existence of non-Maxwellian features in the EEDF, is the presence of errors in the calibration of the ECE interferometer, in particular affecting the comparison of second and third harmonics. In this case, different ratios of $T_{e,X3}/T_{e,X2}$ would be related only to changes in the magnitude of the magnetic field of the various pulses. While no in-vessel calibrations have been possible since 2018, the properties of the instruments were constantly monitored regularly repeating in-lab measurements on a cold (liquid $N_2$) and a room temperature source. These measurements have been used to correct for changing sensitivity of the diagnostic, greatly improving agreement of the ECE profiles with those produced by LIDAR and HRTS. Concerning the measurements in this database, Figure 8a shows that the ratio between LIDAR and ECE measurements (limited to $T_{LID} > 4keV$) is not correlated with the vacuum magnetic field ($B_{vac}$). Figure 8b, instead, shows the ratio between second and third harmonics against $B_{vac}$. Here, it can be observed that, while most of the points for baseline pulses that reach $T_{X3} \sim T_{X2}$ are concentrated at $B_{vac}$ ~3.3 T and higher, similar examples can be found also for cases around ~2.8 T.

## 5 Conclusions and prospects

A very large database of $T_e$ measurements (more than 14000 time points from 246 discharges) was collected from high performance DD and DT pulses performed at JET. These include a variety of plasma conditions and different scenarios, some of which were tested using DT mixture for the first time. These measurements, taken with LIDAR Thomson scattering and an X-mode interferometer, showed discrepancies when high $T_e$ were reached. Differently from what had been observed in the previous studies of this kind [3–5], here the discrepancies were less pronounced and, most importantly, they were not only in the direction of $T_{ECE} > T_{LID}$ for high $T_e$. In fact, a richer phenomenology was observed, including also $T_{ECE} < T_{LID}$, with different scenarios showing different behaviours. Comparison of the peaks of the second and third harmonics of the ECE spectra suggests the presence of non-Maxwellian features of the EEDF for most of the pulses in this data set. Even for plasma conditions where $\tau_{X3}$ was large enough to have $T_{X2} \sim T_{X3}$, in fact, $T_{X2} > T_{X3}$ was observed instead in many pulses.

The possibility that EDF distortions could result in different measurements from ECE and Thomson diagnostics can be investigated using simple models that applies an arbitrary perturbation in velocity space to a Maxwellian EEDF and then calculating the resulting ECE spectrum. Such a study, employing a bipolar distortion of a Maxwellian EEDF, comparing the results of the model with the experimental measurements presented in this database, obtained promising results [24]. However, at the moment, a clear physical cause for EDF distortions which could give rise to these discrepancies has not yet been identified. Several hypothesis have been suggested and are being investigated, including, for example, effects related to the fast ion population (high NBI/ICRH power and DT fusion reactions). The ions could interact with the electron population directly through collisions or through the excitation of MHD modes (such as Alfvén modes) that can then transfer energy to the electrons. The latter possibility is being investigated via gyrokinetic simulations for parameters approaching those of some cases included in the database. The database presented in this work could be further expanded by the inclusion of other diagnostics. For example, equilibrium reconstructions that included constraints from more sources (such as ion and electron temperature and density) would allow better comparisons between diagnostics with different lines of sight, such as HRTS. Furthermore, data from the second ECE interferometer on JET, collecting emission predominantly in O-mode will be added to the database to study also O1 emission. Finally, JET is equipped with an oblique ECE system which collected data for some of the pulses included in this work. Processing these data could result in another way to obtain information on the EEDF.

## 6 Acknowledgements

This work has been carried out within the framework of the EUROfusion Consortium, funded by the European Union via the Euratom Research and Training Programme (Grant Agreement No 101052200 — EUROfusion). Views and opinions expressed are however those of the author(s) only and do not necessarily reflect those of the European Union or the European Commission. Neither the European Union nor the European Commission can be held responsible for them. This work has been part-funded by the EPSRC Energy Programme [grant number EP/W006839/1]. To obtain further information on the data and models underlying this paper please contact PublicationsManager@ukaea.uk. For the purpose of open access, the authors have applied a Creative Commons Attribution (CC BY) licence to any Author Accepted Manuscript version arising.

## References

[1] A.E. Costley, R.J. Hastie, J.W.M. Paul, J. Chamberlain, Phys. Rev. Lett. **33**, 758 (1974)

[2] J. Sheffield, in *Plasma Scattering of Electromagnetic Radiation*, edited by J. Sheffield (Academic Press, 1975), pp. 191–210, ISBN 978-0-12-638750-6, `https://www.sciencedirect.com/science/article/pii/B9780126387506500148`

[3] I. Fidone, G. Giruzzi, G. Taylor, Physics of Plasmas **3**, 2331 (1996)

[4] G. Taylor, R.W. Harvey, Fusion Science and Technology **55**, 64 (2009), publisher: Taylor & Francis _eprint: https://doi.org/10.13182/FST55-64

[5] E. de la Luna, V. Krivenski, G. Giruzzi, C. Gowers, R. Prentice, J.M. Travere, M. Zerbini, Review of Scientific Instruments **74**, 1414 (2003)

[6] E. De La Luna, D. Farina, L. Figini, G. Grosseti, S. Nowak, C. Sozzi, M. Beurskens, O. Ford, T. Johnson, E.c. Jet, in *Electron Cyclotron Emission and Electron Cyclotron Resonance Heating (EC-15)* (WORLD SCIENTIFIC, 2009), pp. 200–207, ISBN 978-981-281-463-0, `https://www.worldscientific.com/doi/10.1142/9789812814647_0026`

[7] V. Krivenski, Fusion Engineering and Design **53**, 23 (2001)

[8] V. Krivenski, 29th EPS Conf. on Plasma Physics, ECA **26B** (2002)

[9] G. Giruzzi, Nucl. Fusion **28**, 1413 (1988), publisher: IOP Publishing

[10] K.V. Beausang, S.L. Prunty, R. Scannell, M.N. Beurskens, M.J. Walsh, E. de La Luna, Review of Scientific Instruments **82**, 033514 (2011), publisher: American Institute of Physics

[11] A.E. White, A.E. Hubbard, J.W. Hughes, P.T. Bonoli, M.E. Austin, A. Bader, R.W. Harvey, Y. Lin, Y. Ma, M.L. Reinke et al., Nucl. Fusion **52**, 063021 (2012), publisher: IOP Publishing

[12] G. Pucella, E. Alessi, S. Almaviva, B. Angelini, M.L. Apicella, G. Apruzzese, M. Aquilini, G. Artaserse, B. Baiocchi, M. Baruzzo et al., Nucl. Fusion **62**, 042004 (2022), publisher: IOP Publishing

[13] F. Romanelli, Nucl. Fusion **53**, 104002 (2013), publisher: IOP Publishing

[14] S. Schmuck, J. Fessey, J.E. Boom, L. Meneses, P. Abreu, E. Belonohy, I. Lupelli, Review of Scientific Instruments **87**, 093506 (2016)

[15] E. de la Luna, J. Sánchez, V. Tribaldos, E.c. Jet, G. Conway, W. Suttrop, J. Fessey, R. Prentice, C. Gowers, J.M. Chareau, Review of Scientific Instruments **75**, 3831 (2004), publisher: American Institute of Physics

[16] R. Pasqualotto, P. Nielsen, C. Gowers, M. Beurskens, M. Kempenaars, T. Carlstrom, D. Johnson, Review of Scientific Instruments **75**, 3891 (2004)

[17] L. Frassinetti, M.N.A. Beurskens, R. Scannell, T.H. Osborne, J. Flanagan, M. Kempenaars, M. Maslov, R. Pasqualotto, M. Walsh, Review of Scientific Instruments **83**, 013506 (2012)

[18] H. Salzmann, K. Hirsch, Review of Scientific Instruments **55**, 457 (1984), publisher: American Institute of Physics

[19] H. Salzmann, J. Bundgaard, A. Gadd, C. Gowers, K.B. Hansen, K. Hirsch, P. Nielsen, K. Reed, C. Schrödter, K. Weisberg, Review of Scientific Instruments **59**, 1451 (1988), publisher: American Institute of Physics

[20] M. Maslov, M.N.A. Beurskens, M. Kempenaars, J. Flanagan, J. Inst. **8**, C11009 (2013), publisher: IOP Publishing

[21] L. Garzotti, C. Challis, R. Dumont, D. Frigione, J. Graves, E. Lerche, J. Mailloux, M. Mantsinen, F. Rimini, F. Casson et al., Nucl. Fusion **59**, 076037 (2019), publisher: IOP Publishing

[22] J. Hobirk, F. Imbeaux, F. Crisanti, P. Buratti, C.D. Challis, E. Joffrin, B. Alper, Y. Andrew, P. Beaumont, M. Beurskens et al., Plasma Phys. Control. Fusion **54**, 095001 (2012)

[23] R.J. Dumont, J. Mailloux, V. Aslanyan, M. Baruzzo, C.D. Challis, I. Coffey, A. Czarnecka, E. Delabie, J. Eriksson, J. Faustin et al., Nucl. Fusion **58**, 082005 (2018), publisher: IOP Publishing

[24] G. Giruzzi, M. Fontana, F.P. Orsitto, E. De La Luna, R. Dumont, L. Figini, M. Maslov, S. Mazzi, S. Schmuck, L. Senni et al., This workshop (2022)

# Current status of ECE system on EAST tokamak

*Hailin* Zhao[1,*], *Ang* Ti[1,**], *Tianfu* Zhou[1], *Zeying* Zhu[1], *Yong* Liu[1], *Bili* Ling[1], *Qing* Zang[1], *Juan* Huang[1], and *Xianzu* Gong[1]

[1]Institute of Plasma Physics, Chinese Academy of Sciences, Hefei 230031, China

**Abstract.** The electron cyclotron emission (ECE) diagnostic on the experimental advanced superconducting tokamak (EAST) has had a major upgrade since 2020, when EAST heating system also went through a significant upgrade, including one NBI system changed from counter-current to co-current (moving from port F to port D), and the antenna and the installation port of LHW and ICRF system have also been changed. The quasi-optical (QO) antenna of P port ECE system has been redesigned, the main purpose of which is to add one oblique ECE view. The angle with respect to perpendicular to the magnetic field is about 10°, which will facillitate measurement of the electron velocity distribution altered by LHW system. The ellipsoidal mirror has also been moved close to the plasma, about 70 cm away from the plasma center, and the poloidal beam waist radius in the plasma has been optimised to be less than 3 cm. The CECE system has also been moved from port G to port C. The frequency coverage of the CECE system has been upgraded to 104-132 GHz by adding one radio frequency (RF) module. Also in the intermediate frequency (IF) module, 8 narrow-band filters have been added to improve the spacial coverage of the system. On port F, a new superheterodyne radiometer with narrow-band filters in IF module has been installed. It consists of eight channels, the radial coverage is about 8 cm, the main purpose of this new system is to study the fine structure of magnetic island.

## 1 Introduction

The electron cyclotron emission (ECE) diagnostic has been widely used in tokamak research since the 1960s[1]. It can provide important information on the time evolution of the electron temperature profile (Te(r)), magnetohydrodynamic (MHD) fluctuation spectra, non-thermal electron behavior and the ECE radiated power loss.

EAST[2] is a fully superconducting tokamak with a non-circular cross-section, The toroidal field can be as high as 2.6 T, the major radius R = 1.85 m, the minor radius a = 0.45 m. EAST tokamak has made an important advance by achieving stable 1056-second steady-state high-temperature plasma operation, setting a record for long-pulse operation in 2021[3].

The ECE diagnostic set has been developed on EAST tokamak for more than 10 years, and includes a conventional ECE system and a correlation ECE (CECE) system.

This paper is organized as follows: In section 2, the heating upgrade on EAST tokamak is introduced. The ECE system front end optical system upgrade is shown in section 3. The CECE upgrade is shown in section 4, a new high resolution ECE system is presented in section 5, and a summary is given in section 6.

## 2 EAST heating upgrade

In order to support the planned physical research on EAST, the auxiliary heating system was upgraded in 2020. The heating system includes neutral beam injection (NBI)[4], lower hybrid wave (LHW)[5] heating and current drive, ion cyclotron range-of frequency (ICRF)[6] heating and electron cyclotron heating (ECH)[7]. A high-power NBI system consisting of two injectors was employed on EAST for plasma heating and current drive. The designed deuterium beam energy is 50–80 keV, beam power is 2–4 MW, and the beam duration is 10–100 s. Each beamline is equipped with two hot cathode ion sources and can be operated individually. The NBI-2 was shifted from port F to port D in 2020, and meanwhile, its injection direction was changed from counter-injection to co-injection. Furthermore, the high power ion source was optimized with better feedback control for stable long pulse beam. The LHW system includes two frequency system, 2.45 GHz and 4.6 GHz. The 2.45 GHz LHW system has a new PAM antenna and was moved from port N to port B. The total power of LHW system is about 6 MW. There were two fast wave heating antennas which are assembled at port I and port B on EAST. In 2020 the antenna at port B was moved from port B to port N, to decrease the effect of pertubation of ICRF on LHW systems. Before the change, when ICRF is turned on, the LHW system coupling is degraded, and the power absorbed by the plasma is also decreased. There are four units of EC systems (No. 1– No. 4), each of which has one gyrotron with a frequency of 140 GHz. An additional two gyrotrons system which have low frequency for low toroidal field is under development. Now the total ECH power can approach 2 MW.

## 3 ECE system front end optical system upgrade

The ECE system of EAST has been running for about 10 years. It includes a heterodyne system and a michelson interferometer system. The heterodyne system can measure the electron temperature with high spatial and temporal resolution, 2 cm and 2.5 $\mu$s respectively. Presently there are 56 channels, with frequency range 97 - 167 GHz. The michelson interferometer system was borrowed from JET, and the frequency range covers 80 - 500 GHz, while the temporal resolution is about 30 ms.

In 2020, taking advantage of the EAST tokamak upgrade, the front end optics of the ECE system in port P was upgraded. The most important change is the adding of a second quasi-optical system, making two, as shown in figure 1. One is for convential ECE systems, another is for oblique ECE measurement. The angle with respect to perpendicular to the magnetic field is about 10°. It will

*e-mail: zhaohailin@ipp.ac.cn
**e-mail: angt@ipp.ac.cn

be used for future studies of high-energy electron distributions generated by LHW system. The beam spot size in the plasma region is less than 25 mm, and the poloidal resolution is improved compared to the old quasi-optical system, which was farther away from the plasma. The ellipsoidal mirror has also been moved closer to the plasma, about 70 cm away from the plasma center, and the poloidal beam waist radius in the plasma has been optimised to be less than 3 cm. After several experiment campaigns, which include wall cleaning, such as lithium coating, silication and so on, the ECE signal intensity is unchanged, this gives confidence to the use of stainless steel mirror as ECE first mirror in future fusion devices such as ITER.

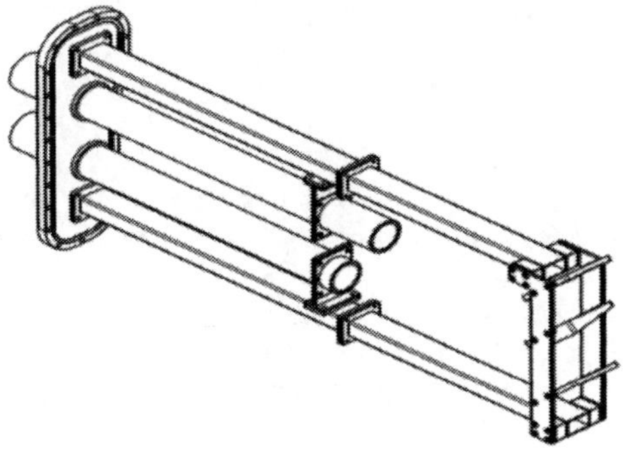

**Figure 1.** Two quasi-optical system for ECE front end

The ECE transmission waveguide is more than 40 meters in length. The direction of the waveguide was also rearranged due to conflict with the water cooling system. One more miter bend was added; the measured transmission loss for 81 to 110 GHz is shown in table 1. The measurement setup used one six-octave frequency to the 13.5 to 18.3 GHz local oscillator (LO), and one horn antenna as a source in the tokamak chamber, with another horn antenna connected to the power meter to measure the transmission waveguide attenuation for different dot frequencies. It can be seen that the attenuation of the 40 meters transmission waveguide for most frequencies is about 10 dB, with some frequencies reaching 14 dB. This roughly factor 10 attenuation of the input power to the transmission line is a significant reduction of signal to the ECE heterodyne receiver. The dominant loss source is believed to be the miter bends so the addition of another one has a deleterious effect. The lesson for future fusion devices such as ITER is to minimize the number of miter bends in transmission lines to keep losses small.

After the front end quasi-optical upgrade and the transmission waveguide direction rearranged, now the ECE signal is smaller than the time before the upgrade. An additional cause of power loss is a reduction of the size of the mirror in the front-end quasi-optical unit. The diameter of the mirror was reduced from 15 cm to 11 cm, thus reducing the surface area by about half. This combined with the extra miter bend loss has the result that the absolute intensity calibration is no longer possible. The voltage difference of the output signal with a chopper as a reference is too small when using hot source inside the tokamak.

**Table 1.** The attenuation of the 40 meters transmission waveguide for different frequencies

| frequency (GHz) | attenuation (dB) |
|---|---|
| 81 | -11.4 |
| 82 | -10 |
| 83 | -11.8 |
| 84 | -7 |
| 85 | -9.3 |
| 85 | -9.3 |
| 86 | -8.3 |
| 87 | -7 |
| 88 | -10.6 |
| 89 | -10.5 |
| 90 | -9 |
| 91 | -9 |
| 92 | -10 |
| 93 | -11.7 |
| 94 | -12 |
| 95 | -11.2 |
| 96 | -10.5 |
| 97 | -11.1 |
| 98 | -12.3 |
| 99 | -11.5 |
| 100 | -9.8 |
| 101 | -12 |
| 102 | -11.3 |
| 103 | -11 |
| 104 | -11.1 |
| 105 | -14 |
| 106 | -13.8 |
| 107 | -12.5 |
| 108 | -12.6 |
| 109 | -10.4 |
| 110 | -13.6 |

The upgrade of the ECE system in Port P cannot be considered a complete success, since the signal intensity of the ECE system is becoming smaller compared with before, but at least experience has been gained, and oblique ECE front end quasi-optical system has been installed, it provides the oppotunity to the future studies of high-energy electron characteristics generated by LHW.

In the next year, an electron temperature gradient scale length ($L_{T_e} = T_e/|\triangle T_e|$) direct measurement system is planned, there will be no need of absolute calibration and the electron temperature profile fitting to get $L_{T_e}$. The conventional method to get $L_{T_e}$ is to fit the electron temperature profile, during this procedure there is plenty of uncertainty. Alternatively, a method has been described to obtain $L_{T_e}$ from a measured change in ECE intensity for a sweep in filter frequency [8]. This comes from the fact that the Te change $\triangle T_e$ is proportional to the change in signal intensity, and the radial position change $\triangle$R is proportional to the change in frequency. By using voltage controlled oscillator (VCO) as LO source, the LO frequency can be preprogrammed in one shot. With a set of swept frequency filters, such as tunable yttrium iron garnet (YIG) filters, $L_{T_e}(R)$ at different radial positions can be obtained.

The current heterodyne system can not measure the electron temperature profile at the low field side when the toroidal field at the magnetic axis is smaller than 1.8 T. In the next year, a 16-channel heterodyne system will be installed, with a frequency range of 81 to 96 GHz, with channel separation of 1 GHz. This system can measure the electron temperature profile at low toroidal field, such as 1.6 T. The low toroidal field on EAST is an attactive regime for low $q_{95}$ and high $\beta_N$ discharges. In this regime, the LHW system and the NBI system can be used, also the ICRF third harmonics heating can be used. In the near future, a new ECH system with a frequency of 105 GHz will be installed. This ECH system can be employed in the lower toroidal field regime.

## 4 CECE system upgrade

A single ECE channel is unable to resolve small-amplitude electron temperature fluctuation above the noise floor due to the inherent thermal noise in standard ECE measurements. By correlating two radially closely spaced radiometer channels, the turbulent fluctuation signal can be retained while the uncorrelated thermal noise is eliminated. This technique, known as correlation ECE (CECE) has been utilized on many machines worldwide, such as Alcator C-Mod[9], DIII-D[10], Asdex-upgrade[11], J-TEXT[12], TCV tokamak[13]. CECE diagnostic provides a powerful technique for overcoming the noise limitations of stand ECE in measuring small amplitude temperature fluctuations. Recently CECE dianostic has been used to measure the electron temperature fluctuation in the improved energy confinement mode (I-mode) discharge in Asdex-upgrade, and interesting phenomenon have been observed[14].

The CECE system has been changed from port G to port C, this port is very close to the 2.45 GHz LHW system, but far away from the 4.6 GHz LHW system. The LHW system can generate strong electromagnetic fields nearby, so the 2.45 GHz system can cause perturbation on the CECE instrument. However, the 2.45 GHz LHW system is seldom used because of the low power absorption at high electron density, so the CECE system can be used without interference in most of the discharges.

Several important upgrades need to be pointed out. The first upgrade is that another radio frequency (RF) module consists of LO and mixer unit has been installed, as shown in figure 2. The frequencies of the two LOs are 102 GHz and 114 GHz respectively, so the CECE system can cover the frequency range from 104 to 132 GHz. In the intermediate frequency (IF) unit, after the low noise amplifier, another 8 channel power divider is used, with eight narrow band filters (250 MHz) are used. The frequencies are 2, 2.2, 3, 3.2, 16, 16.2 ,17, 17.2 GHz. The reason for this is that YIG filter can change frequency from 4 to 18 GHz, and if 8 channel YIG filters are all at high frequcies, such as above 15 GHz, the controller needs more current to drive the YIG filters, and sometimes the current is so large that the YIG filter setting fails. Using these eight channel narrow band filters, more frequency range can be covered.

**Figure 2.** CECE RF unit with two LOs and mixers

The second upgrade is that a new RF transmission line was installed with lower power loss. The RF transmission line transfers the RF signal, which is the output of RF unit. The intensity of the RF signal is usually -10 to -40 dBm, and the frequency range is 2 - 18 GHz. The RF unit is close to the tokamak, just near the waveguide. However, the IF unit which converts the RF signal to the voltage signal is about 4 meters away from the RF unit. This long distance needs transmission line with low power loss, usually the attenuation should be less than 3 dB. The old transmission line has very high attenuation, about 10 dB, and this is not suitable for electron temperature fluctuation measurements, especially when the cross power spectrum is calculated, the high frequency (above 100 kHz) information will be lost.

The third upgrade is the intermediate attenuator between the two low noise amplifiers was properly chosen. In the past, the gain of the two amplifiers was 30 dB and 45 dB respectively, and the attenuator was 2 dB. Most of the time we can see no change in the output signal when the heating power is large. The reason is that the first amplifier was overdriving the second amplifier and this was fixed with a larger intermediate attenuator which is 10 dB. This means that the 1 dB compression point of the second amplifier needs to be considered. In the past, due to the absence of the high frequency wideband source, only the IF unit is tested. At this experiment campaign, we borrowed a noise source, and used it to make a complete system test, including both the IF and RF units. It proves that only with complete test, can the whole system work properly.

The CECE system can measure the electron temperature fluctuation at $\rho$ from 0.7 to 1, this is the region where the electron temperature fluctuation is largest. Especially at the pedestal top, where $\rho$ is about 0.85 to 0.95.

## 5 The high resolution ECE system

The new high resolution ECE system has been installed on port F, very close to the 4.6 GHz LHW system. The window for ECE measurement is very small, the diameter

 https://doi.org/10.1051/epjconf/202327703007

is about 11 cm, and it is near the midplane. This system is intended to detect the 2/1 and 3/2 MHD modes, including tearing mode and neo-classical tearing mode. Accroding to the typical mode location, the LO and filters are chosen. The system has eight channels, and the LO frequency is 118 GHz, and the frequency of eight narrow band filters (250 MHz) is very closely spaced. The frequency can cover 122 to 125 GHz, about 8 cm radially. There are many diagnostics in port F, the space is limited, ECE mirror support installation inside the vacumm chamber is not available. A curvature window and lens antenna are employed to improve the poloidal optical collection capability. The beam radius inside the chamber without curvature window is shown in figure 3, and the beam radius inside the chamber with curvature window is shown in figure 4. It can see that with curvature window, 3dB poloidal beam radius changes from 40 mm to about 20 mm, and is more centralized. Another feature is the use of the second harmonic mixer. This has never been used on EAST before. On future fusion devices, as the toroidal field is much higher, high frequency (above 200 GHz) LO and mixer will be used. Nevertheless, the high frequency LO and mixer is very hard to produce, so second harmonic mixer is a good candidate. Now the whole high resolution ECE system can obtain signal when there is slide away discharge with very low electron density, the output of the conventional ECE system usualy reaches a saturation voltage (above 4 V), the output voltage of the high resolution ECE system is about 1 V. At normal discharge with moderate electron density, the output voltage is very small. This means that the output of the high resolution ECE system is much lower than that of the conventional ECE system, hence there are further improvements needed to obtain an optimally functioning system.

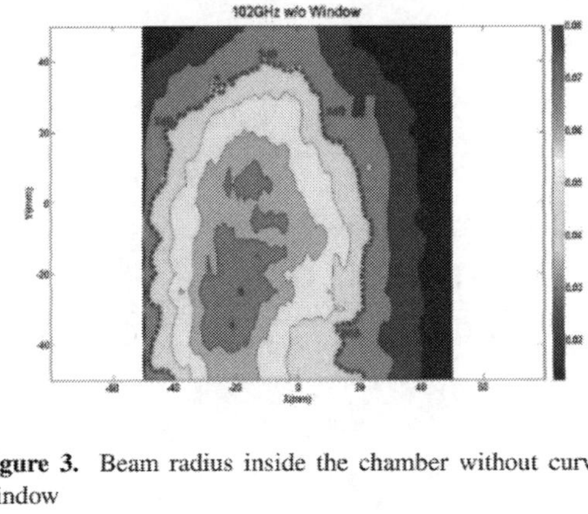

**Figure 3.** Beam radius inside the chamber without curvature window

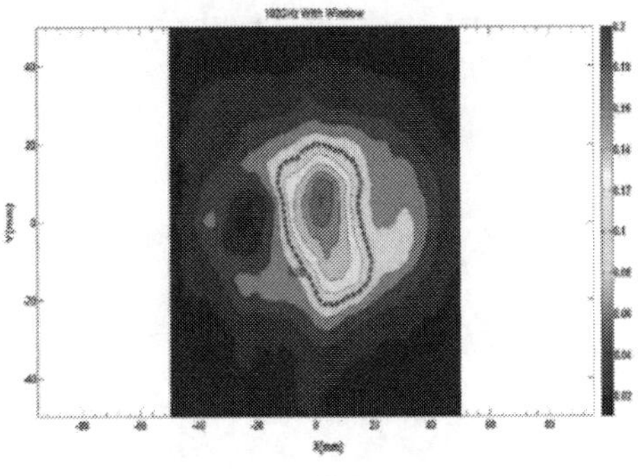

**Figure 4.** Beam radius inside the chamber with curvature window

## 6 Summary

The ECE system on EAST has been upgraded in year 2020. The main upgrade includes the front end quasi-optical system upgrade, comprised of two quasi-optical systems, one is for conventional ECE system, including 56 channel heterodyne system and Michelson interferometer, the other is for future oblique ECE system, which can be used for studies of high-energy electron characteristics generated by LHW. The transmission line also has been rearranged, after the upgrade, the signal intensity is smaller than before, partly because the surface area of the mirror is about half of the old mirror, partly because the transmission waveguide has more miter bends, it causes more power attenuation. The CECE system has been moved to port C and upgraded to 16 channel output, and two LO and mixer units have been installed to cover a wider frequency range. The CECE system is now physically closer to 2.45 GHz LHW system, but farther away from the 4.6 GHz LHW system. As the 2.45 GHz LHW system is less often used, the CECE system can operate without interference in more EAST discharges, and contribute more to physical research. A new high resolution ECE system has been installed in port F, the design of the optical system and the idea of improving the poloidal resolution have been verified, and the ability of the second harmonic mixer to work near the tokamak device is verified. However, to obtain good signal output, more work needs to be done.

In the next year, a $L_{T_e}$ directly measurement system is planned, there will be no need of absolute calibration and the electron temperature profile fitting to get $L_{T_e}$. VCO will be used as LO and by scanning LO frequency, together with the change of YIG filter frequency, $L_{T_e}(R)$ at different positions can be obtained.

Also, one 16 channel ECE system which covers 81 to 96 GHz with 1 GHz separation per channel is on schedule. This will measure the electron temperature profile for $B_T$ smaller than 1.8 T. Low toroidal field discharge is very attractive especially with low $q_{95}$ and high $\beta_N$ .

**Acknowledgments**

This work was supported by the National MCF Energy R&D Program of China under contract Nos. 2019YFE03030002.

## References

[1] Lichtenberg A J, Sesnic S, Trivelpiece A W., Phys. Rev. Lett., **13**, 387 (1964)

[2] Songtao Wu, EAST Team., Fusion Eng. Des., **82**, 463 (2007)

[3] Yuntao Song et al., Sci. Adv., **9**, eabq5273 (2023)
[4] C. Hu et al., Plasma Sci. Technol., **17**, 817 (2015)
[5] Fukun Liu et al., Nucl. Fusion, **55**, 123022 (2015)
[6] Xinjun Zhang et al., Plasma Sci. Technol., **13**, 172 (2011)
[7] Xiaojie Wang et al., Fusion Eng. Des., **96-97**, 181 (2015)
[8] S. Houshmandyar et al., Rev. Sci. Instrum., **92**, 033510 (2019)
[9] N.T. Howard et al., Rev. Sci. Instrum., **85**, 118D11 (2014)
[10] A.E. White et al., Rev. Sci. Instrum., **79**, 103505 (2008)
[11] S.J. Freethy et al., Rev. Sci. Instrum., **87**, 11E102 (2016)
[12] H. Zhou et al., Rev. Sci. Instrum., **89**, 10H105 (2018)
[13] M. Fontana et al., Rev. Sci. Instrum., **88**, 083506 (2017)
[14] R. Bielajew et al., Phys. Plasmas, **29**, 052504 (2022)

EPJ Web of Conferences **277**, 03008 (2023) https://doi.org/10.1051/epjconf/202327703008
*EC21*

# New Q and V-band ECE radiometer for low magnetic field operation on LHD

*Tokihiko* Tokuzawa[1,2*], *Yuki* Goto[1], *Daisuke* Kuwahara[3], *Masaki* Nishiura[1,4], and *Takashi* Shimizu[1]

[1]National Institute for Fusion Science, National Institutes of Natural Sciences, Toki 509-5292, Japan
[2]The Graduate University for Advanced Study, SOKENDAI, Toki 509-5292, Japan
[3]College of Engineering, Chubu University, Kasugai 487-8501, Japan
[4]Graduate School of Frontier Sciences, The University of Tokyo, Kashiwa 277-8561, Japan

**Abstract.** To meet the demand for information on electron temperature fluctuations in low magnetic field experiments in the Large Helical Device (LHD), a new ECE radiometer covering the Q and V bands has been installed. Combination mirrors are installed in the vacuum vessel to focus the beam and efficiently propagate the radiated electron cyclotron waves. Notch filters are used to eliminate stray light from the gyrotron, and a 32-channel heterodyne radiometer is constructed using a filter bank system. As a result, oscillations of electron temperature and both electromagnetic and electrostatic fluctuations were successfully observed.

## 1 Introduction

Electron cyclotron emission (ECE) measurements have been applied to electron temperature measurements in many torus plasma devices confined by magnetic fields. The Large Helical Device (LHD) [1, 2] has been equipped with three types of ECE measurement system (Michelson interferometer, grating polychromator (GPC), and radiometer) to provide electron temperature information since the beginning of the experiment (1998). [3-11]. Recently, plasma fluctuations (such as L-H transitions and MHD bursts, etc.) at a low magnetic field around $B_t = 1 - 1.375$ T have attracted attention at LHD [12-15]. However, the existing ECE radiometers cover frequencies from 50 –150 GHz and are optimized especially for the high frequency range above 100 GHz. At this time, we have developed an ECE radiometer for such low magnetic field strength experiments.

Figure 1 shows the spatial distribution of the characteristic frequency at the LHD horizontal port when the on-axis toroidal field strength ($B_t$) is 1T. Unlike the tokamak, the LHD magnetic field strength distribution has a hill-shape, which allows observation from the outer port in the outer half of the region from the plasma magnetic axis. As shown in Figure 1(b), the extraordinary mode (X-mode) second harmonic EC wave satisfies the optical depth of $\tau > 1$ in the entire plasma region. Since the EC waves in the core region are optically thin but overlap with the third harmonic waves in the periphery, the system was designed with $0.3 < \rho < 1.0$, as the measurement region at this magnetic field configuration, in which the magnetic axis position in the vacuum field $R_{ax} = 3.60$ m, the helical coil pitch parameter $\gamma = 1.2538$, and the ratio of the quadrature field $B_q = 100$ %.

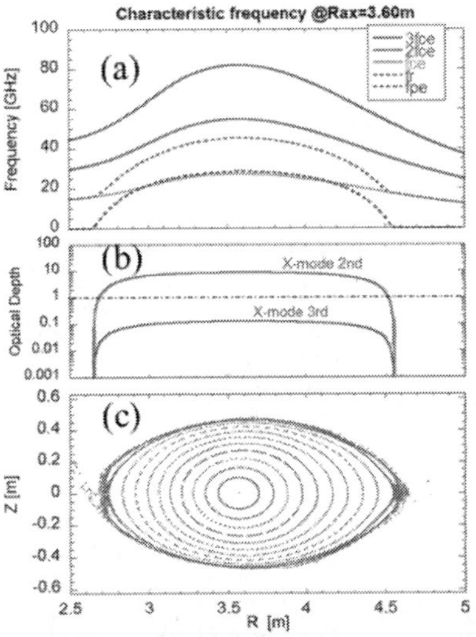


**Fig. 1.** (a) Radial profile of characteristic frequencies in the major radial direction in the LHD horizontal crosssection at $B_t = 1$T and $R_{ax} = 3.60$m. The blue line is the 3rd harmonic $3f_{ce}$, the red line is the 2nd harmonic $2f_{ce}$, the yellow line is the fundamental $f_{ce}$, the green dotted line is the right-hand X-mode cutoff frequency, and the brown dotted line is the electron plasma frequency $f_{pe}$. Here the electron density is calculated as $n_{e0} = 1\text{x}10^{19}$ m$^{-3}$ at the center. (b) Optical thickness at $T_e = 2$ keV, where the X-mode 2nd harmonic $f_{ce}$ exceeds 1 almost everywhere in the plasma confinement region. (c) Closed flux surface per $\rho = 0.1$ and ergodhic region in vacuum.

* Corresponding author: tokuzawa@nifs.ac.jp

This paper describes the diagnostic system, including the focusing optics and heterodyne radiometer circuit, in Chapter 2, and preliminary plasma experimental results in Chapter 3.

## 2 Experimental arrangements of ECE

In the application of the ECE radiometer for this low magnetic field experiment, the system was required to efficiently receive EC waves from the plasma and to detect the received signals with high temporal and spatial resolution. The details of the developed system for each of these items are described below.

### 2.1 In-vessel optical mirrors

The LHD has a complicated vacuum vessel structure for cooling the superconducting coils, and the distance between the port and the plasma is far, as shown in Figure 2. Since it is also difficult to prepare a large vacuum window, we considered installing a focusing optics system inside the vacuum vessel. For this purpose, two aluminium mirrors are used in combination. One mirror (M1) is a concave mirror (500 x 350 mm, focal length of 600 mm) and the other (M2) is a plane mirror (350 x 250 mm). To transmit the EC wave from the plasma to the receiver antenna with these mirrors, the beam width at the vacuum window must be narrowed down to 150 mm or less. The beam radius $\omega$ at each position is obtained from the following Gaussian beam propagation equation,

$$\omega(z) = \omega_0 \sqrt{1 + \left(\frac{\lambda z}{\pi \omega_0^2}\right)^2} \quad (1)$$

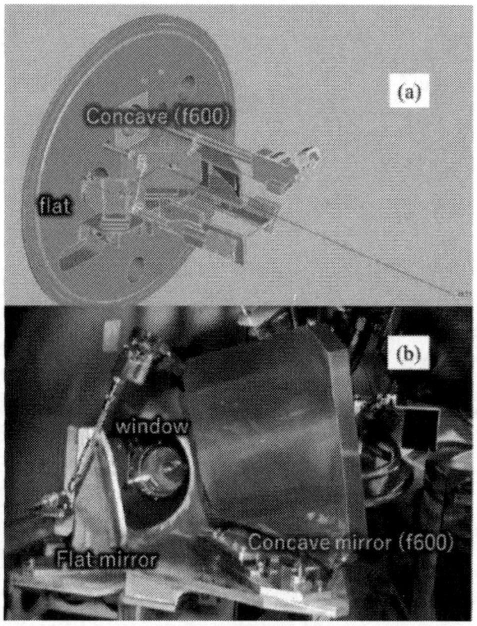


**Fig. 4.** (a) CAD design and (b) photograph of combination mirrors in LHD vacuum vessel.

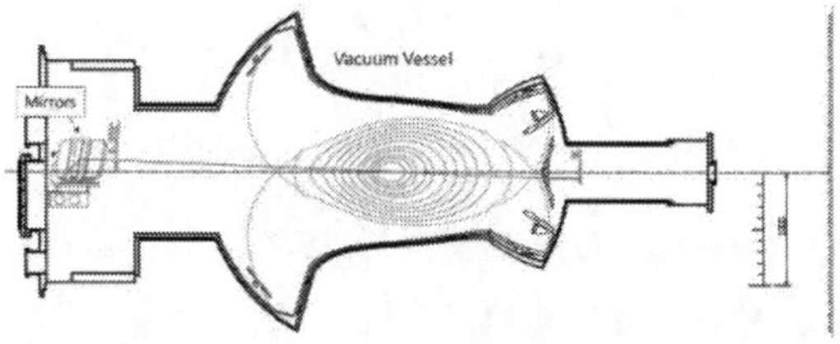


**Fig. 2.** Cross-sectional view inside the LHD 9-O port vacuum vessel, with two mirrors to focus the beam at the plasma center; the distance from M1 to the plasma center is around 2.5 m. The EC waves are detected at an angle of 2.3 degrees, with the line of sight looking slightly upward from the horizontal plane.

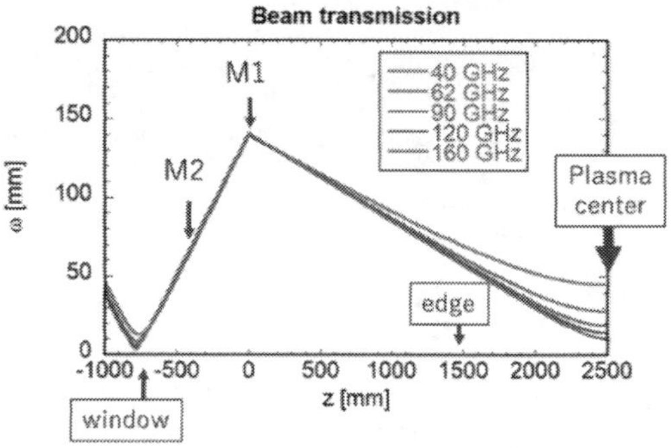


**Fig. 3.** Beam size evolution at each position, determined by Gaussian beam propagation theory. z = 0 for the position of M1.

and is shown in Figure 3. Here, $\lambda$, $\omega_0$ are the wavelength and the beam radius at the beam waist, respectively. $z$ is defined as the distance to the location of the beam waist, and the location of the mirror M1 is set as the reference point. This optical system is expected to have a beam width of 40 mm at 40 GHz, 28 mm at 62 GHz, and 10 mm at 160 GHz at the plasma centre. In the vacuum vessel, the two mirrors are placed in the arrangement shown in Figure 4. To avoid interference with existing equipment, the M1 mirror was placed slightly (115 mm) above the horizontal plane in the vertical direction. However, the change in pitch angle was slight and the field of view was sufficient, so there were no problems with ECE measurements. A shutter was also installed on the vacuum windows to protect them from dirt during wall conditioning.

To evaluate the performance of the focusing optics, we placed a millimeter-wave source at the position of the vacuum window and investigated the beam propagation characteristics. For example, as shown in Figure 5, the beam radius at 1.5 m from the M1 mirror is 79 mm for the 39 GHz millimeter wave and 67 mm for the 150 GHz. These results are in good agreement with the calculated values from Gaussian beam propagation.

https://doi.org/10.1051/epjconf/202327703008

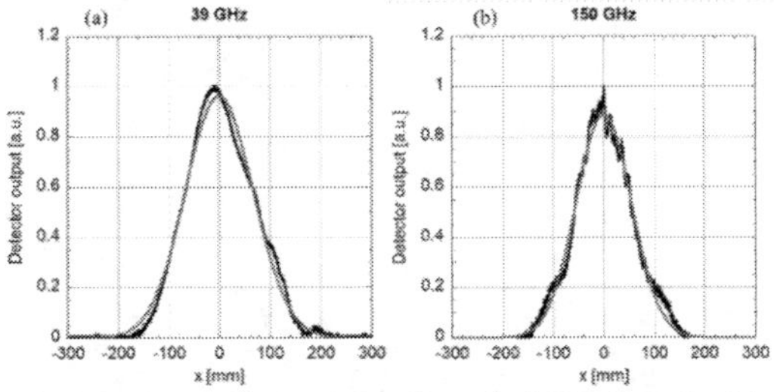


**Fig. 5.** Beam profiles at 1.5 m from the M1 mirror for (a) 39 GHz and (b) 150 GHz millimeter waves. Black lines are measurements, red lines are Gaussian fits. Intensities are normalized to the maximum of the measured values.

## 2.2 Heterodyne Radiometer system

A Q-band and V-band heterodyne radiometer system has been developed. The system consists of two sections which are the radiofrequency (RF) section and the intermediate frequency (IF) one in each frequency band.

### *2.2.1 RF section*

As shown in Figure 6, two bands, Q-band and V-band, were constructed. The EC waves are transmitted through a 2.5-inch corrugated waveguide at a distance of approximately 12 meters from the receiving antenna. After each band is divided and amplified by a low-noise amplifier (LNA) with a gain of 30 dB, the ECE waves are led to a mixer and detected in each band. Both systems use a low-noise DSO oscillator as the local (LO) source for high-sensitivity heterodyne detection, which is multiplied and input to the mixer. The IF signal at the mixer output is amplified by an IF amplifier and led into the next IF section.

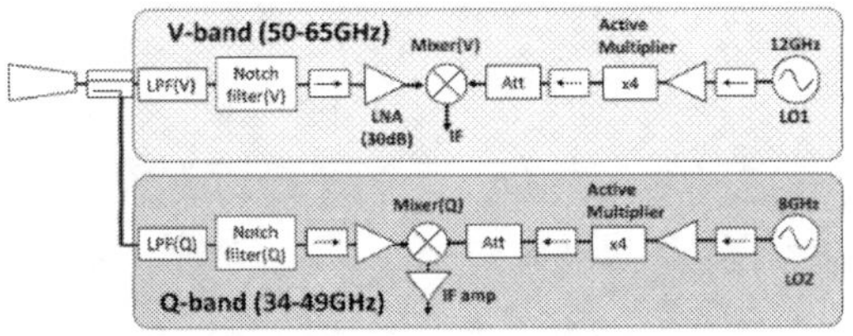


**Fig. 6.** System schematic of heterodyne radiometer RF section. LPF, LNA, and Att represent a low-pass filter, a low-noise amplifier, and an attenuator, respectively. The LO frequencies of the mixer are 48 GHz and 32 GHz for V-band and Q-band, respectively.

56 GHz and 77 GHz gyrotrons are used as second harmonic ECH for plasma ignition and heating in the experimental condition of the magnetic field strengths near $B_t = 1$ T. Due to the high-power output of these gyrotrons, it is necessary to eliminate stray light effects that can damage radiometers and cause saturation of mixers and amplifiers. For this reason, we developed a waveguide cavity-type notch filter and the detailed design values for the Q-band case are given in Ref. 16. On the other hand, a V-band notch filter was also fabricated with the design dimensions, as shown in Table 1. Notch filters (NOTCH (A) and NOTCH (B)) were fabricated separately for each gyrotron frequency, and the waveguides were connected in series so that the two frequencies could be excluded. The performance of the combinate notch filter is shown in Figure 7. The attenuation is 80 dB at 56 GHz and 60 dB at 77 GHz, indicating that a notch filter with sufficient performance could be developed.

**Table 1.** Design parameters for V-band notch filters for 56GHz and 77 GHz.

| | NOTCH (A) | NOTCH (B) |
|---|---|---|
| Center frequency [GHz] | 56.0 | 77.0 |
| Cavity diameter [mm] | 4.12 | 3.0 |
| Cavith length [mm] | 4.546 | 2.62 |
| Cavity gap length distance between two resonators [mm] | 5.726 | 3.42 |
| Slit width [mm] | 0.5 | 0.5 |
| Slit length [mm] | 3.759 | 1.6 |
| Slit depth [mm] | 0.5 | 0.3 |

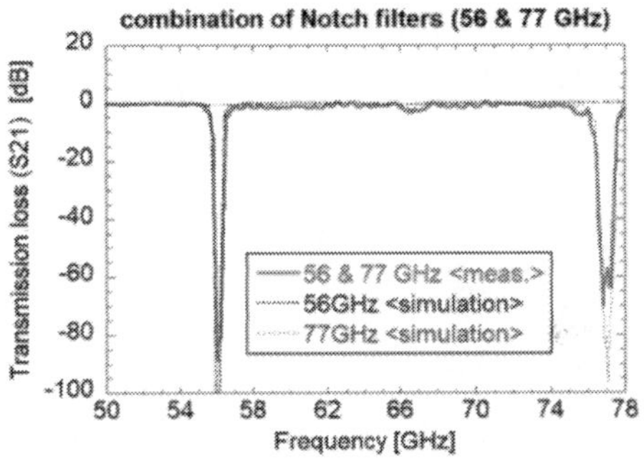


**Fig. 7.** Frequency response of combined notch filters.

### *2.2.2 IF section*

A schematic diagram of the IF section is shown in Figure 8, where a similar system is used for both Q- and V-band to detect each frequency ECE component. The IF signal at the output of the mixer is divided into two parts, and the 2-9 GHz component is amplified by the LNA and enters the filter bank. There it is separated into eight frequency components, discriminated by a BPF with a bandwidth of 800 MHz (described below), and then detected by the Schottky diode detector. The detector output signal is collected throughout the discharge by an ADC (NI PXIe-6378) with 16-bit resolution and 1 MS/s sampling rate. On the other hand, the 10-17 GHz component is down converted to 2-9 GHz by Mixer3 using the 19 GHz signal (LO3), because we want to align the characteristics using a similar filter bank set for signal detection. Thus, each band has 16 frequency channels each, making a total system of 32 channels.

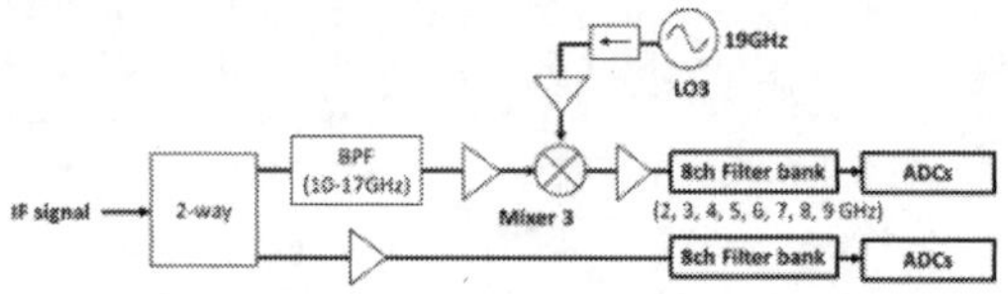


**Fig. 8.** Schematic diagram of IF section.

In the filter bank, each ECE wave is discriminated by a 2-9 GHz BPF. The bandwidth of all BPFs is designed to be $\pm 400$ MHz. The transmission characteristics of the BPFs, measured using a vector network analyser, are shown in Figure 9. It can be seen that each frequency component can be discriminated with the contamination of other frequency components suppressed to -50 dB or more.

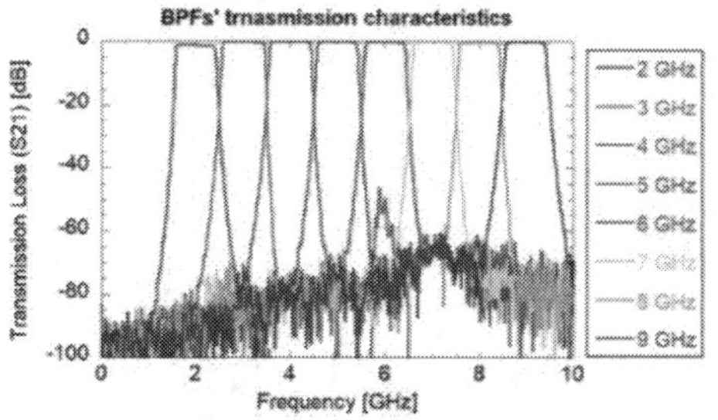


**Fig. 9.** Frequency response of each bandpass filter.

### *2.2.3 Integration test*

The entire heterodyne radiometer system was then tested with a noise source. Instead of EC wave radiation from the plasma, we examined how the signal output of each frequency component from the filter bank changed when a combination of noise sources and attenuators for each band were placed at the antenna locations in the RF section diagram in Figure 6. The strength of the RF signal input to the mixer (from the noise source) was varied by controlling the attenuator from a squeezed state (-32 dB) to a released state (-1 dB). The detector output obtained at that time is shown in Figure 10. Each channel corresponds to 54-57 GHz frequency components. The detector output was found to vary linearly with respect to the change in RF intensity.

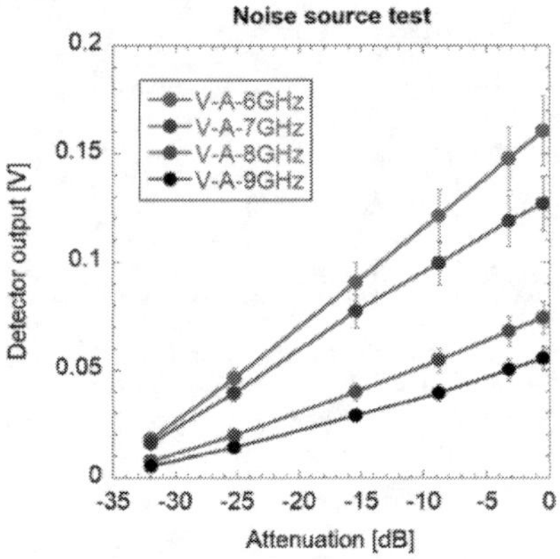


**Fig. 10.** Radiometer integrated test results with a noise source. Change in signal strength from the filter bank when the RF input signal to the mixer was controlled by adjusting the attenuator to vary the strength of the input signal from the noise source. V-A-6, 7, 8, and 9 GHz correspond to BPFs at 54, 55, 56, and 57 GHz, respectively.

## 3 Experimental Observations in LHD

This radiometer is designed for application to plasma experiments under magnetic field conditions from the magnetic field strength of $B_t$ = 1 T to around 1.375 T. As shown in Figure 11, in the so-called inward-shifted configuration with the major radius at the magnetic axis of $R_{ax}$ = 3.60 m, almost the entire region is observable at $B_t$ = 1 T and 1.2 T, and even at 1.375 T, so is the region with the normalized small radius $\rho > 0.5$. In the standard configuration with $R_{ax}$ = 3.75 m, the entire region is observable at 1 T, and even at 1.375 T, the measurement is possible around the rational surface region with $\iota/2\pi$ = 0.5. Here, $\iota$ is the rotational transform.

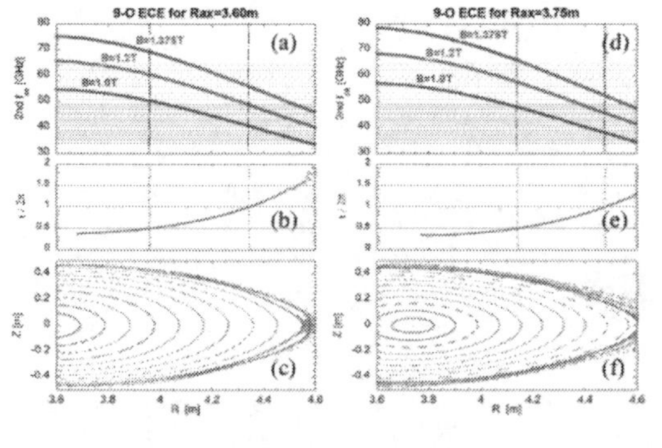


**Fig. 11.** (left) Radial distribution at Rax=3.60 m and (right) $R_{ax}$ =3.75 m. (a, d) Second harmonic cyclotron frequencies at $B_t$ =1 T, 1.2 T and 1.375 T. The orange and the light blue horizontal lines indicate the Q- and V-band measured channels' frequencies, respectively. (b, e) Iota distribution (c, f) Magnetic closed flux surfaces for every $\rho$=0.1.

Some observation examples in plasma experiments are shown below.

NBI power modulation experiments were performed at the magnetic axis of $R_{ax}$ =3.75 m and the magnetic field strength of $B_t$ = 1.375 T. As shown in Figure 12(a), NBI4 and NBI5 which are perpendicular NBI beams were injected with power modulation from 3.4 s to 8.2 s. The time variation of the electron temperature $T_e$ obtained with this radiometer is shown in Figure 12(b). As can be seen from the figure, $T_e$ oscillates significantly in sync with the power modulation. The radial profile of $T_e$ measured by Thomson scattering of the OFF (green) and ON (red) timing of NBI power modulation is shown in Figure 13(a). Also shown are the fitted data of $T_e$ profiles obtained from this data by fitting based on Bayesian statistics [17, 18]. It shows the difference in $T_e$ due to modulation seems to be arisen the region around R=4.5m. Figure 13(b) shows time-expanded ECE radiometer data measured in this region, demonstrating that the phase difference between the R<4.51m region and that outside the R>4.51m region is observed by ECE. In contrast to local heating by ECH, NBI heating is generally considered to affect a wider radial region from the outside of the plasma at the same time. The present measurements revealed that the heating of NBI may be accompanied by localized changes in the plasma. The ECE measurements also show a different response in the core region than the Thomson scattering measurements. Further analysis is needed to integrate and understand these results, and this is an issue for the future.

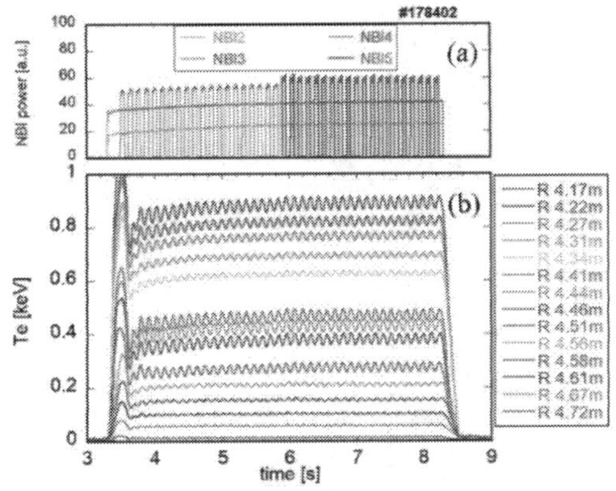


**Fig. 12.** Time variation observation of $T_e$ in NBI power modulation experiments. (a) Incident modulated power of the four NBIs and (b) time variation of the converted signal intensity $T_e$ at each radial position of the ECE.

The radiometer is also expected to observe MHD fluctuations such as Alfvén eigenmodes and electrostatic oscillations. Figure 14 shows an example of the frequency spectrum of the radiometer signal when these oscillations are generated. Comparing the signal strength before the plasma discharge as a noise level, the signal was found to be about one order of magnitude stronger during the discharge. In particular, modes identified as frequency chirp-up reversed shear Alfvén eigenmode (RSAE), geodesic acoustic mode (GAM) oscillations around 20 kHz [19, 20] and these higher harmonics were successfully observed. Since we can now obtain information on various types of temperature fluctuations, we expect to be able to use this information to study their spatial structure and interactions in the future.

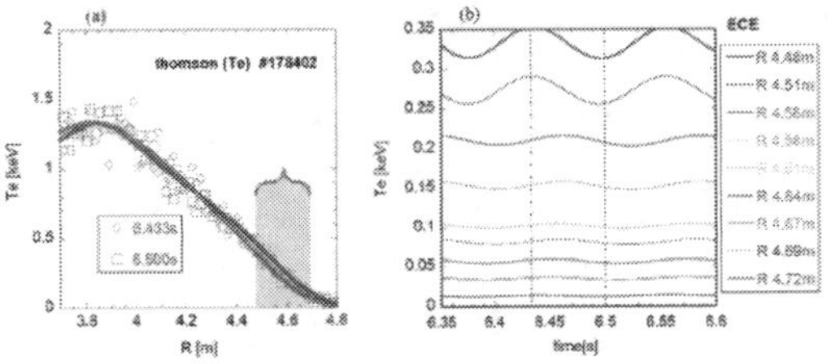


**Fig. 13.** (a) $T_e$ radial profiles measured by Thomson scattering. Green and red solid lines are obtained by fitting. (b) Time variation of electron temperature obtained by ECE in the region shown in blue in (a).

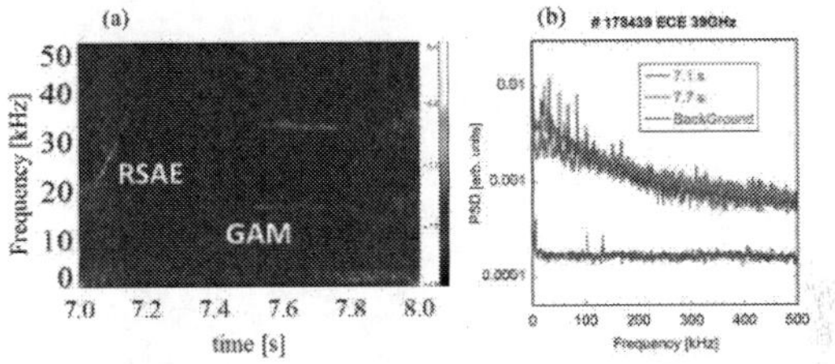


**Fig. 14.** Examples of RSAE and GAM oscillation observations. (a) Spectrogram, and (b) frequency spectra of two timings. The gray line is the estimated background noise for this channel.

## 4 Conclusion

A new ECE radiometer covering the Q and V bands has been developed. The purpose of this radiometer is to observe fluctuations and time variations of electron temperature in low magnetic field experiments at LHD. Due to the limitation of the available port and vacuum window, a focusing mirror designed by Gaussian beam optics was installed in the vacuum vessel for efficient ECE wave transmission. The beam width on the test bench is in good agreement with the estimated value with Gaussian beam propagation. The heterodyne radiometer uses a low-noise local oscillator and has a dynamic range of more than 30 dB when tested with a noise source. To remove the stray lights due to 56 and 77 GHz gyrotrons, a notch filter was created using an oversized waveguide, and more than 50 dB of attenuation was obtained. By implementing this ECE system in LHD, oscillations due to NBI modulation of the radiated ECE waves from the plasma, Alfvén eigenmodes, and GAM oscillations were also successfully observed.

In addition, the LHD data can be accessed from the LHD data repository at https://www-lhd.nifs.ac.jp/pub/Repository_en.html.

## Acknowledgments

The authors are grateful to the LHD experiment group for the excellent operation of the device. This work was partially supported by KAKENHI (Nos. 21H04973 and 19H01880), by a budgetary Grant-in-Aid from the NIFS LHD project, under the auspices of the NIFS Collaboration Research Program (NIFS22ULPP051 and NIFS22KBAP065).

## References

1. A. Iiyoshi et al., Nucl. Fusion **39**, 1245 (1999).
2. Y. Takeiri et al., Nucl. Fusion **57**, 102023 (2017).
3. Y. Nagayama et al., Rev. Sci. Instrum. **70**, 1021 (1999).
4. P. C. de Vries et al., Phys. Plasmas, **7**, 3707 (2000).
5. H. Sasao et al., Fusion Engineering and Design **53**, 153 (2001).
6. Y. Nagayama et al., Fusion Eng. Des., **53**, 201 (2001).
7. K. Kawahata et al., Rev. Sci. Instrum., **72**, 387 (2001).
8. K. Kawahata et al., Rev. Sci. Instrum., **74**, 1449 (2003).
9. Y. Nagayama et al., IEEE Trans. Plasma Sci., **32**, 1716 (2004).
10. T. Tokuzawa et al., Fusion Science and Tech. **58**, 364 (2010).
11. H. Tsuchiya et al., Plasma and Fusion Research, **6**, 2402114 (2011).
12. S. Sakakibara, et al., Nucl. Fusion **57**, 066007 (2017).
13. Y. Suzuki et al., "Improved particle confinement in the L-H transition in the 3D magnetic field", 5th Asia-Pacific Conference on Plasma Physics, 26 Sept-1 Oct, (2021).
14. Wenqing HU et al., Plasma and Fusion Research, **6**, 2402031 (2021).
15. J. Varela et al., Nucl. Fusion **60**, 046013 (2020).
16. M. Nishiura et al., Rev. Sci. Instrum. **92**, 034711 (2021).
17. K. Fujii et al., Rev. Sci. Instrum. **88**, 013508 (2017).
18. K. Fujii et al., IEEE Trans. Plasma Sci. **47**, 3305 (2019).
19. K. Toi et al., Phys. Rev. Lett. **105**, 145003 (2010).
20. T. Ido et al., Nucl. Fusion **51** 073046 (2011).

# Upgrade of the relative calibration methods and Bayesian inference processing for electron cyclotron emission radiometry

*Xin Yu*[1], *Zhongbing Shi*[1], *Zengchen Yang*[1], *Min Jiang*[1], *Wei Chen*[1], *Tianbo Wang*[1], and *Wenan Pan*[1]

[1]Southwestern Institute of Physics, Chengdu 610041, China

**Abstract.** An upgraded local oscillator (LO) hopping calibration method based on a blackbody hot source and a perturbation analysis of the magnetic field difference method are introduced in this work. The blackbody hot source is used to evaluate the difference in the relative coefficients between the two LO hopping frequencies in the same channels. Then the coefficients are obtained by multiplying the LO hopping frequencies coefficients by LO hopping calibration coefficients. In this way, it is more flexible and stable than the in-situ calibration. The magnetic field difference method provides another calibration method to obtain the relative calibration coefficients of the electron cyclotron emission radiometers (ECE). In general, the magnetic field difference method needs two similar shots but with a difference of 2.1% (for HL-2M) in the magnetic field. Meanwhile, there are some errors because of the deviation of detection positions in the same channels between the two shots. For evaluating the calibration errors, the impact of the displacement, $T_e$ perturbation of the core region, and magnetic field difference has been discussed. The result shows that a larger magnetic field difference can improve the accuracy of the calibration. In the end, Bayesian inference has been utilized to evaluate the calibration coefficients and get the most probable calibration coefficients along with its the confidence interval.

## 1 Introduction

ECE radiometry is a powerful $T_e$ and $T_e$ perturbation ($\delta T_e$) diagnostic with high spatial-temporal resolution and it has been established on many tokamaks [1-4]. The calibration of the ECE radiometry is the key step in obtaining the absolute or relative $T_e$ profile. The conventional calibration method is the in-situ calibration using the blackbody hot source with known temperature and it is utilized on many devices such as JET, DIII-D, EAST, and ITER (in the future) [5-8]. However, the blackbody hot source method is complex and inflexible because it requires a long time operation (tens of minutes or even hours) and a large space to operate. Therefore, handy and flexible relative calibration methods like the LO hopping and magnetic field difference method are required [9].

Following the introduction section, an upgraded LO hopping calibration method will be described in Section 2. The perturbation analysis for the magnetic field difference method will be discussed in Section 3. In Section 4, Bayesian inference for the relative calibration methods will be introduced. Finally, Section 5 presents the summary.

## 2 Upgraded LO hopping method

The principle of the LO hopping method is to compare the signal amplitude between the adjacent channels in the same measurement position by hopping the LO frequency. However, the mixer efficiency varies with the LO frequency which results in different initial bias voltages at different LO frequencies. In the strict sense, there are many elements varying with the different LO frequencies such as the antenna gain, transmission loss, and electronic response (including the mixer's efficiency). Therefore, merely comparing the signal amplitude is not complete which will cause prominent calibration errors. The upgraded LO hopping calibration method uses a blackbody hot source to evaluate the response of the measurement channels to different LO frequencies. With this method, the relative $T_e$ can be derived as below:

$$I_i(f_i) \times C_i(f_i) = T_e(f_i) \quad (1)$$

$$I_i(f_{i+1}) \times C_i(f_{i+1}) = T_e(f_{i+1}) \quad (2)$$

$$I_{i+1}(f_{i+1}) \times C_{i+1}(f_{i+1}) = T_e(f_{i+1}) \quad (3)$$

$$I_{bi}(f_i) \times C_i(f_i) = T_b(f_i) \quad (4)$$

$$I_{bi}(f_{i+1}) \times C_i(f_{i+1}) = T_b(f_{i+1}) \quad (5)$$

Then we get :

$$S_i = \frac{C_i(f_{i+1})}{C_i(f_i)} = \frac{T_b(f_{i+1}) \times I_{bi}(f_i)}{I_{bi}(f_{i+1}) \times T_b(f_i)} \quad (6)$$

$$C_{i+1}(f_{i+1}) = \frac{I_i(f_{i+1}) \times C_i(f_{i+1})}{I_{i+1}(f_{i+1})} = \frac{I_i(f_{i+1}) \times S_i \times C_i(f_i)}{I_{i+1}(f_{i+1})} \quad (7)$$

where $I$, $C$, $f$, $I_b$, $T_b$, $S$, and $i$ are the plasma ECE signal amplitude, calibration coefficients, radiation frequency, radiation amplitude of the blackbody hot source, radiation temperature of the blackbody hot source, LO hopping calibration coefficients ratio, and the channel number, respectively. Setting the $C_1$ to 1, the relative calibration coefficients can be derived with Eq. 7.

Normally, the blackbody hot source's in-situ absolute calibration is the conventional calibration method for ECE radiometry. However, the absolute calibration coefficients change slowly over time because of the change in the response of the radiation frequency (RF) amplifier, intermediate frequency (IF) amplifier (mainly caused by active devices), the loss of the transmission line, window coating, and so on. It can be approximatively described as follow:

$$C_i(t,f_i) = C_i(0,f_i) \times RFA_i(t,f_i) \times RFM_i(t,f_i) \times IFA_i(t,f_i) \times IFM_i(t,f_i) \times L_i(t,f_i). \quad (8)$$

In the above, the $C_i(0)$ is the initial coefficient. The $t$, $RFA$, $RFM$, $IFA$, $IFM$, and $L$ are the time, the relative variation coefficients of the RF amplifier, RF mixer, IF amplifier, IF mixer, and the loss of the transmission, respectively. Generally, the variation of the response of those active microwave devices is independent so that the response of the whole working band enhances or abates independently. For a frequency interval of 1 GHz (the interval of IF), the response variable of the active microwave devices and transmission loss can be approximatively considered to be linear. Therefore, Eq. 8 can be simplified as:

$$C_i(t,f_i) = C_i(0,f_i) \times RFA(t) \times RFM(t) \times IFA_i(t) \times IFM_i(t) \times L(t). \quad (9)$$

The relative variation ratio of the calibration coefficients in the two hopping LO frequencies ($S_i$) can be derived as below:

$$S_i = \frac{C_i(t,f_{i+1})}{C_i(t,f_i)} = \frac{C_i(0,f_{i+1})}{C_i(0,f_i)}. \quad (10)$$

This means $S_i$ does not vary with time or vary slowly. It should be noted that $IFA_i$ and $IFM_i$ can vary differently in different channels which induces a difference in the relative coefficients between different channels:

$$\frac{C_i(t,f_i)}{C_i(0,f_i)} \Big/ \frac{C_j(t,f_j)}{C_j(0,f_j)} = \frac{IFA_i(t) \times IFM_i(t)}{IFA_j(t) \times IFM_j(t)} \neq constant. \quad (11)$$

where the $j$ represents another channel. Therefore, the blackbody hot source's in-situ absolute calibration needs to be conducted once in a while, but the upgraded LO hopping relative calibration uses the blackbody hot source just once.

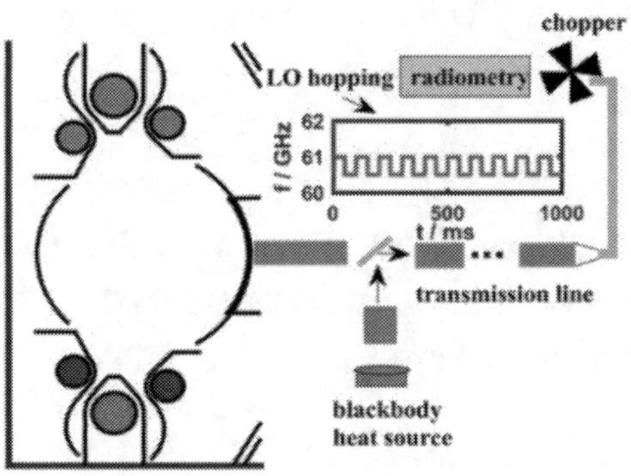


**Fig. 1.** Schematic diagram of obtaining the relative variation ratio of the calibration coefficients in the two hopping LO frequencies.

As shown in Fig. 1, a mirror is utilized to transmit the radiation generated by the blackbody hot source to the radiometry from the transmission line. The radiation signal is hopping by the chopper. Subsequently, by comparing the signal intensity in different LO hopping frequencies, the $S_i$ can be obtained by Eq. 6. The schematic diagram of the upgraded LO hopping calibration is shown in Fig. 2. The measurement positions are constant between two adjacent channels in the two LO hopping frequencies ($LO_1$, $LO_2$). Therefore, the relative calibration coefficients can be derived by Eq. 7 and the $S_i$ can be used to calibrate several times without blackbody hot source evaluation.

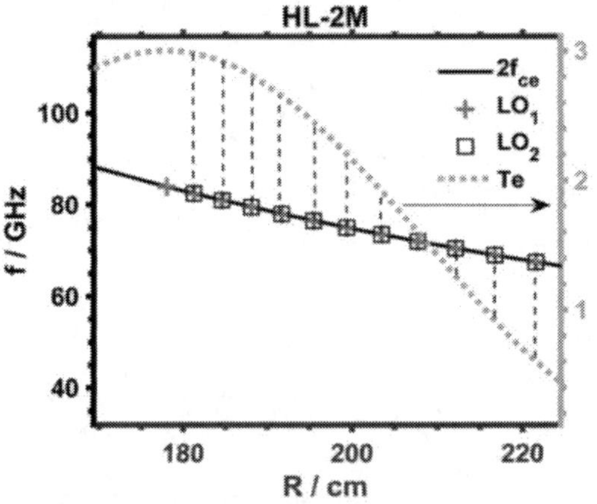


**Fig. 2.** Schematic diagram of the upgrade LO hopping relative calibration.

## 3 Perturbation analysis of the magnetic field difference method on HL-2M

The magnetic field difference method is another relative calibration method for the ECE which is very handy and flexible. Generally, the calibration coefficients are obtained by comparing the signal intensity in the same discharge parameters with an appropriate magnetic field difference. Normally, the appropriate magnetic field difference is dependent on the minimum of the mean of the position deviation ($\delta R_m$) between the adjacent channels with the change of the magnetic field in two shots as shown in Fig. 3. The $B_{t00}$ and $B_{t01}$ are the calibration magnetic fields on the magnetic axis. The relative calibration coefficients can be obtained below:

$$C_{i+1} = \frac{\frac{2I_{Ai}}{C_i(R_{Ai}-R_{Bi+1}} - \frac{I_{Ai}-I_{Bi}}{R_{Ai}-R_{Bi}}}{\frac{2I_{Bi+1}}{R_{Ai}-R_{Bi+1}} + \frac{I_{Ai+1}-I_{Bi+1}}{R_{Ai+1}-R_{Bi+1}})} \quad (12)$$

In the above, the $I_A$, $I_B$, and $R_A$, $R_B$ are the ECE signal intensities and the measurement positions in the two shots, respectively. Unfortunately, the position deviation always exists since the variation of the magnetic field causes different changes in the detected location for the different radial channels. Consequently, the calibration coefficients obtained by this method are not very accurate. Theoretically, by sweeping the magnetic field rapidly, the radiometry can measure the same positions in adjacent channels. In fact, it is difficult for the toroidal magnetic coils, especially for superconducting toroidal magnetic coils.

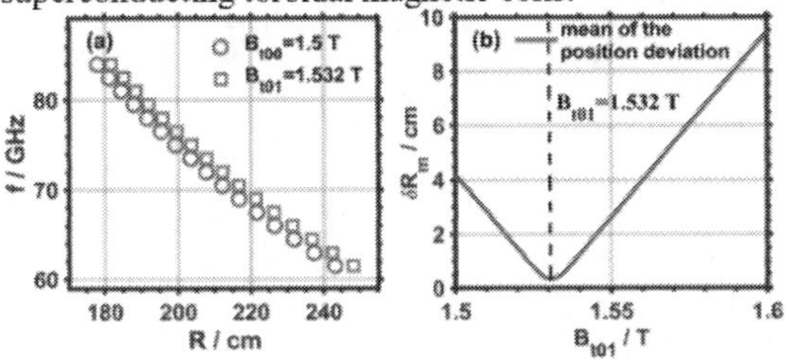


**Fig. 3.** Calibration of multi-channel ECE radiometry by magnetic field difference method. (a) Measured positions at $B_{t00} = 1.5$ T and $B_{t01} = 1.532$ T, (b) The position deviation between adjacent channels ($\delta R_m$) as $B_{t01}$ increases from 1.5 T to 1.6 T.

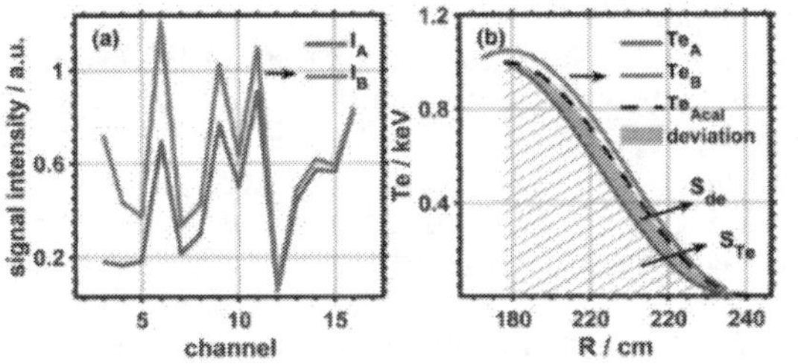


**Fig. 4.** Schematic diagram of the perturbation analysis of the magnetic field difference method. (a) The signal intensity of different channels. (b) Calibration results with the magnetic field difference method.

However, the perturbation analysis shows that the magnetic field difference in which the $\delta R_m$ reaches up to the minimum is not the most appropriate. FIGURE. 4. (a) shows the signal intensities obtained from the preset $T_e$ profiles ($T_{eA}$, $T_{eB}$) in Fig. 4. (b). The $I_A$ is set as a random series and the $I_B$ is obtained by

$$I_B = \frac{T_{eB} \times I_A}{T_{eA}} \quad (13)$$

About the preset profiles, they are fitted using the standard temperature profile model:

$$T_e = T_e(0)\left(1 - \left(\frac{r}{a}\right)^2\right)^{\alpha} \quad (14)$$

where $r$ is the minor radius, $a = 65$ cm, and $\alpha$ is set up to 2.7. The core $T_e$ ($T_{e0}$) and the magnetic axis positions ($R_0$) are set as 1 keV, 178 cm for the $T_{eA}$, and 1.05 keV, 179 cm for the $T_{eB}$, presumably induced by the plasma displacement or other physical problems. The dashed line shown in Fig. 4. (b) is the calibrated $T_e$ ($T_{eAcal}$) profile regarding the $T_{eA}$. The shadowed area is the deviation between $T_{eA}$ and $T_{eAcal}$. Therefore, the error is defined as the ratio of the deviation area to the $T_e$ profile area ($S_{de}/S_{Te}$ as shown in Fig. 4. (b)) and the accuracy is defined as 1- $S_{de}/S_{Te}$. By setting up the $T_{e0A} = 1$ keV, $R_{0A} = 178$ cm, and $B_{t0A} = 1.5$ T, scanning the $R_{0B}$, $T_{e0B}$, and $B_{t0B}$, the calibration accuracy depending on the $R_{0B}$, $T_{e0B}$, and $B_{t0B}$ can be obtained as shown in Fig. 5. (the $T_{e0A}$, $T_{e0B}$, $R_{0A}$, $R_{0B}$, $B_{t0A}$, and $B_{t0B}$ are the $T_{e0}$, $R_0$, and $B_{t0}$ of the $T_{eA}$ and $T_{eB}$).

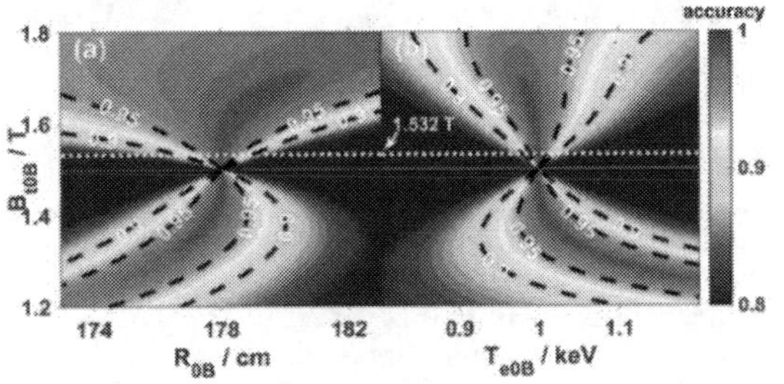


**Fig. 5.** The dependence of calibration accuracy on (a) $R_{0B}$ and $B_{t0B}$, (b) $T_{e0B}$ and $B_{t0B}$ ($R_{0B} = R_{0A}$) where $T_{e0A} = 1$ keV, $R_{0A} = 178$ cm, and $B_{t0A} = 1.5$ T.

FIGURE 5(a) and (b) show the dependence of calibration accuracy on the $R_{0B}$, $T_{e0B}$, and $B_{t0B}$. The thick dashed lines display the contour lines of the calibration accuracy of values 0.95 and 0.9, and the dotted line displays the contour of the $B_{t0B} = 1.532$ T where the mean of the deviation reaches the minimum. The calibration accuracy is more dependent on $T_{e0B}$ and it is difficult for the calibration accuracy to reach 0.95 when the $B_{t0B} = 1.532$ T for both the $R_{0B}$ and $B_{t0B}$. Higher $B_{t0B}$ can expand the range of accurate calibration.

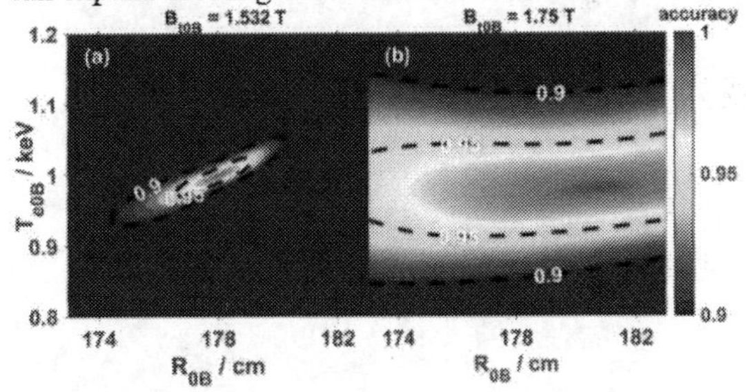


**Fig. 6.** The dependence of calibration accuracy on $R_{0B}$ and $T_{e0B}$ when (a) $B_{t0B} = 1.532$ T and (b) $B_{t0B} = 1.75$ T, where $T_{e0A} = 1$ keV, $R_{0A} = 178$ cm, and $B_{t0A} = 1.5$ T.

FIGURE 6(a) and (b) show the dependence of the calibration accuracy on the $R_{0B}$ and $T_{e0B}$ when $B_{t0B}$ equals 1.532 T and 1.75T. Apparently, 1.532 T is not the most appropriate $B_{t0B}$ for this calibration method because the allowable range is much more narrow than the case of $B_{t0B} = 1.75$ T compared to Fig. 6. (b).

FIGURE 6(b) shows that for the calibration accuracy of 0.95, the $R_{0B}$ and $T_{e0B}$ can vary in the range of 173 - 183 cm and 0.95 - 1.05 keV approximately. In other words, the position deviation and variation of the $T_{e0}$

between the two calibration shots caused by the plasma displacement or the difference of the discharges are allowed to deviate by 5 cm on $R_0$ and 5% on $T_{e0}$ for 95% calibration accuracy, which is acceptable for the control on the tokamak plasma.

The perturbation analysis shows the feasibility of accurate calibration of the $T_e$ profile using this method. It's worth noting that, the accurate calibration range with this method depends on many elements, such as the geometric dimensioning of the plasma, the working magnetic field, and the working frequency band of the ECE radiometry. Therefore, the perturbation analysis needs to be performed before the calibration to get an appropriate magnetic field difference. Due to the convenience and flexibility, the magnetic field difference method is chosen as one of the calibration methods for ECE radiometry on HL-2M to access the calibration coefficients.

## 4 Bayesian inference processing

In order to evaluate the calibration coefficients obtained by the upgraded LO hopping method or magnetic field difference method, the Bayesian inference [10] is utilized to process the calibration data. Normally, the calibration coefficient $C_{im}$ depends on the measured temperature $T_{ei}$, signal intensity $I_{im}$, and the systematic and statistical errors $\varepsilon_i$ (the $i$ and $m$ denote the channel number and calibration number):

$$C_{im} = {}^{T_{ei}}/_{I_{im}} + \varepsilon_i \quad (15)$$

As mentioned above, it can be solved by the Bayesian theory:

$$p(T_e \mid C,\theta) \propto p(C \mid T_e,\theta) \cdot p(T_e \mid \theta) \quad (16)$$

In the above, the $\theta$, $p(T_e \mid C,\theta)$, $p(C \mid T_e,\theta)$, and $p(T_e \mid \theta)$ are the additional information for inferences, posterior probability distribution, likelihood probability distribution, and prior probability distribution. In the Gaussian Process method, the $p(C \mid T_e,\theta)$ and $p(T_e \mid \theta)$ can be formulated by Gaussian Process (GP):

$$p(C \mid T_e,\theta) = \frac{1}{|\Sigma_C|^{\frac{1}{2}}(2\pi)^{\frac{n}{2}}} \exp\left[-\frac{\left({}^{T_e}/_{I}-C\right)^T \Sigma_C^{-1}\left({}^{T_e}/_{I}-C\right)}{2}\right] \quad (17)$$

$$p(T_e \mid \theta) = \frac{1}{|\Sigma_{Te}|^{\frac{1}{2}}(2\pi)^{\frac{n}{2}}} \exp\left[-\frac{\left(T_e-\mu_{T_e}\right)^T \Sigma_{Te}^{-1}\left(T_e-\mu_{T_e}\right)}{2}\right] \quad (18)$$

In the above equation, the $n$ and $\mu_{Te}$ are the number of the channels and expected temperature, respectively. The $\Sigma_c$ and $\Sigma_{Te}$ are the calibration uncertainty and inference uncertainty covariance matrixes, respectively. They are denoted as:

$$\Sigma_c = \begin{bmatrix} C_{e1}^2 & \cdots & 0 \\ \vdots & \ddots & \vdots \\ 0 & \cdots & C_{en}^2 \end{bmatrix}, \quad (19)$$

$$\Sigma_{T_e} = \begin{bmatrix} kf(1,1) & \dots & kf(1,n) \\ \vdots & \ddots & \vdots \\ kf(n,1) & \dots & kf(n,n) \end{bmatrix}, \quad (20)$$

$$kf(i,j) = \sigma_{T_e}^2 \exp\left(-\frac{d_{ij}^2}{2\sigma_l^2}\right), d_{ij} = \left|R_i - R_j\right| \quad (21)$$

In the above, the $C_{ei}$, and $R_i$, $R_j$ are the calibration errors and the measured positions, respectively. The $\sigma_{T_e}^2$ and $\sigma_l^2$ are the hyperparameters obtained by Occam's razor optimization criterion. Consequently, the combination of the prior and likelihood is derived as below:

$$p(T_e \mid C,\theta) \propto p(C \mid T_e,\theta) \cdot p(T_e \mid \theta) \propto \exp\left[-\frac{\left({}^{T_e}/_{I}-C\right)^T \Sigma_C^{-1}\left({}^{T_e}/_{I}-C\right)}{2}\right] \exp\left[-\frac{\left(T_e-\mu_{T_e}\right)^T \Sigma_{Te}^{-1}\left(T_e-\mu_{T_e}\right)}{2}\right] \quad (22)$$

The expected number $(\mu_{T_e}^*)$ and covariance matrix $(\Sigma_{T_e}^*)$ of the prior can be derived as:

$$\begin{cases} \mu_{T_e}^* = \mu_{T_e} + \left({}^{1}/_{I}{}^{T} \cdot \Sigma_C^{-1} \cdot {}^{1}/_{I} + \Sigma_{Te}^{-1}\right)^{-1} {}^{1}/_{I}{}^{T} \cdot \Sigma_C^{-1} \cdot \left(C - {}^{1}/_{I} \cdot \mu_{T_e}\right) \\ \Sigma_{T_e}^* = \left({}^{1}/_{I}{}^{T} \cdot \Sigma_C^{-1} \cdot {}^{1}/_{I} + \Sigma_{T_e}^{-1}\right)^{-1} \end{cases} \quad (23)$$

Experimentally, a magnetic field difference relative calibration is processed with Bayesian inference on HL-2A. The data points are 100 times calibration in a stable discharge period of 200 ms. FIGURE 7 shows the processing result. The shadowed area is the 95% confidence interval and the solid line is the expected value of the highest probability. It can be seen that the maximum probability profile is different from the mean of the original samples and it is more smooth. The maximum uncertainty comes from the core region of the plasma. Bayesian inference provides a method with a confidence interval to determine the calibration coefficients and is helpful to evaluate the calibration accuracy.

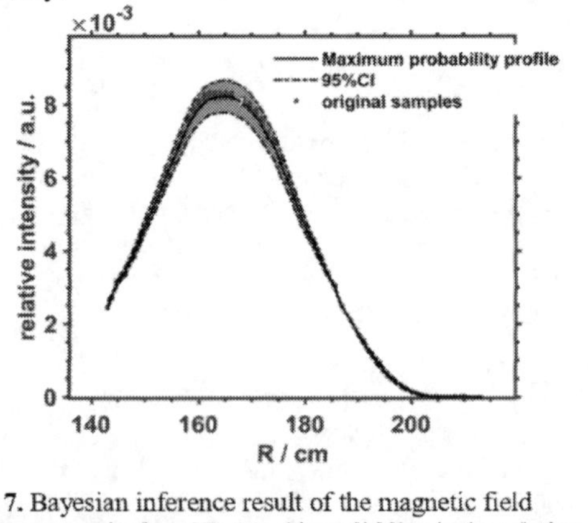


**Fig. 7.** Bayesian inference result of the magnetic field difference method on HL-2A. (the solid line is the relative $T_e$ of maximum probability, and the area between the dash lines is a 95% confidence interval.)

## 5 Summary

An upgraded LO hopping calibration method utilizes the blackbody hot source to evaluate the relative variation ratio of the calibration coefficients in the two hopping LO frequencies. The absolute coefficients can be obtained by simple ratio calculation. The advantage of this method is more flexible and stable. The perturbation analysis of the magnetic field difference method shows that the magnetic field when the mean of the position deviation reaches the minimum is not the most appropriate. By perturbation analysis, the most appropriate magnetic field and the allowable range of errors are obtained. In this analysis, the position deviation and variation of the $T_{e0}$ between the two calibration shots caused by the plasma displacement or the difference of the discharges are allowed to deviate by 5 cm on $R_0$ and 5% on $T_{e0}$ for 95% calibration accuracy. It is acceptable for the control of the tokamak plasma. In addition, the Bayesian inference processing has been simply utilized to evaluate the calibration coefficients. The result shows that the maximum probability profile is different from the mean of the original samples and it is more smooth. The maximum uncertainty comes from the core region of the plasma. Therefore, it can provide a method with a confidence interval to determine the calibration coefficients and helps us to evaluate the calibration accuracy. The experimental verification will be carried out at the end of 2022.

## 6 Acknowledgments

This work is supported by the National Magnetic Confinement Fusion Science Program of China (Grant Nos. 2017YFE0301203), the National MCF Energy R&D Program (No. 2022YFE03060000), the Sichuan Science and Technology Program (Nos. 2021JDJQ0029), and the National Natural Science Foundation of China (Grant Nos. 12175055 and Grant Nos. 12075079).

## References

1. Z. B. Shi, W. L. Zhong, M. Jiang, Plasma Sci. Technol. **20**, 094007 (2018)
2. Z. J. Yang, X. M. Pan, X. D. Ma, B. W. Ruan, R. B. Zhou, and C. Zhang, Rev. Sci. Instrum. **87**, 11E112 (2016)
3. S. J. Freethy, G. D. Conway, I. Classen, A. J. Creely, T. Happel, A. Köhn, B. Vanovac, and A. E. White, Rev. Sci. Instrum. **87**, 11E102 (2016)
4. D. D. Truong, and M. E. Austin, Rev. Sci. Instrum. **85**, 11D814 (2014)
5. S. Schmuck, J. Fessey, T, Gerbaud, et al, Rev. Sci. Instrum. **83**, 125101 (2012)
6. M. E. Austin, R. F. Ellis, J. L. Doane, et al, Rev. Sci. Instrum. **68**, 480 (1997)
7. Y. Liu, S. Schmuck, H. Zhao, et al, Plasma Sci. Technol. **18**, 1148 (2016)
8. G. Taylor, M. E. Austin, A. Basile, et al, EPJ Web of Conferences. EDP Sciences, **147**, 02003 (2017)
9. Z. B. Shi, et al. Rev. Sci. Instrum. **85**, 023510 (2014)
10. M. A. Chilenski, M. Greenwald, Y. Marzouk, et al, Nucl. Fusion, **55**, 023012 (2015)

# Design of Stray Radiation Sensor for ITER ECE Diagnostic

*S. Danani*[1*], *Sheetal Punia*[1*], *Ravinder Kumar*[1], *Hitesh Kumar B. Pandya*[1,2], *Vinay Kumar*[1,2]

[1]ITER-India, Institute for Plasma Research, Gandhinagar 382428, India
[2]Homi Bhabha National Institute (HBNI), Mumbai 400094, India

**Abstract.** The Electron Cyclotron Emission (ECE) diagnostic has a primary role in the measurement of electron temperature profile and electron temperature fluctuations in ITER. This diagnostic shall be exposed to significant power due to unabsorbed Electron Cyclotron Heating (ECH) power in the plasma. The expected stray power loads could be a few tens of watts, and therefore, the protection of millimetre wave components is one of the design challenges of ITER ECE diagnostic. This protection system includes sensors, a band stop notch filter, and a shutter to stop the RF stray radiation from being incident on the sensitive components. The sensors will be positioned along the ECE transmission line, and shall be used for real-time power monitoring of the stray radiation. Here, we describe a novel design of a sensor for monitoring the stray radiation power. This sensor is a Schottky Diode rectenna, known for high-power and high-speed millimetre wave detection capability. It consists of a 2x2 microstrip patch antenna array, a matching circuit, a diode, and a low pass filter. The antenna array is designed analytically and optimized in CST Microwave Studio, for wide reception angle, high gain, and low side lobe levels. Furthermore, the rectifying circuit is optimized using Agilent Advanced Design System (ADS) software to get better rectification and impedance matching of the signal, thereby improving its detection sensitivity. The ADS simulation results show that the detection sensitivity is about 1000V/W for input power of -30 dBm at 170 GHz, thereby achieving the required performance of the sensor.

## 1 Introduction

The Electron Cyclotron Emission (ECE) diagnostic has a primary role in the measurement of electron temperature profile and electron temperature fluctuations in ITER. The supplementary roles of this diagnostic are the measurement of plasma energy, radiated power, run away electron behaviour and edge electron temperature profile. This diagnostic is also

*Corresponding author: (1) suman.danani@iterindia.in
(2) sheetalpunia.iitd@gmail.com

used as a back-up measurement of the ELM temperature transient [1]. The ECE measurement system consists of a front-end radiation collector, high temperature black body calibration sources, transmission lines (TLs), and the radiation measurement instruments (Radiometer and Fourier Transform Spectrometer). The ECE radiation emerges out through primary window and enters into the ex-vessel transmission line system in the port interspace region of Equatorial port plug (EPP) 09. This diagnostic shall be exposed to significant stray radiation power due to unabsorbed Electron Cyclotron Heating (ECH) power in the plasma. Per the latest design of the Diagnostic First Wall aperture (DFW) for 55.F1, the opening area is 0.024 m$^2$. Therefore, for the background power during the plasma start-up, it is approximately 6 kW for 5.5 s. For the non-optimal absorption during plasma flat-top, the power entering the 55.F1 DFW is ~ 48 watts [2]. Therefore, the protection of millimetre wave components is one of the design challenges of ITER ECE diagnostic [3].

The protection system shall include sensors, a band stop notch filter, and a shutter to stop the RF stray radiation from being incident on the sensitive components. Monitoring of the stray radiation at various locations is an important aspect of the stray radiation protection system. The objective of this sensor is to detect the expected ECH loads ~ 48 watts, being coupled into the ECE TL & receiver system, with a fast response time. From the experiences on other machines, the response time should be less than a few milliseconds. Further, it is to be noted that the stray radiation during plasma start-up is higher, so the shutter shall be kept closed in order to protect the ECE diagnostic. These sensors will be positioned along the ECE transmission line, and shall be used for real-time power monitoring of the stray radiation. One location that is being considered to mount the first such sensor, is inside the enclosure of Polarization splitter unit within the ECE beam path. However, work is in progress to finalize the position of first sensor and also number of sensors required and their respective positions along the transmission line.

In this paper, a novel design of a sensor is described for monitoring of stray radiation power. In Section 2, design of various components of sensor is described in detail. The ADS simulation results of the designed sensor are discussed in Section 3. In Section 4, summary and conclusion is given. Section 5 lists some future work anticipated in order to demonstrate the performance of this sensor to satisfy ITER ECE requirements.

## 2 Design of sensor

The sensor is a rectenna based system, known for high-power and high- speed millimetre wave detection capability [4], [5]. A rectenna is a device that captures the RF signal and converts it into DC voltage. A typical rectenna consists of four main components: a microstrip antenna, a matching circuit, a diode, and a low pass filter as shown in Figure 1. The antenna works as a receiver to capture the RF signals. A low-pass filter does the function of harmonic rejection to eliminate power loss due to harmonics. A rectifier is used to rectify the signals by using rectifying diode.

The design of sensor is based on microstrip technology, due to its various advantages over the conventional techniques. The advantages include: compact size, low cost, high integration, and high conversion efficiency. Also, the microstrip patch antennas are versatile in terms of resonant frequency, polarization, impedance, and are omnidirectional. These features are lacking in traditional systems, which are therefore not suitable for our application.

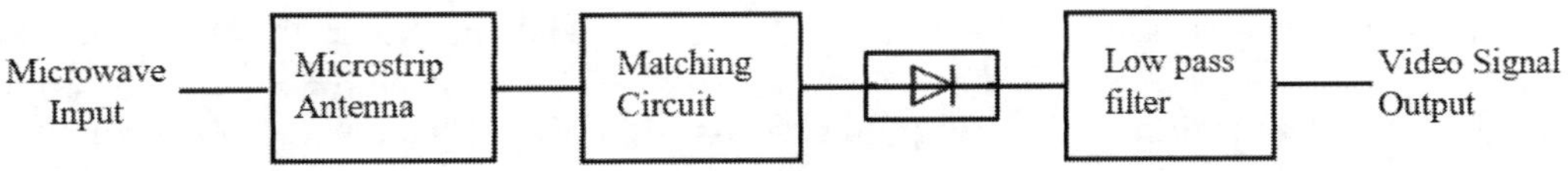

*Figure 1: Schematic of proposed Stray radiation sensor*

Details of various components of rectenna are described below.

## 2.1 Microstrip Antenna

The antenna is a crucial element of the sensor design as it must sense all the stray radiation power entering through the port. The desired characteristics for antenna design are wide reception angle, high gain, large bandwidth, low side-lobe-level and good power handling capability. Fig. 2 shows CST model of a 2×2 patch antenna array designed for the resonant frequency of 170 GHz. The microstrip patch antenna is a single-layer design consisting of four main parts: (i) Patch; (ii) Substrate; (iii) ground plane; (iv) feeding part. The details of antenna design can be found in [6]. The substrate chosen is RO3003 substrate with dimensions 3.55×3.0×0.254 $mm^3$. The substrate thickness is kept small to get higher bandwidth. A symmetric corporate feeding network is used to feed different patches of the antenna array with 50-ohm microstrip lines. The quarter-wave transformer is used to match impedance between the feeding line and the patch to reduce reflection losses. The array provides a large aperture, which makes it capable of detecting extremely weak signals from distant sources with improved gain and radiation pattern compared to a single patch antenna. The antenna array also offers better power handling capabilities than a single antenna element. The characteristics of designed antenna are given in Table 1.

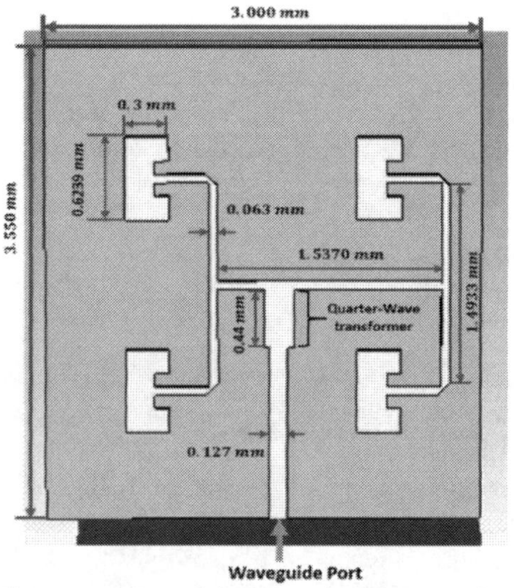


*Table 1: Characteristics of antenna array*

| Parameters | Antenna Array |
|---|---|
| S-Parameter ($S_{11}$) | -50 dB |
| VSWR | ≤ 2 |
| Bandwidth | 12% |
| Gain (IEEE) | 8.78 dBi |
| Side Lobe Level (SLL) | -15.5 dB |
| Angular width (3dB) | $70.5^0$ |
| Total Efficiency (@170GHz) | 79% |

*Figure 2: CST model of inset-fed directional microstrip patch antenna 2×2 array for 170 GHz resonant frequency.*

## 2.2 Low Pass filter

The nonlinear components of rectifying circuits, such as diodes, generate harmonics of the fundamental frequency. These harmonics are reradiated back to the antenna which result in low total efficiency due to the loss in power. To eliminate harmonics reradiating back, filter is adopted. Filter is used to suppress the harmonics and maximize the total conversion efficiency. A low pass filter is designed using ADS software as shown in the Fig. 3. Here, the fourth-order maximum flattening filter passband (3 dB) with an upper frequency limit of

40 GHz and a 20 dB cut-off frequency of 80 GHz is used in designing. The substrate material RO3003 with dielectric constant 3.0 is chosen for the microstrip circuit. The dielectric thickness is $0.254\,mm$, the microstrip line material is copper, and the microstrip line thickness is $35\mu m$. The microstrip layout of the low pass filter is shown in Fig. 4.

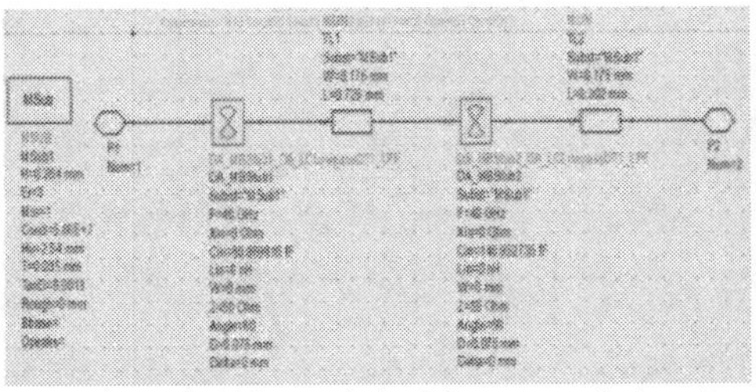

*Figure 3 : Microstrip model with two radial stubs of low-pass filter*

*Figure 4 : Microstrip circuit of the low pass filter*

The ADS simulation results for insertion loss and input impedance of low pas filter are shown in Fig. 5(a) and (b) respectively. The insertion loss is 30.8 dB and 29.1 dB at 170 GHz and 200 GHz respectively, and the harmonic signals are suppressed. Since the diode output port will be connected to the low-pass filter, the input impedance of the designed low-pass filter is first analysed, and is 19.5-j*20.7 Ohm at 170 GHz.

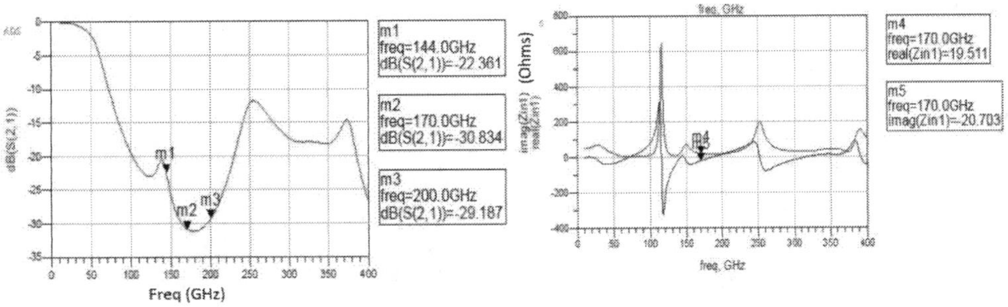


*Figure 5 : The ADS Simulation result at 170 GHz (a) for transmission coefficient (S21) where the minima depicts the maximum power transfer, and (b) for input impedance of the low pass filter where its minimum value signifies its easy passage from source to load.*

## 2.3 Schottky Diode

Diode converts the AC power collected by the antenna to DC. A G-band (110-300GHz) ZBD produced by VDI is used for our application, due to its high cut-off frequency and better I-V characteristics. The response time of detector is 1 μs, which is sufficient for detection of stray radiation power. The minimum chip dimensions are 255×88×43 μm. The diode simulation model for VDI G-Band ZBD in ADS is shown in Fig. 6.

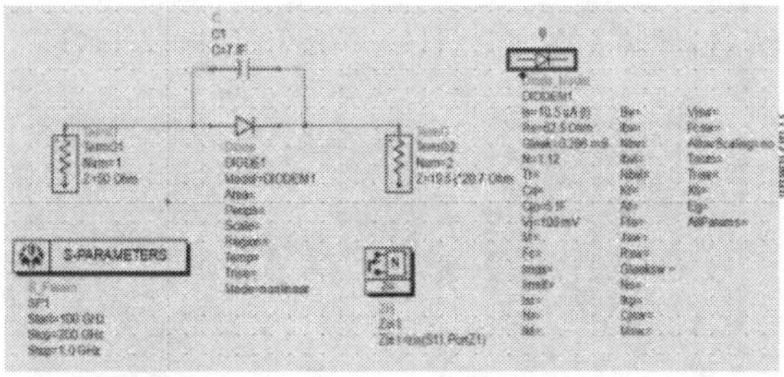


*Figure 6: Diode model of VDI G-band ZBD*

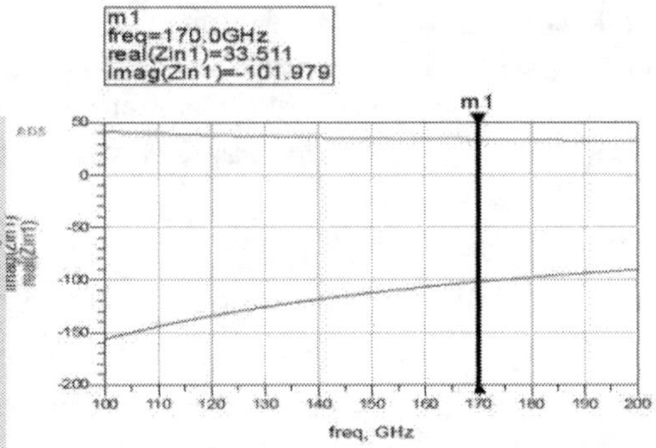


*Figure 7: Diode input impedance simulation results*

The variation of input impedance of the diode with frequency is shown in Fig. 7. Since the gyrotron output frequency in the ECE system is 170 GHz, the input impedance of the diode should be matched to 33.5-j*101.9Ω.

### 2.4 MATCHING CIRCUIT

The matching circuit is used to match the input impedance of the detector to a 50 Ω impedance to reduce the reflection, so that maximum microwave power enters the diode, thus improving the detection sensitivity. An impedance matching method combining a tapered-line match and a series resistor match is proposed in order to attain maximum sensitivity of detector at 170 GHz. The simulation results of the matching circuit when load impedance is set to 33.5-j*101.9 Ω are shown in Fig. 8. The VSWR is 1.1 at 170 GHz. The power delivered to the diode for different input powers is shown in Table 2.

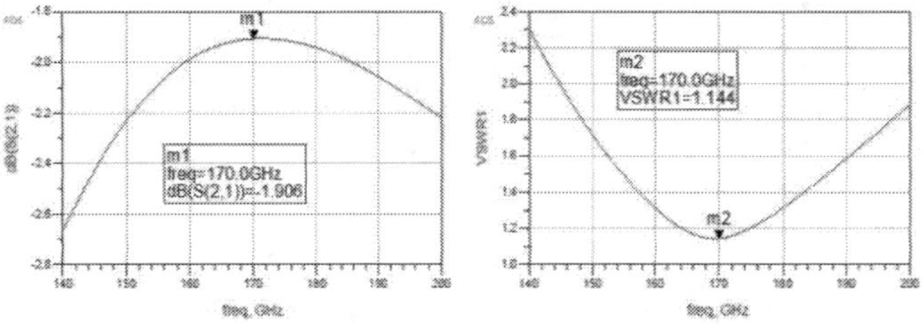


*Figure 8: Simulation result of matching circuit when load impedance is set to 33.5-j*101.9Ω*

Table 2: Power delivered to the diode for different input power

| RF Power (dBm) | Pdel_Watts |
|---|---|
| 10.000 | 0.006 |
| 15.000 | 0.020 |
| 20.000 | 0.064 |
| 25.000 | 0.204 |
| 30.000 | 0.645 |
| 35.000 | 2.039 |
| 40.000 | 6.447 |

## 3 Simulation Results

The rectifying circuit, consisting of the matching circuit, low pass filter and the diode, is then simulated and optimized in ADS and is shown in Fig. 9. The harmonic balance simulation of the detector is carried out.

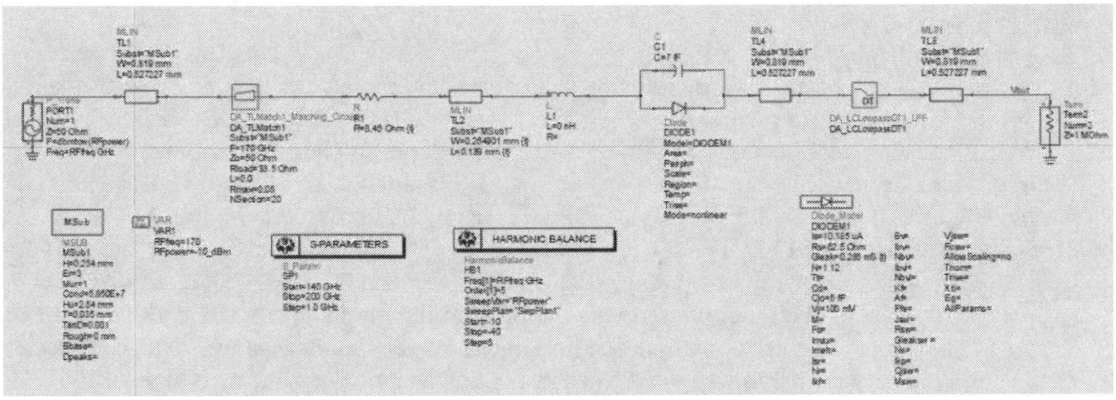


*Figure 9 : The detector circuit and harmonic balance simulation configuration.*

Simulation is done for different input RF power. When the input power is -10 dBm, the output voltage is about 0.07 V, and the detection sensitivity is about 700 V/W. When the input power is -20 dBm, the output voltage is about 0.01V, the detection sensitivity is about 1000V/W; when the input power is -30dBm, the output voltage is about 0.001V, and the detection sensitivity is about 1000V/W. The simulation results when $L = 0$ are similar to the simulation results when $L = 0.05nH$. The simulation results show that the detector has good linearity for a frequency of 170 GHz as shown in Fig. 10.

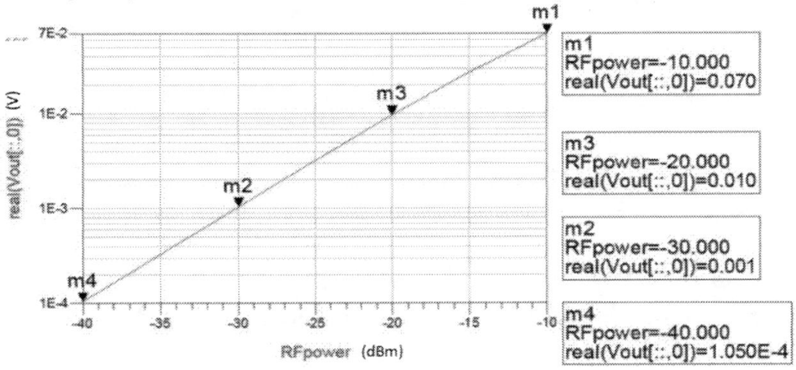


*Figure 10 : The harmonic balance simulation results at 170 GHz*

The effect of variation of RF frequency is also studied. When the input power is set to $-10\ dBm$, the simulation results show that the detection sensitivity is greater than $500V/W$ in the frequency range of $199\ GHz\sim157\ GHz$, which is shown in Fig. 11. The series inductance ($0$ or $0.05nH$) has little effect on the results. Hence, the detector designed is more suitable for use in the ECE system.

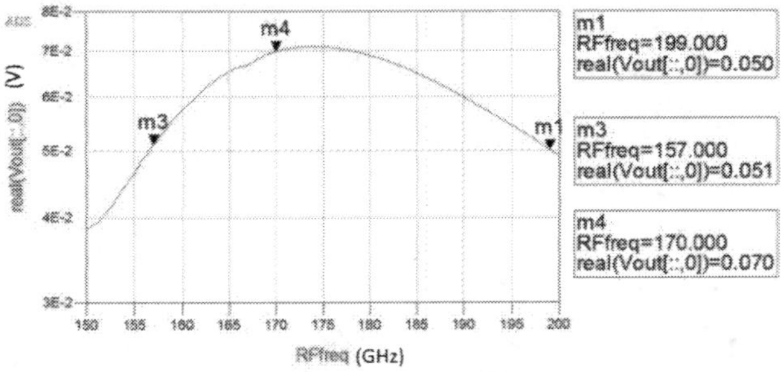


*Figure 11 : The harmonic balance simulation results at -10 dBm.*

## 4 Conclusion

A novel design of a sensor for monitoring the stray radiation power is described in this paper. This sensor is a Schottky Diode rectenna, known for high-power and high-speed millimetre wave detection capability. It consists of a 2x2 microstrip patch antenna array, a matching circuit, a diode, and a low pass filter. A 2×2 microstrip antenna array has been designed and simulated for the resonant frequency of 170 GHz using CST Microwave Studio software. At resonant frequency of 170 GHz, the return loss is -50 dB, angular width is $70.5^0$ and the gain is 8.6 dBi, which meets desired antenna requirements reasonably well. The rectifying circuit, consisting of the matching circuit, low pass filter and the diode, is simulated and optimized in ADS software. The harmonic balance simulation of the detector shows that the detector has good linearity for a frequency of 170 GHz and the detection sensitivity is about 1000V/W for input power of -30 dBm. It is therefore concluded that the above design satisfies all the performance requirements of the sensor and seems to be a possible solution for monitoring of stray radiation power. However, a prototype of the designed sensor shall be developed and tested for its performance. Further optimization of sensor design shall be done if needed, based on the experimental results.

## 5 Future work

The future work involves development of a prototype of the designed sensor and test it for its performance. Also, the mounting position of the sensor inside the ex-vessel TL shall be finalized and a method for its calibration shall be developed.

## 6 References

1. Taylor et al, EPJ Web of Conferences 147, 02003 (2017)
2. Preliminary design proposal of stray radiation protection system, ITER IDM UID PKZCAZ.
3. Johan W. Oosterbeek. et al. "Loads due to stray microwave radiation in ITER", Fusion Engineering and Design 96, 2015
4. A. Etinger et al. "Characterization of a Schottky Diode Rectenna for Millimeter Wave Power Beaming Using High Power Radiation Sources" ACTA PHYSICA POLONICA A Vol. 131 (2017)
5. Maho Matsukura et al. "Instantaneous measurement of high-power millimeter-wave beam for 28 GHz gyrotron", Rev. Sci. Instrum. 90, 024703 (2019)
6. Sheetal Punia, et al "Design and Analysis of a 170 GHz Antenna for Millimeter-wave Applications" *http://arxiv.org/abs/2208.10075*

# Physics Studies for Assessment of Requirement of Displacement Compensation System for ITER ECE Diagnostic

*S. Danani*[1*], *Hitesh Kumar B. Pandya*[1,2], *Pratik Vaghashiya*[1], *Shrishail Padasalagi*[1], *Ravinder Kumar*[1], *W. L. Rowan*[3], *J. P. Ziegel*[3], *Vinay Kumar*,[1,2]

[1]ITER-India, Institute for Plasma Research, Gandhinagar 382428, India
[2]Homi Bhabha National Institute (HBNI), Mumbai 400094, India
[3]Institute for Fusion Studies, The University of Texas at Austin, Austin, TX 78750, USA

**Abstract:** The Electron Cyclotron Emission (ECE) diagnostic has the key function of measuring the core electron temperature profile and electron temperature fluctuation, from the intensity of electron cyclotron radiation emitted from the plasma along the major radius. The ECE diagnostic consists of three main systems: (1) front-end optics, which collects the radiation from the plasma, (2) transmission lines including polarizer splitter unit, which transports the ordinary and extraordinary ECE emission modes separately from the front-end and distributes it to the instrumentation, and (3) detection and analysis instrumentation which is housed at a distance from the tokamak, in the diagnostics building [1]. With its high electron temperatures and harsh environment, ITER presents various challenges for the diagnostic system. One of the most insidious is the misalignment between the in-vessel front-end optics and the ex-vessel transmission line which is caused by vibration of the vacuum vessel during operational and baking phases. Since the electron temperature is inferred from the intensity of the ECE, transient misalignment may lead to poor accuracy in this critical measurement. These displacements are expected to be ~ 15 mm in vertical (z) and horizontal (x) directions, and ~ 5 mm in the toroidal (y) direction. It is important to minimize the effect of these displacements, so that the system maintains alignment during operation, and reliable temperature information is attained. Our objective is to first study the coupling losses due to imperfect coupling of Gaussian beams owing to port plug displacements. Measurements are done to determine the power loss due to coupling of offset beams experimentally. The measured value for coupling loss is ~2 dB at 120 GHz, which is quite high, and it is therefore concluded that a mechanism is needed to compensate for the displacements.

*Corresponding author: suman.danani@iterindia.in

## 1 Introduction

The ECE diagnostic has the key function of measuring the electron temperature profile and electron temperature fluctuation, from the intensity of electron cyclotron radiation emitted from the plasma along the major radius. Other roles of this diagnostic are to obtain information on non-thermal electron populations and the power loss due to ECE [1]. The ECE diagnostic consists of three main systems: (1) front-end optics, which collect the radiation from the plasma, (2) transmission lines including the polarizer splitter unit, which transport the ordinary and extraordinary ECE emission modes separately from the front-end and distributes them to the instrumentation, and (3) detection and analysis instrumentation which is housed at a distance from the tokamak in the diagnostics building. Figure 1 shows the EQ09 port plug, the gas compliant seal, the polarization splitter unit, and the ex-vessel transmission line. Both the polarizer splitter unit and the ex-vessel transmission line are mounted in the Interspace Support Structures (ISS). The front-end optics collects and focuses the ECE radiation from the plasma at D-point, an aperture with diameter 41.75 mm. The emission passing through the aperture is then collected by ellipsoidal mirrors inside the Polarization Splitter Unit (PSU).

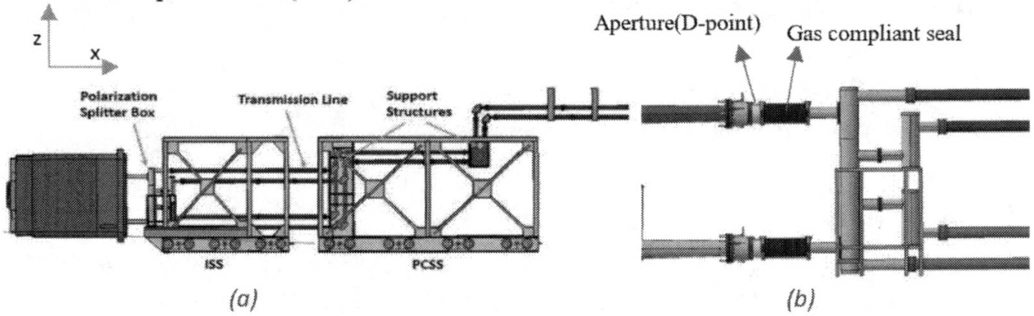


*Figure 1: (a) showing the port plug and Ex-vessel TL including Polarization Splitter Unit, in the ISS and PCSS, (b) showing the gas compliant seal, aperture (D-point) and ex-vessel TL (Polarization Splitter Unit)*

The PSU consists of two Gaussian beam telescopes which consists of three off-axis ellipsoidal mirrors, one flat mirror, and a wire grid polarizer as shown in Figure 2.

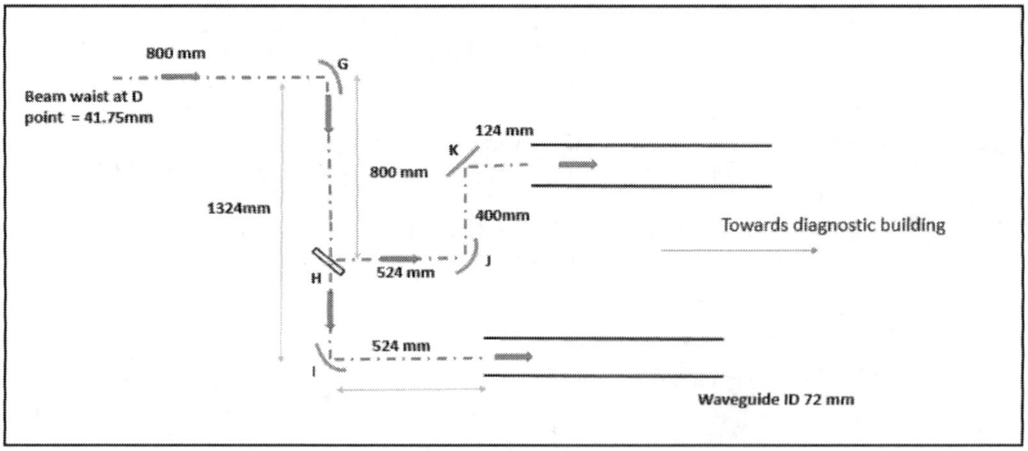


*Figure 2: Optical layout of the Polarization Splitter Unit*

With its high electron temperatures and harsh environment, ITER presents various challenges for the diagnostic system. One of the most insidious is the misalignment between the in-vessel front-end optics and the ex-vessel transmission line which is caused by vibration of the vacuum vessel due to plasma phenomena including vertical displacement events. Also, during operational phase @100°C and baking phase @240°C, there is relative displacement between component on Port plug back flange and interfacing components at D point [2]. Since the electron temperature is inferred from the intensity of the ECE, misalignment may lead to poor accuracy in this critical measurement. These displacements are expected to be ~ 15 mm in vertical (z) and horizontal (x) directions, and ~ 5 mm in the toroidal (y) direction. It is important to minimize the effect of these displacements, so that the system maintains alignment during operation, and reliable temperature information is obtained. An assessment of coupling loss is done for the operational phase, considering above displacement values. Here, the displacements in the operational phase will affect the ECE measurements, so only these displacements have been considered.

## 2 Gaussian Beam Coupling

Our objective is to first study the coupling losses due to imperfect coupling of Gaussian beams owing to port plug displacements. In this paper, different types of Gaussian beam misalignments are discussed and the power loss due to coupling of offset beams is estimated theoretically. Gaussian beam misalignments are mainly due to: (a) Axially aligned beam with offset waist (b) Offset beams (c) Tilted beams, as shown in Figure 3. [3]

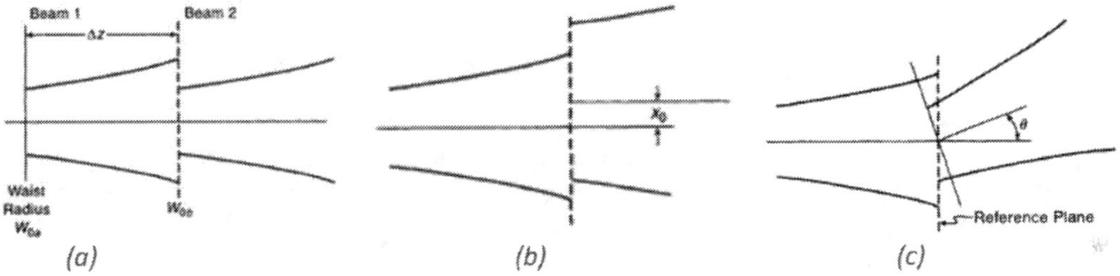


*Figure 3: (a) Axially aligned beam with offset waist, (b) Offset Beams and (c) Tilted Beams*

Here we have considered only the first two cases, as the third case of tilted beam is not applicable.
Axially aligned beam with offset waist: The Gaussian beam is coupled to another beam having the same waist radius, but with a nonzero axial offset.
The coupling loss due to imperfectly coupled beams is given by,

$$K_x = \frac{4}{\left(x + {}^{1}/_{x}\right)^2 + \chi^2 x^2} \qquad (1)$$

Here, $x$ is the ratio of waist radii, given by,

$$x = \frac{w_{0a}}{w_{0b}} \qquad (2)$$

and, $\chi$ is the axial waist offset given by

$$\chi = \left(\frac{\pi\omega_a^2}{\lambda}\right)\left(\frac{1}{R_b} - \frac{1}{R_a}\right) \tag{3}$$

$R$ is the Radius of curvature of the wavefront, given by,

$$R = z + \frac{\left(\pi\omega_0^2/\lambda\right)^2}{z} \tag{4}$$

Offset Beams: Two beams are offset if their axes of propagation are parallel but one is displaced relative to the other. Coupling loss due to offset is given by,

$$K_{offset} = exp\left[-2\left(\frac{x_0}{\delta_{off}}\right)^2\right] \tag{5}$$

with, $$\delta_{off} = \left[\frac{\left(w_{0a}^2+w_{0b}^2\right)^2+\left(\lambda\Delta z/\pi\right)^2}{w_{0a}^2+w_{0b}^2}\right]^2 \tag{6}$$

where $x_0$ is the magnitude of offset and $\delta_{off}$ as the lateral offset distance, at which amplitude coupling is reduced by a factor of $e$.

## 3 Calculation of coupling losses

### 3.1 Axially aligned beams with offset waist

Using equation (1), the coupling loss for axially aligned beams with offset waist is calculated as $K_x = 2.9999 \times 10^{-4}$ dB. Considering the number of times the Gaussian beam couples inside the PSU as 3, the above value is then multiplied by a factor of 3, which gives $K_{xtot} = 0.0009$ dB.

### 3.2 Offset beams

Using equation (5), the coupling loss for offset beams is calculated for the expected vertical displacement of 15 mm, giving $K_{offset}$= 0.43 dB, for f = 115 GHz. This is then multiplied by a factor of 3, which gives $K_{offsettot}$= 1.29 dB. The coupling loss is calculated for the desired frequency range from 110 GHz to 170 GHz, which is the intended range for measurements, and is given in Table 1.

**Table 1.** Theoretical Coupling loss for Offset beams

| Frequency (GHz) | $K_{offset}$ (dB) |
|---|---|
| 110 | 1.27 |
| 115 | 1.29 |
| 120 | 1.32 |
| 125 | 1.34 |
| 130 | 1.36 |
| 135 | 1.38 |

https://doi.org/10.1051/epjconf/202327703011

| | |
|---|---|
| 140 | 1.40 |
| 145 | 1.41 |
| 150 | 1.43 |
| 155 | 1.44 |
| 160 | 1.46 |
| 165 | 1.47 |
| 170 | 1.48 |

From above calculations, it can be seen that the coupling loss is maximum for vertical displacements (case b), and is negligible for radial displacements (case a). Toroidal displacement is negligibly small, and is not significant contributor to the coupling losses.
In order to validate the theoretical values, measurements are done to determine the loss due to vertical displacements.

## 4 Measurement of coupling loss

An experimental mock-up is done to study coupling losses due to vertical displacements, using a prototype polarization splitter unit (PSU). The experimental set-up consists of a D-band source and detector, PSU and a set of corrugated horn antennas for coupling of radiation with the PSU, for input (with source) and output (with detector) gaussian beams, as shown in Figure 4.

*Figure 4: Experimental set-up for measurement of coupling loss*

*Figure 5: Well defined laser spot at both the output ports of PSU after alignment is done*

First, the mirrors inside the PSU are aligned using a diode laser as shown in Figure 5, in order to ensure proper alignment. After alignment is completed, output power is measured at Port 2 (lower port) of the PSU. Experiments are conducted for normal position, when the input beam is perfectly aligned with the PSU, and offset position, when the input beam is 15 mm offset. The ratio of output power measured for normal and offset position gives the coupling loss.
A comparison of theoretical and experimental values for coupling loss is shown in Figure 6. It can be seen from the figure that the experimental and theoretical values match reasonably well. Figure 7 shows the variation of transmission with normalized offset for different

frequencies. It is observed that the transmission decreases with increasing value of normalized offset as expected.

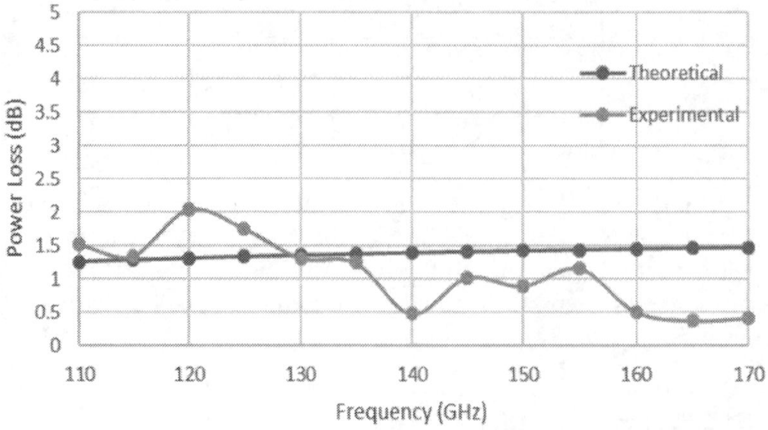


*Figure 6: Comparison of theoretical and experimental values for coupling loss*

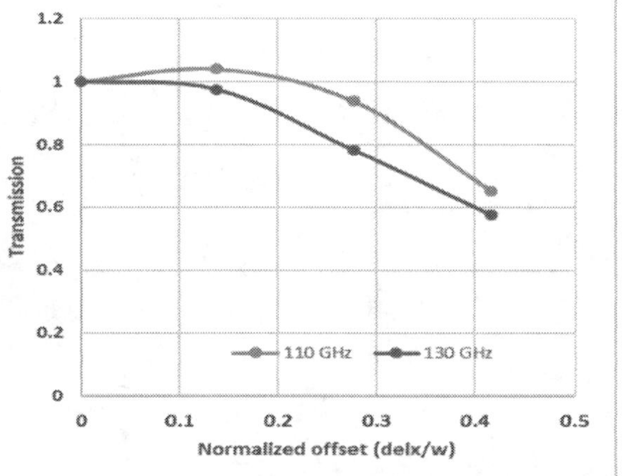


*Figure 7: Transmission vs. normalized offset*

## 5 Summary

It is observed from theoretical and experimental results that coupling loss is high for vertical displacements, with maximum loss ~2 dB at 120 GHz. Based on these assessments, there is a need to offset the displacement against the requirements mentioned above.

During Operation- In order to compensate this displacement of 15 mm in the vertical direction, the PSU will be assembled 15 mm offset (vertically upwards) with the compliant gas seal, during assembly phase at ambient temperature (room temperature). Consequently, at operation temperature @100$^0$C, the front-end optics shall automatically get aligned with the PSU. The deviation expected is ±2 mm which can be addressed by oversized mirrors inside the PSU. However, provision will be made for one-time adjustment after getting the actual window axis position for 100$^0$C.

During Calibration- It is proposed to perform calibration during Restart/Integrated Commissioning, when the machine shall be at operation temperature @ 100$^0$C (without

plasma). The calibration may be performed during short term maintenance phase when the machine will be at operation temperature. The situations will be identical to that in the operational phase (with plasma) and the front-end optics shall remain aligned with the polarization splitter unit. Under these conditions, calibration can be performed reliably and an accurate calibration coefficient can be obtained which can be further used to obtain plasma electron temperature.
Although it seems like at present stage the passive alignment is feasible, to account for other factors related to the port integration and environment, as a matter of the risk mitigation and ALARA, parallel work on alignment mechanism and dedicated tests on the passive alignment are planned and the decision will be made for the Final Design Review.

## 6 Acknowledgements

*Work of WLR and JPZ supported by the U.S. DOE under Contract No. DE-AC02-09CH11466 with Princeton University. All U.S. activities are managed by the U.S. ITER Project Office, hosted by Oak Ridge National Laboratory with partner labs Princeton Plasma Physics Laboratory and Savannah River National Laboratory. The project is being accomplished through a collaboration of DOE Laboratories, universities and industry. The views and opinions expressed herein do not necessarily reflect those of the ITER Organization.*

## 7 References

1. Taylor et al, EPJ Web of Conferences 147, 02003 (2017)
2. Thermal relative displacements of the ITER Vacuum Vessel Port Flanges & Connecting Ducts (VV PHTS temperature effect) IDM UID L9VRZK
3. Goldsmith P F, Quasioptical Systems: Gaussian Beam Quasioptical Propagation and Applications (IEEE Press Series on RF and Microwave Technology), 1998

# Microwave current drive for STEP and MAST Upgrade

*Simon* Freethy[1], *Lorenzo* Figini[2], *Mark* Henderson[1], *Hana* El-Haroun[1], *Bengt* Eliason[4], *Sam* Gibson[1], *Krassimir* Kirov[1], *Alf* Köhn-Seemann[3], *Ivan* Konoplev[1], *Samuli* Saarelma[1], *Ridhima* Sharma[1], *David* Speirs[4], *Roddy* Vann[5], *Helen* Webster[1], *Thomas* Wilson[1] and the STEP team[1]

[1] UKAEA, Culham Science Centre, Abingdon, OX14 3DB, United Kindgom
[2] Istituto per la Scienza e Tecnologia dei Plasmi, Consiglio Nazionale delle Ricerche, 20125 Milano, Italia
[3] University of Stuttgart, IGVP, Pfaffenwaldring 31, 70569 Stuttgart, Germany
[4] University of Strathclyde, 16 Richmond St, Glasgow, G1 1XQ, United Kingdom
[5] University of York, York, YO10 5DD, United Kingdom

**Abstract.** The UK's Spherical Tokamak for Energy Production (STEP) reactor design program has recently taken the decision to use exclusively microwave-based heating and current drive (HCD) actuators for its reactor concepts. This is based on a detailed assessment considering all viable HCD concepts, covering the grid to plasma efficiency, physics applications, technology maturity, integration, maintenance, and costs. Of the two microwave techniques: Electron Cyclotron (EC) and Electron Bernstein Wave (EBW), EC was deemed the lowest risk and EBW is retained as a potential path to a more efficient, higher performing device. To assess the ECCD efficiency, the GRAY beam tracing code was employed to perform detailed scans of the launcher position, toroidal and poloidal launch angle, and frequency over the first 3 cyclotron harmonics. For EBW, GENRAY/CQL3D were used to estimate the CD efficiency, demonstrating promising results. To reduce the physics uncertainties in present models for EBW coupling and current drive, MAST Upgrade will install two dual frequency (28, 34.8 GHz), 900kW, 5s gyrotrons from Kyoto Fusioneering, as part of the MAST Upgrade enhancements package. This will be accompanied by a flexible 2D steering launcher system to allow midplane co- and counter-CD and above midplane launch for co-direction off-axis CD. Coupling efficiency is quantified by measuring the heating induced by reflected (i.e. non-coupled) power to a plate inserted in the reflected beam path. The experiments will also include EBW driven solenoid-free start-up, increasing power and pulse length by a factor of 10 compared to previous MAST experiments. This presentation will discuss the STEP microwave studies and the MAST Upgrade physics design and capabilities.

## 1 Introduction

The STEP (Spherical Tokamak for Energy Production) program is the UK's Spherical Tokamak (ST) reactor design endeavour. In the first tranche (2022-2024), the program aims to produce a concept design for a steady state ST reactor, producing ~100 MWe. As such, the auxiliary current drive system is a critical user of recirculating power. Due to the technological advantages of microwave systems, the present STEP concept uses microwave techniques exclusively for the heating and current drive (HCD) system.

Electron Cyclotron Current Drive (ECCD) will be the primary mechanism for current drive due to the relative maturity and flexibility of the technique. Electron Bernstein Wave Current Drive (EBCD) is adopted as the secondary technique, due to the promise of higher overall current drive efficiency and the ability to operate at high density. Due to the relative uncertainty and reduced flexibility of an EBCD system, two plasma concepts are being developed, one which uses ECCD exclusively and one which uses a combination of EBCD and ECCD, where a balance is found which plays to the relative advantages of the two techniques.

For the EC only concept, the ECCD is responsible not only for the steady state plasma in combination with the bootstrap current, but also for the majority of the plasma ramp-up, as space constraints in the spherical tokamak limit the size of the central solenoid. In the EB+EC concept, ECCD is responsible for the ramp-up as the EB coupling scheme requires a critical density and dictates that the launched beam must follow the magnetic field pitch-angle evolution at the O-mode cut-off. This would require a relatively large steering range and is presently considered too technically challenging. For the steady state phase, where overall efficiency is critical, the EBW is responsible for the majority of the current drive, which is predominantly off-axis. Since the EBW has limited radial access in the STEP concepts to $\rho \gtrsim 0.4 - 0.5$, the required central current drive is provided by ECCD.

EBCD is still relatively immature compared to ECCD. To address the uncertainties of EBCD, MAST Upgrade will install two dual frequency gyrotrons (28 and 34.8 GHz), capable of 900 kW (3s) and 800kW (4.5s). The experiment will measure the EBCD efficiency for the first time on an ST and will provide a facility for testing linear coupling, the effect of density perturbations, non-linear coupling effects as well as solenoid-free EBW start-up experiments. This is paired with a theory and modelling effort for STEP and MAST-Upgrade to develop the physics basis for the use of EBCD on a reactor. Here we describe the system design and the analysis which has set the requirements.

## 2 ECCD optimisation on STEP

As a design principle, we want to minimise the steering required from the ECCD and EBCD launchers. This fact, combined with the fact that microwaves will provide all the auxiliary current through the plasma ramp-up and steady state phases, will in practice mean that a wider range of frequencies will be required on STEP than for a device like ITER in order to reproduce the desired current profile and provide current profile tailoring.

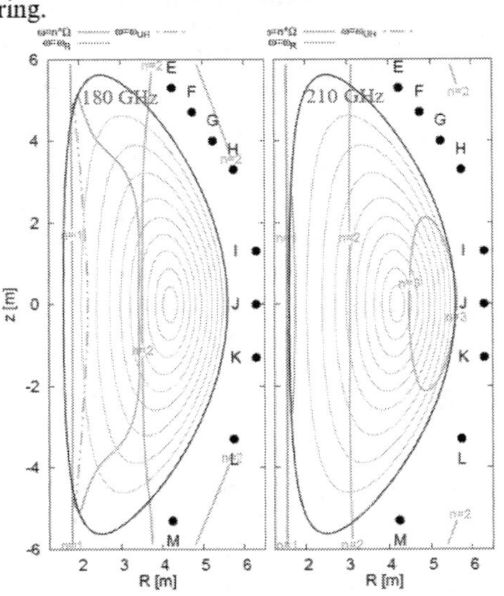


**Fig. 1**: The plasma equilibrium and candidate launch locations for the SPR-045 concept. Harmonics, X-mode cut-off and upper hybrid resonance are shown for 180 GHz (left) and 210 GHz (right)

To optimise the EC configuration for current drive efficiency, a number of candidate launcher positions were selected, as can be seen in Fig. 1. GRAY [1] was then used in single ray configuration to scan the frequency over fundamental, 2[nd] and 3[rd] harmonics (70 – 240 GHz), and toroidal and poloidal angles from each launcher for X and O mode. The result was a rich dataset for the operational space of the ECCD system for that plasma concept [2]. From there, the most efficient configurations were used to fit to the plasma auxiliary current density profile. If the ECCD power assumed by the JETTO simulation is too far from the power required to fit the profile, then another iteration is made.

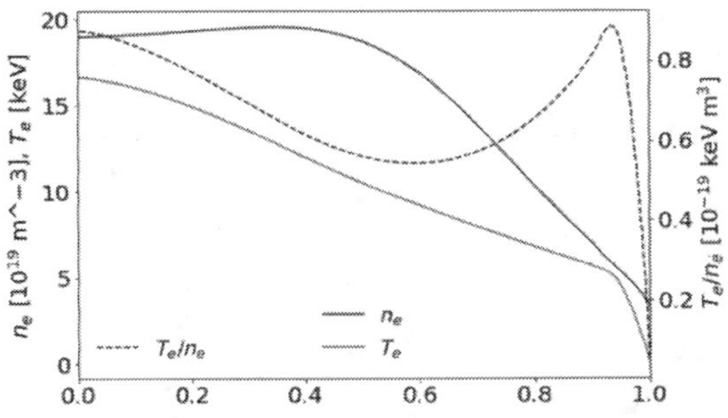


**Fig. 2**: The electron temperature and density profiles for the SPR-045 plasma concept

Fig. 2 shows the density and temperature profiles for SPR-045, the latest ECCD only plasma concept, and Fig. 1 shows the plasma equilibrium and characteristic frequencies for 180 and 210 GHz for this scenario, along with the labelled candidate launcher positions.

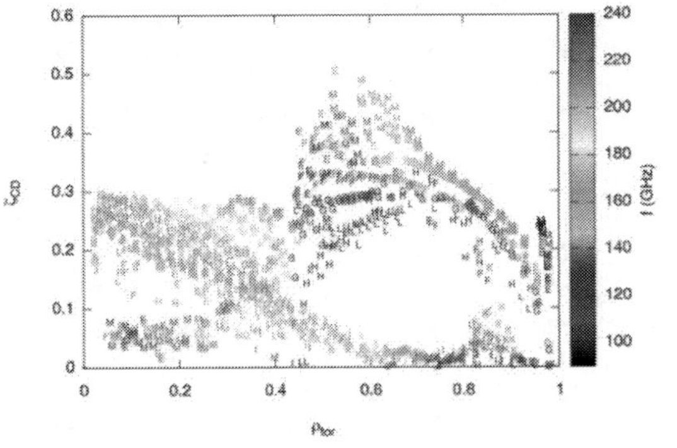


**Fig. 3**: A selection of the configurations and their normalised current drive efficiency. Launch points are represented by letters and the colour denotes the launch frequency

Fig. 3 shows the current drive efficiency at the radial location of the peak of the current density profile. The values are normalised to [3]

$$\zeta = 3.21 \frac{R_m n_{19}}{T_{keV}} \frac{I}{P} \tag{1}$$

This normalisation is useful for comparing the relative efficiency of ECCD and EBCD in different scenarios. The efficiency $\zeta$ is comparable to the optimal values found for DEMO [4], and the most efficient launchers over a wide range of radii are E, F, G and M. For more details of the approach and the resulting shape of this curve, please see [2].

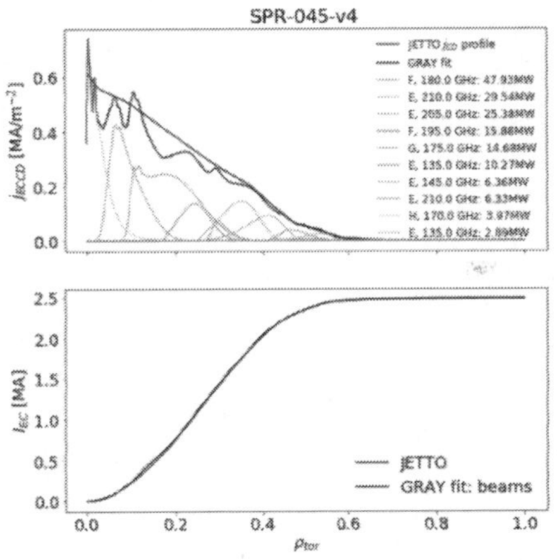


**Fig. 4**: (a) The current density profile from integrated modelling (JETTO) and GRAY fit to that profile. (b) The cumulative integral of the current density for GRAY and JETTO

The current density and cumulative current profiles were fit to the required auxiliary current and current density profiles using a non-negative least squares (nnls) algorithm and the result for this for SPR-045 is shown in Fig. 4. This leads to a realistic estimate for the total EC power required, as well as the power required in each configuration. Configurations requiring more than 1MW in power are listed in the plot legend of Fig. 4. In total in this scenario, 166 MW is required to drive 2.5 MA of auxiliary current, for a total plasma current of 20.5 MA. The bootstrap fraction is calculated self-consistently within JETTO with the assumptions for density and thermal transport. However, since for the ST the transport is not well understood and there is limited empirical data, changes in these assumption could lead to large variations in the recirculating power. The exposure of the design to this risk is less for a more efficiecnt current drive methodology, hence the need to develop EBCD.

Modellling for the plasma ramp-up is not yet complete and will be presented at a later date. However, the ramp-up will take place at much lower density than the steady state phase, resulting in higher real terms efficiency from the ECCD system. Then follows a transition from EC dominated current drive to bootstrap dominated current drive at higher density.

## 3 EBCD on STEP

The STEP EBW concept is still under development, so here we present the analysis approach and some preliminary results of this analysis. A low-field side (LFS) coupling to the EBW via the O-X-B process was chosen, as coupling from the high-field side (HFS), such as was demonstrated in [5], is limited in density by the left-hand X-mode cut-off, presenting too strict a density limitation, and the beam must pass through the fundamental harmonic in order to reach the upper hybrid resonance, where the conversion to EBW occurs. For the plasma parameters envisaged for the STEP flat-top, a high percentage of this power would be absorbed at the fundamental harmonic, making this impractical. Direct X-B tunnelling from the LFS is also not appropriate as the overall coupling efficiency is very sensitive to the density gradient and is only efficient at $k_0 L_n$ values far below what is expected for any of the STEP concepts.

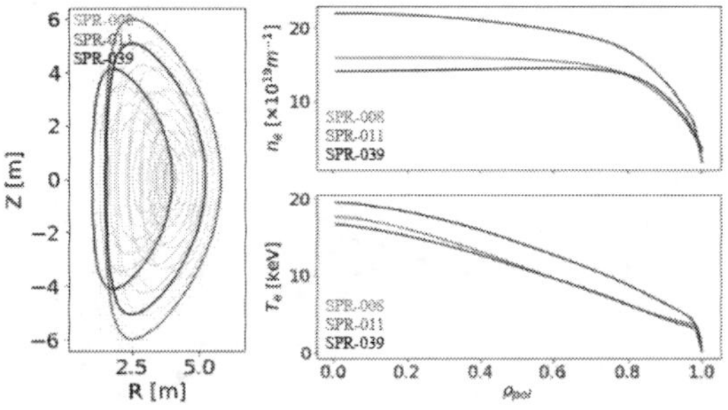


**Fig. 5**: Equilibria, density, and temperature profiles for SPR-008, SPR-011 and SPR-039

Initially we're interested in what the expected current drive efficiency would be for perfect coupling, over what radial range we can expect to drive current, and then whether or not the linear coupling scheme looks feasible for the STEP concepts. More advanced analysis which looks into the potential impact of non-linear effects and density perturbations will be examined later in combination with the MAST Upgrade EBW system.

### 3.1 Penetration and current drive efficiency

For the assessment of current drive efficiency of EBCD in these scenarios, GENRAY [6] was used for the ray-tracing component and CQL3D [7] for the radiative transport and current drive calculations. For each of the concepts, the EBWs are launched between the fundamental and second harmonics as normally access at higher harmonics is not possible as the presence of an O-mode cut-off is required. The launch angles are restricted to the critical angles required for coupling and the height and frequency are scanned over the relevant ranges. Several concepts were analysed and Fig. 5 shows equilibria, density, temperature profiles for three of these. For more a more detailed account of the study, see [8].

Exceptional current drive efficiencies were found, as shown in Fig. 6, with normalised efficiencies up to 1.5, three times higher than the ECCD in a similar region. This makes EBCD a very promising avenue and would significantly lower the re-circulating power requirements for the reactor.

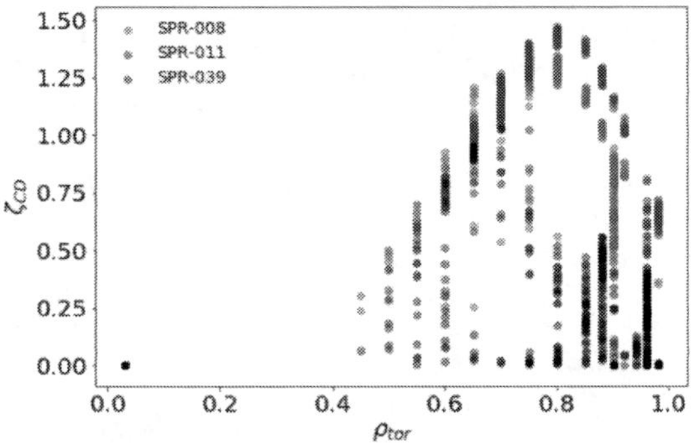


**Fig. 6**: normalised current drive efficiencies for the three different plasma concepts

Access to radii less than $\rho_{tor} \simeq 0.5$ is limited primarily by a combination of Doppler broadening of the resonance and the non-monotonic B-field profile typical of the high-β ST reactor [8]. The radial restriction was approximately the same across the ST concepts we examined. The limited radial range for EBW means that ECCD is required for core current drive to avoid the formation of a current hole in the plasma centre, which could give rise to MHD instabilities. This in turn sets a limitation on both the lowest and highest operating densities. True high-density operation, without ECCD, can only be reasonably achieved by increasing the EBW penetration and this can be done by operating at lower temperature. In principle, a lower temperature, higher density operating point would be preferable, but it remains to be seen if such a scenario can be self-consistently achieved.

### 3.2 Coupling estimates

The EBWs are coupled to from the low field side of the plasma via the O-X-B scheme [9]–[11]. For a plane wave approaching a plasma slab at the critical angle, this conversion is 100%, for increasing deviations from the critical angle, a tunnelling effect determines the rate at which the coupling efficiency falls off. We can estimate this from an analytical formula developed by Mjølhus [12], in combination with ray-tracing. This effect determines not only the steering angle sensitivity, but also the acceptable divergence of a gaussian beam, where finite divergence takes power away from the critical wavevectors into less efficient ones. The beam divergence effect can be quantified by the following [13]:

$$< \tau_{OX} > \simeq \left(1 + 3\frac{\kappa}{z_R} + 2\frac{\kappa^2}{z_R^2}\right) \quad (2)$$

Where $\kappa = \pi L_{n\sqrt{\omega_c/2\omega}}$, $z_R = k_0 w_0^2/2$ and $w_0$ is the beam waist radius. It is possible for back-tunnelling to occur between the slow and fast X-mode [14], limiting the efficiency of the coupling at very high density gradients. This is not a concern for these plasmas however as the density gradients are too shallow to allow this to happen.

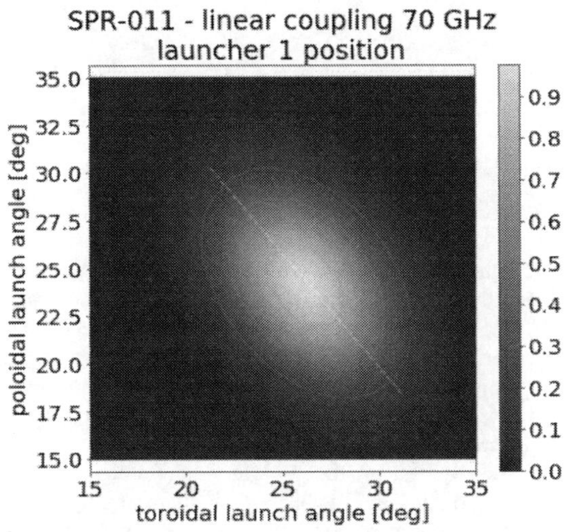


**Fig. 7**: An example of the coupling window for SPR-011 at 70 GHz, for a launching mirror placed 1 m from the plasma surface at the mid-plane. Calculations are for a single ray.

The EBW plasma scenario is still under development, but we present here some example calculations of SPR-011 at 70 GHz, which is the lowest frequency considered for this scenario and is likely a best case for EBW coupling. Fig. 7 shows a coupling window for a hypothetical launcher placed 1 m away from the plasma surface at the plasma midplane. There is a $\pm 1^o$ tolerance for 95% coupling, which is well inside present launcher capabilities. The density scale length, $L_n = n/\nabla n$, is estimated from a density profile which is calculated by EUROPED [15] to be consistent with the peeling ballooning stability limit for this plasma. Fig. 8, shows the coupling efficiency as a function of the increasing beam waist (decreasing divergence) at 70 GHz. We consider the density scale length uncertain at this stage, so we also show the performance for ½ and 2 times the nominal gradient estimated from EUROPED. We can see for this scenario that a 15 cm beam is sufficient to keep the coupling above 99% and larger beams mitigate the uncertainty in $L_n$ to some degree.

In reality, this calculation misses the plasma curvature effects and the variation of the polarisation across the beam. These effects will limit the maximum conversion efficiency and will likely cause the curves in Fig. 8 to roll over. In order to estimate this effect, as well as the effects of density perturbations, we plan on performing 3D full wave simulations.

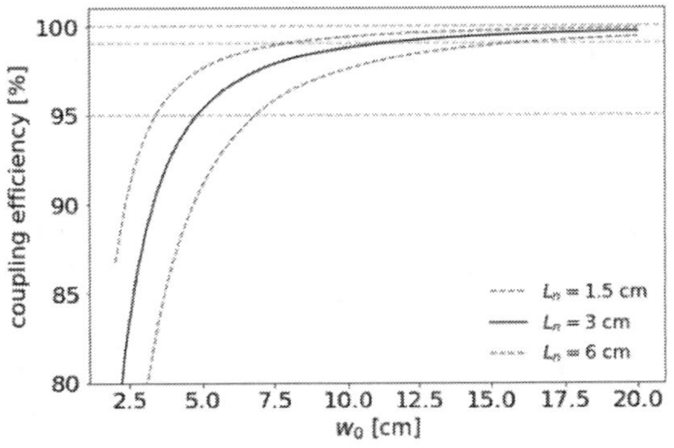


**Fig. 8**: The transmission coefficient as a function of the beam waist size. Larger beams have lower divergence and thus more efficient coupling.

## 4 EBCD system on MAST Upgrade

There are many physics challenges to be addressed before EBCD can become the default choice for current drive for STEP. Although there has been experimental evidence of efficient EBCD of $\zeta_{CD} \simeq 0.4$ on the stellarator [11], a more extensive validation of the current drive models for EBWs on STs is required. Furthermore, there has never been an experimental validation for the predictions of highly efficient current drive on STs of $\zeta_{CD} > 1$ [13], [16], which are also seen in our calculations for STEP. It has been shown that high-efficiency coupling from the LFS to EBW can be achieved [17], [18], but the effect of density perturbations and non-linear effects require more extensive investigation.

To provide an experimental basis for the use of EBCD on the spherical tokamak, an EBCD system will be installed on MAST Upgrade. The system consists of two dual-frequency gyrotrons (28/34.8 GHz), each capable of 900kW for 3 seconds or 800kW for 4.5 seconds. The aim of this system is to: provide an experimental test of the EBCD efficiency on an ST, examine open issues regarding the LFS coupling scheme, (such as the linear coupling behaviour and turbulence [19]), examine collisional damping and non-linear effects [20] and extend the original MAST experiments into EBW based solenoid-free start-up. An engineering design of the system can also be found at this workshop [21].

### 4.1 Coupling efficiency

In order to operate an EBW system on a modern experimental tokamak, it is necessary to show that the power can be efficiently coupled to the plasma and that the stray radiation can be kept to a reasonable minimum.

In order to calculate the coupling efficiency for MAST Upgrade, we need to account for the possibility that the slow X-mode, which would normally convert 100% to the EBW, instead tunnels out of the plasma to the fast X-mode branch. This is different to STEP as the MAST Upgrade scale lengths are smaller and the frequency is lower. The efficiency of this process is given by [14], [22]:

$$\tau_{XX} = e^{-\pi\eta} \qquad (3)$$

$$\eta = \left(1 - N_{\parallel}^2\right)^{3/2} \left[\frac{\omega_c L_n}{c\alpha}\left(\sqrt{(1+\alpha^2)} - 1\right)^{1/2}\right]_{UHR} \qquad (4)$$

where $\alpha = \omega_p/\omega_c$, and the term in the square brackets is evaluated at the upper hybrid resonance. We also note the use of a square root for the $\alpha$ term in curved brackets, rather than the square found in [22]. The total reflectivity is then given by

$$R_{tot} = 1 - R_{XX} < \tau_{OX} >, \qquad (5)$$

where $R_{XX} = 1 - \tau_{XX}$.

For this study, three separate full wave codes were employed, FFW, IPF-FDMC and EMIT-2D [19]. Initially, calculations were done in the plasma slab at 28 GHz. For the slab, $k_0 L_n = 25$, f = 28GHz, $B_0$ = 0.85 T and is directed parallel to the density contours.

As a numerical check of equation (2), the beam waist size was scanned and the reflectivity was calculated with IPF-FDMC, the results of which are shown in Fig. 9. As can be seen from the Figure, the reflectivity from the full wave simulations is lower than predicted by the analytical result and it is suspected that this is due to the 2D nature of the full-wave simulations. This result however confirms the design choice to minimise the divergence of the launched O-mode. A scan of the phase front curvature at fixed divergence was also carried out, by altering the position of the beam focus, but only a negligible effect was seen in the coupling efficiency.

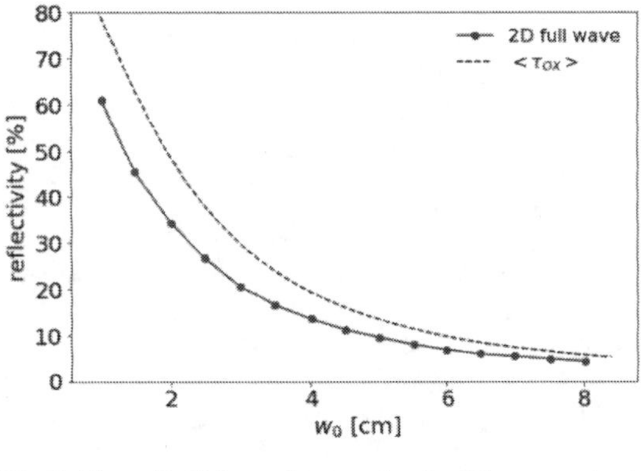


**Fig. 9**: The reflectivity vs beam waist size. The approximate formula in Equation (1) is plotted for comparison

To guide the frequency selection, we then scanned $L_n$ and calculated the coupling efficiency. Fig. 10 shows the reflectivity from the full wave simulations, $1 - \tau_{OX}$, for this scan in blue. The green solid line is the reflectivity from Equation 2, the red dashed line indicates the transmission coefficient for the X-mode tunnelling, given in Equation 3 and the solid line indicates $R_{tot}$. It can be seen from the figure that there is a minimum in reflectivity for $k_0 L_n \simeq 5$, as the tunnelling from the slow to fast X-mode begins to dominate the reflectivity, and this corresponds well with the analytical estimate shown. We see good agreement between the analytical and full wave calculations on the position of the minimum, however the full wave simulations underestimate the reflectivity at higher $k_0 L_n$ due to their 2D nature. 3D full wave simulations are in progress to check the analytical results presented here.

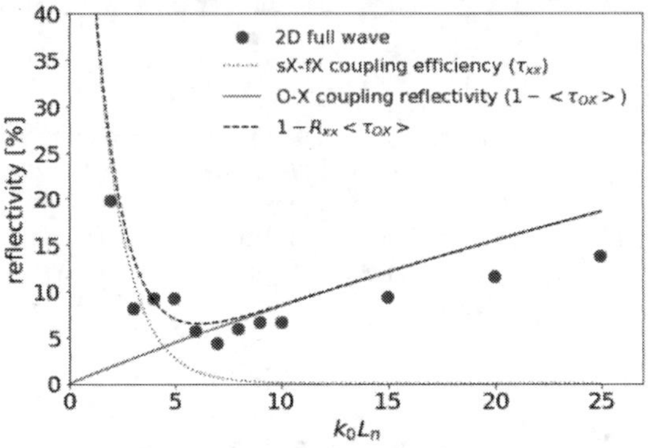


**Fig. 10**: The reflectivity as a function of density scale length. For small $L_n$, the reflectivity increases due to back-tunnelling of the slow X-mode to the fast X-mode

As realistic edge density gradients were not available for MAST Upgrade at the time the analysis was performed, example MAST shots were used. Fig. 11 left shows three example density profiles for H-mode, and low and high-density L-modes. The three dashed lines represent the cut-off densities for a fundamental harmonic (17.25 GHz), a second harmonic (35 GHz) and an intermediate frequency (28 GHz). As you can see from the figure, the fundamental harmonic causes a mode conversion at the bottom of the steep gradient region, and it is ambiguous as to whether this conversion would take place in the scrape off layer. Aside from an unpredictable gradient, the role of density perturbations in this region is likely much greater than for higher frequencies. Thus, it is preferred to have a second harmonic system.

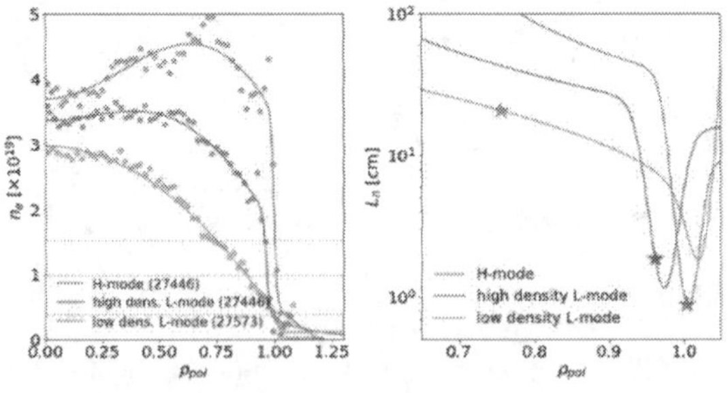


**Fig. 11**: Left: density profiles for H-mode, high-density L-mode and low-density L-mode. Dashed lines indicate cut-off densities for 17.25, 28 and 34.8 GHz. Right: density scale lengths for these profiles. Stars represent scale lengths at 34.8 GHz in each case.

The right panel of Fig. 11 shows the gradient scale lengths for the example profiles and the stars denote the scale lengths for 34.8 GHz for the three examples. Fig. 12 shows a similar figure to Fig. 10, but for MAST parameters for 34.8 GHz, $w_0 = 5.5$ cm. One can see that the high-density L-mode and H-mode are either side of the minimum in reflectivity, where the low-density L-mode shows a reflectivity of around 25%. Fig. 12 also shows that the steep gradients in H-mode may increase the reflectivity, necessitating an approach similar to that outlined in [23].

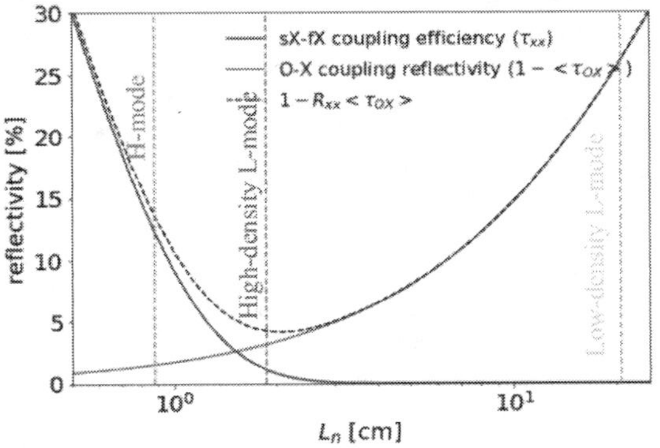


**Fig. 12**: The analytical reflectivity as a function of scale length for MAST Upgrade parameters. Dashed vertical lines represent the expected scale lengths at the cut-off position for 34.8 GHz in H-mode, high-density, and low-density L-modes.

To catch the reflected beams, the system is designed with "interceptor plates". These are graphite plates placed in the path of the reflected beams and will be designed to scatter the reflected beam over a wide angular range, thus reducing the power density of any stray radiation in the vessel. Fig. 13 shows the approximate positions of these plates for the midplane launchers (similar plates are designed for the upper launchers) and the incident and reflected beams are shown at the optimal coupling angle for a plasma current of 800kA and 2000kA. The plates are also designed to work over the full range of plasma radii anticipated. The plates will be monitored with infrared cameras to measure the heating and this will be the primary diagnostic of reflected power (and thus coupling efficiency) for both machine protection and scientific purposes.

## 4.2 Current drive efficiency

In order to set the power requirements for the system, a critical factor was that a robustly measurable current could be driven. Initially the target was to be able to drive 10% of the plasma current and this was later supported with integrated modelling and forward modelling of the Motional Stark Effect diagnostic to find the ultimate sensitivity limitations. GENRAY and CQL3D were again used to predict the current drive efficiency. These predictions were performed before MAST Upgrade first plasma, so TRANSP [24] predictive design points were used. Fig. 14 shows the density and temperature profiles as well as the characteristic frequencies at the plasma midplane.

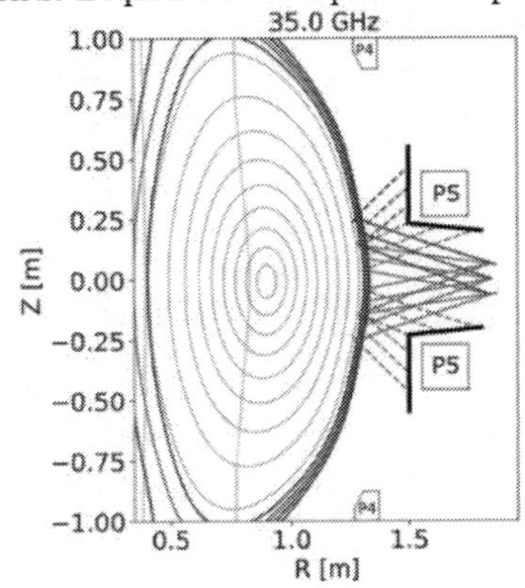


**Fig. 13**: A schematic plot of the poloidal cross-section of the beams from the midplane launcher for $I_p$ = 800kA (green) and $I_p$ = 2000 kA (blue). The dashed lines represent the beam reflected from the plasma after coupling. Thick black lines represent "interceptor plates".

A second harmonic system was preferred as the coupling is far more reliable at these frequencies, as shown in Section 4.1. From a launcher positioned on the machine midplane, it is possible to take advantage of a symmetry to drive co-current and counter-current at the same radial location, by coupling power into either the upper or lower coupling window. With two independent midplane launchers, one can drive any combination of co-current, counter-current, or balanced current (heating), greatly increasing the experimental flexibility of the system to disentangle heating and current drive effects. Furthermore, to drive current far off-axis without changing the gyrotron frequency, a pair of launchers are placed above the midplane (70 cm). For a detailed presentation of the launcher design please see [21].

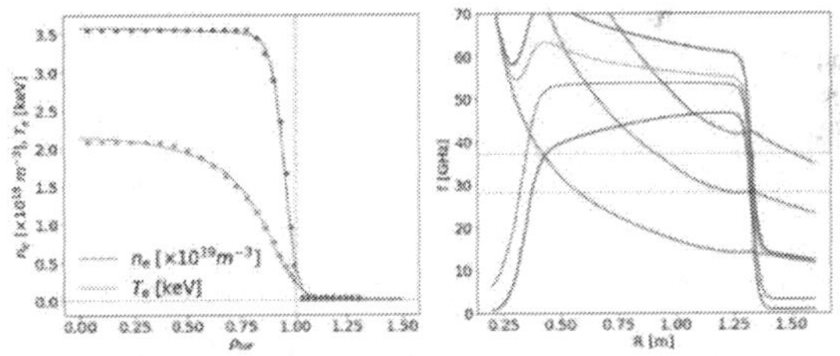


**Fig. 14**: Left: density and temperature profiles for MAST Upgrade TRANSP prediction C1a. Right: the characteristic frequencies of this scenario.

Current drive efficiencies between 0.1 A/W ($\zeta$ = 0.43) and 0.14 A/W ($\zeta$ = 0.6), shown in Fig. 15, were found close to the plasma axis, with the difference only being the plasma radius. EBW closer to the plasma centre appeared to gain a boost in CD efficiency. Since the plasma current in these scenarios is 1.2 MA, the requirement is then to have at least 1.2 MW of power coupled into the plasma as EBW.

## 5 Solenoid free start-up

It was shown on MAST that EBW, launched from the high field side, could provide highly efficient solenoid-free plasma start-up [5], [25]. This is of great interest to the STEP programme as the solenoid in the STEP concepts is very limited in size and can only produce around 1MA at the plasma start-up. Any assistance that the heating system can provide would greatly assist the start-up phase. The MAST Upgrade experiments plan to increase both the power and time of the microwave radiation by up to a factor of ten, to further explore this novel technique. Similar calculations were performed for the off-midplane launcher, finding lower normalised efficiency of $\zeta = 0.3$, shown in Fig. 16.

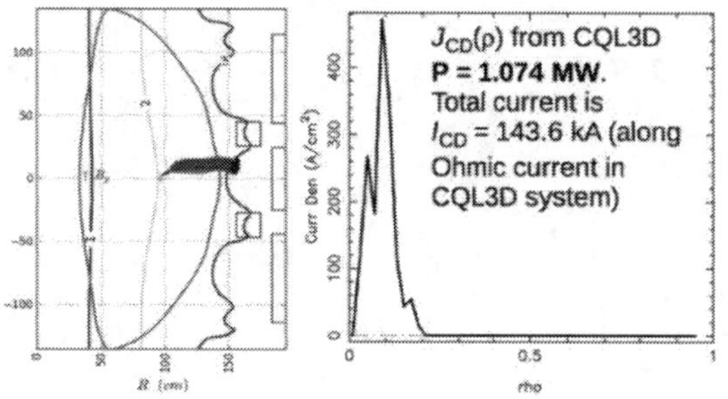


**Fig. 15**: Left: O-mode and EBW rays at 37 GHz. Right: the current density as calculated by CQL3D. The efficiency is 0.14 A/W and the corresponding $\zeta$ is 0.63

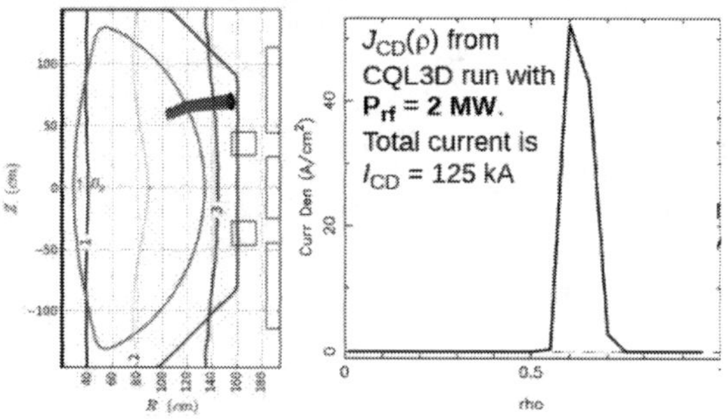


**Fig. 16**: off axis efficiency from a 70 cm high launcher. The efficiency is 0.063 A/W and the corresponding $\zeta$ is 0.3

## 6 Summary

In summary, the STEP spherical tokamak reactor will use microwave techniques for all heating and current drive requirements. ECCD is anticipated to be the primary source of auxiliary current drive due to its maturity and flexibility. However, EBCD is a planned opportunity due to its promise of high current drive efficiency at high density. GRAY was used to optimise the launch configuration for SPR scenarios finding efficiencies comparable to similar optimisations for ITER and EU-DEMO. EBCD was analysed using GENRAY and CQL3D and current drive efficiencies of around 3 times higher than ECCD were found, but this is limited to off-axis current drive as the penetration is limited by Doppler broadening and relativistic effects. The coupling to EBW was analysed using known analytical expressions, and the coupling was found to require large diameter beams to maintain high efficiency. The role of non-linearities and density perturbations remains to be investigated.

To provide a firm experimental basis for the use of EBCD, and solenoid free start-up for STEP, a 1.8 MW EBCD system will be installed on MAST Upgrade. The system is designed to give co-, counter- and balanced current from the midplane launcher, and could achieve greater than 10% of the total plasma current based on efficiency estimates using GENRAY/CQL3D. A launcher above the midplane will give off-axis deposition at the same frequency and magnetic field. A high coupling efficiency is expected, >96%, and interceptor plates are placed in the path of the reflected beam to decrease the power density of the stray radiation and serve as a measure of the coupling efficiency.

## References

[1] D. Farina, "A Quasi-Optical Beam-Tracing Code for Electron Cyclotron Absorption and Current Drive: GRAY," *Fusion Sci. Technol.*, vol. 52, no. 2, pp. 154–160, Aug. 2007.

[2] L. Figini, "Electron Cyclotron Current Drive Efficiency in the STEP Device," *EPJ Web Conf.*, p. under review, 2022.

[3] T. C. Luce *et al.*, "Generation of Localized Noninductive Current by Electron Cyclotron Waves on the DIII-D Tokamak," *Phys. Rev. Lett.*, vol. 83, no. 22, pp. 4550–4553, Nov. 1999.

[4] E. Poli *et al.*, "Electron-cyclotron-current-drive efficiency in DEMO plasmas," *Nucl. Fusion*, vol. 53, no. 1, p. 013011, Jan. 2013.

[5] V. Shevchenko, Y. Baranov, M. O'Brien, and A. Saveliev, "Generation of Noninductive Current by Electron-Bernstein Waves on the COMPASS-D Tokamak," *Phys. Rev. Lett.*, vol. 89, no. 26, p. 265005, Dec. 2002.

[6] A. P. Smirnov and R. W. Harvey, "The GENRAY Ray Tracing Code," 2003. [Online]. Available: https://www.compxco.com/Genray_manual.pdf . [Accessed: 01-Jul-2022].

[7] R. . Harvey and M. . McCoy, "The CQL3D Fokker-Planck Code," 2015. [Online]. Available: https://www.compxco.com/cql3d_manual.pdf. [Accessed: 01-Jul-2022].

[8] T. Wilson, "Electron Bernstein Wave (EBW) current drive profiles and efficiency for STEP," *EPJ Web Conf.*, p. under review, 2022.

[9] J. Preinhaelter and V. Kopecký, "Penetration of high-frequency waves into a weakly inhomogeneous magnetized plasma at oblique incidence and their transformation to Bernstein modes," *J. Plasma Phys.*, vol. 10, no. 1, pp. 1–12, Aug. 1973.

[10] A. Mueck *et al.*, "Demonstration of Electron-Bernstein-Wave Heating in a Tokamak via O–X–B Double-Mode Conversion," *Phys. Rev. Lett.*, vol. 98, no. 17, p. 175004, Apr.

2007.
[11] H. P. Laqua, H. Maassberg, N. B. Marushchenko, F. Volpe, A. Weller, and W. Kasparek, "Electron-Bernstein-Wave Current Drive in an Overdense Plasma at the Wendelstein 7-AS Stellarator," *Phys. Rev. Lett.*, vol. 90, no. 7, p. 075003, Feb. 2003.
[12] E. Mjølhus, "Coupling to Z mode near critical angle," *J. Plasma Phys.*, vol. 31, no. 1, pp. 7–28, Feb. 1984.
[13] J. Urban *et al.*, "A survey of electron Bernstein wave heating and current drive potential for spherical tokamaks," *Nucl. Fusion*, vol. 51, no. 8, p. 083050, Aug. 2011.
[14] A. K. Ram and S. D. Schultz, "Excitation, propagation, and damping of electron Bernstein waves in tokamaks," *Phys. Plasmas*, vol. 7, no. 10, p. 4084, 2000.
[15] S. Saarelma *et al.*, "Self-consistent pedestal prediction for JET-ILW in preparation of the DT campaign," *Phys. Plasmas*, vol. 26, no. 7, p. 072501, Jul. 2019.
[16] G. Taylor *et al.*, "Efficient generation of noninductive, off-axis, Ohkawa current, driven by electron Bernstein waves in high β, spherical torus plasmas," *Phys. Plasmas*, vol. 11, no. 10, pp. 4733–4739, Oct. 2004.
[17] G. Taylor *et al.*, "Efficient coupling of thermal electron Bernstein waves to the ordinary electromagnetic mode on the National Spherical Torus Experiment," *Phys. Plasmas*, vol. 12, no. 5, p. 052511, May 2005.
[18] H. P. Laqua, V. Erckmann, H. J. Hartfuß, and H. Laqua, "Resonant and Nonresonant Electron Cyclotron Heating at Densities above the Plasma Cutoff by O-X-B Mode Conversion at the W7-As Stellarator," *Phys. Rev. Lett.*, vol. 78, no. 18, pp. 3467–3470, May 1997.
[19] A. Koehn-Seemann, "Numerical investigations of the O-X mode conversion process in MAST Upgrade," *EPJ Web Conf.*, p. under review, 2022.
[20] M. Senstius, "Nonlinear degradation of O-X-B in MAST Upgrade," *EPJ Web Conf.*, p. under review, 2022.
[21] H. Webster, "MAST Upgrade Microwave Heating and Current Drive System - Engineering Design Overview," *EPJ Web Conf.*, p. under review, 2022.
[22] A. . Ram and A. Bers, "Excitation and emission of electron cyclotron waves in spherical tori," *Nucl. Fusion*, vol. 43, no. 11, pp. 1305–1312, Nov. 2003.
[23] H. Igami, M. Uchida, H. Tanaka, and T. Maekawa, "Polarization adjustment of incident electromagnetic waves for optimal mode-conversion to electron Bernstein waves," *Plasma Phys. Control. Fusion*, vol. 46, no. 1, pp. 261–275, Jan. 2004.
[24] "TRANSP users pages." [Online]. Available: https://transp.pppl.gov/index.html.
[25] V. F. Shevchenko *et al.*, "Long Pulse EBW Start-up Experiments in MAST," *EPJ Web Conf.*, vol. 87, p. 02007, Mar. 2015.

# ITER ECH&CD Control System: architecture, interfaces and status of development.

*Giuseppe* Carannante[1*], *Mario* Cavinato[2], *Katarina* Cindric[2], *Peter* De Vries[1], *Federico* Felici[3], *Martino Giordano* Ferrari[1], *Giuseppe* Ferrò[1], *Mark* Henderson[4], *André* Neto[2], *Melanie* Preynas[1], *Matthias* Reich[5], *Filippo* Sartori[2] and *Luca* Zabeo[1].

[1] ITER Organization, Saint-Paul Lez Durance, France
[2] Fusion for Energy, Josep Pla 2, Barcelona, 08019, Spain
[3] Swiss Plasma Center - EPFL, Lausanne, Switzerland
[4] UKAEA, Culham Science Centre, Abingdon, OX14 3DB, United Kingdom
[5] Max-Planck-Institut für Plasmaphysik, Garching, Germany

**Abstract.** The ITER ECH&CD system is designed to inject 20 MW of millimetre-wave at 170 GHz into the vacuum vessel. The system is composed of many sub-systems, namely High-Voltage Power Supplies (HVPS), Gyrotrons, Transmission Lines (TL), Ex-vessel Waveguides (EW), Launchers. It is the role of the EC Plant Controller (ECPC) to integrate all the Sub-system Control Units (SCU), to prepare the system for operation and to execute the real-time requests coming from the plasma control system. The ECPC also implements plant level protection functions involving more than one sub-system and it interfaces with the ITER Central I&C. This paper gives an overview of the EC system and a description of the control system development focusing on the architecture and the interfaces. Control and protection functions are presented together with a functional allocation to better define interfaces and responsibilities. The preliminary design of the interface with the Plasma Control System to implement advanced control functions is also presented.

## 1 Introduction

The ITER Electron Cyclotron Heating and Current Drive (EC H&CD) system is being designed to supply up to 20MW of power at 170 GHz. This power is generated from a set of 24 RF Sources (or Gyrotrons) and the associated High Voltage Power Supplies (HVPS) and transmitted to the plasma via Transmission Lines (TL), Ex-Vessel Waveguides (EW) and two different types of Launcher: one Equatorial (EL) and four Upper Launchers (UL), as shown in **Fig. 1**. HVPSs and RF sources are placed in Building B15 (or RF Building); transmission lines run along the walls of Building B13 and reach 5 different openings in Building B11 to arrive to the 5 launchers (see **Fig. 1** for more details).

The HVPS and RF Sources are grouped into 12 modular sets, such that each system can be operated independently of the status of the other modules.

Inline switches in the TL allow the power generated by a single gyrotron to be directed toward different launchers, as explained in Section 2. This allows the EC system to be more flexible and to implement different control functions.

The EC system is used for a variety of heating and current drive (HCD) applications and it requires a complex control system, consisting of a Plant Controller (ECPC) and a series of Subsystem Control Units (SCUs). The control system is used to monitor the EC system status and drive the actuators (power, steering mirrors, switches etc.) following the requests coming from the Plasma Control System. The ECPC is also in charge of plant level protection functions and of interfacing with the ITER Central I&C.

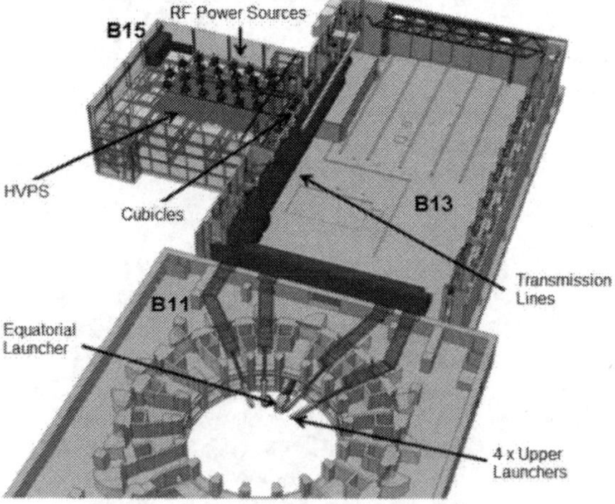


**Fig. 1.** ITER ECH&CD plant layout.

As it can be seen in **Fig. 2**, the procurement strategy of the EC system is quite complex and involves 5 different domestic agencies. To make the integration easier, the ECPC was introduced as a coordinator of the entire plant. Clear interfaces and responsibilities were

* Corresponding author: giuseppe.carannante@iter.org

agreed upon during the system conceptual design review in 2013 [1] and further detailed during periodical workshops.

The installation phase for the first subsystems started in the beginning of 2022 and the first commissioning is expected to start by the fall of the same year. The installation and commissioning will continue for several years, in which the plant will be in a number of different configurations. The ECPC needs to cope with this complication and to allow parallel operation of different subsystems at the same time by implementing the required functionalities and ensuring a safe operation. This is achieved by means of operation modes and tasks, as it will be explained in the next sections.

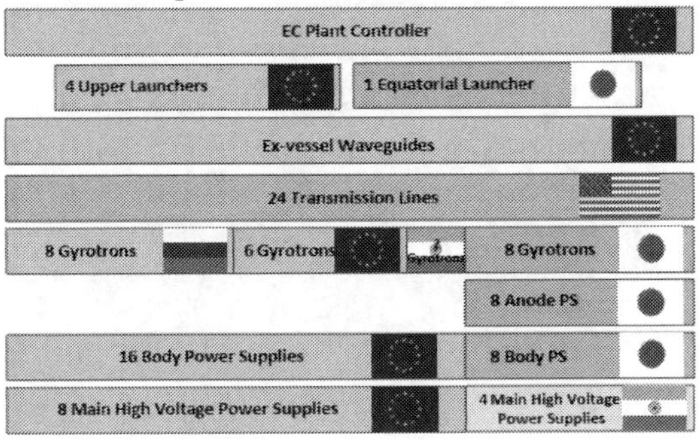


**Fig. 2.** EC system procurement strategy.

## 2 System layout

The EC system is quite distributed and its operation is rather complex because of the different functions that can be performed. To allow a higher flexibility in the implementation of the plasma control functions, the generated power can be redirected toward different launchers in real-time during the pulse.

In particular, 24 transmission lines arrive to the EL and 8 lines arrive to each of the ULs. This means that the transmission lines have 24 entries on the gyrotron side and 56 entries on the launcher side. To allow moving the power from one launcher to another, inline switches are placed in the TLs. While all the ULs have 8 input transmission lines, only 4 gyrotrons can be connected to UL 12 and UL 13 and 8 gyrotrons can be connected to UL 15 and UL 16, more details can be seen in **Fig. 3** for the UL 12 and for the UL 15.

Moving the power from one launcher to another requires a series of activities. The RF generation needs to be paused during the movement, this is achieved by switching off the HVPS. This operation may require changing the mirrors position, the polarisers, the TL switches, the isolation valves and the isolation shutter valves. The entire switching sequence is performed in less than 3 seconds.

The EC operation is carried out from the ITER control room; for local commissioning or special operation requiring the operators to be close to the EC equipment, a temporary control room is installed in the RF building.

### 2.1 Confinement barriers and safety functions

From the vacuum point of view, the EC system can be separated in 3 different sections, namely the Gyrotrons, the TL and the Launchers.

The gyrotron is kept under vacuum by internal ion pumps controlled by the gyrotron SCU.

The output window of the gyrotron is connected via a Matching Optic Unit (MOU) to a TL that is comprised of many different components, as shown in **Fig. 4**, and goes up to the other vacuum boundary represented by the diamond window in B11. This section is kept under vacuum by a dedicated loop implemented by the ITER Vacuum System.

The EW and the Launchers are then connected to the vacuum vessel.

The diamond window is the boundary with the EW, the continuation of the TL in the B11 toward the launchers. The diamond window is a first confinement barrier and it is protected by an isolation valve that is closed in case of potential loss of vacuum. This valve is controlled by the Vacuum System and the Central Safety System (CSS) and requests to open it or close it can be issued by the ECPC only for operation or maintenance purposes.

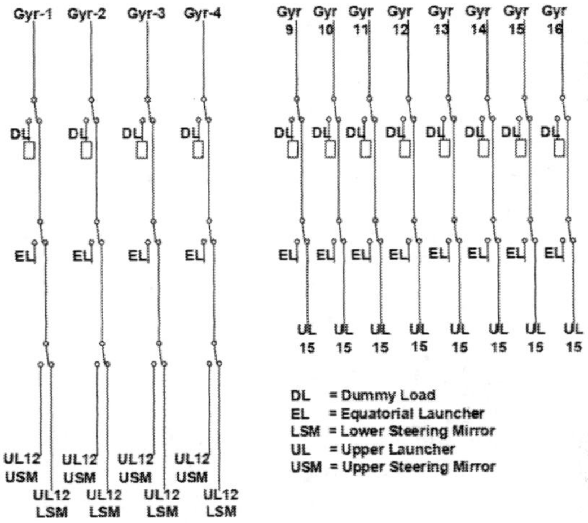


**Fig. 3.** Transmission Lines inline switches configuration for UL in Port12 and UL in Port15.

Additional confinement barriers are present at the Port Cell and the Gallery boundaries and are protected by isolation shutter valves (see **Fig. 4**). These failsafe pneumatic valves are controlled by the EC control system and by the CSS via two solenoid valves in series on the same compressed air circuit. The safety actuation is always performed by the CSS.

Every time the CSS closes one of the isolation valves, it also sends a request to the ECPC to trip all the gyrotrons connected to that line, in order to reduce the risk of damage of the gyrotron itself, the diamond window and other RF components. This action is not classified as a safety function but as an investment protection.

Additional safety functions are defined to isolate the cooling water and gas lines crossing the safety

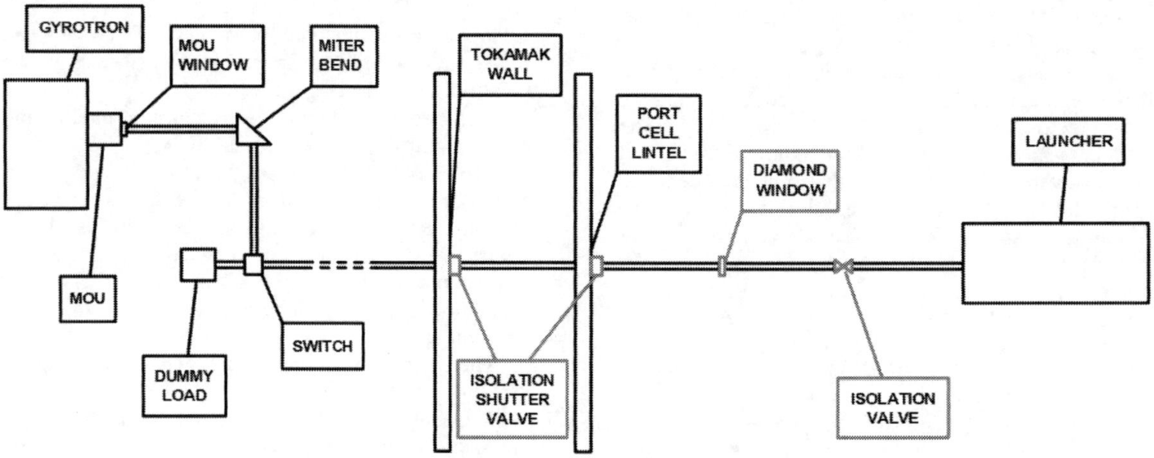


**Fig. 4.** Simplified layout of a single Transmission Line.

boundaries, to detect loss of cooling of the EW components and to detect arcs in the diamond window.

Occupational safety functions are also identified, related to electrical and radiological risks. The access to high voltage area is controlled in the HVPS with both passive and active protection functions. To protect against X-rays produced by the gyrotrons during operation, the entire gyrotron area is fenced and the access is controlled by the Plant Safety System. Specific procedures are put in place to avoid the operation if personnel is present inside the area.

Since the implementation of these safety functions has been transferred to the Central Safety System and it is not in the scope of the EC Control System, the details of the functions and their implementation goes beyond the scope of this paper.

## 3 ITER Central I&C

The ITER Central I&C is composed of different layers of control and protection systems:

- The Central Safety System (CSS): Provides plant-wide nuclear (CSS-N) and occupational safety (CSS-OS) functions.
- The Central Interlock System (CIS): Provides plant-wide investment protection functions.
- CODAC System: Provides overall plant systems coordination, supervision, plant status monitoring, alarm handling, data archiving, plant visualization (HMI) and remote experiment functions.
- CODAC networks: A set of networks providing the physical and logical interconnection between CODAC System and plant systems I&C.

The functions of different CODAC networks include distribution of configurations, status, commands and events via the Plant Operating Network (PON), distribution of timing via the Time Communication Network (TCN), fast real-time communication via the Synchronous Databus Network (SDN), data storage functionalities via the Data Archiving Network (DAN).

The Plasma Control System is implemented in the same CODAC infrastructure and has a direct interface to the ECPC via SDN.

## 4 Staged approach

A staged approach has been introduced in 2016 with the re-baselining exercise. It introduced the different operation phases and the requirements for each phase of the plant systems.

The ITER Staged Approach has three non-nuclear operating phases prior to DT operation. The first phase (First Plasma) will include 7 MW of installed EC power coming from a set of 4 HVPSs, 7 gyrotrons, 7 TLs and 1 UL in Port 16. Due to the efficiency of the various components involved in the transmission, the power delivered to the plasma is 5.8 MW.

The first plasma will be carried out with polarizers and launcher mirrors in a fixed position. The total EC on-pulse length for each FP pulse shall be less or equal to 300 ms. This requirement was set for the design of the mirror of the First Plasma Protection Components and it is now a constraint for the first plasma. This pulse length of 300 ms is the total integrated time for the EC to be on in each ITER pulse; it can be delivered in a single time window (of 300 ms) but the EC power can also be modulated over the entire FP pulse, lasting up to 1 s. Only the UL in Upper Port 16 will be used and all the 7 gyrotrons will be connected to it. The cooling water system for the TL in B13, EW and UL will not be functional.

These reduced functionalities make the control system for first plasma much easier, although the current architecture and interfaces are already compliant with more complex functions.

The following phase will be the Pre-Fusion Plasma Operation-1 and will have the entire EC system installed and functional and the operation will be extended to 100 seconds. Even if some of the advanced PCS control functions, like NTM control, will not be needed in this phase, the required interfaces will already be

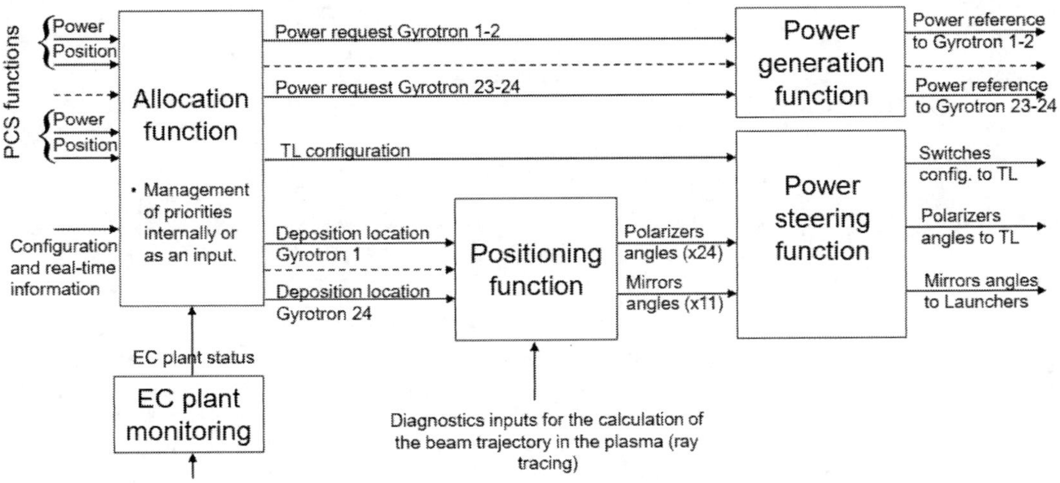


**Fig. 5.** EC system functional analysis.

implemented in the EC control system to allow sufficient time to test and debug them before the next operation phase.

Since different control and protection functionalities are needed before each plasma campaign, the staged approach has also been applied to the control system. Both the hardware and the software of the EC Plant Controller will be updated in stages, going from the simple tools and simulators needed to allow the commissioning of the subsystems (Stage 1) to the complete system integrating all the subsystems and implementing the full PCS interface

- Stage 1: a portable system used to perform the Factory Acceptance Tests (FAT) of the SCUs and the Site Acceptance Tests (SAT) if the final configuration is not available on site;
- Stage 2: a temporary system to support the SAT of HVPS, Gyrotrons and TL on an RF dummy load, implementing all the partial control and protection functions and simulating all the required signals coming from missing systems;
- Stage 3 (First Plasma): a system to support integrated commissioning of the entire EC system and the operation for first plasma;
- Stage 4 (Full system): a system to support advanced operation and the complete interface with central I&C.

The FAT Tools will be used in 2022 for the first HVPS commissioning.

Stage 2 has been already delivered to IO and successfully went through the SAT. It will be installed in its final location in B15 as soon as the access will be granted and it will be used for the commissioning of the first gyrotron by the fall of 2023.

The next step is to go through the FDR of the stage 3 and work toward the additional requirements coming from the parallel operation of multiple gyrotrons.

## 5 Functional analysis

The functional analysis of the EC control system has been performed for the system conceptual design review in 2013 and refined with the design evolution of the different components.

Protection and control functions have been firstly identified and then allocated to the various components in the control system architecture.

Apart from internal faults of the single subsystems like loss of control or loss of communication, the protection functions listed in **Table 1** have been identified and allocated to the different SCUs.

The main functions needed to control the power generation and deposition are summarized in **Fig. 5**. The functional allocation and the interface with the PCS will also be discussed.

**Table 1.** List of protection functions.

| Protection function | SCU |
|---|---|
| Arc inside the gyrotron tube (detected as an overcurrent in HVPS) | HVPS |
| Arc on gyrotron output window and MOU (detected as a light flash on a fibre optic cable) | Gyrotron |
| Mode jump in gyrotron | Gyrotron |
| Cooling fault at gyrotron | Gyrotron |
| Arc in the RF dummy load | TL |
| Cooling fault at dummy load and TL | TL |
| Polarizers control malfunctioning | TL |
| Arc on diamond window (vessel side) | EW |
| Cooling fault at EW | EW |
| Cooling fault at diamond window | EW |
| Arc inside launcher | Launcher |
| Cooling fault at launcher | Launcher |
| Steering mirror control malfunctioning | Launcher |
| Poor vacuum in MOU, transmission line, RF dummy load | Vacuum system |
| To execute stop requests from PCS | ECPC |
| To execute stop requests from CIS | ECPC |
| To execute stop requests from CSS | ECPC |

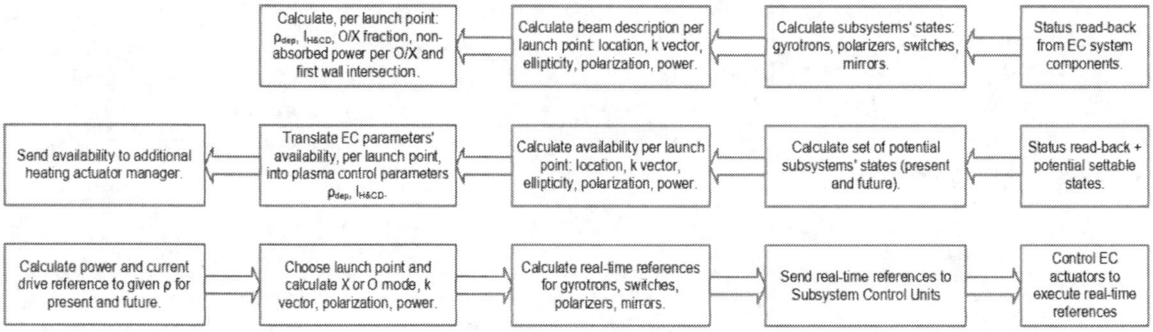


**Fig. 6.** Information flow for Plasma Control Functions between ECPC and PCS.

The final goal of the EC control system is to generate the required amount of power and deposit it at the required location in the vacuum vessel. Many power and location references can be generated at the same time because many different launching points are available: 3 launching points in the EL and 2 launching points in each UL. This allows the EC system to be used for more than one plasma control function at a time.

Given the fact that different mirrors can access different but overlapping areas in the plasma, the first thing to do is to allocate the power and location references to specific launching points. Then, the polarisation and the launching mirror position have to be computed. Once all these references are available, the control system needs to guarantee that the HVPSs, gyrotrons, TL switches, TL polarisers and the launching mirrors are all controlled in real-time to execute them.

The allocation of functions between PCS and ECPC has been performed trying to make use of all the features of the system while minimising the duplication of information, interfaces and functions. The problem has been divided into three flows of information: the knowledge of the current power and deposition location, the knowledge of the power availability in the present and in the near future, the control of the power deposition for the different plasma control functions.

These three flows are represented, from top to bottom, in **Fig. 6**. The red blocks represent the functions allocated to the ECPC while the blue blocks are the ones implemented by the PCS. The arrows represent the information flow, not necessarily implemented in a physical interface since more than one block can reside in the same hardware component. The red/blue arrow is the interface between the two control systems and it is implemented by the fast real time network SDN.

The first function (first line in **Fig. 6**) implemented by the ECPC is to interface with the EC subsystems to retrieve the EC status, to calculate the full beam characterisation per each launching point and send it to the PCS, which in turn translate this into relevant plasma information.

The second ECPC function is to calculate the potential subsystem state in terms of: accessible location, k vector, ellipticity, polarisation, power. The potential state is the one that can be reached in the near future (3 seconds maximum) by rearranging the available power. This information is crucial to the PCS to optimize the use of the EC and the other additional heating systems [2, 3, 4]. The parametric representation of the potential EC state is an open point and will be further investigated during the design phase. The PCS role for this function is again to translate the EC potential state into relevant information for the plasma, mainly in terms of power for heating and current drive and of deposition location ($\rho_{dep}$).

These first two functions complete the first two lines in **Fig. 6**. The direction of the arrows change in the last line, since it represents the information flow generated by the PCS. For each control function, the PCS firstly generates a desired power request and deposition location, then assigns the reference to a specific launching point with the related information (k vector, polarisation) and finally sends it to the ECPC. It is the role of the ECPC to translate this information into low level references for the gyrotrons, TL switches and polarisers, launcher mirrors and to make sure that the references are followed in real-time.

PCS also takes into account the constraints from the EC to move the power from one launcher to another and may decide to pre-emptively allocate power to a specific area to be able to cope with potential control needs. As an example, a specific flux surface could be tracked and EC power pre-allocated to it if an NTM is likely to occur.

One of the functions for which the EC system is the main actuator is the stabilisation of Neoclassical Tearing Modes (NTM). This function puts very stringent requirements on the control system to be able to modulate the RF generation synchronously with the NTM rotation up to 5 kHz. Since the interface with the PCS is much slower (the expected update rate is between 100 Hz and 1 kHz), different solutions are currently under investigation. One solution is that PCS calculates the NTM frequency and phase and communicates it to the ECPC at the actual communication rate, the ECPC would then implement a phase locked loop to generate the triggers for the gyrotrons. This solution would work under the assumption that the phase shift of the islands rotation is substantially slower than the ECPC-to-PCS communication. Another option is to add a high speed interface with a dedicated diagnostic to the ECPC system. Two possible diagnostics are the Magnetics diagnostics and the inline ECE [5]; the latter would use one of the beam lines as ECE channel. As of today, there is no high speed link to the Magnetics diagnostics and the inline ECE is not in the ITER baseline design. Hence, the issue is still open and it will have to be addressed in the future.

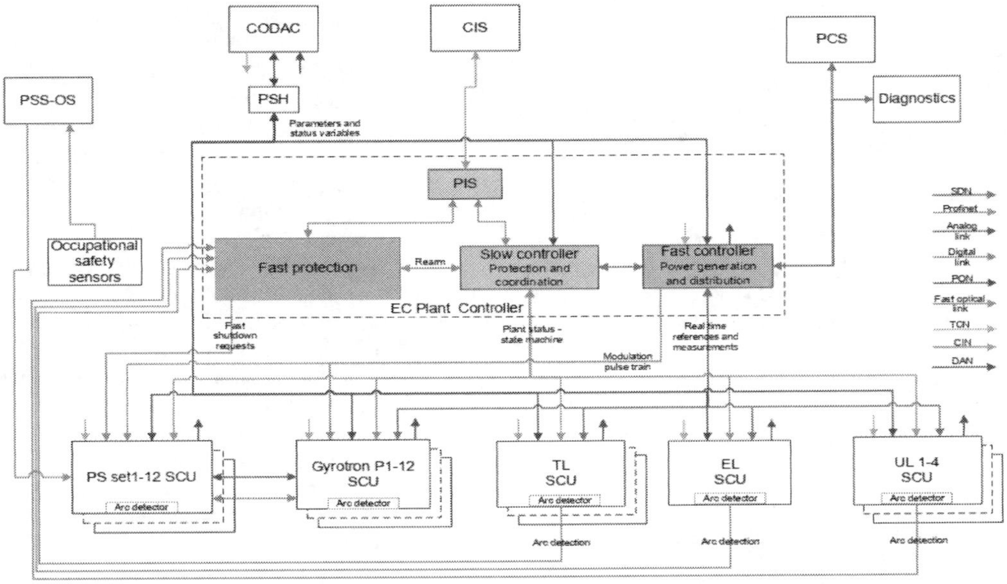


**Fig. 7.** EC control system architecture.

## 6 Architecture

The control system architecture is depicted in **Fig. 7**. Each of the subsystems mentioned above is independently controlled by an SCU. Two gyrotrons are grouped in one SCU because they are connected to the same HVPS set, hence they are considered as one operation unit. The ECPC acts as master of the SCUs, supervising the operation of the entire plant and providing the interface to the ITER central I&C. The ECPC includes a Plant Interlock System (PIS). The occupational safety functions are implemented by a Plant Safety System for Occupational Safety (PSS-OS), managing electrical and radiological hazards in building B15. The implementation of the PSS-OS is delegated to the Central Safety System and it is not included in the scope of the EC control system.

Control, protection and safety functions are always implemented in the lowest possible layer of the control hierarchy, which means that functions involving only one subsystem are implemented in the SCUs; functions involving more than one subsystem are implemented in the ECPC.

CIS and CSS implement functions involving external plant systems and they can inhibit the operation of the EC system in every operation mode.

Communication among different controllers is via fibre optic cables whenever possible, including network interfaces. Standard ITER networks are used to communicate among EC control system devices and with the Central I&C.

Each SCU has a direct connection to PON for parameters setting and to report back the states.

## 7 Operation principles

The EC plant is composed of many subsystems and a high flexibility is required in its operation. To allow the system to be flexible enough, a complex operation has been introduced in which the system is broken down in groups that can work together or in a segregated way.

For each SCU and for the ECPC, the segregation defines if it is under the supervision of a higher level controller (Synchronous control) or if it is controlled by an operator via a CODAC HMI (Asynchronous control).

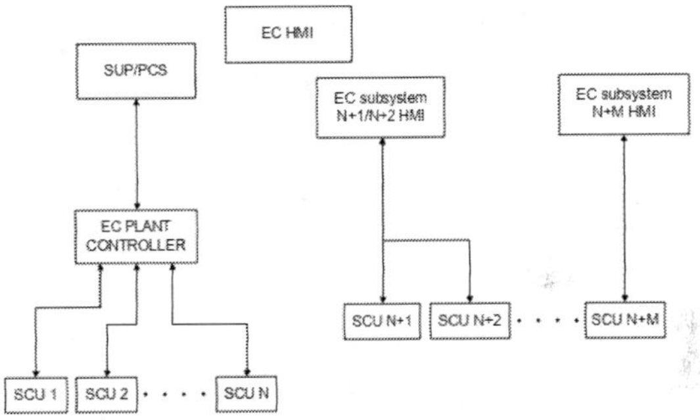


**Fig. 8.** Subsystems operating with different segregation modes.

When the ECPC is in Synchronous control, the requests coming from the Central I&C are then translated into subsystems requests and sent to the SCUs that are Synchronous. The Central I&C is represented either by the SUP (ITER supervisor, configuring the plant systems and controlling their operating states) or by the PCS. In **Fig. 8**, the first N subsystems are in Synchronous operation and M subsystems are in Asynchronous operation (directly controlled via their HMIs).

When the ECPC is in Asynchronous operation, only the requests coming from an operator are executed. Even if the ECPC is Asynchronous, its operation can include one or more subsystems, which will be in Synchronous operation with the ECPC.

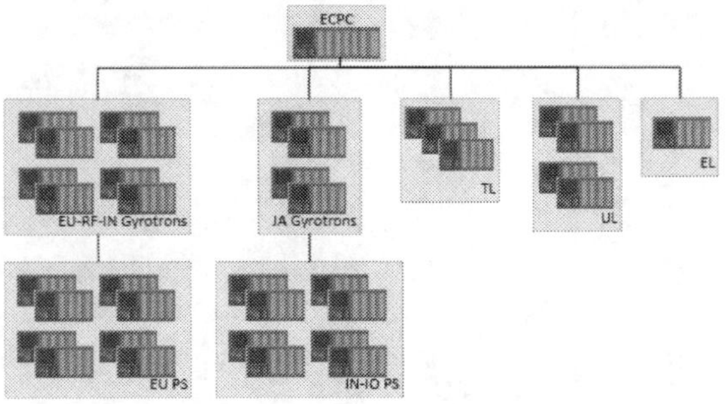


**Fig. 9.** EC control system hierarchy.

When a given subsystem (HVPSs, gyrotrons, TLs, EWs, launchers) is in Asynchronous mode, an operator can directly control the subsystem by sending requests via the CODAC HMI.

The supervisory hierarchy inside the EC system is shown in **Fig. 9**. The supervisory system of the HVPS SCU is the correspondent gyrotron SCU; for all the other subsystems, the supervisor is the EC Plant Controller.

**Fig. 10** shows the source of the commands in a generic SCU in the different segregation modes. The interface with the HMI is always active for monitoring but the commands are only accepted when the system is in Asynchronous control.

It is worth noting that only the HVPSs have a local mode, in which the system is operated with a local HMI (on the HVPS control cubicle); this mode is only used for the HVPS commissioning or maintenance on an electrical dummy load.

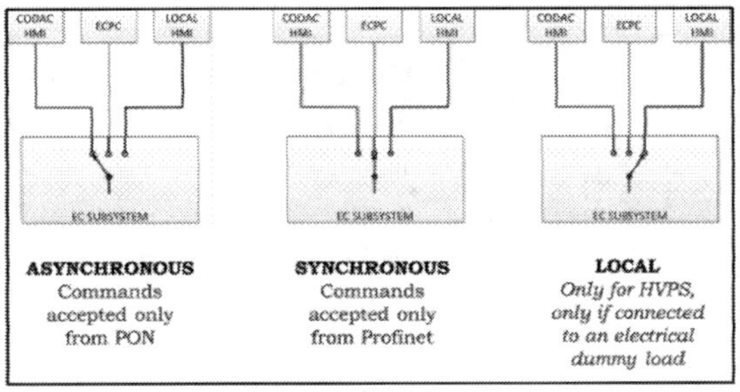


**Fig. 10.** Segregation and source of commands for the EC subsystems.

Each one of the 12 sets of HVPSs, gyrotrons, TLs and of the 5 launchers can perform a different operational task at the same time.

Tasks can involve a single subsystem, a group of subsystems or the entire EC system. Many tasks can be performed at the same time; the supervision and the plant protection is ensured by the ECPC and the Central I&C.

Taking the gyrotron as an example, the following tasks can be performed:

- Gyrotron commissioning in free air: the transmission line is not connected and the beam pattern at the output of the gyrotron matching optic unit is checked. This operation is better carried out in the temporary control room in the RF building because frequent access to the gyrotron area might be required. This operation is only needed for the first gyrotron commissioning or maintenance. Absorbing material and a target to collect the thermal footprint is needed. Special safety provisions are put in place.
- Alignment of beam in transmission lines: same as the previous task but a section of transmission line of a given length is added. This operation is only needed for the first TL commissioning or maintenance.
- Gyrotron operation on dummy load: transmission line is connected up to the RF dummy load. This operation can be carried out from the main control room or from the temporary control room. This operation is needed for the commissioning of gyrotrons or other components.
- Gyrotron operation on a short pulse dummy load in different locations: to measure TL efficiency and calibrate the RF power delivered to the plasma for each gyrotron, a short pulse dummy load needs to be mounted in different locations (at least at the end of the line in B13 or B11).
- Gyrotron conditioning: full details in Section 7.1.
- Commissioning of the control system: the control system of each of the subsystems can be tested without transmitting RF power to the vessel. This operation is needed for first commissioning and after each control system upgrade; it can be carried out from the main control room or from the temporary control room in B15.

### 7.1 Gyrotron conditioning

The standard gyrotron preparation requires ramping up the filament current and waiting for the emitter temperature to be stabilised before being able to perform a long pulse. This time is needed for the emitter ring to reach the thermal operating point. Since there is no direct measurement of the emitter temperature, very short pulses are performed to check the status of the gyrotron by looking at the beam current.

During the pulse, the cool down effect on the emitter, due to the beam current, needs to be compensated by acting on the filament power. Different solutions exist to adapt the filament power during the pulse, based on feedback or feedforward control.

After a pulse, the emitter state is unknown and it depends on the pulse length and on the control algorithm chosen to stabilise the beam current. This means that there is a transient after each pulse, lasting several minutes, during which the gyrotron might not be ready for another pulse.

When the EC system is used in an ITER pulse, all the power needs to be ready as soon as possible to allow the plasma control system to perform its functions. For this reason, gyrotron conditioning will have to be performed during the pulse as a parallel activity. Two mechanisms are needed to perform this function: an RF dummy load that can always be used if needed; the ability of the control system to perform a conditioning task to one or several gyrotrons, while the rest of the system is potentially performing different tasks. The control system needs to accommodate for this need and allow parallel operation of different subsystems as explained at the beginning of the section and in **Fig. 8**.

Another reason to perform conditioning is when an unpredicted behaviour is observed. As an example, if a gyrotron is having frequent arcs during a pulse, an asynchronous/conditioning task can be selected for that particular gyrotron so that an operator can try conditioning it on a dummy load to be able to use that gyrotron again in the same or in the next pulse.

For the sake of completeness, it is worth mentioning that having a reliable model of the gyrotron might solve the problem of knowing the emitter state at all times. A modelling activity has been started in 2018 by F4E [6], to predict the gyrotron behaviour and to be able to design more robust controllers. The activity is still ongoing with different goals: refine the model of the gyrotron, validate the model inside the pulse and between pulses (when the beam current is off), design a controller stabilising the beam current during the pulse and controlling the emitter temperature outside the pulse. The results of the modelling activity will not be presented in this work and will be the subject of a future publication.

## 8 Conclusions

The EC control system is a complex and distributed system, it has many interfaces and includes control, protection and safety functions.

An EC Plant Controller was introduced in 2013 to control and supervise the EC system and to provide an interface layer with the Central I&C.

The system went through a Conceptual Design Review in 2013, where the main requirements were agreed among all the stakeholders. Many workshops and subsystems design reviews have been then organised to further clarify the requirements and follow up the design of the ECPC and of the SCUs. All the EC control system stakeholders actively participated to these meetings, together with the responsible officers of all the relevant interfacing systems.

The system is now approaching a new phase with the installation of the HVPS and Gyrotrons in 2022. By the fall of 2022, the first commissioning of the HVPS procured by F4E will start.

The installation and commissioning activities will continue for a long period until the First Plasma and then the rest of the systems will be installed for the next plasma operation phase. The ECPC will adapt to the different phases by providing the required functionalities and by ensuring a safe operation.

It is important to note that the EC control system involves 5 different DAs and it is the results of a big distributed effort of all the parties including the ITER Organisation, the DAs and fusion laboratories around the world.

*The views and opinions expressed herein do not necessarily reflect those of the ITER Organization.*

## References

1. G. Carannante, M. Cavinato, F. Gandini, G. Granucci, M. Henderson, D. Purohit, G. Saibene, F. Sartori, C. Sozzi, User requirements and conceptual design of the ITER Electron Cyclotron Control System, Fusion Eng. Des., **196-974**, 420-424 (2015)
2. F. Wagner, A. Becoulet et al, Plasma Phys. Control. Fusion, **52**, 124044 (2010)
3. M. Henderson et al, Physics of Plasmas, **22**, 021808 (2015)
4. A.R. Polevoi et al, Reassessment of steady-state operation in ITER with NBI and EC heating and current drive, Nucl. Fusion, **60** (2020)
5. H. van den Brand, W.A. Bongers, J.K. Stober, W. Kasparek, D. Wagner et al, Inline ECE measurements for NTM control on ASDEX Upgrade, Nuclear Fusion, **59**, (2018)
6. N. Badodi, A. Cammi, A. Leggieri, F. Sanchez, L. Savoldi, Energies, **14**, 2068 (2021)

# The ECRH-Power Upgrade at the Wendelstein 7-X Stellarator

H. P. Laqua[1], K. A. Avramidis[2], H. Braune[1], I. Chelis[3], G. Gantenbein[2], S. Illy[2], Z. Ioannidis[2], J. Jelonnek[2], J. Jin[2], L. Krier[2], C.Lechte[3], A. Leggieri[4], F. Legrand[4], S. Marsen[1], D. Moseev[1], H. Oosterbeek[1], T. Rzesnicki[2], T. Ruess[2], T. Stange[1], M. Thumm[2], I. Tigelis[5], R. C. Wolf[1] and the W7-X team[6]

[1] Max-Planck-Institute for Plasma Physics, Greifswald, D-17491, Germany
[2] IHM, Karlsruhe Institute of Technology (KIT), Karlsruhe, D-76131, Germany
[3] IGVP, Stuttgart University, D-70569 Stuttgart, Germany
[4] Microwave & Imaging Sub-Systems, THALES, Vélizy-Villacoublay, F-78141, France
[5] Department of Physics National and Kapodistrian University, Athens, GR-15784, Greece
[6] The W7-X team is the list of authors in Thomas Sunn Pedersen *et al* 2022 *Nucl. Fusion* **62** 042022

**Abstract.** The existing ECRH system at W7-X consists of 10 gyrotrons, with output power levels ranging from 0.6 MW up to 1.0 MW each at a frequency of 140 GHz, quasi-optical transmission lines and microwave launchers at the plasma vessel. Compared to other large fusion experiments, W7-X has a relatively low power-to-volume ratio. However high heating power is particularly necessary for achieving high plasma beta values, where the improved confinement of fast ions, one of the optimization criteria of W7-X, can be examined. It is therefore necessary to expand the ECRH systems in several consecutive steps. It is planned to increase the number of gyrotron positions from 10 to 12 and at the same time to evolve the gyrotron output power in several development steps from 1 MW to nominal 1.5 MW and, finally, up to 2 MW. At the same time, the transmission lines will also be upgraded for 2 MW operation. A special effort is also made to improve the reliability of the system by the fast control system,

## Introduction

The ECRH system for stellarator Wendelstein7-X (W7-X) had been designed about 22 years ago [1]. Nevertheless, it has reliably fulfilled its tasks during the first experimental campaigns. After the successful results of the past campaigns [2],[3], the need for a modernisation and power upgrade has arisen, resulting in the strategy for the performance upgrade presented here.

## Motivation

The existing 140 GHz ECRH system at W7-X consists of 10 gyrotrons with output power levels ranging from 0.6 MW up to 1.02 MW each at a maximum " reliable" total power of 8 MW, quasi-optical transmission lines and microwave launchers at the plasma vessel. Up to 7.5 MW with a transmission efficiency of 94% were delivered to the plasma. Compared to other large fusion experiments, W7-X has a relatively low power-to-volume ratio. However high heating power is particularly necessary for achieving high plasma beta values, where the improved confinement of fast ions, one of the optimization criteria of W7-X, can be examined. In addition, with higher heating power, one expects the achievement of improved confinement regimes, such as the H-mode.

For the further experimental campaigns, two different operational scenarios are planned for achieving the scientific goals of W7-X.

In the long-pulse scenario, plasma operation for up to 30 minutes with reactor-relevant power fluxes of up to 10 $MW/m^2$ on the plasma-facing components is to be demonstrated. To achieve this, ECRH with an absorbed power of 10 MW is required. For this purpose, two additional gyrotron positions will be equipped and fitted with 2 gyrotrons of the 1.5 MW power class.

In the high beta scenario, plasma confinement at reactor-relevant pressures will be demonstrated. For this, however, plasma heating powers of 20-30 MW are required, which can be achieved together with the NBI for plasma discharge length of up to 10 s. For this an ECRH expansion up to 18 MW is envisaged. Because of the more favourable scaling properties for the normalised plasma pressure $\beta$, a lowering of the magnetic field to 1.7 T and the operation of the ECRH at the third harmonic X-mode (X3) is also planned.

## General strategy

The strategy for ECRH expansion is based on previous experience with the ECRH systems at W7-X and other fusion plants. The main findings from this will be highlighted below.

- The construction of non-modified series gyrotrons over many years prevents progress in gyrotron technology. At the end of a long delivery process, the last series gyrotrons are already outdated. This is especially the case when a number of gyrotrons requires a delivery time of 5-10 years. In the past neither our gyrotron producer nor the host team could deliver, install, test and commission more than 1 gyrotron per year.

- The ECRH system must be considered as a whole. The weakest components determine the performance. The individual components must be matched to each other as optimally as possible. This already applies to the building of the ECRH installation and the access space to the experiment. The effort and cost of ECRH performance is far greater than the cost of providing adequate installation space, so economizing on installation space does not make sense.

- All components should be equipped with sufficient safety margin and reserves for a later power expansion since there must be technology progress expected.

- The control system in a long-pulse or continuous wave experiment is fundamentally different from that in a pulsed experiment. Here, the control system must analyse the interlock signal in real time and, if necessary, enable a quick restart. It is also important to categorise the interlocks and carry out appropriate reactions, such as stop or warning messages.

- The complexity of the system must be minimized. Every unnecessary actuator, sensor, cooling circuit reduces the reliability of the system. For this reason, an existing system should always be checked for the possibility of simplifying or omitting components, sensors, etc. These findings can then be implemented when planning the expansion.

## Gyrotrons

The strategy for increasing ECRH power is based on utilising the existing infrastructure for optional upgrades, which has been already planned in the initial installation. First of all, the two reserve positions F1 and F5 will be equipped with two new 140 GHz CW gyrotrons of the 1.5 MW class. For each position, the entire periphery, such as water cooling and transmission lines, will of course be built, taking into account previous experience. For example, the remote control of all transmission mirrors was abandoned because it is only needed once during the adjustment phase and is then switched off. Manual adjustment is possible without any problems but needs more time. The ECRH launchers were designed from the outset to operate with 12 beams. Also, all transmission components are already designed for a power of 2 MW, although so far they could be tested up to 1 MW only. The high-voltage supply at the IPP can supply 12 gyrotron positions with 3.3 MW of electrical power each. For pulses of less than 10 s, even twice the electrical power is possible.

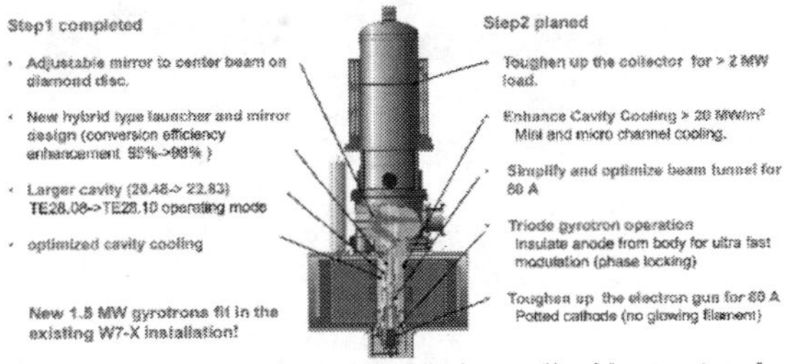


**Fig. 1.** Improvements in gyrotron design realised in step 1 and planned for step 2.

The basic strategy is first to fill the vacant positions and then to replace the outdated, weaker gyrotrons with more modern, more powerful ones. The constraint is that the new gyrotrons should be as compatible as possible with the old installation, so that the system retains a high degree of modularity and parts from one position can be used in the others. This is also extremely labour and cost efficient and reduces the development risk for new components.

Therefore, as shown in Fig.1 a new gyrotron has been developed together with KIT and Thales, which is based on the successful 1 MW W7-X design and is fully compatible with the existing system. The original TE28.08 cavity operation mode was changed to the TE28.10 mode, which increase the cavity surface by 25%, The cavity cooling was also optimised even further, which made it possible to increase the output by 50% power to 1.5 MW. Furthermore, the more than 20-year-old design of the internal mode converter was optimised, increasing the mode conversion efficiency from the gyrotron mode to the HE11 output mode from about 95% to 98%. Finally, the output coupler mirror opposite to the diamond window was improved to be externally adjustable, which should enable perfect centering of the microwave beam at the diamond window in order to minimize thermal stresses due to asymmetric loads. The new cavity and microwave optic was successfully tested in advance in an almost identical short pulse gyrotron, where the 1.5 MW operation at 45 % efficiency was demonstrated in the millisecond range [4].

This development project, which is close to completion with the delivery of the industrial gyrotron to the KIT test facility, is part of a four-stage development program for increasing the ECRH power at W7-X up to 18 MW by 2030 as shown in Fig.2. The industrial gyrotrons are to be improved step by step in permanent cooperation with KIT and Thales. In the second step, which is currently being prepared, 3 more 1.5 MW CW gyrotrons will be purchased. Their critical components, however, will be successively developed to the 2 MW power class level, so that a 2 MW prototype gyrotron can be built in the third step. If this is successful, 3 more CW gyrotrons of the 2 MW class would be ordered in a fourth step, whereby successive improvements should also be possible. In this process the risks are shared between consumers and industry. This strategy will promote European gyrotron development and ensure that the most modern gyrotrons are always available and acquired.

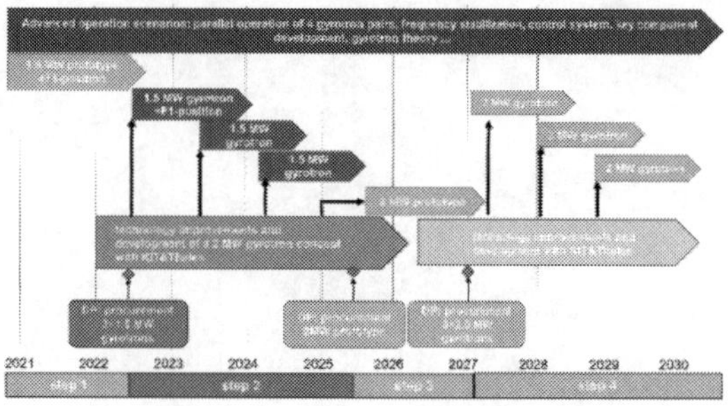


**Fig. 2.** Four Step Strategy for the ECRH-upgrade at W7-X. Step 1 and step 2 are already approved.

## Transmission line

In line with the increase of the number of gyrotrons from 10 to 12, the transmission line must also be expanded by two more beams. However, this is only necessary for the individual beam sections. The multi-beam section is already designed for 12 beams and does not require any additional adjustment, since it was already aligned for the existing beams. .A new microwave beam only needs to be adjusted on the beam combining mirror and it is automatically imaged on the beam dividing mirror. In addition, the new mirrors were designed with the proven technology of a stainless steel base body with cooling channels on which a 2 mm thick copper layer is electroplated. The launcher in the plasma vessel was already built for 12 beams.

In addition to increasing the number of beams, their transmission capacity must also be raised to the increased beam power of 1.5 and later 2 MW. The cooling capacity of the mirrors was already designed for 2 MW beam power, but the previous experimental campaigns showed that problems due to arcing at certain points in the transmission line already occurred below 1 MW. This was particularly problematic in the summer months with high absolute humidity. The problem mainly occurs between mirror no. 13 and no. 14 in the transmission line where a tiny vertical focus exists. Its waist varies between 11 mm and 20 mm depending of the beam line

Here the electric field strength is strongest and the convection air flow is in the unfavourable direction parallel to the beam. Therefore, the design of the mirror 13/14 combination was modified in such a way that the beam waist at the focal point was enlarged. This was initially implemented for the two new beam lines, but can be also implanted in the existing beam lines if necessary. To further reduce arcing, the transmission line was equipped with a powerful air drying system that blows the pre-cooled dry air into the critical area. With the help of a corner reflector, which was installed directly in front of the diamond vacuum window on the plasma vessel, the complete transmission line could be tested with maximum power as shown in Fig. 3. The beam of the A gyrotron was transmitted to the corner-reflector via the A beam line and then sent back to our long pulse microwave load via the F mirror line. This situation simulates the transmission of 2 out of 3 beams to a launcher and the transmission of the maximum gyrotrons power of 900 kW could be shown. Unfortunately, the transmission for higher powers can only be verified when such gyrotrons are available. The arrangement also allows the exact determination of the transmission losses because the beam passes through the transmission path twice. In this special special experimental arrangement the calibration error of the calorimetric load vanishes, since we measure the power with the same load, but with different beam paths, the direct path from the gyrotron to the load and the long path over twice the transmission line length. This resulted in a transmission efficiency of about 94%, which corresponds to the theoretical value and confirms previous measurements, where the corner-reflector was positioned at the end of the multi beam section [5].

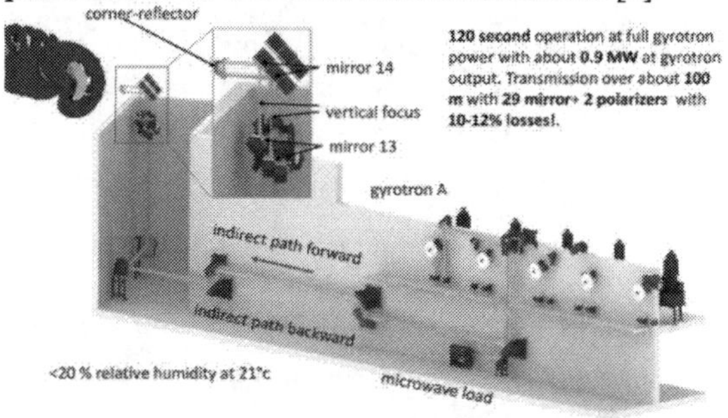


**Fig.3.** Sketch of the full power tests over **twice the length** of the transmission line with the **A beam** to the corner-reflector and via the **F beam** line back to the microwave load. The relative humidity was set below 20% at 21°c.

## Control

The control system is the crucial component for a safe and reliable ECRH system. However, the requirements for safety and reliability can be contradictory. In particular, the fast interlock system must react in sub-milliseconds in the event of an error and switch off the gyrotron. In long pulse or continuous wave mode, however, it is possible to analyse the error automatically and immediately and switch the gyrotron on again if possible. An example of this is the mode recovery system shown in Fig. 4, which detects a mode loss within 0.2 ms and lowers the acceleration voltage until the operation mode settles again. Then the voltage is raised again to a value just below the original set-point. From there, the voltage is ramped up to the nominal value in a much slower ramp. In this way, the gyrotron samples its maximum power range [6].

After the operation phase OP1.2b, the interlocks that occurred were statically evaluated as shown in Fig. 5 and appropriate improvements were made. The main cause of gyrotron shutdowns was arcing in the transmission line. These should be reduced by the new air conditioning. In addition, the arc detection is currently being changed from light sensors to the much faster RF sensors. Faster detection should also allow faster restarting. Other issues are often problems with certain sensors, e.g. water flow monitoring, which often cause false alarms. Here a systematic renewal is taking place.

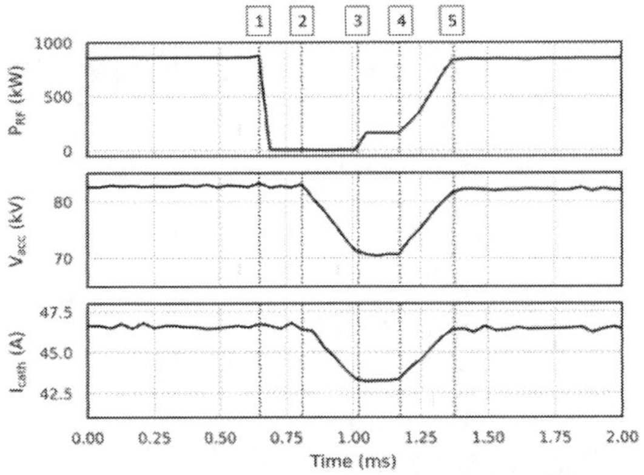


**Fig. 4.** Operation diagram of the mode recovery system [2]. 1) Mode loss 2) Detection of the mode loss and rapid decrease of the acceleration voltage by 10 kV. 3) Reestablishment of the gyrotron mode. 4) Rapid ramping up of the acceleration voltage to just below the nominal value. 5) Slow voltage ramp up to the nominal value.

Another future-oriented development is the external frequency stabilisation of the gyrotrons [7]. With the help of a phase lock loop (PLL), the frequency of the diode-type gyrotrons is controlled with the acceleration voltage as shown in Fig.6. This enables highly precise scattering experiments with CTS diagnostics . However, if it were also possible to lock the gyrotron in phase, completely new possibilities would open up for ECRH operation. All the gyrotrons would then emit coherently and it would be possible to merge many microwave beams into one. It would also be possible to switch the direction of the sum beam with the phase of the individual beams [8].

**Interlocks in OP1.2b**

| Interlock / Gyrotron | A1 | B1 | C1 | D1 | E1 | A5 | B5 | C5 | D5 | E5 | Summe |
|---|---|---|---|---|---|---|---|---|---|---|---|
| RF (mode loss) | 20 | 10 | 17 | 23 | 5 | 2 | 1 | 2 | 3 | 10 | 88 |
| I-Body | 5 | 5 | 8 | 9 | 14 | 6 | 3 | 2 | 1 | 16 | 69 |
| Overcurrent | 4 | 0 | 17 | 5 | 8 | 1 | 5 | 0 | 5 | 2 | 47 |
| Arc Gyrotron | 1 | 8 | 0 | 0 | 0 | 3 | 0 | 0 | 0 | 0 | 12 |
| Arc M13 (ECRH tower) | 20 | 6 | 16 | 21 | 17 | 13 | 17 | 15 | 22 | 21 | 168 |
| Sweep PSU | 6 | 9 | 1 | 1 | 1 | 7 | 7 | 0 | 0 | 3 | 30 |
| Cooling | 0 | 2 | 0 | 0 | 4 | 4 | 8 | 1 | 4 | 4 | 27 |
| Communication | 3 | 1 | 3 | 2 | 8 | 0 | 3 | 0 | 1 | 1 | 22 |
| Summe | 54 | 41 | 57 | 61 | 57 | 36 | 34 | 20 | 41 | 57 | |

**Fig. 5.** List the interlocks of the individual gyrotrons that occurred in operation phase 1.2b and their causes.

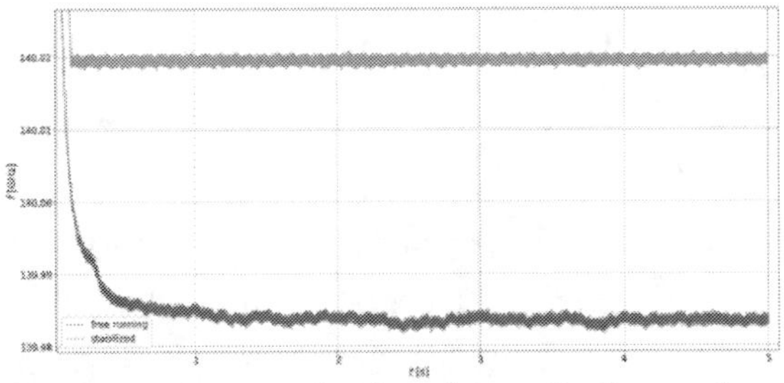

**Fig. 6.** Frequency as a function of time of a free running gyrotron in blue and a frequency stabilized gyrotron in orange.

## Summary and conclusions

The strategy for the expansion of the ECRH system at W7-X has been presented, based on the experience gained in the construction and operation of the existing ECRH facility. The gyrotron development will be carried out in a four-step process, which guarantees a permanent improvement of the tubes. In addition, the entire ECRH facility will be optimised to meet the higher requirements for power generation and transmission. This means that in ~2030, 12 fully equipped microwave beam lines with a high level of reliability and a total output of around 18 MW (4x1MW+4x1.5MW+ 4x2MW) are foreseen.

## References

[1] Erckmann et. al. Fusion Science and Technology, **52** (2007), 291.
[2] Beidler et al., Nature **596** (2021), 221.
[3] T. Sunn Pedersen et al., Nucl. Fusion **62** (2022), 042022.
[4] K. A. Avramidis et al., Fusion Engineering and Design **164** (2021), 112173.
[5] T. Stange, et al, EPJ Web of Conferences **157** (2017), 02008.
[6] F. Wilde et al., Fusion Engineering and Design **148** (2019), 111258.
[7] L. Krier, et al., https://doi.org/10.1109/IRMMW-THz50926.2021.9566847
[8] V. Erckmann, W. Kasparek et al.: 15th IEEE International Vacuum Electronics Conference (IVEC 2014), 2014, Monterey, CA, USA, 3.3.

## Acknowledge

This work has been carried out within the framework of the EUROfusion Consortium, funded by the European Union via the Euratom Research and Training Programme (Grant Agreement No 101052200 — EUROfusion). Views and opinions expressed are however those of the author(s) only and do not necessarily reflect those of the European Union or the European Commission. Neither the European Union nor the European Commission can be held responsible for them.

# MAST Upgrade microwave heating and current drive system – engineering design overview

*Helen* Webster[1*], *Simon* Freethy[1], *Mark* Henderson[1]; *Joe* Allen[1], *Ottone* Caretta[1], *Shail* Desai[1], *James* Holden[1], *Dariusz* Maszczyk[1], *Elias* Melidis[1], *Hendrik* Meyer[1], *Ewan* Muir[1], *Asanka* Munasinghe[1], and *Santhosh* Surendran[1]

[1]United Kingdom Atomic Energy Authority, Culham Centre for Fusion Energy, Culham Science Centre, Abingdon, Oxon, OX14 3DB, UK

**Abstract.** MAST-Upgrade (MAST-U) is undergoing several enhancements to deliver increased performance and functionality. One such enhancement is the design, development, and implementation of an Electron Bernstein Wave (EBW) Heating and Current Drive (HCD) System. The MAST-U EBW System aims to provide experimental data for model validation, and to provide a greater understanding of EBW physics and its capabilities. The MAST-U EBW System provides up to 1.8 MW of microwave power into the plasma, through a system comprising: high voltage power supplies; two gyrotrons; evacuated transmission lines; a steerable in-vessel launching system; and associated control and ancillary systems. The gyrotrons from Kyoto Fusioneering have a 0.9 MW output power capability at the dual frequencies of 28GHz and 34.8GHz, allowing start-up and current drive studies to be carried out at their respective optimum frequencies. Additional diagnostics, termed interceptor plates, are proposed to sit in the path of the first reflection. These will measure the reflected power from the plasma, to both act as an interlock if the reflected power is too high, and provide key information on the coupling efficiency.

## 1 Introduction

MAST-Upgrade (MAST-U) is undergoing several enhancements to deliver increased performance and functionality. One such enhancement is the addition of a 1.8MW, 4.5s Electron Bernstein Wave Heating and Current Drive System (EBW H&CD), operating at 28GHz / 34.8GHz.

The primary aim of the system is to provide an experimental basis for the use of EBW Current Drive on a spherical tokamak, in support of the STEP (Spherical Tokamak for Energy Production) programme, examining open issues, and providing data on experimentally achieved coupling efficiencies to assist with model validation. The system is designed to provide a high degree of flexibility, particularly in the launching system, to provide a broad experimental data set. Additionally, the system provides the capability to extend earlier MAST experiments into EBW based solenoid free start-up. More details of the motivation for this system can be found in these conference proceedings [1].

The two roles of the system (solenoid free plasma start-up and high-density current drive) have differing requirements. The first of these requires excitation of Electron Bernstein Waves at the fundamental electron cyclotron frequency, and the latter at the second harmonic.

The power requirements of the system are defined by the need to drive a clearly measurable current. The high-level requirements are shown in Table 1, along with the design parameter values chosen to meet these requirements. This paper provides an overview of the engineering design at the current preliminary design phase.

## 2 System Overview

The system designed to meet these requirements is termed the MAST-U EBW System. The system external to the MAST-U vessel comprises: high voltage power supplies; two dual frequency gyrotrons; evacuated transmission lines; a long pulse dummy load; and associated services such as water cooling and vacuum pumping. Inside the MAST-U vacuum vessel, a steerable in-vessel launching system provides flexible options via two separate launchers – off-axis and on-axis. Finally, additional diagnostics termed "interceptor plates" sit in the path of the first reflection.

The off-axis launcher has a small steering range, and can operate in co current drive only, whereas the on-axis launcher can operate in the co and counter plasma current directions (independently for each beam).

In the present system design, with the gyrotron output power offset by the transmission and launcher losses, a power to the plasma of up to 1.6MW is predicted. A summary of the design values against the requirements is shown in Table 1.

The system is located primarily within the existing MAST-U building. Co-location of the high voltage power supplies and gyrotrons in the same area minimises the cable length between them, reducing cable stored energy and minimising risks of gyrotron damage, in case of a

* Corresponding author: helen.webster@ukaea.uk

gyrotron arc event. The location of the gyrotrons within the building allows for a minimal length and number of bends in the transmission line route to minimise power losses.

**Table 1.** System requirements and design values

| | Parameter | Requirement | Design Value |
|---|---|---|---|
| **Heating and Current Drive** | Frequency | Second harmonic, 35-37GHz | 34.8GHz |
| | Power (to the plasma) | >1200kW | 1500-1600kW |
| | Launch | Launch to achieve on and off axis current drive | Two on-axis beam paths, Two off-axis beam paths |
| **Solenoid Free Start-Up** | Frequency | Fundamental, 28GHz | 28GHz |
| | Power (to the plasma) | >200kW | 650-750kW |
| | Launch | X mode, to provide co-current drive | Quasi-high field side launch from centre column polarising tile (see Section 6.4) |

## 3 Power Supplies

The High Voltage Power Supplies are designed to meet the gyrotron requirements; for a triode type gyrotron this consists of cathode, body and anode power supplies. For cost and space optimisation the cathode supply is shared between the two gyrotrons, rated for 110A, -55kV.

Initially, modulation can be carried out using the cathode supply at 1kHz. Future modifications are proposed to use a switching system on the anode power supplies to enable independent modulation. Examples of the desirable power modulation are shown in Figure 1. Shown here are examples of maintaining a constant average power whilst changing the maximum instantaneous power, allowing non-linear effects in the EBW coupling scheme to be investigated.

The power supplies are designed to keep the arc energy low (<10J), and output over/under shoot, ripple and stability to acceptable levels. The power supplies have an overall duty cycle of 1/60, equating to a 5 second pulse every 5 minutes (with a higher intra-pulse duty cycle).

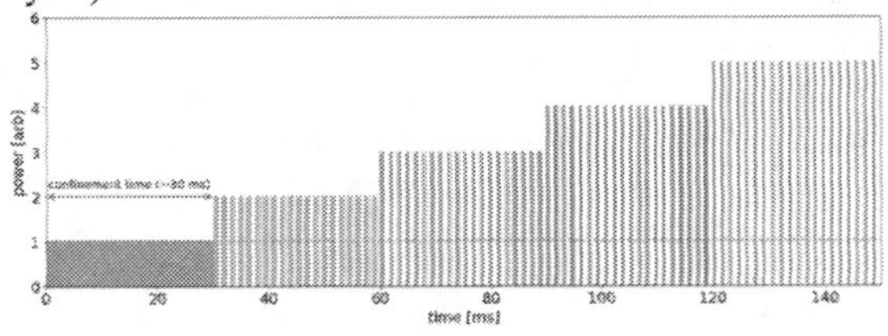


**Fig. 1.** Example of desirable modulation capability

## 4 Gyrotrons

The gyrotrons are supplied by Kyoto Fusioneering, and include the gyrotron tubes, cryogen-free Superconducting Magnets (SCMs), Matching Optics Unit (MOU), and low voltage power supplies.

The gyrotrons have two ratings, 800kW for 4.5 seconds, and 900kW for 3 seconds (see Table 2). These power values are coupled into the transmission line after the MOU in the $HE_{11}$ mode. The power output can be varied from <100kW to the values specified above.

The gyrotrons are dual frequency, at 28GHz and 34.8GHz, allowing start-up and current drive studies to be carried out close to their respective optimum frequencies (more detail of frequency selection can be found in these conference proceedings [1]). The MOU is designed to couple into an $HE_{11}$ waveguide at both frequencies without modifying the MOU optics; the mirrors are optimised for an average of the two frequencies.

One full power and pulse length dummy load is provisioned, shared between the two gyrotrons. The function of this is to absorb the power from the gyrotrons for testing and commissioning. It contains appropriate instrumentation on the cooling circuit to obtain calorimetric measurements.

**Table 2.** Operating parameter combinations for the gyrotrons

| Frequency (GHz) | Power (kW) | Pulse Length (s) |
|---|---|---|
| 28.0 / 34.8 | 800 | 4.5 |
| | 900 | 3 |

## 5 Transmission Line Design

The transmission line connects the MOU output to the interface with the ports on the MAST-U vessel. There is one transmission line per gyrotron.

Transmission of power from the gyrotron to the MAST-U vessel is achieved through the use of cylindrical corrugated waveguide with an internal diameter of 88.9mm, and associated mitre bends. 88.9mm is chosen to provide a balance between losses at the chosen frequency, and the size of the line to integrate into an existing building (whilst maintaining typical or standard diameters). The line is evacuated to high vacuum due to the need to avoid breakdown arising from the high power of the beam, and also the need to share a vacuum with MAST-U.

The transmission lines perform additional functions as follows:

- Deviate power to a calorimetric load.
- Provide polarisation control of the output beam.
- Monitor the forward power to detect potential mode jumps in the gyrotron and ensure microwaves are being transmitted as expected.
- Deviate power to either upper or mid-plane launcher.
- Provide isolation from the tokamak vacuum (when not operating into tokamak).
- Provide pumping access for maintaining vacuum in the line.
- Provide electrical isolation of the transmission line where required.

https://doi.org/10.1051/epjconf/202327704004

These functions are met by the design shown in Figure 2. Each line contains an elliptical and a plane polariser to give full polarisation control of the output beam, and several switches to provide the deviation to the load, or to the two launch options. There is no window at the interface to the MAST-U vessel, instead, gate valves provide the vacuum isolation when required. Vacuum pumping is carried out on the MOU, dummy load, and on or near the isolating gate valves on the interface with the MAST-U vacuum vessel. Electrically insulating breaks are located at the MOU interface and MAST-U vessel interface.

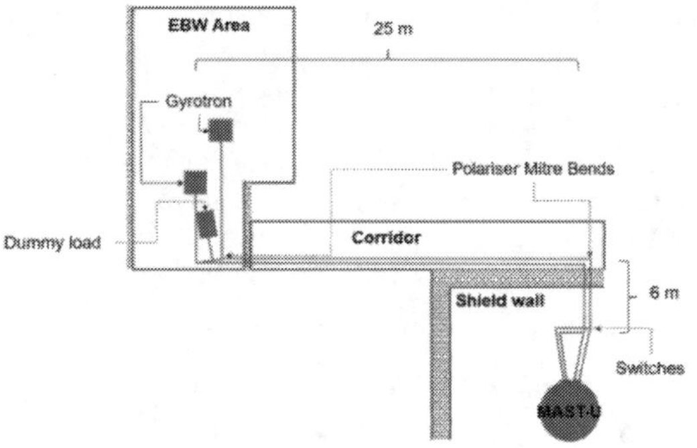


**Fig. 2.** Illustrative layout of the transmission line showing key components (top view)

The transmission lines achieve a calculated 85-90% transmission efficiency. The route is optimised to reduce the number of mitre bends, since at the given frequencies, each bend contributes a 2-2.5% loss [2].

## 6 Launcher Design

The primary function of the launcher system is to direct the microwave energy towards the plasma as required for EBW coupling. To meet the system requirements, the direction of the beam needs to be adjustable in both the toroidal and poloidal directions. The launcher system needs to be able to launch the beams on and off axis. To achieve this, the system has two launchers, termed the off-axis / upper launcher, and the on-axis / mid-plane launcher.

### 6.1 On-axis / Mid-plane Launcher

The requirement that the on-axis mirrors can be used to send beams in the co- and counter-plasma current direction requires placement of the final mirror at the midplane (later referred to as M2 mirrors). The symmetry of this location gives equal magnitude current in both directions, allowing a balanced, zero current configuration (one co- and one counter-).

The two midplane mirrors each have to provide a capability to steer the beam in one of two principal directions, corresponding to two plasma locations where the best coupling of the microwaves is expected for each of co- and counter-current drive. The steering is also required to be flexible for a range of plasma parameters, and to cover a range around the optimum coupling locations. This gives a steering range as shown later in Figure 6.

The mirrors are designed to achieve a large beam waist (>50mm) after the final steering mirror, which minimises the beam divergence (<4°) for optimum EBW coupling, and also to achieve the start-up beam requirements (see Section 6.4).

The launching mirrors are designed to be large enough to transmit >99% of the power which lands on them to maximise the transmission efficiency to the plasma edge and to minimise stray radiation inside the vessel, whilst not restricting diagnostic lines of sight. The mirror size is dictated by the 28GHz beam, which has a larger beam width.

The mid-plane launcher consists of two fixed mirrors delta-M1 and epsilon-M1, and two steering mirrors, delta-M2 and epsilon-M2; these are shown in Figure 3, with delta beams on the left as shown in the image. The image shows the expected range of output beams.

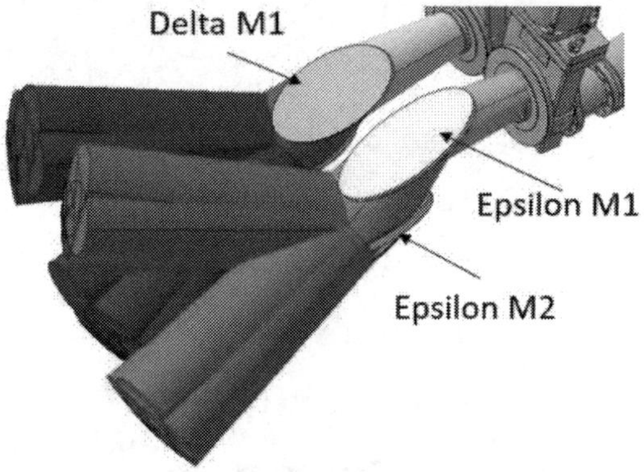


**Fig. 3.** CAD model of on-axis launcher showing the two beams, and a range of output beams in the co and counter plasma current directions, and for a range of plasma parameters.

### 6.2 Off-axis / Upper Launcher

The requirement for the off-axis mirrors is to send beams in the co-current configuration only. Achieving counter-current would have required a separate set of mirrors, which was discounted for the scope of the initial system due to the cost, complexity, and port availability. The beams need to be launched with a given beam diameter, and at a given toroidal angle, whilst avoiding clashes with other in-vessel equipment.

The off-axis design uses a four-mirror system. The design uses three fixed mirrors per beam, and one steering mirror for each beam (four mirrors for each of the delta and epsilon beams).

Figure 4 shows the design of the beam path. The mirrors are numbered in sequence from the gate valve. M1, M2 and M3 are fixed mirrors, and M4 is a steering mirror. The primary function of the M1 and M2 mirrors, which are ex-vessel, is to focus and cross the beams so that they pass through the port cleanly. The M3 mirror provides the final focussing, and the M4 mirror is flat, and provides the final positional adjustability and launch, with

https://doi.org/10.1051/epjconf/202327704004

the M4 mirrors positioned to ensure the beams enter the plasma with equivalent poloidal launch geometry.

The M4 steering mirror only requires a small steering range, since it is only required to achieve co-injection (the small steering range is then to accommodate different plasma configurations).

This design also requires an additional "vacuum chamber" added to the port flange. This accommodates the M1 and M2 mirrors, and provides a larger surface area for the waveguide interface, allowing for the gate valves to be positioned and appropriately spaced.

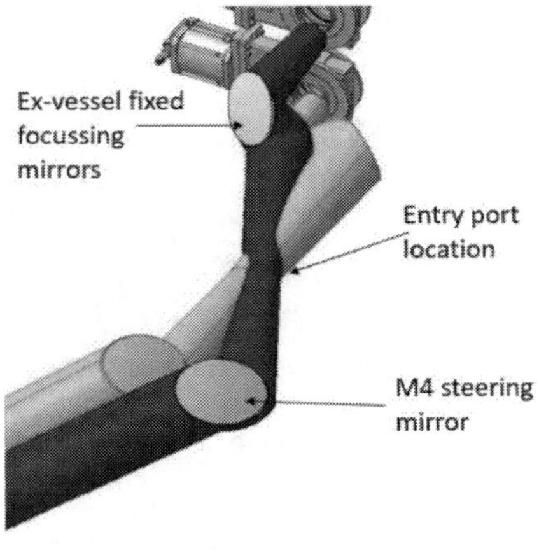


**Fig. 4.** Off-axis launcher geometry showing four mirror system with beams crossing through the entry port into the vessel

### 6.3 Launcher Engineering Design

The on-axis and off-axis launchers described in the previous sections need to be realised into an engineering design. Key considerations at this design stage are the choice of mirror material and beam steering mechanism, hence these aspects are considered in the following sections.

The mirror material chosen is copper coated graphite. This combination provides maximum reflection via the high electrical conductivity of the copper, whilst minimising electromagnetic forces (see Section 6.3.1).

#### *6.3.1 EM Force Calculations*

Analysis was undertaken to determine the likely electromagnetic forces acting on the mirrors, as this also impacts the mechanical design. The analysis uses a main formula for forces experienced by a conducting, rectangular plate in a changing magnetic field, with vector manipulation used to align the magnetic field data with the orientation of the EBW System mirrors. The methodology considers three origins of forces, from the changing magnetic field due to:

- Standard fluctuations in the MAST-U field
- The intra-pulse steering of the mirrors (for the steering mirrors only)
- Disruptions

It is concluded from the preliminary analysis results that the mirror needs to be made primarily of a material such as graphite with a lower electrical conductivity, to ensure that the forces remain low. This is particularly significant for the steering mirrors, where a motor will be required to hold the mirror in position. This eliminated the use of pure copper, or copper alloy, as a mirror material, and also gave a maximum acceptable copper coating thickness.

#### *6.3.2 Thermal Analysis*

Preliminary thermal analysis was undertaken to determine an appropriate material choice. This is based around the use of copper as the main reflective surface due to the high electrical conductivity leading to higher microwave reflection. The incident beam, modelled as a Gaussian, was simulated as an incident heat flux with appropriate absorption parameters. Multiple pulses were simulated in accordance with the expected duty cycle to investigate the ratchetting effect. A structural assessment was subsequently performed to determine the stresses induced by the thermal expansion.

Using the data obtained from the simulations, and evidence of prior experience from ASDEX [3], it was found that using a graphite body mirror with a thin (<100μm) copper coating on the mirror face would be suitable. The simulations found that when using conservative values, including accounting for additional factors such as plasma radiation, the temperatures and stresses reached were acceptable. Figure 5 shows the maximum temperature reached after several pulses when the ratchetting effect has stabilised for a copper coated (front face only) graphite mirror.

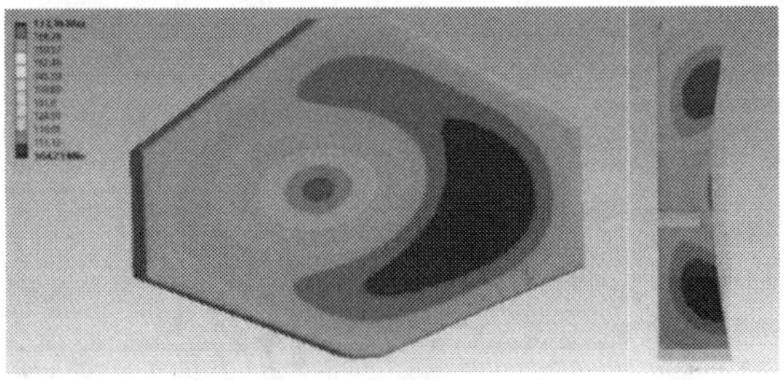

**Fig. 5.** ANSYS model of a simplified mirror geometry, showing the maximum temperature (~170°C) reached after ratcheting with a graphite mirror with a copper coated front face. Left – front view, right - cross section view.

#### *6.3.3 Mechanical Design*

Mirrors M4 (off-axis) and M2 (on-axis) are steerable mirrors. The steering mechanism design is for an externally actuated pushrod-based system for both launchers. The system consists of:

- External motors
- Linear actuators with vacuum feedthrough through ConFlat flange
- Stainless steel pushrods
- Stainless steel mounting brackets
- Vacuum compatible bearings and joints

The mid-plane launcher requires a large steering capability, with the anticipated range shown in Figure 6. Ideally, the rotation point is also about the front face.

Therefore, a tip-tilt style system is proposed which provides an effective rotation point around the front face as required. This consists of 3 pushrods each with a ball joint and sliding joint at the mounting plate. The mirror attaches to the mounting plate.

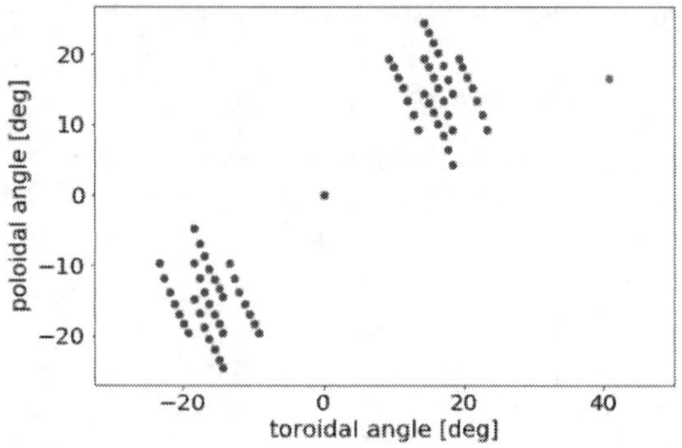


**Fig. 6.** Anticipated steering range for mid-plane launcher. Two groups of points represent co- and counter- directions; points cover range of plasma parameters.

### 6.4 Start-up

The start-up design aims to achieve a quasi-high field side launch (high field side launch is not achievable with MAST-U geometry). This approach involves launching a beam in O-mode through the plasma towards a polarising tile on the centre column, which changes the polarisation from O-mode to X-mode, and the wave is then absorbed by the plasma.

The design for the system makes use of an existing polarising tile already installed on the MAST-U centre column, shown in Figure 7. The tile was designed for 28GHz, so the use of the dual frequency gyrotron enables the tile to be used without modification.

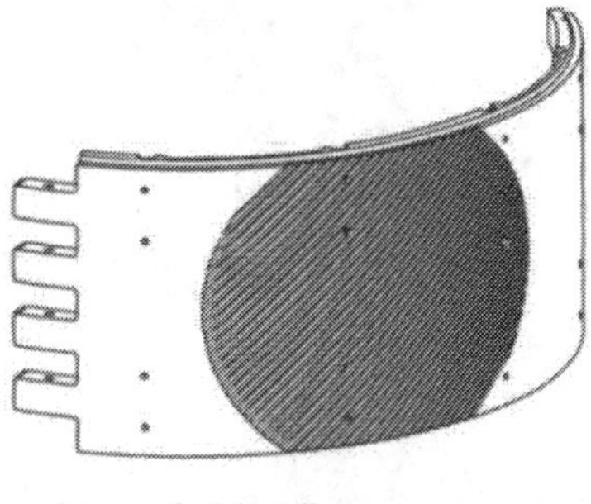

**Fig. 7.** Centre column polarising tile

In order for the start-up beam to reach the polarising tile, the design uses the on-axis launcher, and the delta beam. The steering mirror delta-M2 can achieve an additional position, using the steering mechanism, which directs the beam towards a final fixed mirror (mirror "H") opposite the polarising tile. That mirror focusses the beam onto the polarising tile. In the initial system, only one beam is used for start-up.

The final mirror is designed to achieve a given spot size on the centre column polariser tile. The tile has a fixed radius of polarising pattern, and hence the fixed mirror size and focussing is optimised to give a beam less than or equal to the given pattern size. This is illustrated in Figure 8, which shows the beam propagation through the in-vessel system. The beam path chosen, min w_CC, minimises the spot size on the centre column tile.

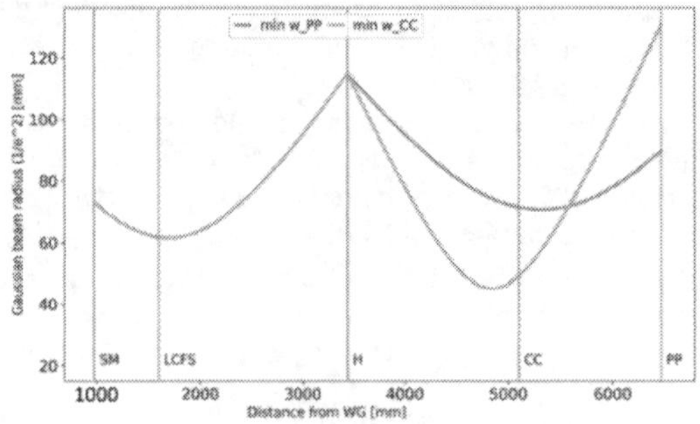


**Fig. 8.** Approximate beam envelope ($1/e^2$ power radius) for start-up. SM – steering mirror; mirror H, final fixed mirror; CC – centre column tile; PP - low field side Protection Plate.

## 7 Interceptor Plates

The microwave beams have a maximum theoretical coupling efficiency of 96% [1]. The coupling is sensitive to the incident angle of the microwaves with respect to the magnetic field and the plasma surface, and the polarisation of the microwave beam; if any of these parameters is not optimal, the coupling efficiency falls, with an extreme of 100% reflection in the worst case. Additionally, the spatial distribution of the reflected power is of scientific interest, since this can be used to support model validation.

To provide this required information on the coupling efficiency, a component which intercepts the microwave beam as it reflects from the plasma is proposed. These are termed "interceptor plates" and sit in the path of the predicted first reflection of the microwave beam over a range of parameters.

The functions of the interceptor plates as described above can be summarised as follows:

- Provide diagnostics on the amount of energy escaping the plasma, with sufficient information on the reflected beam.
- Provide diagnostics for a fast response interlock to protect the machine in case the power fails to couple with the plasma.
- Protect the poloidal magnets and other in vessel components from being damaged by the reflected microwave beams.
- Diffuse the remaining beams so that any further in vessel interactions with the scattered microwaves are reduced to a relatively benign energy density.

Figure 9 shows the approximate location of the interceptor plates around the MAST-U P5 coils. These capture the reflected beams for a range of plasma currents and plasma radii. The plates are designed to "wrap" around the coils to ensure they have minimal impact on diagnostic lines of sight, and to provide good protection for the coils themselves. Interceptor plates are provisioned for each of the beam launch options – off-axis, on-axis co, on-axis counter.

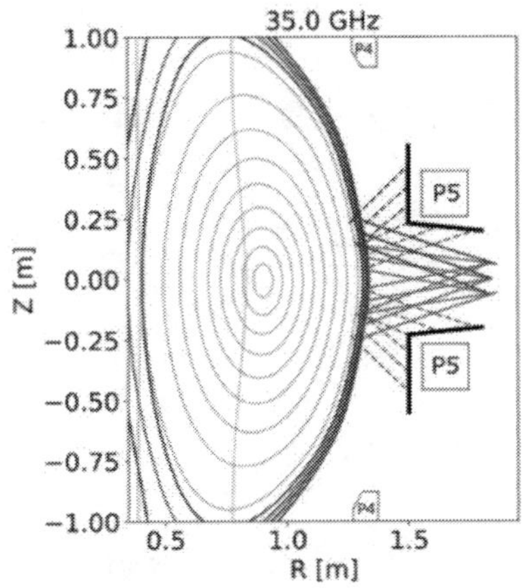


**Fig. 9.** A schematic plot of the poloidal cross-section of the beams from the midplane launcher for Ip = 800kA (green) and Ip = 2000 kA (blue). The dashed lines represent the beam reflected from the plasma after coupling. Thick black lines represent "interceptor plates" [1].

The concept design for any given interceptor plate is shown in Figure 10 and consists of two graphite plates, >10mm thick. These plates have a scattering / diffusive machined surface to achieve the function of reducing the overall stray power density within MAST-U. One of the graphite plates is inclined at ~5° to the horizontal. The graphite plates are attached to a strap which fits around the MAST-U in-vessel coil, with feet to provide adjustment. The graphite is chosen for its thermal and moderately low microwave absorption properties.

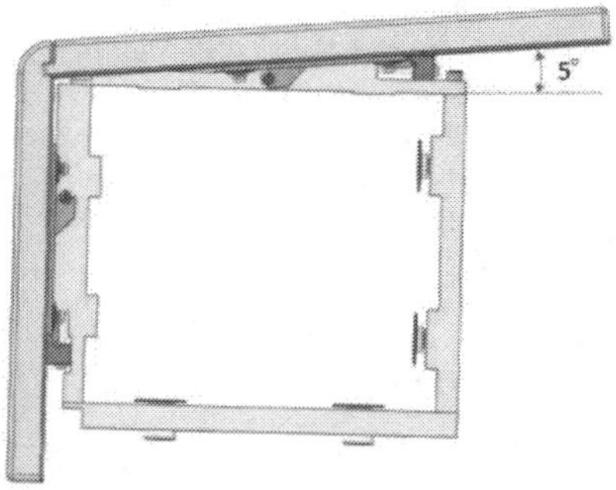


**Fig. 10.** Concept of an interceptor plate

The plates achieve their diagnostic capabilities via two methods. Firstly, Infra-Red camera(s) are positioned to view the surface of the plates (this drives the need for the nearly-horizontal graphite plate to be angled). By directly viewing the front surface, a fast response is achieved, and hence this can be used as both an interlock (high rate of change of temperature viewed with camera means that coupling is too low and the system shuts down) and a diagnostic.

A second measurement method, embedded thermocouples, is used for redundancy. The thermocouples are embedded to a depth of ~3mm from the front surface of the plates, which gives a fast enough response for both an interlock, and some limited diagnostics.

## 8 Control and Ancillaries

The EBW System as described is augmented by a supporting infrastructure. The most significant elements are the control system, and the cooling system.

The philosophy for the control system is for a distributed system, to support staged testing and commissioning, including separation of initial testing into a dummy load from integrated testing into the MAST-U tokamak. The control system is divided into five main sub-systems:

- Personnel Access Safety System, providing personnel protection.
- Launch Control Sub-System, controlling set up of transmission line and launch parameters.
- Power Control Sub-System, controlling power supplies and gyrotron to achieve desired power output.
- Operation Management Sub-System, providing slower timescale plant control and monitoring.
- Asset Protection Sub-System, providing primarily fast protection for the gyrotrons.

The control system is designed to be remotely operable and combines primarily commercially available solutions into a robust and coherent design.

The primary function of the cooling system is to cool the gyrotron and dummy load during and after pulses, and provide sufficient heat rejection capability to maintain the gyrotron inlet water temperature below 35°C.

The EBW cooling sub-system comprises multiple circuits with separate pumping all fed from a buffer vessel. The buffer vessel is connected to a wider MAST-U plant cooling system, and stores the heat during a pulse, before slowly exchanging the water with the MAST-U cooling system between pulses. The MAST-U cooling system is then responsible for the overall heat rejection through an existing cooling tower system.

## 9 Conclusion

This paper has reported the key design specifications, principal components, key challenges and innovative features of the MAST-U EBW System at the preliminary design phase. The System achieves a high degree of experimental flexibility through the modulation capability and range of power inputs, and the variety of in-vessel launch options, hence providing significant support to model validation, and information on EBW physics and its capabilities in spherical tokamaks. The MAST-U EBW System is now entering the detailed design phase.

This work has been part-funded by STEP, a UKAEA programme to design and build a prototype fusion energy plant and a path to commercial fusion.

## References

1. S.Freethy, EPJ Web Conf., (this conference)
2. J.L.Doane, Fusion Sci. Tech. **53**, 159-173 (2008)
3. F.Leuterer et al., Fusion Sci. Tech. **55** 31–44 (2009)

# Status of the ITER ECRH&CD control system development

*Giuseppe* Carannante[1*], *Mario* Cavinato[2], *Katarina* Cindric[2]; *Martino* Ferrari[1]; *Giuseppe* Ferrò[1]; *Andre* Cabrita Neto[2] and *Filippo* Sartori[2]

[1]ITER Organization, Saint-Paul Lez-Durance, France
[2]Fusion for Energy, Josep Pla 2, Barcelona, 08019, Spain

**Abstract.** The ITER ECRH&CD system is designed to inject 20 MW of millimetre-wave at 170 GHz into the vacuum vessel. The system is composed of many sub-systems, namely High-Voltage Power Supplies (HVPS), Gyrotrons, Transmission Lines (TL), Ex-vessel Waveguides (EW), Launchers. It is the role of the EC Plant Controller (ECPC) to integrate all the Sub-system Control Units (SCU), to prepare the system for operation and to execute the real-time requests coming from the plasma control system. Plant level protections are also implemented by the ECPC, in charge of ensuring the safe operation of the plant, while optimizing the power availability. While control and protection functions are always pushed to the lower possible controller able to implement them, the operational requirements and flexibility of the system make it impossible to fully segregate many functions, since each gyrotron is connected to at least two different launching mirrors. To simplify the SCUs' development and to respect the responsibility boundaries imposed by the procurement strategy, all the functions involving more than one sub-system are implemented in the ECPC, which exposes a single operational interface towards the ITER Central I&C. The status of the control system development is presented in this work.

## 1 Introduction

The ITER Electron Cyclotron Heating and Current Drive (ECH&CD) System is designed to inject 20 MW of millimetre-wave at 170 GHz into the vacuum vessel. The system is being procured by 5 domestic agencies, Europe (F4E) [1], Japan (JA-DA), India (ITER-India), Russia (RFDA), United States (USIPO), plus the ITER Organization (IO). The EC system is composed of different sub systems, each one with its own local controller called Subsystem Control Unit (SCU). The main sub-systems are the High Voltage Power Supplies (HVPS), RF sources (Gyrotrons), Transmission Lines (TL), Ex-vessel Waveguides (EW) and Launchers. Due to the complexity of the ITER EC system, the number of the contributing parties and the staged installation [2], the Electron Cyclotron Plant Controller (ECPC) was added to the architecture. The main ECPC functions are the integration of all the sub-systems, the system preparation for operation and the execution of the real-time requests coming from the Plasma Control System (PCS). Even though coordination and supervision of the subsystems are the essential roles of the ECPC, the most important function is to provide protection at plant level, ensuring safe operation while fully optimizing the power availability.

The ECPC together with the SCUs forms the Electron Cyclotron Control System (ECCS). Control and protection functions are always pushed to the lower possible controller able to implement them. To facilitate the SCUs' control system development and to reduce the risk of re-implementation of components, all the functions involving more than one sub-system are implemented in the ECPC.

The expected operation scenarios of the EC System are classified as those that are synchronous and those that are asynchronous to the ITER pulse. As high availability of the EC System is required, the ECCS will allow a part of the EC plant to be operated synchronously on the plasma, while another part of the plant will be operated asynchronously (e.g. for commissioning and maintenance). The coordination of the ECPC with ITER Central Instrumentation and Control (I&C) and SCUs is done using a set of hierarchical state machines and sequences, detailed in sections 3 and 5.

The ECPC development was divided into stages as it will firstly be used only for the acceptance and integration tests of the SCUs, and later also for the plasma operation:

- Stage 1: Transportable system able to verify the correct functionality of the SCUs interfaces to the ECPC
- Stage 2: Gyrotron Commissioning control system able to implement integrated protection of a partial EC plant and support SAT (Site Acceptance Tests) of a Gyrotron, HVPS and TL SCUs
- Stage 3: First Plasma system
- Stage 4: Fully functional system implementing complete control techniques

Stage 1 and 2 control systems were successfully developed, manufactured, and delivered to ITER in December 2021. The delivered system architecture of the ECPC supports testing of all the interfaces and functionalities of the SCUs. The first use-case of the developed ECPC control system is the SAT campaign

* Corresponding author: giuseppe.carannante@iter.org

https://doi.org/10.1051/epjconf/202327704005

of the EUDA HVPS planned to start at the beginning of 2023. During the development of the EUDA HVPS system, a conclusion was reached that no market available components were able to satisfy the specifications of the Gyrotron to HVPS interface (for references and measurements). Therefore, bespoke modules (section 6) were developed.

Even though first plasma just requires the ability to schedule the gyrotron operation for short pulses, many advanced functionalities are already present in the following plasma campaign, e.g. switching gyrotrons to different launchers in real-time, modulating the gyrotron power to synchronize with the rotation of the neoclassical tearing mode (NTM), automatic conditioning of gyrotrons. These advanced functions require all the systems to minimise the preparation time and to accept real-time references. The need to condition a gyrotron for a long pulse is one of the weakest points in this context. A research activity is being conducted to develop a model-based control of the gyrotron heater to maintain it always prepared for long pulse operation.

## 2 Functions and architecture

The implementation was based on the ECCS design presented on the Conceptual Design Review (CDR) in November 2013. All the control and protection requirements were identified and allocated to the different controllers of the I&C architecture.

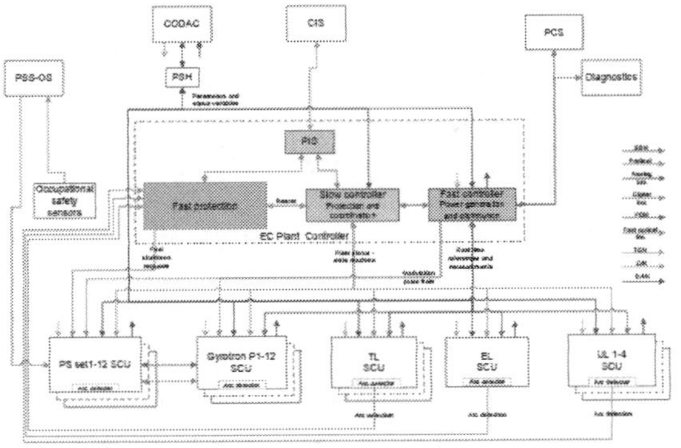

**Fig. 1.** ECCS Architecture.

The main functions of the ECPC as per [3] are the following:

- Manage EC subsystems parameters and waveforms
- Allow operation of each single EC subsystem
- Coordinate the state machines of the EC plant
- Control the EC plant following real-time requests and references coming from the PCS
- Implement fast and slow protection functions (the choice of components given in section 2.2)
- Publish to CODAC (Control, Data Access and Communication) signals to be monitored

The ECCS architecture, with the ECPC components marked in the dashed box, is given on **Fig. 1**. A brief description of the components follows.

### 2.1 ECCS Networks and interfaces

Given the uncertainties in the project, reducing the number of physical interfaces and at the same time allowing for increasing and changing them has been attained by adopting standard network interfaces, like Synchronous Databus Network (SDN) among fast controllers and Profinet among PLCs. Moreover, the interfaces towards CODAC, as shown in [4], had to be respected. CODAC System and Networks provide overall plant systems coordination, supervision, plant status monitoring, alarm handling, data archiving, plant visualization (HMI) and remote experimental functions.

The networks which are used in the ECCS and their functionalities are:

- Plant Operating Network (PON): slow data monitoring; parameters setting; local clock synchronization in slow controllers; state machine management of the ECPC and SCUs when in asynchronous mode
- Data Archiving Network (DAN): delivery of acquired fast data to the ITER data storage system (e.g. gyrotron current measurements)
- Time Communication Network (TCN): local clock synchronization in EC fast controllers
- SDN: delivery of real-time data between plant systems; fast communication to exchange reference, scheduling and diagnostic information between EC fast controllers
- Profinet: connects all EC slow controllers; allows SCU state machine and protection functions management in synchronous operation

The majority of the SCUs have all the listed network interfaces. However, the HVPS SCU also has an interface towards the PSS-OS (Plant Safety System for Occupational Safety), as it includes control of high voltage components which have to be operated in a way compliant with occupational safety.

In addition to the CODAC services and internal EC connections, the ECPC interfaces with the following ITER central I&C services:

- The Central Safety System (CSS): plant-wide nuclear (CSS-N) and occupational safety (CSS-OS)
- The Central Interlock System (CIS): plant-wide investment protection

Moreover, the ECPC will implement an interface towards the PCS for real-time power requests.

### 2.2 Protection functions implementation

The Slow Controllers (PLC) and the Fast Protection Controller (FPGA) are used to implement the protection functions, based on the following criteria:

- PLCs are used for protection functions that require reaction times of >50 ms.
- FPGAs are used for protection functions that require faster reaction times.
- Optical fiber is the preferred solution for fast protection functions in order to improve the speed, noise reduction and isolation

In addition to the implementation of the plant level slow protection functions (e.g. over-temperatures, lack of flows), the Plant Slow Controller configures the Fast Protection Controller protection functions and it is therefore the master of the plant protection.

The Plant Fast Protection Controller sends shutdown requests to the HVPS SCUs based on the configured protections. After receiving the shutdown request, the HVPS will remove the output voltage in less than 10 μs.

### 2.3 Control functions implementation

The Plant Fast Controller is the master of the plant operation, it defines which SCUs operate synchronously and manages the state machines of the synchronous subsystems. The Plant Fast Controller interfaces with the Central I&C to receive requests which are then translated into SCU commands and waveforms. It will schedule the EC operation by assigning the available Gyrotrons and Launchers to different control functions. It will compute in real-time the power references for the Gyrotrons, the polarization for the TL and the mirrors positions for the launchers.

The Plant Fast Controller receives from the Plant Fast Protection Controller a packet containing the information of the faults and events that have happened in the plant. These events are then timestamped using the central TCN time and are published to the Central I&C on DAN.

The Plant Slow Controller relays the state machine commands and statuses between the Plant Fast Controller and the SCU Slow Controllers, which manage the SCUs' state machines.

## 3 State machine and operation management

The orchestration and operation synchronization between all the EC plant systems is based on a set of state machines and sequences.

The ECPC state machine, given on **Fig. 2** is the highest node in the hierarchy of state machines. It interfaces to Central I&C and coordinates all the systems of the plant which are to be operated synchronously. To define which systems are to be operated synchronously, the concept of mode is introduced. A mode will indicate which SCUs are expected to be slaves of the ECPC. The evolution through the states of the ECPC state machine will automatically trigger actions on the state machines of the SCUs which will then trigger sequences of SCU sub-systems (e.g. auxiliary power supplies). The conditions of each ECPC state and sub-state transitions are linked to the states of the SCUs defined for a given mode. Plant protections will automatically be updated with the selection of a different mode.

Each ECPC state has three sub-states:

- Progress – a transition has been started and is ongoing in one or more SCUs. A command is issued to the SCU to start a sequence. The SCU sequence then issues commands to SCU sub-system sequences
- Stable – the transition is finished and the conditions for a given state have been fulfilled. SCU states are compliant
- Error – the transition which was started has failed or the condition for the stable sub-state has been lost

The main ECPC states are:

- Idle –checking that all the services have been started and that communication is established between the different ECPC controllers
- Configure Protections – a state in which new definitions of the plant protections can be transferred
- Standby – SCUs required for the mode are in Standby state (e.g. gyrotron auxiliaries in operation)
- Ready – SCUs are ready to receive a trigger to start a pulse
- Online – power generation according to received inputs. Trigger signals sent to SCUs

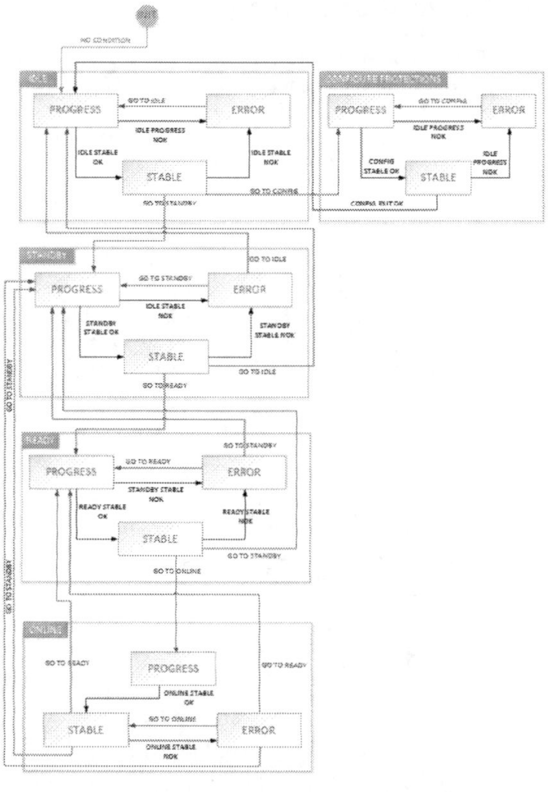


**Fig. 2.** ECPC State Machine.

An example of the state machine coordinated behaviour follows. One of the standby modes definitions of the GCU is normal operation of one gyrotron. Normal operation means that all the auxiliary sub-systems of the gyrotron are required to be functioning. The GCU will receive a command from the ECPC to prepare the system for Standby. The GCU will automatically issue commands, in a given order, to the different gyrotron sub-systems. When all the sub-systems are functioning, the GCU condition for Standby will be fulfilled. This information will then be transmitted back to the ECPC and the state of the ECPC state machine will be updated accordingly.

A prototype of both the ECPC and the Gyrotron SCU were deployed and tested at the FALCON test facility at the Swiss Plasma Centre (SPC) [5].

## 4 Frameworks and libraries

To ease the development process, the implementation of the control system is based on frameworks and libraries. This facilitates debugging and application maintenance, and it presents a time-saving, a more efficient and cleaner way of implementing. When possible, the frameworks already in place in other ITER systems were re-used, and when not, custom solutions for the EC system were developed.

The Fast Controller development is based on the MARTe [6] framework. MARTe is a C++ modular and multi-platform framework for the development of real-time control system applications. The main advantages of the framework are the separation between the platform specific implementation, the environment details and the user code, as well as the software quality assurance process of the components. This framework is used in many ITER projects and it is spread throughout the entire fusion community.

The Slow Controller integration with instrumentation uses the Unified Control Library (UCL) objects. UCL has been developed as part of a joint IO and F4E effort to facilitate the development, commissioning, and maintenance of industrial ITER systems (e.g. Vacuum system, Cooling Water System). The UCL objects are divided into families (e.g. input/output, alarm, control) where each of them covers a certain layer within the application. The benefit of the UCL objects is that they provide standardized interfaces towards the field, as well as towards CODAC.

## 5 Gyrotron controller

The SCU with the most complex set of controllers and interfaces is the Gyrotron Controller SCU (GCU), managing a pair of gyrotrons. The GCU is defined with a Standby and a Ready mode which describe the set of sub-systems that should be operational in a given state. The modes present degrees of freedom as the plant can easily be reconfigured for different types of operation, ranging from gyrotron conditioning on a dummy load to firing with the pair of gyrotrons. System protections are automatically updated based on the selected Standby and Ready modes.

The GCU Fast Controller implements a set of state machines, linked to the preparation of the system and the real-time interface and operation. It receives the trigger from the Plant Fast Controller to start the power on the HVPS. It interfaces to the HVPS SCU to which it sends analogue references and digital triggers and acquires the HVPS measurements and statuses. Measurements and references are received and sent using analogue optical cards (described in section 5).

The GCU Slow Controller is the master of the SCU control system and responsible for the preparation of all the auxiliary power supplies of the gyrotron pair, the HVPS state machine and the GCU Fast Controller state machine. The GCU Slow Controller implements a set of configurable slow protections which are defined for the different modes of the operation of the GCU (e.g. operation with one or two gyrotrons). It configures the fast protections of the system in the GCU Fast Protection Controller.

The GCU Fast Protection Controller gathers protection information from the plant, e.g. arcs, and issues a fast shutdown petition to the HVPS used for the operation of the pair of gyrotrons.

The GCU Fast Controller implements the real-time control of the gyrotron. It receives requests to switch on and off the gyrotron from the Plant Fast Controller and propagates this request to the HVPS. This architecture allows to switch on the gyrotron power after only few milliseconds from the request.

### 5.1 Universal Gyrotron Controller framework

A Universal Gyrotron Controller (UGC) framework was developed to provide the mechanisms to manage all types of ITER Gyrotron Controllers (SCUs) using a variety of hardware platforms. UGC is a software framework used to perform the following functions:

- Manage Gyrotron auxiliaries
- Manage HV power supplies
- Implement protection functions
- Implement the standardized ECPC-SCU interface
- Implement the standardized CODAC interface

The software architecture of the SCU identifies multiple layers interconnected by a number of internal buses, as shown on **Fig. 3**. Starting from the bottom layer, the Device Access Layer, contains all the software that is necessary to handle any peripheral. This ranges from IO cards to complex digital/analogue interfaces to auxiliary devices.

The main components of the Device Access layer are the following blocks:

- Basic step – performs a single action on the auxiliary device, e.g. sets one output on the PLC
- Level 0 sequence – description of a fixed set of functions, e.g. switching on of a power supply
- Communication block – standard interface block for the different communication protocols, e.g. RS232, MODBUS
- Subsystem Status monitor – checks and publishes the different states of a subsystem
- HVPS reference generator and data acquisition – provides an interface of the Fast Controller to the HVPSs. Sends voltage and current references, trips and commands. Receives measurements as well as statuses. Measurement acquisition and reference generation are done via Hotlink modules (described in section 6)

The layer above the Device Access layers consists of SCU Process Blocks, which are components that process data that is read from or written to one of the internal buses. The main blocks are:

- Command Manager – pre-processing of the different sources of sequence commands and priorities. Issues individual commands to each level 0 sequence of a subsystem
- State Manager – implements a state machine to manage the transitions between the subsystem manual/auto and segregation states

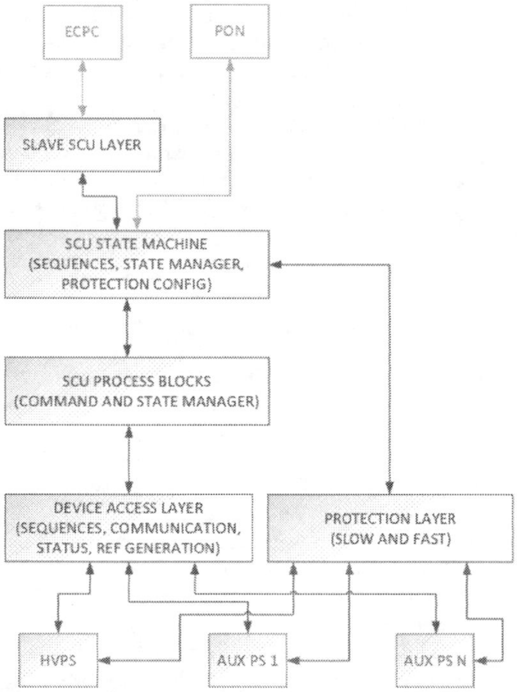


**Fig. 3.** UGC layers.

Above that, we find the SCU State Machine layer. This is used to control the SCU operation logic with a cluster of state machines and sequences. The main state machine of the SCU manages the overall operation of the SCU. It prepares the system outside of a pulse and manages the operation during a pulse. Another SCU state machine defines if the SCU is in asynchronous operation (management done through PON) or if it is in synchronous operation (slave to ECPC, as described in section 3). The command manager uses this state to filter and issue the commands to the bottom layer.

The SCU state machine takes into consideration the current state of the whole plant, the state machine and the commands coming from the command manager. When a condition is fulfilled, the state of the state machine will be changed.

The top layer is the Slave SCU Layer. It contains the description of the SCU behaviour toward the ECPC.

An additional function implemented in the UGC is the gyrotron protection. The protection layer implements this function by issuing trips to the different power supplies based on a defined set of protections. The protections are configured for different modes of operation of the plant, and when changing from one mode to another, the corresponding protections are loaded automatically. The protection functions can be executed either in the Slow Controller or in the Fast Protection Controller (see Section 2.2). The processing of the input information in the Fast Protection Controller is done in stages (as per figure **Fig. 4**):

- Input stage – synchronizes the digital inputs to the system clock through a chain of flip flops. Analyses each input (positive/negative logic; pulse train) and generates an alarm for further stages
- Logic function block stage – performs logical function operations on the alarms coming from the input stage or the logic function block stage
- Output stage – selects which alarms coming from the input stage or the logic function block stage should issue a fast shutdown signal to the HVPS

The Slow Protection Controller implements the same stages and can process both digital and analogue inputs.

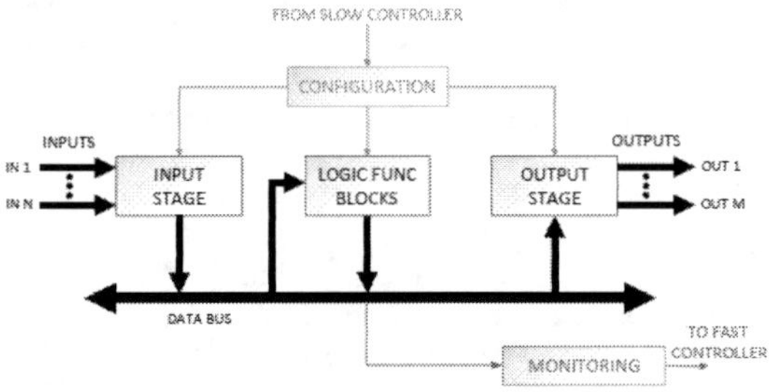


**Fig. 4.** Fast Protection Controller.

## 6 Hotlink modules

The requirements on the reference generation and data acquisition for the HVPS were such that no "off the shelf" IO card available on the market was adequate. There was a need for a solution which provides a high speed interface with optical isolation; is compatible with the National Instruments Compact RIO (cRIO) chassis and compatible with the solution of the HVPS supplier. Therefore, Hotlink (HOTLink® family of data communication products provides a simple and low-cost solution to high-speed data transmission) modules, were designed and manufactured. The main application of the modules is to implement a high speed analogue interface between the HVPS SCUs and the Gyrotron Controller SCU.

The CRIO-Hotlink TX and CRIO-Hotlink-RX modules are respectively a high speed analogue to optical converter and a high speed optical to analogue converter.

The modules have three interfaces (as shown on figure **Fig. 5**):

- Analogue interface – two differential inputs and two single ended output channels
- Optical serial interface – transmission / reception of 200 Mbit digital data

- Digital parallel interface – 8bit parallel bus compatible with the cRIO chassis

The modules can be configured for full speed of 5 MSa/s for single channel transmission or 2.5 MSa/S for double channel mode.

The Hotlink RX operation mode is to receive data from the optical interface and to send data both to the analogue output channel and the digital interface. The Hotlink TX operation modes are to transmit data:

- acquired from the analogue input channel to both the optical and the digital interface;
- read from the cRIO 8bit digital bus to the optical interface and the analogue outputs.

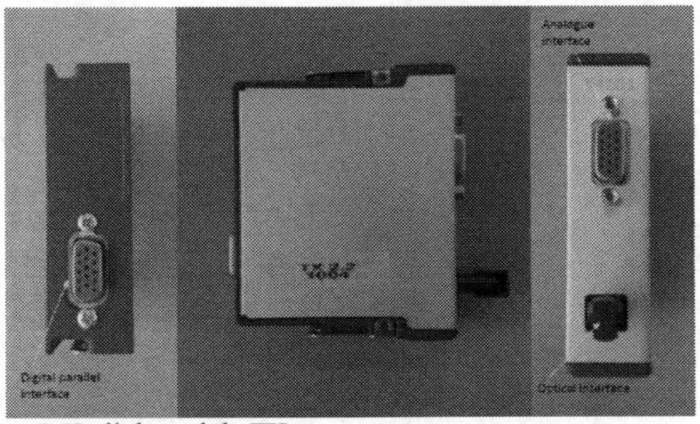


**Fig. 5.** Hotlink module TX.

An example of the interfaces and modules operation modes is given on figure **Fig. 6**, where a pair of Hotlink modules is in the following setup: a signal is generated on the oscilloscope and fed into the analogue input of the TX module. That signal is then transmitted on the optical interface of the TX module to the optical interface of the RX module and also on the analogue output of the TX module. The analogue output of the TX module is displayed on the oscilloscope as CH1. The acquired optical signal on the RX is transmitted on the analogue output of the module and displayed as CH2 on the oscilloscope.

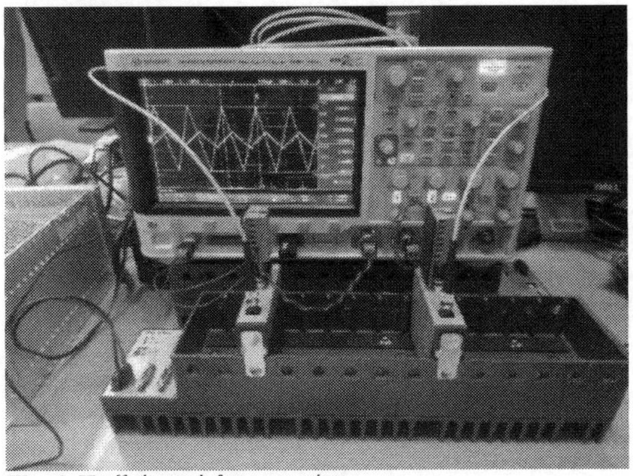

**Fig. 6.** Hotlink modules operation.

## 7 Future activities

One of the biggest challenges of the gyrotron long pulse operation is how to keep the gyrotron ready. Nowadays, the gyrotron has to be conditioned with a series of short pulses before it reaches conditions which are suitable for a long pulse campaign. In order to maintain the gyrotron ready, the state of the gyrotron heater should be known. To be able to estimate the state, a model-based control of the gyrotron heater is being investigated.

Other future activities include the start of the SAT of the EUDA HVPS in the beginning of 2023, the SAT of gyrotrons in the second half of 2023 and the preparation of the first plasma control system final design review which is planned for 2024.

## 8 Conclusion

A description of the current status of the ECCS control system was presented. The ECCS system consists of the Plant Controller and different SCU subsystems. Both the ECPC and the SCU controllers are a cluster of Fast, Slow and Fast Protection Controllers, where each implements a given set of functions. The interfaces and networks used in the control system comply with the ITER standards. The coordination of the ECPC with Central I&C and SCUs is done using a set of hierarchical state machines and sequences. To define which SCU systems are operated synchronously to the pulse and which asynchronously, the ECPC mode is introduced. Software standardization is achieved with the usage of frameworks. Hardware of the ECCS is based on "off-the-shelf" components. However, the HVPS reference generation and measurement acquisition requirements were such that bespoke modules, Hotlink modules, were designed and manufactured.

*The views and opinions expressed herein do not necessarily reflect those of the ITER Organization.*

## References

1. F. Albajar, G. Aiello, S. Alberti et al, "Status of Europe's contribution to the ITER RC system", EPJ Web of Conferences 87 (2015)
2. T. Gassmann, B. Beaumont, et al, "Integration of IC/EC systems in ITER", Fusion Eng. Des., 85, 1245–1249 (2010)
3. G. Carannante, M. Cavinato, F. Gandini, G. Granucci, M. Henderson et al., Fusion Eng. Des., 96-97, 420-424 (2015)
4. ITER Organization, "CODAC Architecture", https://www.iter.org/mach/codac/architecture
5. D. Fasel, S. Alberti, J. Dubray, T. Goodman et al, Fusion Eng. Des., 146, 1942-1946 (2019)
6. A. C. Neto, F. Sartori et al., "An agile quality assurance framework for the development of fusion real-time applications," 2016 IEEE-NPSS Real Time Conference (RT), 1-7 (2016)

# Development of the Multi-Beam Transmission Line for DTT ECRH system

*Saul* Garavaglia[1*], *Alex* Bruschi[1], *Francesco* Fanale[2,3], *Gustavo* Granucci[1], *Alessandro* Moro[1], *Paola* Platania[1], *Afra* Romano[2,3], *Stefan* Schmuck[1], *Alessandro* Simonetto[1] and *Espedito* Vassallo[1]

[1]Institute for Plasma Science and Technology – CNR, via Cozzi 53, 20125 Milano, Italy
[2]ENEA, Fusion and Nuclear Safety Department, C. R. Frascati, via Fermi 45, 00044 Frascati, Italy
[3]DTT S.C. a r.l., via Fermi 45, 00044 Frascati, Italy

**Abstract.** The DTT tokamak, whose construction is starting in Frascati (Italy), will be equipped with an ECRH system of 16 MW for the first operation phase and with a total of 32 gyrotrons (170 GHz, ≥ 1 MW, 100 s), organized in 4 clusters of 8 units each in the final design stage. To transmit this large number of power beams from the gyrotron hall to the torus hall building a Quasi-Optical (QO) approach has been chosen by a multi-beam transmission line (MBTL) similar to the one installed at W7-X Stellarator. This compact solution, mainly composed of mirrors in "square arrangement" shared by 8 different beams, minimizes the mode conversion losses. The single-beam QOTL is used to connect each gyrotron MOU output to a beam-combiner mirror unit and, after the MBTL, from a beam-splitter mirror unit to the ex-vessel and launchers sections located in the equatorial and upper ports of 4 DTT sectors. A novelty introduced is that the mirrors of the TLs are embodied in a vacuum enclosure, using metal gaskets, to avoid atmospheric absorption losses and microwave leaks. The TL, designed for up to 1.5 MW per single power beam, will have a total optical path length between 84 m and 138 m from the gyrotrons to the launchers. The main straight section will travel along an elevated corridor ~10 m above the ground level. The development of the optical design reflects the constraints due to existing buildings and expected neutron flux during plasma operation. In addition, the power throughput of at least 90% should be achieved.

## 1 Introduction

The Divertor Tokamak Test (DTT) is a superconducting device ($B_T$=6 T, $I_P$=5.5 MA, $R_0$=2.19 m, a=0.7 m, pulse duration of 100 s, 1 pulse per hour) under construction in ENEA Frascati (Rome, Italy) [1]. DTT will study suitable solutions for the power exhaust issue in conditions relevant for the future fusion device DEMO [2]. For this purpose, a key role is assigned to the additional heating systems; DTT will be equipped with ECRH, ICRH and NBI for a total power of 45 MW. The ECRH system will provide 16 MW for the first operation phase and a total of 32 gyrotrons (170 GHz, ≥ 1 MW, 100 s); a joint procurement with F4E for the first 16 gyrotrons is started and the acceptance of the first pre-series gyrotron is foreseen in the second half of 2023. The ECRH system is linked with several physics tasks including the core heating and MHD mode control [3]. The effective capability of the ECRH system to fulfil the requirements has been verified with the use of beam tracing GRAY [4, 5] considering the reference DTT Single-Null full power scenario [6].
The development strategy is based on the implementation of automatic operations, high reliability, high modularity and reduced maintenance. Existing and assessed technology (exploiting ITER and W7-X experience) is preferred according to the ambitious schedule for the first plasma in 2028. The architecture of the system (see Fig. 1) consists of 4 clusters, each one equipped with 8 gyrotrons. The gyrotron output Gaussian beams are transmitted by an evacuated Quasi-Optical (QO) single/multi-beam transmission line (SBTL/MBTL) towards two antennas, one upper and one equatorial, placed in the same DTT sector.

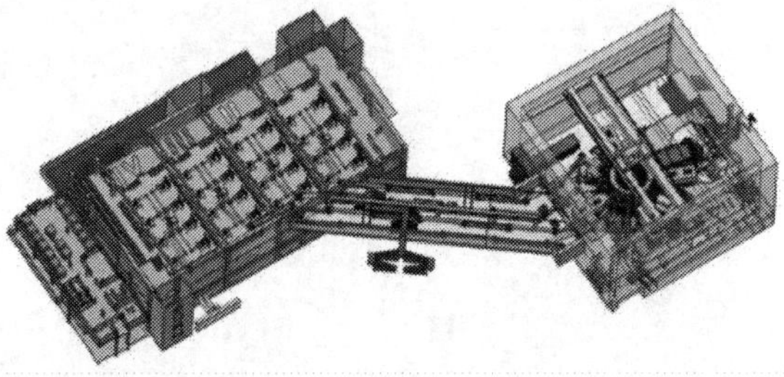

**Fig. 1.** 3D CAD model of ECRH building (left), THB (right) and the joining bridge (centre).

The ECRH building is organized on three different levels: the basement hosts the cooling system and the electrical substation, the ground level carries 16 gyrotrons High Voltage Power Supplies (HVPS)

* Corresponding author: saul.garavaglia@istp.cnr.it

groups, and the first level accommodates 4 gyrotron clusters. The 4 MBTLs run from the ECRH building to the Torus Hall Building (THB) over a bridge, penetrate the bioshield of the THB at 2.6 m above the equatorial plane of the machine and are connected to the upper and the equatorial ports of the tokamak sectors 12, 14, 16 and 18 dedicated to ECRH [7]. This paper reports the updated description and the solution adopted for the transmission line (TL) of the ECRH system, together with the wave propagation assessment to validate the model.

## 2 TL Conceptual design

The TL concept is based on the QO propagation of Gaussian beams at 170 GHz under vacuum, realized by a set of mirrors in a confocal arrangement. For the most part of the path, the TL exploits the multiple beam option with 8 $TEM_{00}$ Gaussian beams, separated at the input plane, sharing a set of oversized parabolic focusing and flat mirrors, similar to the W7-X stellarator TL [8]. In order to minimise the transmission losses and arc risk, TL mirrors are proposed for the first time embodied in a vacuum enclosure using large metal seals. The design requirements are a target transmission efficiency of 90% and a power handling capability of 1.5 MW per single beam allowing any future power gyrotron upgrade. The design of the whole TL is a trade-off between the mirrors dimension (depending on the beam radius on the mirror surface) and the distance between two focusing mirrors. The main MBTL unit is composed by 4 identical dog-legs, including each a focusing mirror and a flat mirror at 90° reflections (Fig. 2).

**Fig. 2.** 3D CAD model of MBTL basic concept, composed by 8 reflecting mirrors (4 focusing and 4 flat mirrors) with 90° reflections.

The focusing mirrors are in square configuration in order to compensate the aberrations due to the off-axis reflection of the beams [9]. The length of this TL section is chosen to cover the straight corridor (Fig. 3). The wave polarisation is selected at the input of each MBTL to minimise the ohmic losses due to mirror reflections. The TL is divided in three parts (Fig. 3): the first connects each gyrotron of one cluster to the combiner mirror at the geometrical centre of the cluster. The second part is the MBTL section that is from the combiner mirror of each cluster to the splitter mirror in the THB. In the last part the eight beams are divided into a bundle of six beams directed inside port 3 (equatorial) and two beams inside port 2 (upper), housing the Equatorial Antenna (EA) and the Upper Antenna (UA) respectively.

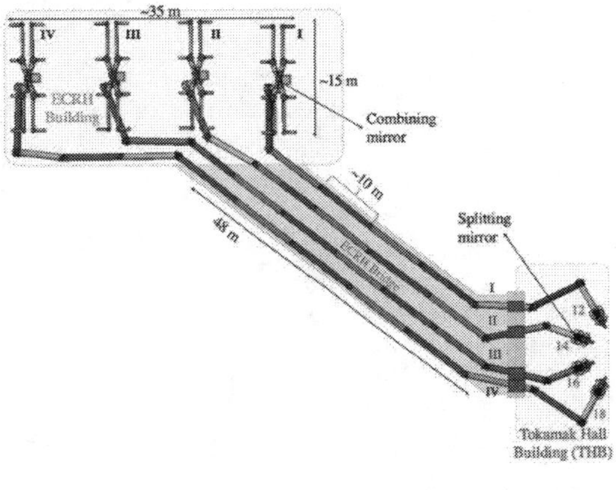


**Fig. 3.** Top view of TLs, including the SBTL in the ECRH building (top left), the MBTL along the bridge (centre) and the penetrations in the THB with the connections to the DTT dedicated sectors (bottom right).

### 2.1 Single-Beam TL in ECRH building

The design of the single-beam transmission line (SBTL) in the ECRH building considers several requirements including the minimisation of the space occupied by the 4 clusters, the minimization of the number and the types of required mirrors, the choice of a beam size small enough to reduce the vacuum enclosure and the number of vacuum components. For these reasons, the 8 gyrotrons of each cluster are arranged into 2 rows of 4 items each, with a 5 m distance between two gyrotrons to avoid mutual magnetic disturbance due to the superconducting magnets (Fig. 4). The 8 output beams are refocussed and collected together in the centre of the cluster on a combining mirror achieving a two axes symmetric cluster arrangement. The created bundle of beams is then reflected downwards to propagate inside the MBTL.

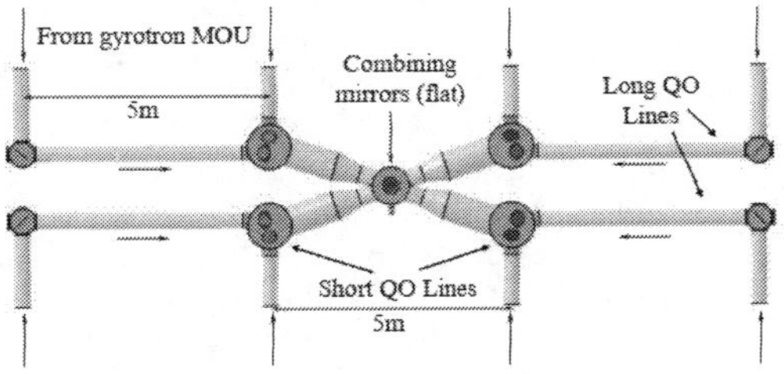


**Fig. 4.** Top view of the 3D CAD model of SBTL arrangement for one cluster in the ECRH building.

This geometry reduces the number of the matching optics types to one for the 4 nearby gyrotrons (short) and one for the remaining gyrotrons (long), maximising the modularity. Each path is composed by one (short) or two (long) single focusing mirrors respectively and one plane polarising mirror to set the favourable input polarisation for the MBTL to minimise the ohmic losses. For each line, the last mirror before the combining mirror is movable to reflect the incident beam towards the combining mirror or the rotable plane mirror placed in the geometrical centre of the cluster above the combining mirror. This rotable mirror directs the beams, one at a time, toward a RF load close to the combining unit through a second fixed focusing mirror located below it as shown in Fig. 5.

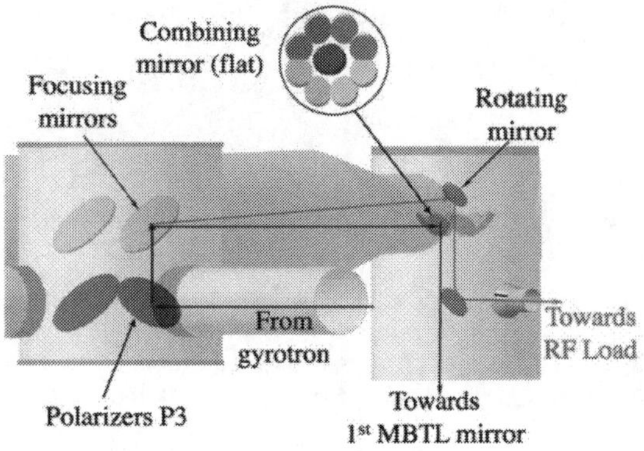


**Fig. 5.** Detail of the connection between SBTL and RF load. The beam coming from the gyrotron can be tilted by a focusing mirror to reach, on top in the centre of the combining mirror, the rotating plane mirror. This mirror is designed to redirect the beam vertically down where the fixed mirror reshapes the beam and directs it into the load (red arrow).

## 2.2 Multi-Beam TL

The MBTL consists of a straight section running along the corridor. At its ends, the connections have been realized in telescopic arrangement to reach the combining mirror under the gyrotrons cluster centres on one side and the splitting mirror in front of the dedicated ECRH tokamak sectors on the other side (Fig. 3). The 8 beams are arranged on the vertices of an ideal octagon (Fig. 5 top) and start (and end) parallel to run in the MBTL spatially separated. They propagate alternatively crossing or parallel to each other between two focusing mirrors on which they have maximum size and minimum power density. The MBTL design considered several constraints including the positions of the pillars and walls of existing buildings, the minimum clearance to be left around the TLs for maintenance, and the use of the same ex-vessel structure in front of the DTT sector. A particular attention has been dedicated to the bio-shield penetration in the THB to limit the neutron streaming in accordance with the restriction rules [10]. In the corridor, the mirror units are positioned at ~10 m distance whereas outside the corridor the distance between focusing mirrors has been reduced and adapted depending on the various paths. Only five types of mirrors with four different dimensions are necessary for the four clusters.

## 2.3 Single-Beam TL in THB

The last section of the TL matches the beams coming from the MBTL with two types of antennas: two beams are redirected to the upper antenna in port 2 and six beams to the equatorial antenna in port 3. The ex-vessel single-beam optics in the THB is the same for all 4 clusters (Fig. 6). The splitting mirror is located in front of the corresponding tokamak sector at about half height between port 2 and port 3. Only one focusing mirror is included to match each single beam at the splitting mirror to the one at the input of waveguide (WG) before the launchers. Two polarising mirrors in z-shape optical path are also included in order to optimise the coupling between the launched wave and the plasma in real time. Three optics configurations have been found: one for the two beams launched from port 2 (yellow, Fig. 6 left), one for the three beams of the top launchers (green) and one of the bottom launchers of port 3 (magenta).

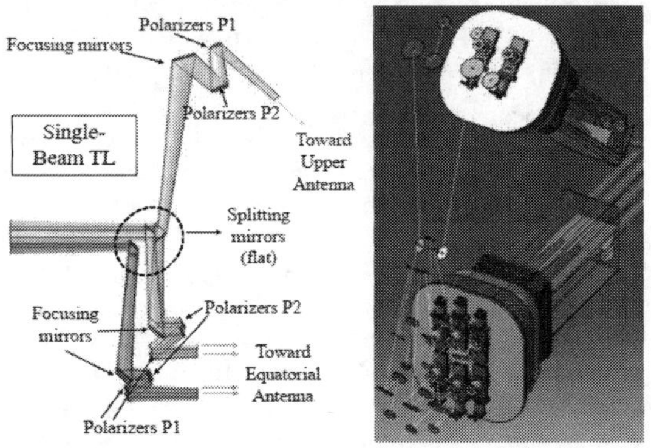


**Fig. 6.** Left: SBTL design in the THB. Right: upper and equatorial launchers and ex-vessel transmission lines.

## 2.4 Mirrors

The design criteria of the MBTL mirrors are driven by the lowest temperature increase of the reflectors during the RF discharge and the minimum surface deformation. The latter seems to depend critically on the temperature inhomogeneity due to the heat load. For this reason, a set of different combinations of materials and cooling solutions has been evaluated. The present reference is a copper body of 20 mm thickness on top of which a planar cooling channel layer is obtained, and a cover copper layer of a minimum thickness of 4 mm. The mirror is water cooled through a double spiral circuit with semi-circular channels section. The mirrors have an elliptical shape and the dimensions, sized to allow the reflection of 99.97% of power, are between ~1.1 × 0.8 m for the MBTL and ~0.35 × 0.25 m for the SBTL. Table 1 reports the number of mirrors for one representative beam of each of the 4 TLs, the total number of mirrors for the MBTL, the path length (considering short and long lines in the SBTL section) and the volume.

**Table 1.** Number of mirrors, length and volume of TLs.

| TL | MBTL mirrors | Total mirrors | Length [m] | Volume [m$^3$] |
|---|---|---|---|---|
| I | 17 | 77 | 84-90 | 68 |
| II | 17 | 77 | 89-95 | 71 |
| III | 21 | 81 | 104-110 | 78 |
| IV | 25 | 85 | 130-137.5 | 87 |

### 2.5 Vacuum system

One of the novelties of the system is the evacuation of the QOTL. The vacuum envelope is a cylindrical chamber that contains the mirrors and the related structures at ~1 mPa of target pressure. The material of the enclosure is stainless steel (a lighter option in Aluminium is under evaluation) with a diameter of 800 mm (300 mm) and thickness of 6 mm (3 mm) for the MBTL (SBTL). A preliminary design of the pumping system was carried out as a function of volume to be pumped out and of surface of the vacuum chamber. Two pumping units are planned at the ends of each TL near the gyrotron clusters and close to the THB in the corridor. The DTT vessel volume is ~100 m$^3$ at a pressure of ~$10^{-2}$-$10^{-3}$ mPa (~10 mPa during the discharge). At present, the connection between the vacuum vessel and TL will be regulated only by an all-metal gate valves without the presence of a diamond windows as used in other tokamaks. This valve is normally closed and is opened only during the ECRH pulse. An analysis of the vacuum system dynamics [11] revealed the need to drastically reduce the conductance between the TL and vacuum vessel not to compromise plasma operations. In fact, at the simultaneous aperture of the 32 gate valves, a gas flow from the tokamak of the same order of the gas injected in the DTT vacuum vessel to sustain the plasma density is expected. For this reason, a section of corrugated WG 63.5 mm diameter, ~1.5 m length has been included in the beam path at the transition between the SBTL enclosure and the launcher internal part to marginalise the impact on the tokamak gas puffing. However due to the uncertainties on the parameters used in the simulations, the variability of DTT scenarios and different operating modes of the ECRH system, a back-up solution with the later use of a diamond windows will remain possible. The optical design has consequently been made compatible for the matching with the WG insertion.

### 2.6 Propagation assessment

In order to investigate the TL design model, evaluations of the propagating power density and losses have been performed with the use of GRASP [12]. After the four TLs have been modelled in the GRASP environment, the beam propagation along the TL has been simulated. The following results refer to the ideal case, obtained considering the TL optically aligned and undeformed mirror surfaces. The coupling losses are low as expected in the modular straight part of the MBTL (0.16%) and higher outside (between 2% and 2.35% depending on the TL). The spillover is of the order of 0.05% for all the TLs due to the mirrors size choice which means that the radiated power not intercepted by the mirrors and enclosed in the vacuum chamber is of the order of a few kW per line. The ohmic losses, depending on the polarisation selected at the input, range between 2.2% and 4.0% including a factor of 1.3 for the mirror's roughness. The total loss is estimated to be between 4.7% and 6.3% for the shortest/longest TL, when SBTL mirrors are included in the analysis. The additional coupling loss due to the transition of WG ($TEM_{00}$-$HE_{11}$-$TEM_{00}$) in the optical path introduces an extra maximum contribution of 4%, which can be reduced either by tuning the waveguide length and beam waist or by adding an optimized $HE_{11}$-$TEM_{00}$ taper at least on one end. To reach a more realistic and all-inclusive estimate of losses, two analyses are still ongoing. The first is the inclusion in the GRASP model of the deformation effects of the mirrors surface due to the thermal load that can introduce effects on the beam alignment. The second analysis calculates the required tolerances in the alignment of the mirrors with a software based on Monte Carlo analysis of the geometric optics [13].

## 3 Conclusion

Progress in the development of the MB TL of DTT's ECRH system is presented. The layout of the system has been consolidated with a modular architecture based on clusters. The design of the Single/Multi-Beam QO TL is extensively studied including the novel solution to evacuate the TL. The next efforts will be dedicated to complete and validate the TL conceptual design including the deformation effects in the electromagnetic analysis with an improved mirror cooling to complete the evaluation of the impact on the losses.

## References

1. R. Martone et al., ISBN 978-88-8286-378-4 (2019)
2. R. Ambrosino et al., Fus. Eng. and Des. **167**, (2021) 112330
3. G. Granucci et al., Fusion Eng. Des. **122** (2017) 349–355
4. D. Farina et al., Fusion Sci. Tech. **52**, 154 (2007)
5. B. Baiocchi et al., this conference (2022)
6. I. Casiraghi et al., Nucl. Fusion **61** (2021) 116068
7. S. Garavaglia et al., Fusion Eng. Des. **168** (2021) 112678
8. V. Erckmann et al., Fusion Sci. Technol. **52** (2007) 291–312.
9. L. Empacher, W. Kasparek, IEEE Trans Antennas Propag. **49** (2001) 483–493
10. A. Colangeli et al., Fus. Eng. and Des. **171**, (2021)112690
11. S. Garavaglia et al., Journal of Vacuum Science and Tech. B *under publication*
12. GRASP, Technical Description, TICRA (www.ticra.com)

13. A. Simonetto, ISTP internal report FP20/05 (rev. Apr. 2022)

# Gyrotron anode voltage control in EAST ECRH system

*Weiye Xu*[1*], *Handong Xu*[1†], *Jian Zhang*[1], *Fei Guo*[1], and *Haozhang Sun*[1]

[1]Institute of Plasma Physics Chinese Academy of Sciences, 350 Shushanhu Rd., Hefei 230031, China

**Abstract.** Four diode-type gyrotrons are used in the EAST ECRH system now. The anode is one of the main components of the gyrotron. We can control the output power of the gyrotron by changing the anode voltage. An anode voltage control system for ASIPP anode power supplies was developed based on the ethernet CompactDAQ chassis NI-cDAQ-9185. The anode voltage can be controlled up to 30 kV, and the maximum current is 100 mA. The maximum modulation frequency is 5 kHz at a 50% duty cycle.

## 1 Introduction

There are now four diode-type [1] gyrotrons [2] used in the EAST ECRH system [3]. According to the latest naming rules of the gyrotrons in EAST, the #1, #3, and #4 gyrotrons are from GYCOM [4], and the #2 gyrotron is from CPI. The output frequency of the #1, #2, and #3 gyrotrons is 140GHz. The output frequency of the #4 gyrotron is 105GHz or 140GHz [5]. A separate anode power supply is required for each gyrotron. The magnitude of the anode voltage affects the microwave power output by the gyrotron, as well as the electric-wave conversion efficiency.

The anode power supplies for #1 and #4 gyrotrons are from GYCOM. The parameters of the anode power supply are shown in **Table 1**. The range of output voltage is 5-35 kV. The resolution of the output voltage is 10 V. The maximum output current is 100 mA. The maximum pulse repetition frequency is 0.1 Hz. The pulse duration is between 10 ms and 1000 s. The remote control of the GYCOM anode power supply is realized via RS-485.

The anode power supplies for #2 and #3 gyrotrons were developed in ASIPP. A new anode power supply with 5kHz modulation capability was also developed in ASIPP [6]. The maximum modulation frequency is 5 kHz at a 50% duty cycle. The parameters of the ASIPP anode power supply are shown in **Table 2**. The maximum anode voltage is 30 kV, and the maximum current is 100 mA. We have developed the anode voltage control system for #2 and #4 gyrotrons based on the ethernet CompactDAQ chassis NI-cDAQ-9185.

**Table 1.** The parameters of the GYCOM anode power supply for #1 gyrotron and #4 gyrotron.

| | |
|---|---|
| **Input Voltage** | 380V ±10% (Three-phase) |
| **Input Frequency** | 50~60Hz ±2% |
| **Work** | Pulse |
| **Output Voltage** | 5~35 kV |
| **Output Voltage Resolution** | 10 V |
| **Maximum Output Voltage** | 35 kV |
| **Maximum Output Current** | 0.1 A |
| **Threshold Current** | 0.2 A |
| **Rising Edge of Output Voltage** | <10 ms |
| **Falling Edge of Output Voltage** | <10 μs |
| **Pulse Length** | 0.01s~1000s |
| **Minimum Interval Between Two Pulses** | 2s |
| **Maximum Pulse Repetition Rate** | 0.1 Hz |

**Table 2.** The parameters of the ASIPP anode power supply for #2 gyrotron and #3 gyrotron.

| | |
|---|---|
| **Input Voltage** | 220VAC |
| **Input Frequency** | 50Hz |
| **Output Voltage** | ≤30kVDC |
| **Output Current** | ≤100mA |
| **Ripple** | ≤0.5% |
| **Rising Edge of Output Voltage** | ≤10μs |
| **Falling Edge of Output Voltage** | ≤10μs |
| **Pulse Length** | 0~1000s |

## 2 GYCOM anode power control

The GYCOM anode power supply can be controlled through serial communication. The high-voltage power supply is equipped with an RS485 interface, which needs to be connected to the switch using the RS485-TCP conversion module, and finally realizes the communication with the host computer. The network architecture of GYCOM anode power control is shown in **Fig. 1**. When communicating, it is necessary to write the command message that the power supply can recognize into the serial port, and the power supply

* Corresponding author: xuweiye@ipp.cas.cn
† Corresponding author: xhd@ipp.cas.cn

responds and outputs the required voltage and current. When the power supply is turned off, the new value will not be saved. The format of the communication protocol message is shown in **Fig. 2**. The control program is written based on Labview [7], which can realize functions such as voltage setting and threshold setting. The device number of the voltage controller is N=1, and the device number of the current controller is N=2. The length in bytes refers to the byte length of the command code, parameter, and check code. The parameter is a hexadecimal string (with low byte first and high byte last) that converts the voltage and current values from decimal floating-point numbers. The check code is a character string obtained by performing Cyclic Redundancy Check (CRC) on the device number, byte length, command code, and parameters and inverting them. The GYCOM anode power supply can set the output voltage value, but it cannot be adjusted in real time. The host computer control interface of the GYCOM anode power supply is shown in **Fig. 3**. We have tested the anode voltage and the test results for the #1 anode power supply are shown in **Table 3**.

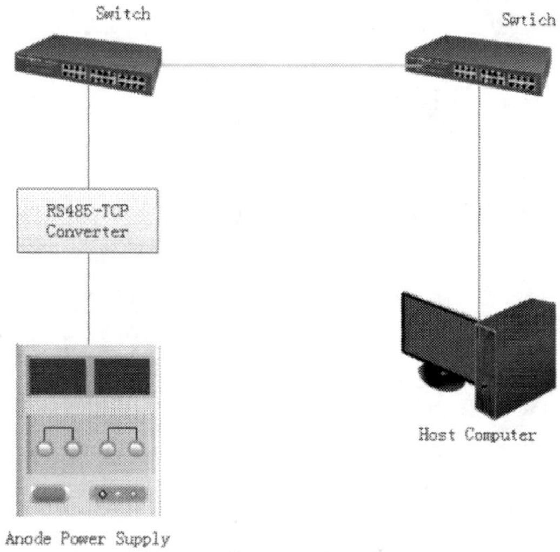


**Fig. 1.** The architecture of GYCOM anode power control.

**Fig. 2.** Communication protocol message format for the GYCOM anode power supply.

**Table 3**. Comparison of reference voltage setting value, read value, and the actual output value of #1 anode power supply.

| Reference voltage setting value (kV) | Reference voltage read value (kV) | Actual output value (kV) |
|---|---|---|
| 20 | 19.961 | 19.97 |
| 21 | 20.968 | 20.96 |
| 22 | 21.966 | 21.96 |

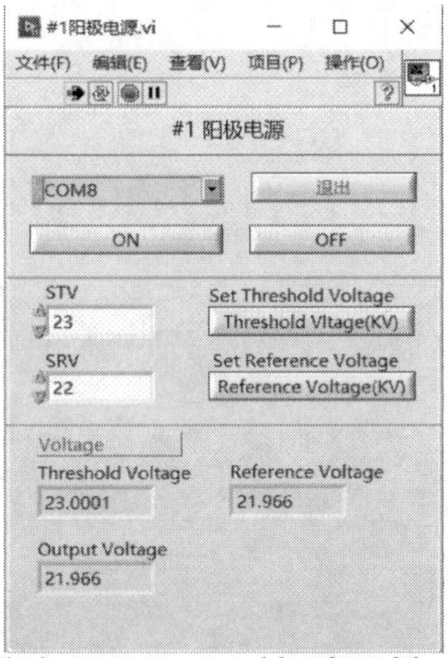


**Fig. 3.** The host computer control interface of the GYCOM anode power supply.

## 3 ASIPP anode power control

Two 30kV/100mA high voltage anode power supplies for #2 and #3 gyrotrons have been developed in ASIPP. In order to realize the real-time control of the anode voltage of the gyrotron, a new anode power supply with 5 kHz modulation capability has also been developed in ASIPP. It will be used for the next gyrotron in ASIPP. The power supply structure can be divided into five parts, rectifier, GTR (giant thyristor) linear compensator, inverter, high-frequency transformer and switch power supply (SPS) [6].

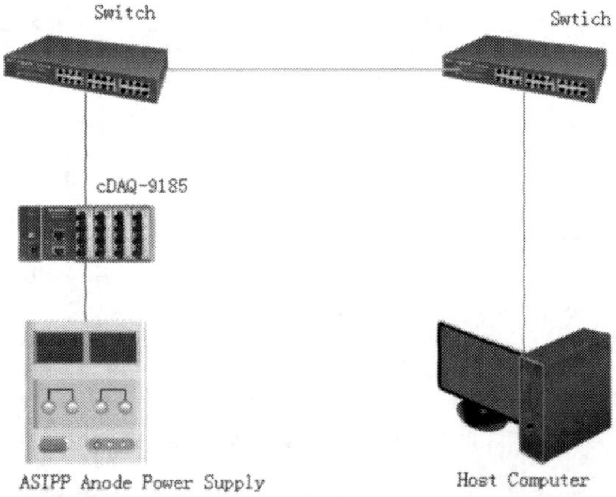


**Fig. 4.** The architecture of ASIPP anode power control.

We have developed the anode voltage control system for #2 and #3 gyrotrons based on the ethernet CompactDAQ chassis NI-cDAQ-9185. This anode voltage control system can also be used for the new anode power supply with 5 kHz modulation capability. The architecture of ASIPP anode power control is shown in **Fig. 4**. The cDAQ-9185 is used to provide a reference voltage to the anode high voltage power supply. The cDAQ-9185 communicates with the host computer through Ethernet.

The control program interface is shown in **Fig. 5** and **Fig. 6**. We developed two control modes, one is the DC mode, and the other is the modulation mode. In the modulation mode, in addition to the modulation output, we can also control the anode power supply to output other waveforms such as sine wave, triangle wave, etc.

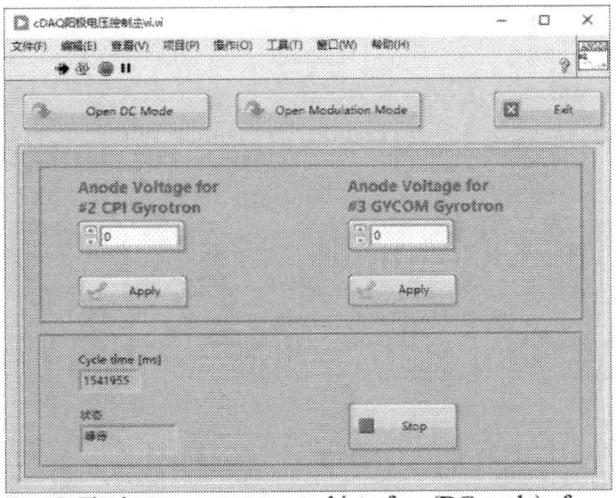


**Fig. 5.** The host computer control interface (DC mode) of ASIPP anode power supply.

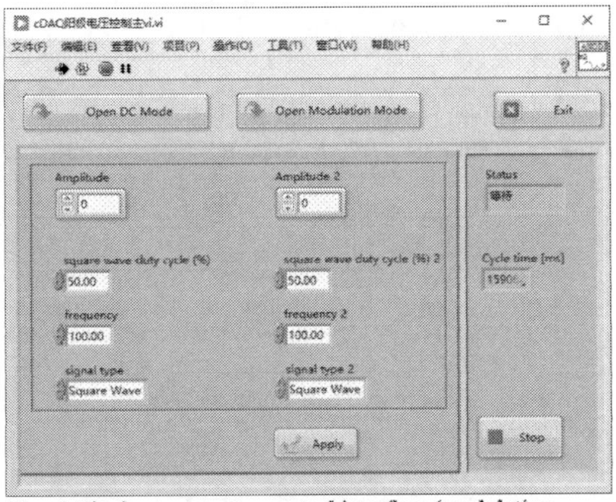


**Fig. 6.** The host computer control interface (modulation mode) of the ASIPP anode power supply.

The new anode power supply with 5 kHz modulation capability has been tested. **Fig. 7** shows the modulation output waveform. The modulation frequency is 5 kHz, the duty ratio is 50 %, and the modulation depth is 100 %.

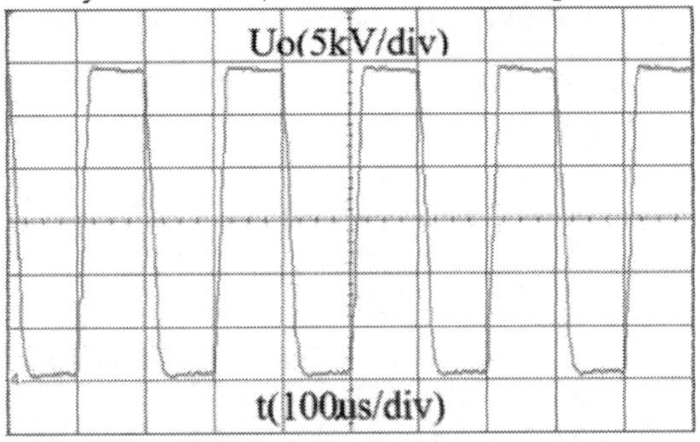


**Fig. 7.** The modulation output waveform of the ASIPP anode power.

# 4 Effect of anode voltage on gyrotron output RF power

## 4.1 Variation of gyrotron output power with anode voltage when the cathode voltage is constant

The anode voltage is an important parameter that affects the gyrotron output power [8]. We have tested the corresponding change in gyrotron output power [9] by controlling the anode voltage.

The changes of the #1 gyrotron output power with the anode voltage (here refers to the voltage of the anode relative to the ground) are shown in **Fig. 8** and **Fig. 9**. It can be seen that as the anode voltage increases, the dummy load power increases, the collector power decreases, and the efficiency improves.

Increasing the anode voltage will have the same effect as increasing the cathode voltage, the electron injection velocity will increase, the energy will increase, and the Larmor radius will increase. Compared with only increasing the cathode voltage, when the anode voltage is increased, the voltage difference between the anode and the collector (that is, the deceleration voltage) becomes larger, so that the drift speed of the electrons in the interaction area is reduced, and the electrons can be in the interaction area for a longer time, the efficiency increase is more obvious. However, the anode voltage should not be too large, because an excessively large anode voltage leads to an excessively large Larmor radius, which may not satisfy the interaction conditions, resulting in a decrease in efficiency [10].

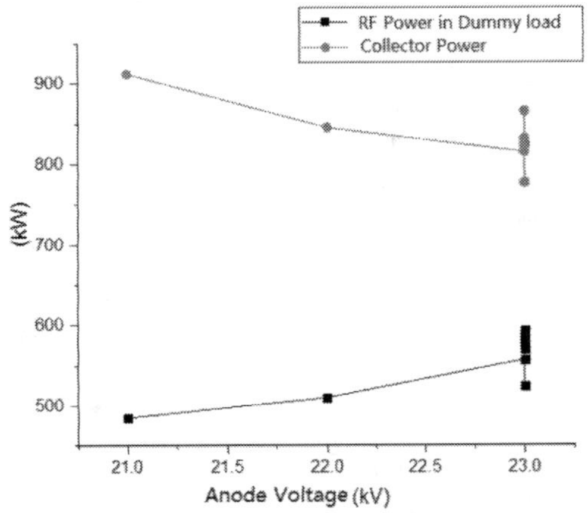


**Fig. 8.** Changes in the output power of the gyrotron when the magnetic field current, filament primary voltage, filament primary current, and cathode voltage are 50.55A, 39.7V, 34.7A, and -44kV, respectively.

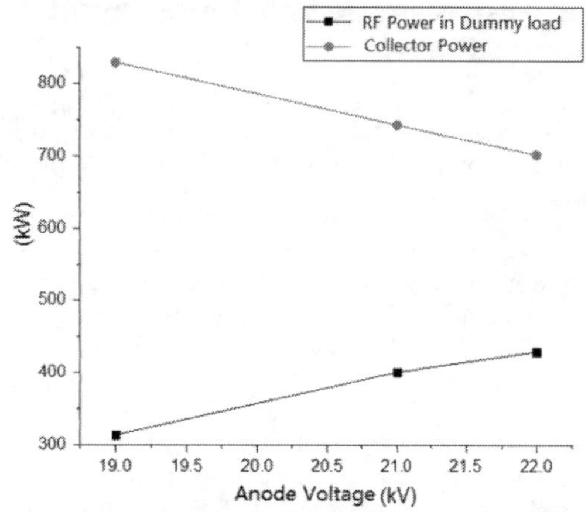


**Fig. 9.** Changes in the output power of the gyrotron when the magnetic field current, filament primary voltage, filament primary current, and cathode voltage are 50.55A, 39.1V, 34.4A, and -42kV, respectively.

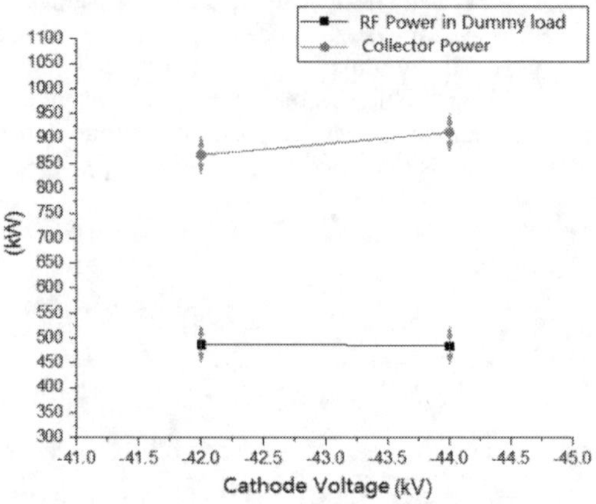


**Fig. 11.** Changes in the output power of the gyrotron when the magnetic field current, filament primary voltage, filament primary current, and beam voltage are 50.55A, 39.7V, 34.7A, and 65kV, respectively.

### 4.2 Variation of gyrotron output power with cathode voltage when beam voltage is constant

When the beam voltage (the sum of the absolute value of the anode voltage and the cathode voltage) is constant, as the absolute value of the cathode voltage increases and the anode voltage decreases, the dummy load power is almost unchanged, the collector power increases, and the efficiency becomes lower, see **Fig. 10** and **Fig. 11**. Since the beam voltage does not increase, the electron injection velocity does not change and the Larmor radius does not change, but due to the drop in the anode voltage, the voltage difference between the anode and the collector (that is, the deceleration voltage) decreases, so that the drift speed of electrons in the interaction region increases, and the time for electrons to stay in the interaction region becomes shorter, resulting in lower efficiency.

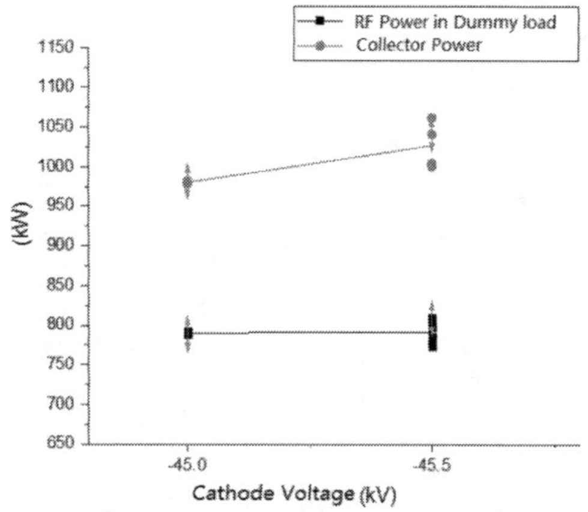


**Fig. 10.** Changes in the output power of the gyrotron when the magnetic field current, filament primary voltage, filament primary current, and beam voltage are 50.4A, 40.9V, 35.28A, and 68.5kV, respectively.

## 5 Conclusions

Four diode-type gyrotrons are now used in the EAST ECRH system. We can control the RF output power of the gyrotron by changing the anode voltage. The control system of the GYCOM anode voltage was developed based on RS485 communication. The control system of the ASIPP anode voltage was developed based on the ethernet CompactDAQ chassis NI-cDAQ-9185. The control program was developed based on Labview. The ASIPP anode voltage can be controlled up to 30 kV, and the maximum current is 100 mA. A new anode power supply with 5 kHz modulation capability has also been developed in ASIPP. It will be used for the next gyrotron in ASIPP. Using anode voltage control, we experimentally tested the relationship between the output power of the gyrotron and the anode voltage. The test results can be reasonably explained by theory.

## Acknowledgements

This work was supported in part by the Performance Improvement Project of EAST, Hefei Comprehensive National Science Center, by the National Key R&D Program of China under Grant No. 2017YFE0300401, and by the HFIPS Director's Fund under Grant No. YZJJ2020QN10.

## References

1 I. Pagonakis, S. Alberti, K. Avramidis, F. Legrand, G. Gantenbein, J. Genoud, J.-P. Hogge, S. Illy, Z. Ioannidis, P. Kalaria, B. Piosczyk, S. Ruess, T. Ruess, T. Rzesnicki, M.-Q. Tran, T.-M. Tran†, M. Thumm, I. Vomvoridis, and J. Jelonnek, EPJ Web Conf. **203**, 04011 (2019).

2 G. S. Nusinovich, M. K. A. Thumm, and M. I. Petelin, J Infrared Millim Te **35** (4), 325 (2014).
3 H. Xu, X. Wang, J. Zhang, F. Liu, Y. Huang, J. Shan, W. Xu, M. Li, J. Lohr, Y. A. Gorelov, J. P. Anderson, Y. Zhang, D. Wu, H. Hu, Y. Yang, J. Feng, Y. Tang, B. Li, W. Ma, Z. Wu, J. Wang, L. Zhang, F. Guo, H. Sun, X. Yan, and E. Team, presented at the 20th Joint Workshop on Electron Cyclotron Emission and Electron Cyclotron Resonance Heating, 2019 (unpublished).
4 A. G. Litvak, G. G. Denisov, and M. Y. Glyavin, IEEE Journal of Microwaves **1** (1), 260 (2021).
5 H. Xu, W. Xu, D. Wu, M. Li, X. Wang, L. Zhang, J. Lohr, J. Doane, J. P. Anderson, Y. A. Gorelov, J. Wang, Y. Hou, W. He, and T. Zhang, Fusion Eng Des **164**, 112222 (2021).
6 J. Zhang, Z. Gao, H. Sun, Y. Zhou, J. Wang, and Y. Huang, Fusion Eng Des **153**, 111479 (2020).
7 W. Xu, H. Xu, F. Liu, F. Hou, and Z. Wu, Fusion Eng Des **113**, 119 (2016).
8 W. Xu, H. Xu, F. Liu, X. Wang, Y. Yang, and J. Zhang, Ieee T Plasma Sci **47** (12), 5251 (2019).
9 W. Xu, H. Xu, F. Liu, J. Wang, X. Wang, and Y. Hou, Plasma Science and Technology **19** (10), 105602 (2017).
10 V. A. Flyagin and G. S. Nusinovich, P Ieee **76** (6), 644 (1988).

# 1.5 MW CW RF Loads for Gyrotrons

*Lawrence* Ives[1] , *Thuc* Bui[1], *Thomas* Habermann[1], *George* Collins[1], *David* Marsden[1], *Jeffrey* Neilson[2], *Tim* Horn[3], *and Chris* Rock[3]

[1]Calabazas Creek Research, Inc., 690 Port Drive, San Mateo, CA 94404 USA
[2]Lexam Research, 10 Black Fox Way, Redwood City, 24062 CA US
[3]North Carolina State University, 915 Partners Way, Raleigh, NC 27695 USA

**Abstract.** A new series of MW-class, RF loads were developed that can dissipate power levels exceeding 1.5 MW at frequencies from 28 GHz to 180 GHz. The new loads are designed to reflect less than 0.25% of the input power and operate continuously (CW). Stainless-steel and anodized aluminum versions were developed. The stainless steel version is designed to meet requirements for nuclear facilities, such as ITER, while the aluminum version is capable of power levels exceeding 2 MW CW. The aluminum version is also lighter and less expensive. This paper describes the design and capabilities of the loads.

## 1 Introduction

Electron cyclotron resonant heating (ECRH) is an important mechanism for increasing the temperature of tokamak plasmas. In addition to plasma heating, RF power is used for current drive and suppression of plasma instabilities. Gyrotrons provide the RF power for these and other applications and generate hundreds of kilowatts to more than 1.5 MW of long pulse or continuous wave (CW) power. Several organizations provide MW-class gyrotrons producing 1 MW or more from 110 GHz to 170 GHz. ITER specifications indicate gyrotrons producing 1.5 MW will eventually be required, and 2 MW sources are envisioned for DEMO.

Development and operation of gyrotrons requires loads to dissipate the RF power. Calabazas Creek Research, Inc. (CCR) developed the first load capable of dissipating more than 1.25 MW CW in the late 1990s [1]. CCR delivered these loads to Europe, Japan, Korea, and the United States and several are still operational. These loads accept RF power in a Gaussian mode transmitted in waveguide or free space.

In 2012, CCR completed development of 2 MW CW loads supporting RF input in 60.3mm or 63.5mm waveguide. In addition to supporting higher power, these loads reduced reflected power to less than 1% [2]. Three loads were delivered to QST in Japan, and one was used to test the 1.9 MW, 110 GHz gyrotron at JT-60A [3]. The loads have dissipated tens of thousands of pulses of 1 MW or more, and no arcs or thermal failures have been reported. In fact, CCR is not aware of any gyrotron test that was prematurely terminated by a CCR load in over twenty years of operation.

Several issues have been addressed over the years. The 1.25 MW loads suffered from persistent water leaks around rotating seals. This issue was eliminated with improve seals in the mid-2000s. These loads also exhibited more than 5% reflected power, which was addressed in the 2 MW loads. The 2 MW loads also incorporated ferromagnetic seals that increased performance and reliability.

**Fig. 1.** 1.5 MW CW RF load

The new load, shown in Fig. 1, was designed to improve reliability and reduce cost by eliminating rotating components and leveraging additive manufacturing to improve performance. The development also focused on dramatically reducing reflected power and accommodating increased mode

https://doi.org/10.1051/epjconf/202327704008

impurities in the RF input. Estimated reflected power is less than 0.25%.

CCR developed two, new, load versions. Many facilities, such as ITER, place strict limits on materials for vacuum and wetted surfaces. CCR's 1.25 and 2 MW load cylinders were fabricated using anodized aluminum to leverage the relatively low weight and high thermal conductivity. This material is prohibited by many fusion research facilities, so CCR developed a load primarily fabricated from stainless steel (SST) and copper. Consequently, this load is significantly heavier and more expensive than previous loads. The reduced thermal conductivity of SST also reduces the maximum power capability to 1.5 MW CW. To accommodate other users, CCR developed an anodized aluminum version capable of higher power at reduced cost and weight. This version also leverages additively manufactured (AM) components not currently qualified for nuclear facilities. The following sections describe design and fabrication of the loads and provide simulations of the RF and thermomechanical performance.

## 2 Mechanical design

The principal innovation in CCR loads is active distribution of the RF beam to uniformly distribute the power and eliminate arcing. All loads capable of MW operation incorporate overmoded cavities where the power is absorbed. This results in constructive and destructive interference from reflected RF waves. When constructive interference occurs near a load surface, it causes excessive RF heating and subsequent outgassing. The associated high electric field can initiate an arc that damages the surface or terminates gyrotron operation. CCR's concept eliminates this occurrence by continuously sweeping the power. The effectiveness of this approach has been confirmed over twenty years of operation where no load arcs or thermal damage have been reported.

The 1.25 MW loads used a rotating mirror at the downstream end of the load to sweep the power; however, this reflected the RF wave back toward the input, resulting in significant reflected power. This was essentially eliminated in the 2MW loads, which used a launcher at the input to sweep the beam. The launch angle was chosen such that the input was shadowed from reflected power by the back side of the launcher. This reduced reflected power from the load to approximately 0.7%, as measured by QST, and confirmed by General Atomics. This achieved the specification that reflected power be less than 1%.

To increase performance and reliability and reduce cost, CCR explored alternatives to rotating components for the RF sweeping. While effective, the cost and complexity of ferromagnetic seals was problematic. The new design uses a bellows to support an internal reflector that sweeps the RF power without rotating. The reflector is revolved around the input in the configuration shown in Fig 2.

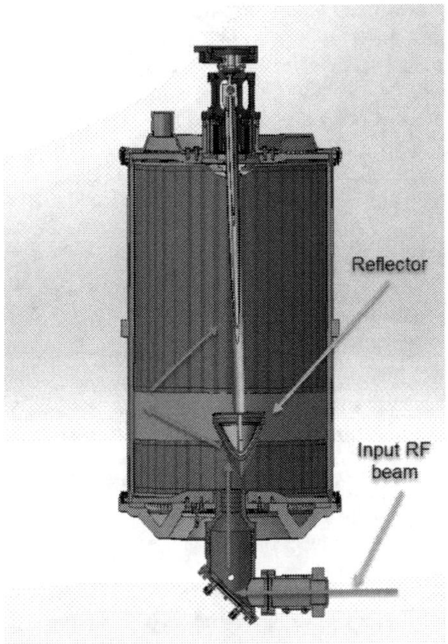


**Fig. 2.** Conceptual mechanical design

This approach provides several advantages. Probably the most important is cost. The ferromagnetic seals in the 2 MW loads cost more than 4,000 USD and requires precision support structures. The bellows and its support components cost less than 1,500 USD. Because the bellows is deflected less than 1.5°, the stresses are far below yield levels, implying unlimited lifetime. The reduced complexity also lowers assembly cost.

CCR tested this design using the setup shown in Fig 3. The assembly operated more than 4,000 cycles over 315 hours. This is almost three times the required lifetime specified by ITER. It was still operating nominally when testing was terminated due to schedule constraints. The horizontal test configuration was chosen to maximize stresses on the assembly.

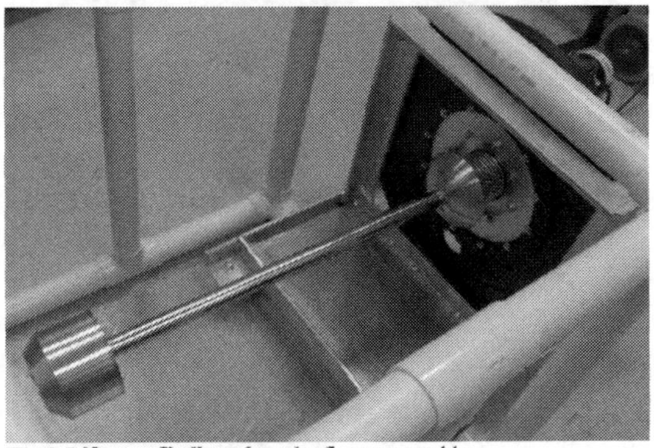

**Fig 3.** Life test of bellows-based reflector assembly

## 3 Input coupler

Reflected power is a critical issue for gyrotron users, as excessive levels can impact gyrotron operation and potentially result in catastrophic failure. A key task was reducing reflected power, hopefully below measurable levels. The design incorporates a focusing miter bend that accepts the input beam from corrugated waveguide and focuses it through a reduced size aperture, as shown in **Fig. *4***. The corrugated input waveguide terminates at the flange where the RF beam is launched and expands as a Gaussian, free space mode.

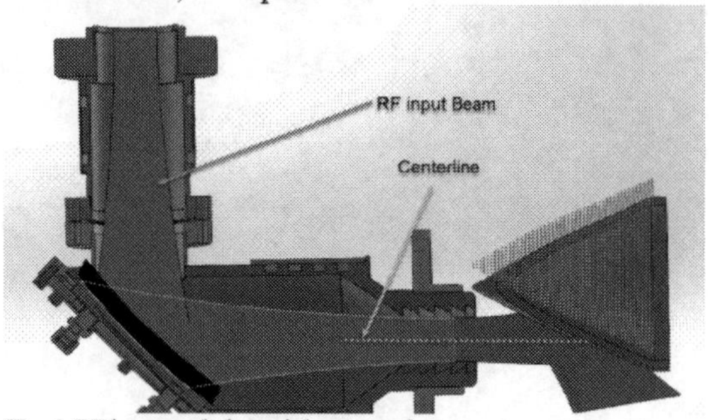


**Fig. 4.** RF beam path through input coupler

The input coupling system was designed using Surf3d. **Fig. *6*** shows the RF beam path from the end of the corrugated transmission line, past the miter bend, through the final aperture to the revolving reflector, and through two bounces on the load cylinder walls. The simulation also provides the power density profiles for thermomechanical analyses and cooling system design.

CCR also investigated load operation in the presence of non $HE_{11}$ input modes. The concern is that power in these modes may be trapped in the input coupler before entering the main load. Any RF power propagating into the cylindrical section of the load will be scattered and absorbed. For proper operation, a significant amount of non-$HE_{11}$ power must be transmitted into the main load by the miter mirror. If this is achieved, there is high confidence that low reflected power can be achieved, even with significant input mode impurities. CCR simulated non-$HE_{11}$ modes, specifically, the $HE_{12}$, $HE_{13}$,

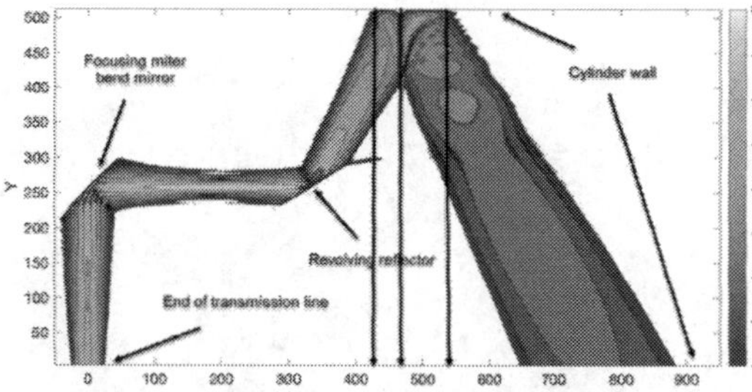


**Fig. 5.** Surf3d simulation of RF beam path into load

$HE_{21}$, $HE_{22}$, and $HE_{23}$. These are the closest modes to the $HE_{11}$ and the most likely to constitute non-$HE_{11}$ power.

Fig. 5 shows the field intensity for the $HE_{12}$ mode. In this configuration, the RF power is launched from a corrugated waveguide at -14 cm in a vertical direction. It impacts the focusing miter mirror and is reflected toward the aperture at location zero on the axis. The far-right boundary is a non-reflecting surface representing the cylindrical section of the load. The aperture at location zero is 20 mm diameter. The reflected power for each mode is determined by numerical integration of **E** x **H** over a circle with the same diameter as the corrugated waveguide and at a position just below the excitation field plane and normalized to the input power to get fractional power reflection (FPR). For the $HE_{12}$ mode the FPR was still quite small at 0.004. In addition, the total power reflected (RPF) was calculated. These results are shown in Table 1. To get total reflected power, the power in each higher mode must be multiplied by the FPR for that mode and summed with reflection phase included. The input mode composition is not known a priori, but assuming 10% of input power is distributed evenly in the five higher order modes and they all add in phase (an unlikely occurrence representing an upper bound in reflection amplitude) the total reflected power would be 0.25%. Since the higher order modes attenuate faster, it is unlikely the power in higher order modes would be equal to the lower order modes, thus it is likely the reflected power would be substantially less since the lower order modes have substantially lower FPR coefficients. The total power reflected (and not coupled to the input waveguide) for assumption of 10% of input power is distributed evenly in the five higher order modes is 1%.

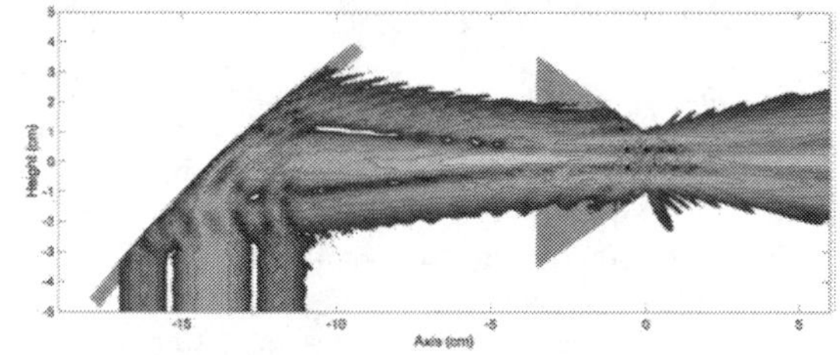


**Fig. 6.** Field intensity for $HE_{12}$ mode excitation. The incident mode is reflected off focusing mirror, and a small part is reflected back by cone funnel on the axis at 0 cm.

These calculations used a 20 mm diameter aperture. From Fig. 6, one can observe that the maximum radial extent of the $HE_{12}$ fields is less than 20 mm. Consequently, a 40 mm diameter aperture would result in total transmission of this mode into the load cylinder. This would also drastically reduce FPR for the other, higher order modes. The actual load aperture is approximately 44 mm diameter, which further reduces or eliminates reflected power from higher order modes.

*Table 1. Fractional Reflection for non-HE11 modes*

| Mode | FPR | Total Power |
|---|---|---|
| $HE_{11}$ | 0 | .003 |
| $HE_{12}$ | .004 | .042 |
| $HE_{13}$ | .047 | .143 |
| $HE_{21}$ | .002 | .019 |
| $HE_{22}$ | .009 | .0443 |
| $HE_{23}$ | .063 | .286 |

There are two issues associated with the higher order modes. One is the level of reflection back into the input waveguide, and the other is the power that fails to enter the waveguide and is dissipated in the coupler region. The analysis above indicates total reflected power for

10% non-$HE_{11}$ modes would be 1%. This is within the cooling capability of the input coupler.

CCR anticipates less than 0.25% of the RF power will be reflected into the transmission line and sufficient cooling is available to dissipate this absorbed power. Note that any higher mode power coupled back into the input waveguide will be rapidly dissipated due to waveguide loss. Consequently, the corrugated waveguide input should be well cooled.

The heat load on the miter bend mirror exceeds that of all other surfaces and is typically the component limiting the power capability of the load. It is fabricated from copper chrome zirconium (CuCrZr), which supports higher power densities than copper. It is also well cooled. For 1.2 MW CW operation, a 26 LPM flow rate results in a vacuum wall temperature of 122° C and wet wall temperature of 73° C. The peak stress is approximately 1/3 the yield stress. Higher coolant flow will allow higher power operation. Provided sufficient cooling, it is anticipated that 2 MW CW operation can be supported.

**Fig. 7.** 3D printed miter mirror assembly

It should be noted that the power density on the mirror is reduced by expansion of the Gaussian beam in the input section of the load. Consequently, the power density will be less than on miter bends in the incoming transmission line. The power delivered to the load will most likely be determined by transmission line components rather than load components.

**Fig. 8.** Solid model of input coupler for 170 GHz operation with 63.5 mm waveguide

ITER specifications prohibit braze joints in the vacuum envelope, so the initial mirror was explosively bonded to the stainless steel support plate. North Carolina State University (NCSU) used additive manufacturing (AM) to build this assembly by 3D printing CuCrZr directly on the support plate. An advantage with this technique, in addition to dramatically lowering cost, includes integration of cooling channels closer to the vacuum surface with increased thermal transfer to the coolant. This could allow higher power operation than available with explosive bonding. Surface finishes achieved with AM are currently unacceptable for most RF applications due to the inherent roughness. This is avoided by machining the reflective surface following printing. The roughness remains on the interior cooling channels, but this enhances heat transfer at appropriate flow rates. Fig. 7 shows a miter mirror assembly built using AM. The machined copper reflecting surface exhibits a mirror-quality finish while the sides show the inherent roughness from printing.

The load design incorporates several features to reduce reflected power. First, the revolving reflector shadows the input from RF waves propagating parallel to the input axis. Stepped walls at the aperture will reflect any incident power back into the cylindrical section of the load. Secondly, only power that impacts the miter mirror at a precise angle will be focused into the input transmission line. Given the input path for reflected power, the probability of this is quite low. Power incident on the mirror at the wrong angle will impact the walls of the stainless steel, smooth bore waveguide, which is well cooled. Only power entering the corrugated waveguide parallel to its axis can propagate back to the gyrotron.

**Fig. *8*** shows the input coupler model for 170 GHz operation. The assembly includes ports for vacuum pumping and an arc detector. All components are water cooled.

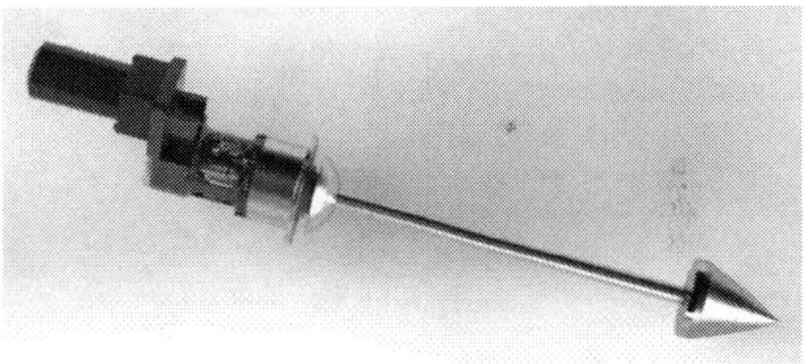

**Fig. 9.** Solid model of the reflector assembly

## 4 Reflector assembly

RF power enters the load cylinder through the input assembly and impacts the revolving cone of the reflector assembly, shown in Fig. 9. The cone revolves around the input coupler every two seconds and evenly distributes the thermal loading while preventing arcing from constructive interference. Water cooling is provided through the support shaft, which also scatters the RF power around the load interior. The cone surface is shaped to maximize axial and azimuthal scattering of the RF beam. An electric motor revolves the cone using a chain drive. CCR has used the same model electric motor for more than twenty years, and none has ever failed. The drive gear is interlocked to ensure the cone is properly revolving prior to injection of RF power.

The full power of the RF beam impacts the cone, but the area of impact moves around the cone surface as it revolves around the input. Consequently, the primary concern is cyclic fatigue. CCR used the incident power density obtained from Surf3d to analyze the thermomechanical performance using a cyclic profile. Fig. 10 shows the relevant temperatures as a function of time for a 30 LPM coolant flow rate. The peak stresses are well below the 44 MPa yield stress, implying unlimited lifetime due to thermal fatigue.

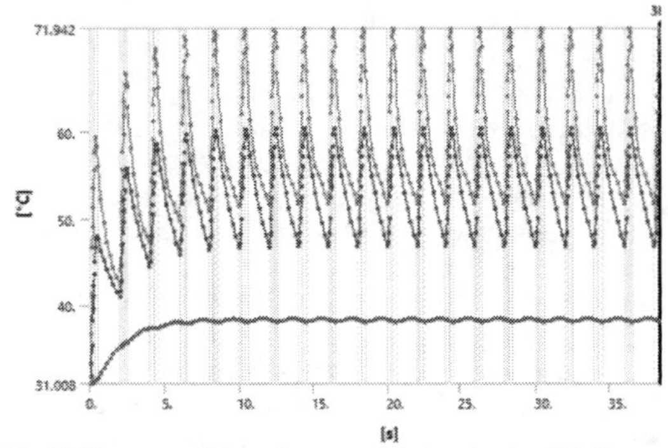


**Fig. 10.** Vacuum wall (green), average (blue), and wet wall (red) temperatures for reflector cone

## 5 Load Cylinder

The input coupler and reflector assemblies are common to both the stainless steel and aluminum versions of the load cylinder assembly. The primary function of the cylinder assembly is to absorb the input RF power and transfer the energy to the coolant. This assembly includes the high flow cooling circuit and provides the primary vacuum envelope for the load.

**Fig. 11.** Stainless steel inner cylinder

The inner cylinder assembly consists of the inner cylinder, shown in Fig. 11, with end plates electron beam welded (EBW) to the ends. The interior of the cylinder includes RF loss material to increase absorption in regions of low to moderate power density. CCR uses plasma sprayed PP-146 compound to increase the RF loss in these regions [4]. This material is 97% titanium dioxide. The exterior is grooved to enhance heat transfer to the coolant. The end plates support the input and output coolant manifolds for the primary coolant and provide the interfaces to the input coupler and reflector assembly. Fig. 12 shows the stainless steel version, which weighs approximately 300 kg. Forgings for the load cylinders, both stainless steel and aluminum, are centrifugally spun, which avoids stresses that can cause warping after machining.

**Fig. 12.** Stainless steel load cylinder assembly

The primary cooling circuit consists of manifold end plates and an outer cylinder. These components are attached to the inner cylinder using bolts and O-ring seals.

CCR analyzed the thermal performance using power densities determined from the Surf3d simulation shown in **Fig. *6*** and the two second periodicity of the rotating beam. The polarization with highest RF loss was used in all cases. Fig. 13 shows the power density profile at the wall for 1.5 MW operation, and Fig. 14 shows the maximum inner wall temperature, which peaks below 62°C. Note that the cylinder reaches steady state operation after approximately twenty seconds. The peak stress is less than 170 MPa, well below the yield stress

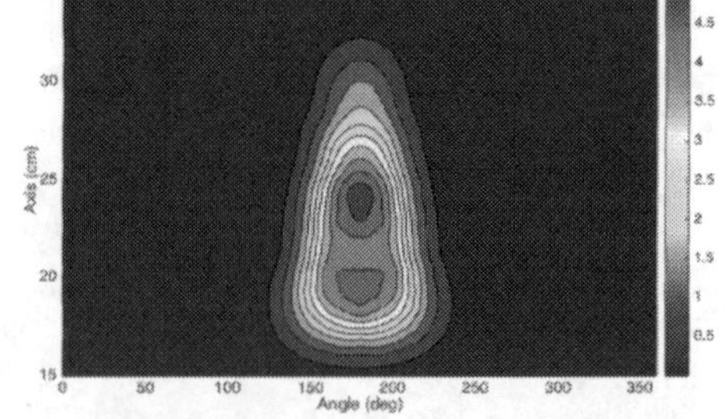


**Fig. 13.** RF power density on cylindrical wall for 1.5 MW input power

of 262 MPa. This indicates the load can operate indefinitely (CW) under these conditions.

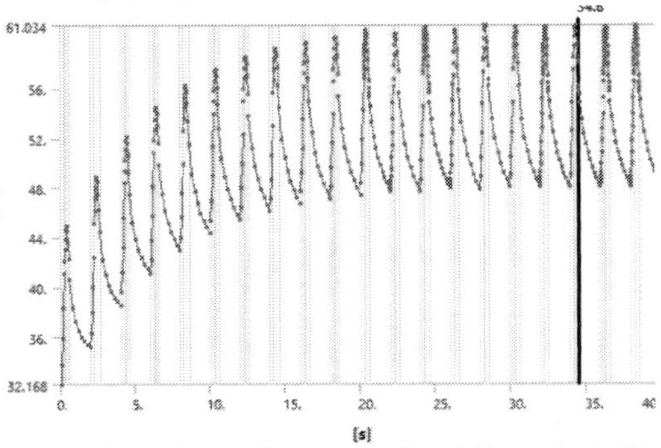


**Fig. 14.** Maximum inner wall temperature through forty seconds of operation

The anodized aluminum version of the assembly is slightly longer than the SST version due to thicker, e-beam welded end plates. The outer cylinder and end plates are also anodized aluminum. The weight of this assembly is less than 200 kg and lower cost than the stainless steel version. The increased thermal conductivity allows operation to more than 2 MW CW. The thermal analysis was presented in reference [2].

## 6 Diagnostics and control system

CCR developed a control system based on programmable logic controllers (PLCs) that monitors the system, controls load operation, and uses temperature and flow sensors to calculate and display dissipated power in real time. The system includes the associated plumbing, sensors, and cables. It also provides an interlock signal to confirm that the load is operating correctly and a digital interface for remote control and data acquisition. Specifically, the system provides the following:

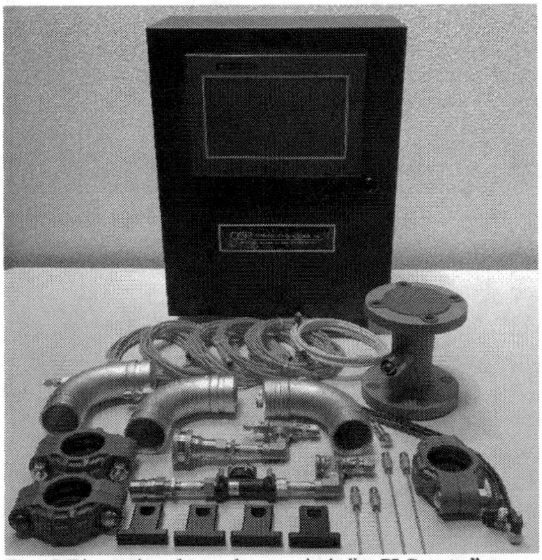

**Fig. 15.** Diagnostic and control system, including PLC controller, plumbing, and sensors

- Controls reflector motor,
- Measures reflector revolution rate,
- Measures reflector, input coupler, and main coolant flow rates,
- Monitors inlet and outlet coolant temperatures,
- Calculates and displays average and peak RF power for short pulse, long pulse or CW operation,
- Provides interlocks for main coolant flow and reflector revolution,
- Includes emergency Off button
- Provides digital I/O (Modbus) for remote monitoring and control

A front panel touch screen provides user control and display of operational and performance parameters. Fig. 16 shows a photo of the PLC controller, temperature and flow sensors, cables, and a portion of the plumbing hardware included with the system. A sample display is shown in Fig. 16.

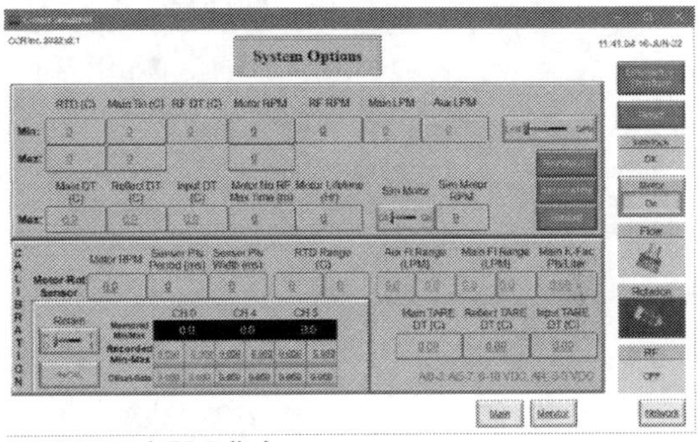


**Fig. 16.** Sample PLC display

## 7 Summary

Calabazas Creek Research, Inc. is continuing to develop high power RF loads for the ECE-ECRH and gyrotron community. The new loads offer improved capability, increased reliability, and lower cost. Both stainless steel and anodized aluminum load cylinders are available with the aluminum version offering higher power operation, significant weight reduction, and reduced cost. The diagnostics and control system provides temperature, flow, and dissipated power data in real time and remote data acquisition and control.

## References

1. R. L. Ives, M. Mizuhara, R. Schumacher, R. Pendleton, “Development of a High Power CW Waterload for Gaussian Mode Gyrotrons,” IEEE Trans. Plasma Sci., Vol. 27, No. 2, April 1999.
2. R.L. Ives, M. Mizuhara, t. Kobayashi, S. Moriyama, g. Collins, P. Borchard, and Jeff Neilson, “Design and Operation of a 2-MW CW RF Load for Gyrotrons,” IEEE Trans. Electron Devices, Vol. 61, No. 6, June 2014.
3. T. Kobayashi, QST-Japan, Private communications, May 2022.
4. Available from Aimtek, Inc., 201 Washington Street, Auburn MA 01501 USA

# MISTRAL campaign in support of W7-X long pulse operation

*Johan W.* Oosterbeek[1,*], *Mathias* Stern[1], *Harald* Braune[1], *Karsten* Ewert[1], *Matthias* Hirsch[1], *Frank* Hollmann[1], *Andreas* Killinger[2], *Heinrich.P* Laqua[1], *Philippe* Maquet[3], *Stefan* Marsen[1], *Venancio* Martínez-García[2], *Dmitry* Moseev[1], *Frank* Noke[1], *Sunil* Pak[4], *Arne* Reintrog[1], *Torsten* Stange[1], *Robert C.* Wolf[1], *Rafael J.* Zubieta-Lupo, and *the* W7-X team[1]

[1]Max-Planck-Institut für Plasmaphysik, Teilinstitut Greifswald, Wendelsteinstraße 1, 17491 Greifswald, Germany
[2]University of Stuttgart, Institut für Fertigungstechnologie keramischer Bauteile Allmandring 7b, D-70569 Stuttgart
[3]Bertin Technologies, 155 rue Louis Armand, CS 30495, 13593 Aix-en-Provence Cedex 3, France
[4]ITER Organization, Route de Vinon-sur-Verdon, CS 90 046, 13067 St. Paul Lez Durance Cedex, France

**Abstract.** Following two initial campaigns [1], Stellarator Wendelstein 7-X (W7-X) has now completed the construction phase by installation of active cooling of all plasma facing components. The machine is presently commissioned for the next campaign (OP2) aiming at 1 GJ per pulse, e.g. 100 s at 10 MW, eventually aiming at 18 GJ, e.g. 1800 s at 10 MW. The key heating system is the Electron Cyclotron Resonance Heating (ECRH) system, consisting of 10 gyrotrons with power per gyrotron ranging from 0.6 MW up to 1.0 MW at 140 GHz. A phased upgrade of the installation is in progress with the addition of 2 gyrotrons and the development of 1.5 MW and 2.0 MW gyrotrons, such that at the end of the upgrade 4 gyrotrons will be available in each power class of 1.0, 1.5 and 2.0 MW [2]. The increased ECRH power, combined with O2 and X3 heating schemes at high densities, will lead to increased microwave stray radiation. This is non-absorbed microwave power that diffuses inside the vessel and is incident on all in-vessel components including vacuum windows and attached diagnostic systems. A fraction of the stray radiation is absorbed by resistive or dielectric losses of these components, leading to thermal loads that scale with stray radiation levels and pulse length. At W7-X a high power microwave stray radiation launch facility 'MISTRAL' is available that is used to qualify in-vessel components for use at specified microwave surface power densities [$Wm^{-2}$]. This paper reports on MISTRAL campaigns in 2020 - 2021 for testing of stray radiation loads during OP2 in W7-X, as well as on an EUROfusion program assessing stray radiation loads on ITER components. A dedicated, absolutely calibrated, caloric load was developed for the campaign to obtain measurement of stray radiation power levels as well as to conveniently expose samples. Amongst other we report on shielding concepts using metal enclosures combined with microwave absorbing coatings and dielectric heating of vacuum windows.

## 1 Introduction

Electron Resonance Cyclotron Heating (ECRH) or Electron Cyclotron Current Drive (ECCD) are a key actuators in magnetically confined fusion experiments. In optically thick plasmas the absorption of the microwaves is very high but heating scenarios, or plasma conditions, with reduced optical thickness may apply. A technical issue is a practical limit to the accuracy of the launching angle and polariation of the microwave beam. Both operational and technical conditions may lead to an inevitable fraction of non-absorbed microwave power that, after many passes through the vessel and the plasma, diffuses into microwave stray radiation. This power is incident on the vessel interior resulting in thermal loads but also propagates through ports interfering with diagnostic operation. These problems are actively worked on in the community and in this paper we report on recent work using the MISTRAL stray radiation exposure facility: reduction of stray radiation, response of microwave absorbing coatings and stray radiation exposure of vacuum windows.

*e-mail: hans.oosterbeek@ipp.mpg.de

## 2 MISTRAL

The Microwave Stray Radiation Exposure Facility 'MISTRAL' is a large resonator in which a microwave stray radiation field is generated to qualify components for stray radiation loads in the vessel [3–6]. It is a tube with length 2 m and a diameter of 1.5 m. A gyrotron Gaussian beam at 140 GHz is injected through a sapphire vacuum window into the vessel where reflects off a mirror with two inclinations: a large inclination directing the beam to the vessel circumference such that the beam makes a trajectory around the inner circumference (Fig. 1a), and a small inclination directing the beam towards the long side of the tube such that it also propagates along the vessel (Fig. 1b.) On each bounce off the vessel wall some of the beam power is scattered towards the centre of the vessel resulting in a microwave stray radiation field in the centre region of the vessel. The reflecting mirror is in fact V-shaped such that two beams are generated, one travelling clockwise, as shown in Fig. 1a), and one travelling counter clockwise. By adjustment of gyrotron settings and duty cycle the CW input power can be set up to 10 kW. The high wall reflectivity combined with modest size apertures leads to a surface power density of p = $(55 \pm 5)\,\mathrm{kWm^{-2}}$. This was measured with a caloric load which was developed for the work described here and is located near the centre of the vessel closure plate when viewing the face in Fig. 1a). A cartoon of the caloric load arrangement is shown in Fig. 2. Samples such as vacuum windows or filters may be mounted in the sample holder. In the interior Teflon hoses with a steady water flow are mounted. The low refractive index of Teflon, combined with the high absorption of water and the reflecting cone at the far end, ensure a very low reflection coefficient, of the order of 2%. In the cone a pyrometer is mounted to monitor the temperature of samples in the sample holder. More details are available in [7].

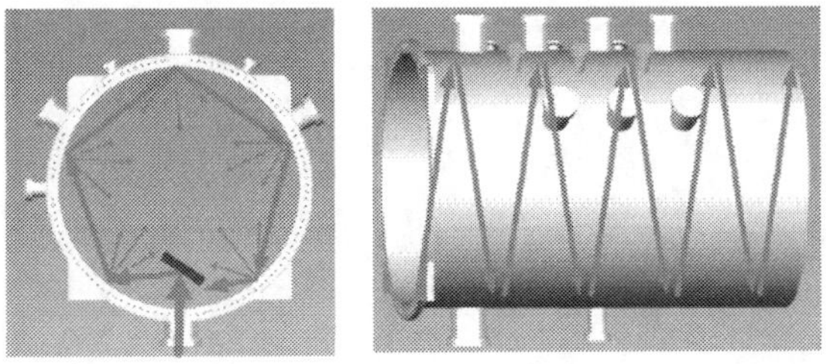

a) b)

**Figure 1.** Conversion of collimated beam to stray radiation.

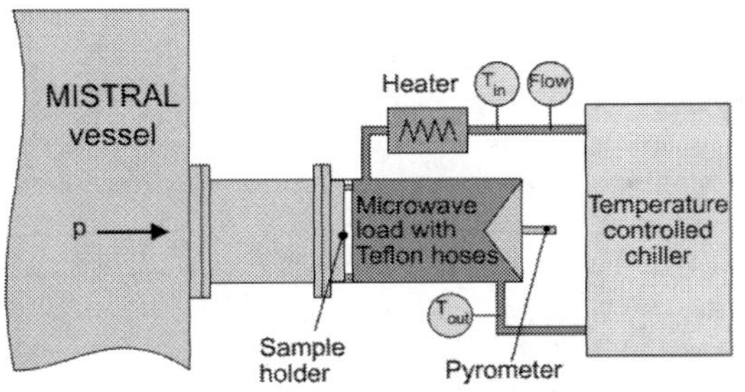


**Figure 2.** Caloric load attached to MISTRAL DN150 port.

## 3 Reduction of stray radiation

Full metallic shielding of components against stray radiation is not always possible. Effort is then directed towards using low-loss components, providing sufficient thermal sink and measures to reduce stray radiation levels. In this section, the reduction of stray radiation in a sub-volume by limiting the input stray radiation level, combined with the use absorbing coatings, is discussed. A detailed model for analysing coupled cavities is found in [8, 9].

### 3.1 Surface power densities in coupled cavities

Often the task is to reduce stray radiation levels inside a sub-volume in the vessel or in a system attached to the vessel. The most basic model is that as depicted in Fig. 3.

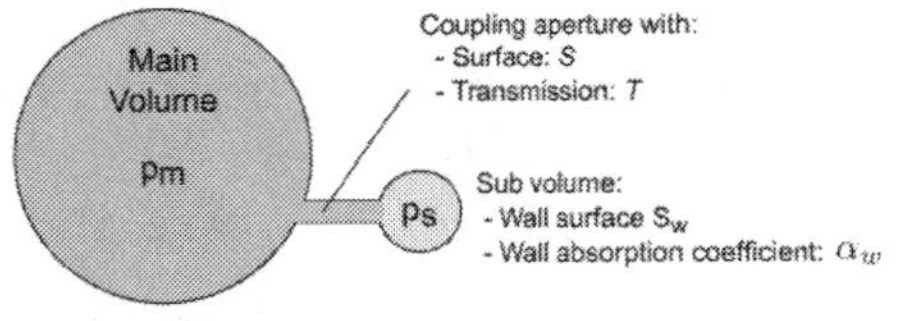


**Figure 3.** Cartoon of coupling between a main volume with surface power density $p_m$ and a small sub-volume with surface power density $p_s$.

The large volume is considered to be a constant source of stray radiation with surface power density $p_m$. The sub-volume is shown attached to the vessel, such as a diagnostic, but it could also be a structure inside the vessel such as an ECW launcher or a divertor enclosure. An intuitive measure is to reduce the surface power density in the sub-volume by reducing the coupling: making the aperture surface $S$ smaller or by placing a filter in the aperture with a low transmission coefficient $T$ for microwaves. But this may not necessarily lead to a reduction of the surface power density inside the sub-volume as in equilibrium the surface power density is determined by the power that flows into the sub-volume AND the losses inside the sub-volume. This is best seen using the power balance:

$$p_m \cdot S \cdot T = p_s \cdot S \cdot T + p_s \cdot S_w \cdot \alpha_w \Leftrightarrow \tag{1a}$$

$$\frac{p_s}{p_m} = \frac{S \cdot T}{S \cdot T + S_w \alpha_w}, \tag{1b}$$

with the symbols are defined in the sketch in Fig. 3. It is seen that in the limit of loss-less walls the surface power density in both cavities is the same. But placing absorbing material in the sub-cavity allows to significantly lower the surface power density in the sub-cavity. This is illustrated in Fig. 4. This does not go at the cost of massive dissipation as the product ST can be lowered by e.g. a reduced aperture or a filter. A practical application of this is the stray radiation protection of the pressure gauges on the MISTRAL. The gauges are mounted on the pumping lines to the turbo pumps. A wire mesh was placed over the pumping aperture and a microwave absorbing coating was placed inside the port, see the photo in Fig. 5. For the selection of the mesh 3 commercially available meshes of different wire and mesh size (unit: lines per inch) were purchased and tested in the sample holder. Four measurements were taken: 1) no mesh, and 2-4 ) with meshes. The mesh transmission is found by normalising the calorimeter power with mesh to the measurement with no mesh. The results are shown in Fig. 6. For use in the MISTRAL the 24 lines/inch mesh was selected as it resulted in low enough stray radiation levels, being mechanically strong and not impeding the pumping speed too much.

### 3.2 Absorbing coating

Stray radiation levels can be lowered by dissipation of the microwaves. The losses can be introduced by application

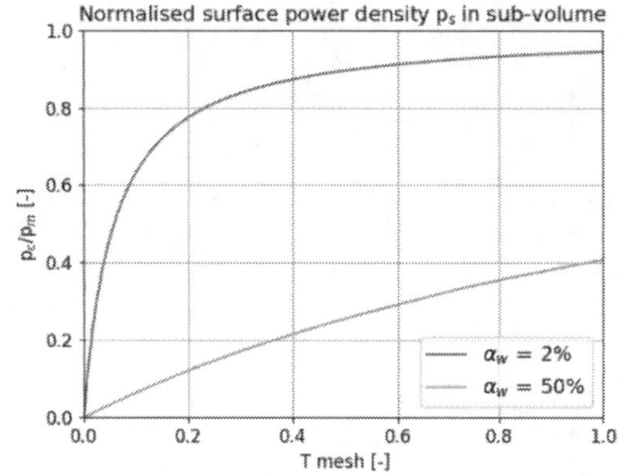


**Figure 4.** Reduction of stray radiation by using a wire mesh filter with transmission T and for two different wall absorption coefficients. While a coating with up to 75% absorption was demonstrated, 50% is shown in the example as usually not all walls are, or can be, coated.

**Figure 5.** Protection of the vacuum gauges in the pumping port. The photo is take outside the MISTRAL with the vacuum pumps removed. At the far end of the port (vacuum side) the mesh is seen. The insert is a microwave absorber: an aluminum sheet coated with a microwave absorbing layer, refer to section 3.2.

of a microwave absorbing coating on metal surfaces [10–12]. This is practical as a thin layer can applied with Atmospheric Plasma Spraying onto walls or pipes, providing a heat sink at the same time. Another advantage of a coating in the mm-wave region is that the thickness can be set such that the electric field reflection coefficient on the surface is minimised, leading to a maximum in absorption. This is caused by multiple internal reflections inside the layer. The resonance occurs under the condition $d = m\frac{1}{4}\lambda$, with $d$ the coating thickness, $m$ an odd integer $(1, 3, \cdots)$ and $\lambda$ the wavelength inside the dielectric. This is the vacuum wavelength $\lambda_0$ divided by the square root of the relative permittivity. The reflection can also be maximised, leading to a minimum in the absorbed fraction. The anti-resonant condition is $d = n\frac{1}{2}\lambda$, with n an integer $(1, 2, \cdots)$. Absorption is quantified using plane wave propaga-

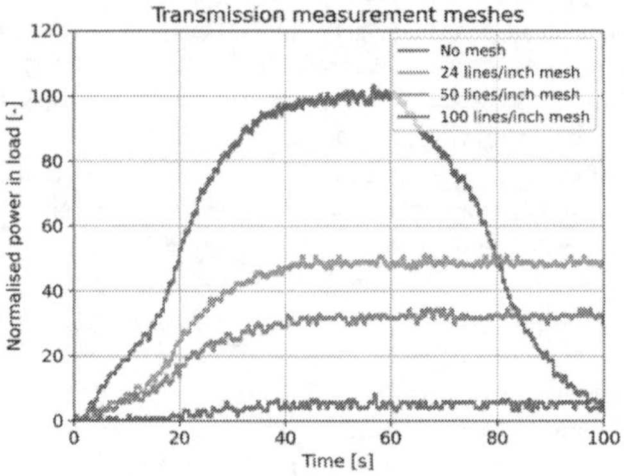


**Figure 6.** Insertion loss of wire meshes using the caloric load. MISTRAL not at vacuum: to avoid possible arcing the calibration pulse (blue) was limited to one minute.

tion: $E = E_0 e^{j\omega t - \gamma z}$ where $\gamma = \alpha + j\beta$ is the propagation constant with $\alpha$ the attenuation constant and $\beta$ the phase constant $2\pi/\lambda$. As the electrical field propagates into the coating it exponentially decays: $E(z) = E_0 e^{-\alpha z}$, while the power decays as $P(z) = P_0 e^{-2\alpha z}$. The attenuation constant is determined by the losses in the coating. These are found by the current densities in the coating [13]:

$$J = j\omega\epsilon'\left(1 - j\left[\frac{\sigma_a}{\omega\epsilon'} + \frac{\sigma_s}{\omega\epsilon'}\right]\right)E \tag{2}$$

The three terms left to right are: i) the loss-less displacement current density, ii) a loss current density induced by the alternating field ($\sigma_a$), and iii) a loss current density caused by a finite static conductivity ($\sigma_s$). In the expression $\epsilon'$ is the real part of the complex permittivity, which is defined in $\epsilon = \epsilon' - j\epsilon'' = \epsilon_0(\epsilon_r' - j\epsilon_r'')$. The imaginary part $\epsilon''$ is defined by the relation $\sigma_a = \omega\epsilon''$. It is common to express the dielectric losses by the loss tangent $\tan\delta_a = \epsilon''/\epsilon'$. One then obtains for the ratios between the square brackets: $\tan\delta_a + \sigma_s/\omega\epsilon'$. These are physically different: the measure for dielectric losses and the measure for ohmic losses, however, with the MISTRAL at a single frequency, they cannot be distinguished [14] and they lead to a net heating rate ($\Delta T/\Delta t$) of the coating. In experiment and modelling at the MISTRAL therefore the two terms are replaced by a single measure for conduction losses: $\tan\delta$, noting that this is not necessarily the dielectric loss tangent as quoted for ceramics.

The coating response is evaluated by illumination with a plane wave. The propagation constant $\gamma$ of the wave is obtained by using both Maxwell equations and by substitution of a plane wave solution. This results in:

$$\gamma = j\omega\sqrt{\mu_0\epsilon_0\epsilon_r'}\sqrt{1 - j\tan\delta} \tag{3}$$

Following the methodology for a vacuum window as in [7], plane wave propagation inside the coating is modelled by considering the layer to be a lossy ceramic but now terminated with a reflecting metal wall. See the schematic in Appendix A. For instruction only normal incidence is shown but the full model includes oblique incidence. An additional change, following Hentrich [12], is the using the complex permittivity in computation of the refracted angle in the layer as opposed to taking only the real part. This has been applied too for computation of the layer impedance. This is required as, contrary to low-loss vacuum windows, the imaginary part of the complex permittivity may no longer be much smaller as the real part.

The coating used in this work is an $Al_2O_3/TiO_2$ coating [10–12], mixed to ratio 87% / 13%. For this coating Hentrich [14] extracted a $0.25\lambda$ resonance thickness of 165 $\mu m$ at 140 GHz. MISTRAL measurements (section 4.3) showed an absorption coefficient of 67% at a thickness of 150 $\mu$m. Using the data a fit was made using a complex permittivity with $\epsilon_r' = 11.2$ and $\tan\delta = 0.13$, see Fig. 7. In the experiment a coating with 300 $\mu$m thickness showed 30% absorption. This is in broad agreement with the model. Evaluation of the substrate surface resistance $R_s$ shows this to be an insensitive parameter in the model: as long as a metal is used, the response is the same within fractions of a percentage point.

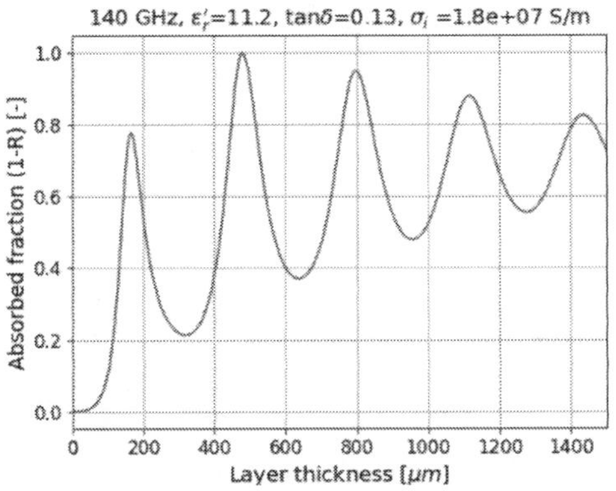


**Figure 7.** Modelled absorbed fraction (1-R) in coating fitted to experiment data. The maxima and minima are caused by multiple internal reflections. At larger thickness these fade and the absorption converges to 1-$R_i$, with $R_i$ the initial reflected fraction off the boundary. The model shows even larger absorption at next two maxima. These are not experimentally verified but result from larger bulk absorption (longer optical path) coinciding with maxima due to multiple internal reflections.

The model was next used to investigate the response at oblique incidence. This shows, besides the effect of the Fresnell equations, little variation with incidence angle. This is explained, as also noted by Hentrich [12], by only small differences in optical path due to the small refracted angels given the high relative permittivity.

The stray radiation response is now found following the methodology in [7]. In short: the coating is exposed to isotropic stray radiation which is modelled by an angular spectrum of incidence angles with a normal distribution and considering both polarisations. The normal distribution is important as large incidence angles are more pronounced as small incidence angles, which may be understood by considering the rays to be incident through a virtual dome placed over the coating. The modelled isotropic stray radiation response of the coating is shown in Fig. 8. In the plot the x-axis shows the maximum considered incidence angle $\theta_i$, for example: $\theta_i = 20°$ means an isotropic stray radiation response when considering angles up to 20°. The plot shows the absorbed fraction for the resonant layer and the anti-resonant layer. For each layer the stray radiation response for either polarisation is shown (perpendicular and parallel). For a layer, the average of both polarisations predicts the isotropic stray radiation response. The data for minima and maxima at normal incidence has been taken from Fig. 7. The predicted isotropic stray radiation response as function of considered range of incidence angles shows a flat response, except for a modest effect when including the very large incidence angles.

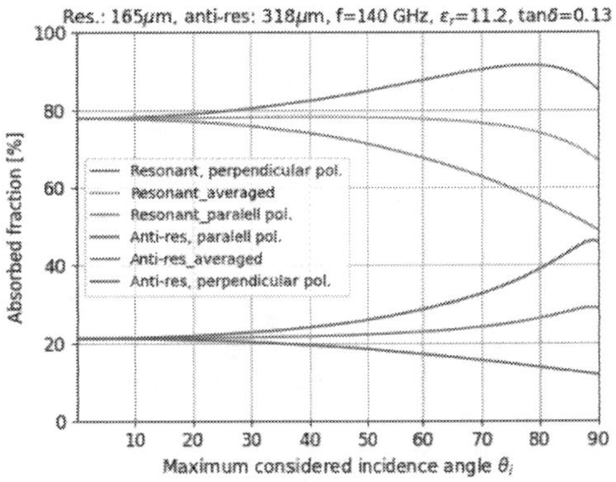


**Figure 8.** Modelled isotropic stray radiation response with input data from Fig. 7. Upper traces: resonant, 165 $\mu$m, lower traces: anti-resonant, 318 $\mu$m. The absorbed fractions are plotted as a function of considered range of incident angles, e.g. $\theta_i = 0°$ is normal incidence while $\theta_i = 20°$ includes all incidence angles up to 20°. For each thickness the response to perpendicular and parallel polarisation is plotted. The averages of these (label 'averaged') is the overall isotropic stray radiation response.

These modelling results together with absorbed fractions of stray radiation power measured (section 4.3) are consistent with a scan over fixed angles by Hentrich [12]. This allows to conclude that the absorbed fraction of power using stray radiation exposure is not significantly affected by the angular distribution of the stray radiation.

Measurement and modelling was all at 140 GHz in this work. It is worth noting that in case the dielectric heating is dominant, the coating absorption at frequencies not too far off, say at 170 GHz, will be the same as long as the layer thickness is adapted to this new frequency too. This may be seen as follows. In case the loss tangent is no too excessive one can approximate $\alpha$ by $\pi\sqrt{\epsilon_r'}\tan\delta f/c$ [7]. The attenuation goes as $e^{-2\alpha d}$, but as $d = 0.25(1/\sqrt{\epsilon_r'})(c/f)$, the scaling with frequency and layer thickness cancels out.

## 4 Vacuum Windows

The work on vacuum window exposure during the MISTRAL campaigns in 2020 and 2021 addressed three topics: i) stray radiation loads in vacuum windows, ii) exposure of ITO-coatings on vacuum windows, and iii) prototype test of ITER window stray radiation sensors.

### 4.1 Stray radiation loads in vacuum windows

The isotropic stray radiation response of vacuum windows is reported in [7]. The work is summarised here.

Vacuum windows are critical components, yet considerable thermal loads may result due to dielectric heating by microwave stray radiation. For low-loss windows the absorbed fraction of power can be computed accurately using the loss tangent and the permittivity. But two questions are frequently raised: i) what is the loss tangent at microwave frequencies? and ii) what is the isotropic stray radiation response? To address these questions the permittivity and loss tangent at 140 GHz for a number of commercially available windows were measured using a collimated beam at low power in the laboratory [15, 16]. The stray radiation response was next measured and modelled at high power in the MISTRAL.

The outcome of the work is that, for a window in a typical port geometry, i.e. by excluding the very large incidence angles, the single pass absorbed fraction may be taken as a upper limit by isotropic stray radiation, while applying a safety margin $(1+\epsilon_r')/2\sqrt{\epsilon_r'}$. As the single pass absorption is easily calculated using the permittivity and loss tangent, laboratory data is now sufficient to get the high power load.

### 4.2 ITO-coated windows

In order to reduce the stray radiation transmitted through a vacuum window an ITO-coating may be applied. This is an Indium Tin Oxide layer which has a surface resistance inversely proportional to the ITO-conductivity and the layer thickness. Note that the surface resistance has unit $\Omega/\square$ as it is defined with the wave incident on a symmetrical surface [17]. Using data extracted from coatings in use at W7-X, the surface resistance is plotted as function of layer thickness using $R_c = 1/(\sigma_c \tau)$ [17], with $\sigma_c$ the thin layer ITO conductivity and $\tau$ the layer thickness, see Fig. 9. Then, using transmission line theory, the reflected fraction of power due to an impedance step from air → ITO is computed, see Fig. 10. Exposure of ITO-coated windows in the MISTRAL confirm that relatively thick coatings, having a surface resistance 20 $\Omega/\square$ or less, show very high stray radiation reflection, comparable to reflection off a metal surface. As a consequence, there is very little absorption inside the ITO coating. On the other hand, very thin coatings with high surface resistance show high transmission and high ITO-induced absorption, resulting in poor attenuation of stray radiation and excessive window heating rates. A practical issue is that an ITO-coating close to the plasma may be eroded by it. The coating must then be on the air-side of the window. While this still gives the large reduction in transmission for stray radiation, it results in double pass absorption in the window which may now cause a considerable thermal load.

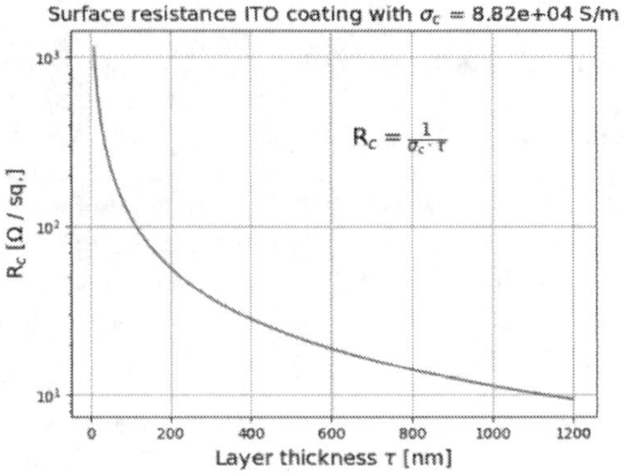


**Figure 9.** ITO surface resistance as function of layer thickness.

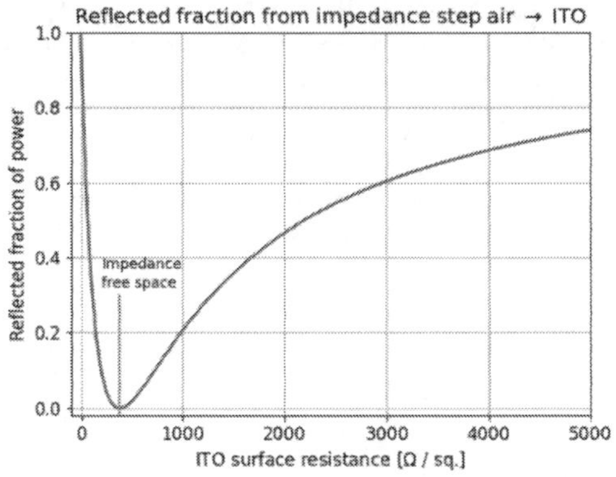


**Figure 10.** Reflection due to impedance step air → ITO-layer.

### 4.3 ITER stray radiation sensors

A novel development at ITER [18] are stray radiation window sensors. These are Molybdenum rings with a microwave absorbing coating that are placed at the edge of the window aperture. A prototype test was made at the MISTRAL, the arrangement is shown in Fig.11. At three locations of the ring the temperature is measured with an embedded thermocouple. The absorbing coating is as described in section 3.2. Figure 12 shows the heating rate for a ring with an almost resonant layer thickness (150 $\mu$m), for a ring with an almost anti-resonant layer thickness (270 $\mu$m) and for a ring with an anti-resonant layer (300 $\mu$m). In the experiment the MISTRAL stray radiation exposure was set to p = $(53 \pm 5)$ kWm$^{-2}$. The heating rate is equated to the power flowing into the body and the heat capacity: $\Delta T/\Delta t = (p \cdot A \cdot S_c)/(m \cdot c_v)$, with A the absorbed fraction, $S_c$ the coated ring surface, m the mass and $c_v$ the specific heat capacity. The absorbed fractions are solved giving: $A_{150}$ = 67% and $A_{270} \approx A_{300}$ = 30%, both with relative

**Figure 11.** ITER window sensor mounted on MISTRAL port.

error of 10%. Having found the absorbed fractions in situ, the sensors are calibrated and can be used to provide the incident surface power density from the measured heating rate. At modest temperature linear scaling can be used but at elevated temperatures the heating rate is no longer linear as radiation becomes important.

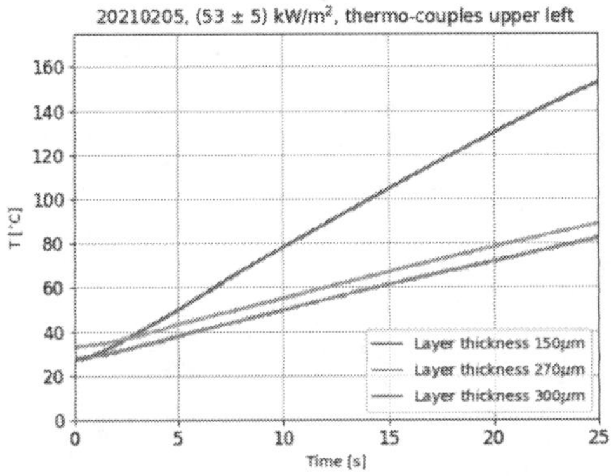


**Figure 12.** Heating rate of the ITER window sensors

*This work has been carried out within the framework of the EUROfusion Consortium, funded by the European Union via the Euratom Research and Training Programme (Grant Agreement No 101052200 - EUROfusion). Views and opinions expressed are however those of the author(s) only and do not necessarily reflect those of the European Union or the European Commission. Neither the European Union nor the European Commission can be held responsible for them.*

## References

1. R.C. Wolf et al., Performance of Wendelstein 7-X stellarator plasmas during the first divertor operation phase, Phys. Plasmas 26, 082504 (2019).
2. H.P. Laqua et al., The ECRH-Power Upgrade at the Wendelstein 7-X Stellarator, this conference.
3. H.P. Laqua et al., *ECRH Heating Scenarios and In-Vessel Components at the Wendelstein7-X Stellarator*, 28th EPS Onference on Contr. Fusion and Plasma Phys. Funchal, 18-22 June 2001. ECA Vol. 25A (2001) 1277-1280.
4. M. Hirsch et al., *The Impact of Microwave Stray Radiation to In-Vessel Diagnostic Components*, in: International Conf. on Fusion Reactor Diagnostics, Varenna, 9 - 13 September, 2013, AIP Conference Proceedings 1612, 39 (2014)
5. D. Hathiramani et al., *Microwave stray radiation: Measures for steady state diagnostics at Wendelstein 7-X*, Fusion Engineering and Design 88 (2013) 12321235
6. H.J. Hartfuss et al., *Problems with ECRH stray radiation in Wendelstein 7-X*, 30th EPS Conference on Contr. Fusion and Plasma Phys., St. Petersburg, 7-11 July 2003 ECA Vol. 27A, O-3.2C
7. J.W. Oosterbeek et al., *Microwave stray radiation losses in vacuum windows*, accepted for publication, Fusion Engineering and Design, 2022
8. H.P. Laqua et al., *Distribution of the ECRH stray radiation in fusion devices*, Proceedings of the 14th Joint Workshop on Electron Cyclotron Emission and Electron Cyclotron Resonance Heating, Santorini, Greece, 9-12 May 2006
9. D. Moseev et al., *Stray radiation energy fluxes in ITER based on a multiresonator model*. Fusion Engineering and Design 172 (2021)
10. Miriam Floristán, et al., *Development and testing of 140 GHz absorber coatings for the water baffle of W7-X cryopumps*, Fusion Engineering and Design 86 (2011) 1847-1850.
11. Andreas Killinger et al., *Plasma Spraying of a Microwave Absorber Coating for an RF Dummy Load*, MPDI Journal, Coatings 2021, 11, 801
12. Andreas Hentrich et al., *Resonant Atmospheric Plasma-Sprayed Ceramic Layers Effectively absorb Microwaves at 170 GHz*, Journal of Infrared, Millimeter, and Terahertz Waves (2022)
13. C.A. Balanis. *Advanced Engineering Electromagnetics*. John Wiley and Sons, 1986
14. Private Communication, Andreas Hentrich, Burkhard Plaum, EC21, 20-24 June 2022, ITER
15. F. Königer, *Measurement system for the precise determination of dielectrical properties in the mm-wave range based on hemispherical open resonators*, Frequenz 43 (1989)
16. I. Danilov, R. Heidinger, *New approach for open resonator analysis for dielectric measurements at mm-wavelengths*, Journal of the European Ceramic Society 23 (2003), 2623-2626
17. Samual. Y. Liao, *Microwave devices and circuits*, Prentice Hall, 1990, Third Edition
18. Private communication, Rafael Zubieta, Philippe Maquet, Sunil Pak

# Appendix A: Model of coating response to incident plane wave

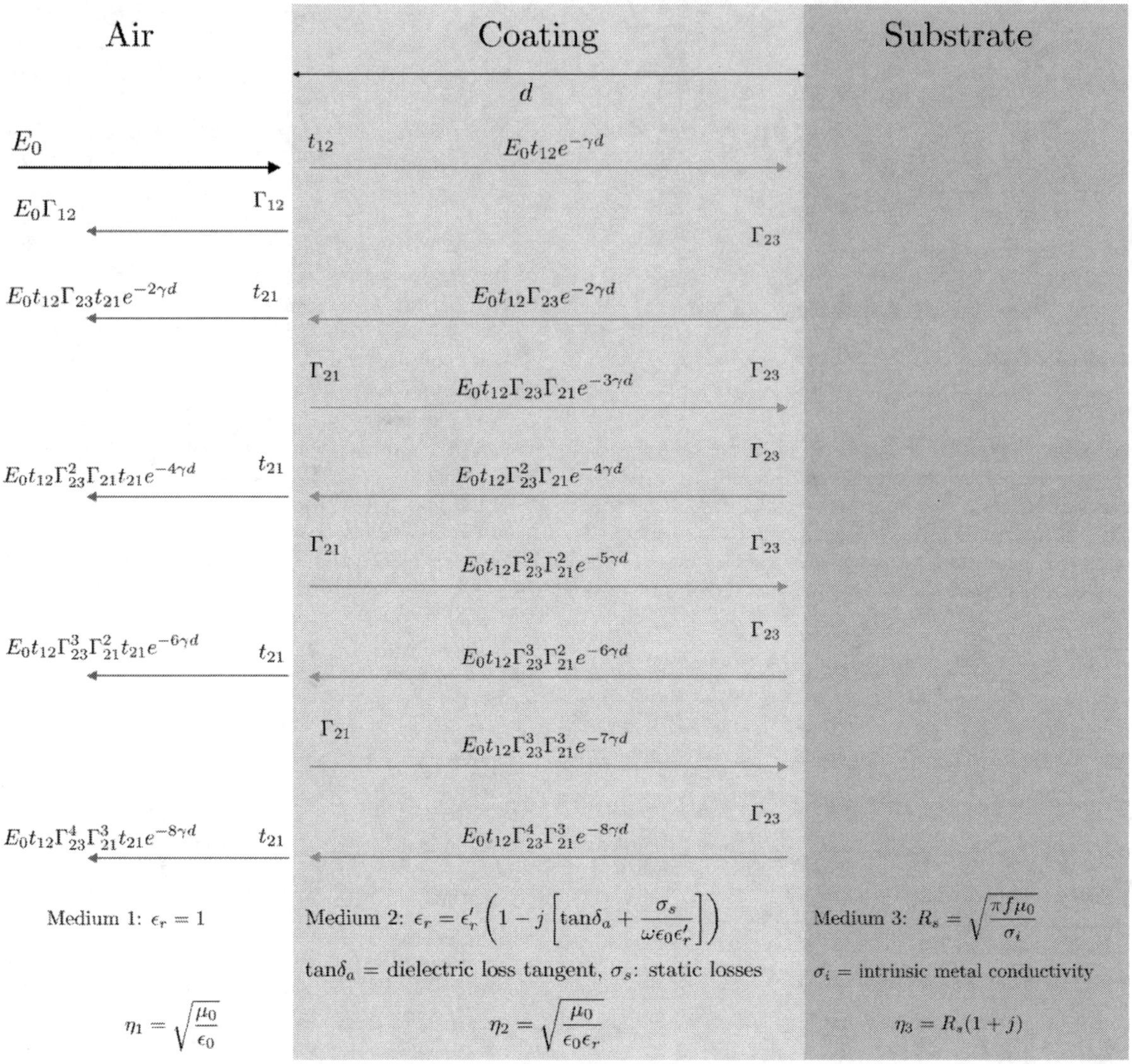


$\gamma$ = propagation constant $= j\omega\sqrt{\mu_0\epsilon_0\epsilon_r'}\sqrt{1 - j\left(\tan\delta_a + \frac{\sigma_s}{\omega\epsilon_0\epsilon_r'}\right)} = \alpha + j\beta$. MISTRAL work: $\left[\tan\delta_a + \frac{\sigma_s}{\omega\epsilon_0\epsilon_r'}\right] = \tan\delta$

$$\Gamma_{12} = \frac{\eta_2 - \eta_1}{\eta_1 + \eta_2} \qquad \Gamma_{23} = \frac{\eta_3 - \eta_2}{\eta_2 + \eta_3} \qquad \Gamma_{21} = \frac{\eta_1 - \eta_2}{\eta_1 + \eta_2} \qquad t_{12} = \frac{2\eta_2}{\eta_1 + \eta_2} \qquad t_{21} = \frac{2\eta_1}{\eta_1 + \eta_2}$$

$$\Gamma_{sum} = \Gamma_{12} + t_{12}t_{21}\sum_{n=1}^{n=\infty}\Gamma_{21}^{(n-1)}\Gamma_{23}^{(n)}e^{-2n\gamma d}$$

Fractional reflected power: $R = \Gamma_{sum}\Gamma_{sum}^*$  Fractional absorbed power: $A = 1 - R$

*Drawing from:*

Constantine A. Balanis, *Advanced Engineering Electromagnetics.* John Wiley and Sons, 1986
Paul F. Goldsmith, *Quasioptical Systems*, IEEE Press, 1998

# High power mm-wave loss measurements of ITER ex-vessel waveguide components at the FALCON test facility at the Swiss Plasma Center

*Timothy P.* Goodman[1,*], *Humberto* Torreblanca[1], *Cinta* Marraco Borderas[1], *René* Chavan[1], *Avelino* Mas Sanchez[1], *Anastasia* Xydou[1], *Mario* Cavinato[2], and *Katarina* Cindric[2]

[1]Ecole Polytechnique Fédérale de Lausanne (EPFL), Swiss Plasma Center (SPC), CH-1015 Lausanne, Switzerland
[2]Fusion for Energy, Josep Pla 2, Torres Diagonal Litoral B3, E-08019 Barcelona, Spain

**Abstract.** Many future fusion devices will rely heavily, if not solely, on electron cyclotron (EC) heating subsystems to provide bulk heating, instability control (neoclassical tearing mode (NTM) stabilization), and thermal instability control. Efficient use of the installed heating power (gyrotrons) requires low-loss transmission of the power over 100s of meters since the mm-wave sources need to be installed where the stray magnetic field has a small amplitude. Transmission lines are used to propagate the mm-wave power over this long distance. Quasi-optical techniques (mirrors) are used at W7X and are planned for DTT, for example. Guided components are installed at DIII-D, TCV and elsewhere and are planned at JT60SA and ITER. High power test facilities exist to evaluate the power transmission of assemblies of guided components (transmission lines). The European test facility FALCON was setup by Switzerland and Fusion for Energy (F4E) in Lausanne Switzerland at the Swiss Plasma Center (SPC) in the Ecole Polytechnique Fédérale de Lausanne (EPFL). Operations are funded through a framework contract with F4E. SPC operates the facility. Two ITER-class 170GHz gyrotrons are housed within the facility and used to evaluate the thermal behaviour of components provided by various ITER partners. Loss measurements are presented for miter bends and waveguides of several materials at two different diameters. The results are used to model the expected losses in the ITER ex-vessel waveguides (EW) of all five EC launchers.

## 1 Introduction

The use of electron cyclotron (EC) heating at the fundamental through 3rd harmonic resonance in fusion experiments [1–6] and future devices [7–9] is well established and is expected to expand significantly to also include electron Bernstein wave heating [10]. EC power is transmitted over 10s or 100s of meters from the gyrotron sources to the launching antennas via quasi-optical mirrors or evacuated over-moded corrugated transmission lines (TLs). In the latter, at high power, power losses are measured to exceed the theoretical losses associated with the lowest loss $LP_{01}$ mode. While the fractional losses are still low, the designs of cooling systems for multimegawatt plant systems depend on the expected losses. Experimental confirmation of the heat transfer coefficients (HTC) between cooling structures and the objects that are being cooled is commonly required. In the same way, confirmation of the power losses within the objects is needed.

Loss measurements in high-power, long-pulse experiments have been reported in, for example, [11–13]. Infrared temperature measurements are made of relatively long TLs. Uncooled waveguides (WGs) store the heat during the pulse, dependent on the thermal mass of the components. Calorimetry is based on the total change in temperature of the components from the start to the end of the mm-wave pulse (when linear) and/or the integral of the flow and water temperature difference when cooled. The latter is used for the mitre bend (MB) mirrors.

Experiments, at QST (formerly JAEA) in Japan, using General Atomics, 63.5mm inner-diameter, corrugated aluminum alloy 6061-T6 WG and MBs, found an exponential decay of the temperature away from the MBs and enhanced losses far from the MBs, even when considering the higher losses resulting from some $LP_{11}$ mode content in the transmitted beam [14–16]. The importance of reflections from the high-power mm-wave calorimetric loads used during transmission measurements has been cited in previous tests [11] and demonstrated recently [13]. Reflections lead to higher heating of the TLs over long distances, but the reflected power level is difficult to quantify. Simulations of the Ohmic attenuation losses from the EC beam including the expected mode content of the ITER TL between the gyrotron MOU and the diamond window of the EW are consistent with the ITER transmission requirements [17]. The ITER loss specification for the TL is 0.05%/m and reported values have ranged from 0.012%/m to 0.028%/m at in inner diameter $\varnothing$=63.5mm (the Ohmic losses scale as $\varnothing^{-3}$ [18], so they are expected to be 2 × larger in the $\varnothing$=50mm waveguide used in the ITER TL).

In this paper we discuss loss measurements of new waveguide manufactured by a European consortium under contract with Fusion For Energy. They were de-

*e-mail: timothy.goodman@epfl.ch

signed according to the ITER specifications for the ex-vessel waveguides (EW) of the ITER launchers (4 upper launchers (UL) - mostly dedicated to neoclassical tearing mode (NTM) stabilization and other current profiling techniques - and one equatorial launcher (EL) for central heating and current drive). The waveguides are uncooled and are made of aluminium alloy 6061-T6 (Al), stainless steel 316L (SS) and CuCr1Zr copper alloy. They have 50mm inner-diameter, are corrugated and evacuated, and incorporate integrated couplings that house double spring-energized all-metal seals to meet the stringent ITER vacuum requirement. For the experiments reported here, only the inner seal is used. (The outer seal creates a service-vacuum space that allows automated vacuum leak testing, at ITER.) In addition to the straight waveguide sections, tests of CuCr1Zr mitre bends are presented.

The ITER EC section has allocated certain levels of loss (specified as transmission efficiencies or transmitted power minimums) to various portions of the EC plant to meet the overall efficiency required by the ITER project. The EW subsystem must have 98% transmission efficiency. The power losses presented here are consistent with this requirement when used to calculate the estimated losses of the EW. Additionally, the data is used to design the EW cooling system to ensure stable, reliable, operation even when fully powered. The theoretical loss rate for the Al waveguides is 0.0048%/m; but, as in previous experiments [11, 16], the measured loss is nearly an order of magnitude higher.

## 2 Measurement layout

The FALCON experimental zone is housed in one hall of the Swiss Plasma Center. The zone is delimited by a protective wall preventing human access during operation of the gyrotron, transmission line and dummy loads. The available space for testing equipment is (width x length x height) 10m x 12m x 6m.

Each of the two gyrotron towers houses a 1MW-class ITER-like gyrotron operating at 170GHz. An evacuated, 8.534m-long, 63.5mm-inner-diameter, Al TL with 2 MBs, manufactured by General Atomics (GA), connects the MOU of a Thales Electron Devices (TED) gyrotron [19, 20] to a spherical load developed by IFP-CNR [21] to allow long-pulse (1000s), high-power testing (950kW demonstrated) . This is the so-called EU TL; it is 1.9m off of the false floor of the testing area.

A GYCOM gyrotron occupies the second tower and its MOU is connected to a similar TL that starts at a similar height as the EU TL, then zig-zags down to 0.81m off of the false floor via 3 MBs to bypass the support tower of the TED gyrotron, and after ≈2.2m bifurcates in an RF switch (180°, or 90° bend) graciously provided by GA. The 90° branch of the switch has been used to test a GA RF load [22]. The through branch (180°) connects to a downtaper from 63.5mm to 50mm inner diameter. This is used for ITER UL component testing and is the so-called ITER TL. Figure 1 shows a panaramic view of the TLs. Only the TED gyrotron and its MOU are seen at the left; the GYCOM gyrotron and MOU are hidden from view. The devices under test (DUTs) start with an MB manufactured by SPC [23] on the lower TL in Fig. 1. Further information about the facility is found in [12].

From the input of this MB, eighteen different combinations of components have been installed over the past 3 years, allowing comparisons of the behaviour of components to each other. The longest series of components in the ITER TL is 9.584m and is terminated by a GYCOM RF load. Shorter combinations were terminated by a second spherical dummy load (20 cooling channels per hemisphere as opposed to 16-channels for the IFP-CNR load, so-called RFL.20 and RFL.16, respectively). All loads measure the same RF power for the same gyrotron operational parameters within the error bars of the measurements: approximately ±3%.

The rest of this paper is divided into sections addressing different types of power losses in a typical transmission line system. Section 3 discusses the mitre bend, in which the beam is reflected from a mirror to affect a change in direction. Mitre bends at FALCON incorportate thermocouples in the mirrors to diagnose losses and beam alignment (see [23]). Section 4 details experiments made using stainless steel waveguides. These are particually lossy and have low thermal diffusivity, allowing visualization of the underlying heat flux in a way not possible with typical waveguide materials. Losses in waveguide far from miter bends, which act as a source of higher order modes near cut-off (here refered to as very high order modes VHOMs), yield a measure of the nearly constant baseline heat flux; these are discussed in Section 5. Load reflections are also discussed in Section 5 before conclusions drawn from this work are summarized in Section 6.

## 3 Mitre Bends

Two mitre bends have been manufactured by SPC. Both MB mirrors are equipped with a pair of thermocouples (TCs) that are 1mm below the reflecting surface and are 4mm on either side of the mirror center. Four additional TCs are embedded in the MB body. In addition, calorimetry is made separately on the mirror and body water cooling circuits. Measurements of the losses in both the mirror and body of the MB confirm that the two MBs are identical within experimental accuracy when installed at the same location in the ITER TL (i.e. as the first MB). This allows them to be used as measurement devices when placed at different locations in the TL: Any differences in losses are due to the TL configuration and not the component itself.

The GYCOM load must be oriented vertically. The load and its required preload are, together, 2.5m tall; therefore, a second miter bend is positioned below the load (see Fig. 2). In this TL, the first MB reflects in the E-plane and the second, under the load, in the H-plane. According to theory, mirror losses are half as large in the H-plane for these 90° MBs [24]. As described in [23], the RF beam (and therefore the heat flux) is not centered on the first MB, so there is a significant temperature difference at the two TC locations (≈50±6°C); however, at the location of the second MB the temperature difference is reduced (≈25°C

https://doi.org/10.1051/epjconf/202327704010

**Figure 1.** The FALCON test facilities 2 TLs: EU TL (upper in 63.5mm inner diameter Al) and ITER TL (lower with switch and downtaper - covered in green tape - to 50mm inner diameter multi-material).

**Figure 2.** The GYCOM preload and load (stainless steel) positioned above an SPC MB (CuCr1Zr) connected by a 0.34m-long CuCr1Zr waveguide covered in red electrical tape and preceded by a 2m-long stainless-steel waveguide convered in black electrical tape.

average, see Fig. 3) and we conclude that the beam is more centered.

As described above, it has been verified that these TCs and their acquisition chain have an identical response to the same input RF beam [23]. Figure 3 shows the difference in temperature (relative to the start of the pulse) of each thermocouple in the MB mirrors for a 15-minute-long, 0.51MW RF pulse. A fast step in acceleration voltage in the gyrotron causes a step in RF power at ≈10s and the fast response is seen on the TCs ($1/e$ response time ≈0.5s [23]). The slow rise over time, starting before ≈200s, is due to re-circulation of warm cooling water and the oscillations at the end of the pulse are due to the feedback control of the heat exchangers in the closed loop cooling system. From independent water temperature measurements of the MB mirror and body, in which the time-changing input water temperature is subtracted, calorimetry shows that the total power loss in the MB as-

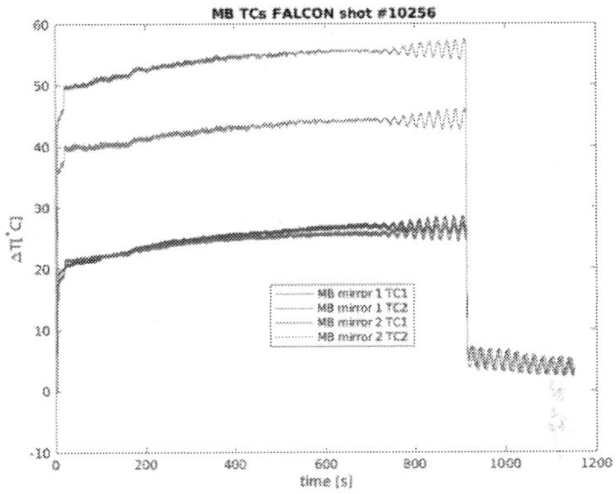


**Figure 3.** The temperature difference traces of the thermocouples from two mitre bends, one in E-plane (higher temperatures) and one in H-plane (lower temperatures). The beam is off-center on mirror 1 and more centered on mirror 2. The average temperature of mirror 1 is approximately twice that of mirror 2, as expected.

sembly is 0.35% of the transmitted power; with 22% of that power lost in the MB body.

## 4 Stainless Steel WGs

Two stainless steel (SS) WGs are available to compare to WGs made of other materials. They both show significant and identical axial variations in temperature on a spacial scale less than their 1.2m length when positioned at the same location in the TL. This was an unexpected result since neither the Al nor the CuCr1Zr WG showed such variations when positioned at the same location in the TL. Because of the very low heat diffusivity in SS, the WGs can be used as a diagnostic. Despite the relatively short TL, these WGs permit measurements "far" from a MB as described below.

It is known that the beam in the TL is not a pure $LP_{01}$, but is a mix of several low order $LP_{mn}$ modes as shown in Fig. 5 of Ref. [23] and indicated by the off-set of the beam on the MB. Having noticed that the distance between the temperature peaks in the spacial pattern corresponds to several of the half-beat-wavelengths of the measured modes in the TL, experiments were carried out to change the mode mixture and measure potential changes in the temperature pattern in the SS WGs. To this end, the mirror alignment in the MOU was changed in several steps, while under vacuum. By measuring the power losses in the MOU and the temperature difference between TCs in the MB, it was verified that the original alignment could be recovered after the experiment.

Figure 4 shows measurements of the temperature rise in the MOU cooling water for 3 shots; for the original mirror alignment, a misaligned mirror and the recovered mirror alignment. Power losses are listed in the legend. The mirror was misaligned in steps to lower the difference between the mirror TCs temperature. Falcon shot 8935 has a more-central beam alignment on the MB mirror; the temperature difference between the TCs was lowered from ±13% to ±4.4%. However, the MOU losses increased concomitantly.

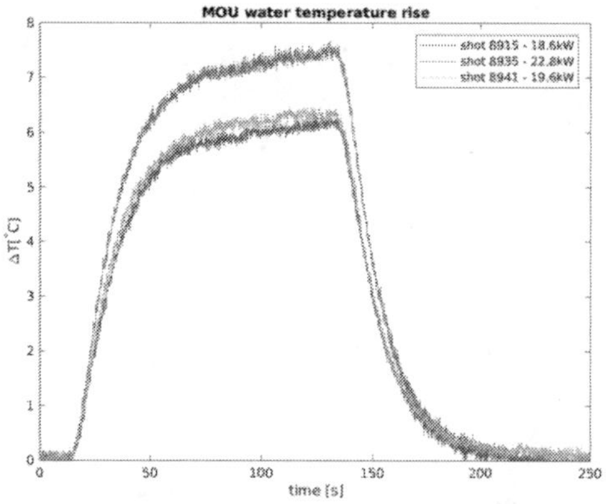


**Figure 4.** Temperature increase in the MOU cooling water for the original mirror alignment (#8915), misaligned mirror (#8935) and recovered position (#8941). The associated power losses are given in the legend.

The axial temperature profiles of the two SS WGs positioned imediately after the first MB are shown in Fig. 5. The relatively massive coupling between the two WGs is seen as a deep notch in the temperature curves. The data is taken from an Infra-red camera viewing both the front side and back side (using a large reflector plate). Several TCs are placed along the WGs to cross-calibrate the IR images. The temperature is higher in the misaligned case and the shape of the profile is different. The next highest order mode ($LP_{11}$) is excited by angles in the input beam coupled from the MOU. It has a beat wavelength with the $LP_{01}$ mode of 3.14m. Significant changes in the RF heating pattern can be expected at 1/4 of this length (0.79m).

The short-scale-length temperature variations are superimposed on a curve that decreases monotonically away from the MB (at z=0 in Fig. 5). Previous investigators have modeled this as an exponential decay in the heating due to VHOMs, modes near cut-off, that are generated in the MB. Here we model the WG heating with a

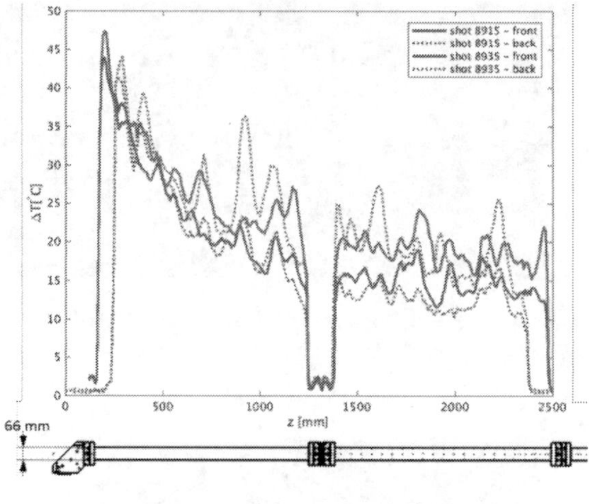


**Figure 5.** The change in the axial temperature profiles for the original (8915) and misaligned shots (8935) shown in Fig. 4 for both the front (solid) and back (dashed) side. SS WGs show the change in the TL heating pattern when the MOU mirror is misaligned. The relatively high thermal mass coupling between the waveguides appears as a notch centered at ≈1300mm, and at each end. (N.B. The drop in $\Delta T$ at each end of the "back" side, shifted with respect to the front side, is due to the size of the reflector - the back is no longer visible to the camera at the extremes.) A sketch of the miter bend and stainless steel waveguides with couplings is shown below the graph.

constant baseline heating plus heat flux decaying exponentially with z. The model parameters are the result of fitting the data from several shots. Figure 6 compares the measured and modelled temperature profile for the longer FALCON shot 8942 (pulse lengths are typically limited by the maximum allowed WG temperature - set here to 80°C to protect the aluminum-sheathed, spring-loaded, metallic vacuum seals).

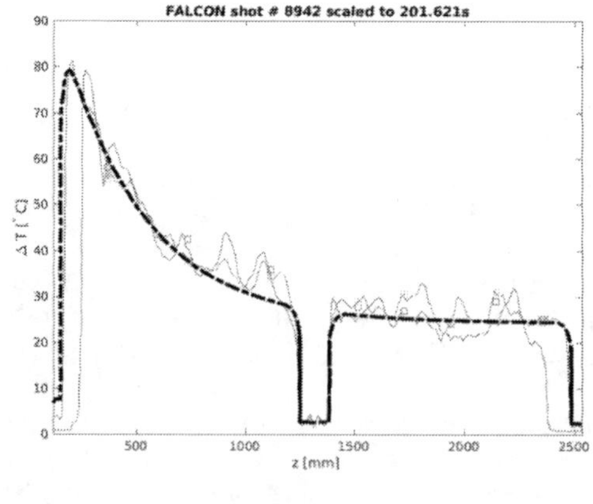


**Figure 6.** The simulated temperature difference resulting from a simple off-set exponential heat flux model (dashed line), compared to the measured axial temperature (blue-front, red-back, red squares - TCs) for a longer pulse in which the MOU mirror was slightly readjusted to better match the original alignment.)

The decay length in the model is found to be 0.37m, whereas the combined length of the two SS WGs is 2.5m; thus, all of the power associated with the VHOMs is absorbed in these two WGs. This forward propagating VHOM power constitutes 1/4 of the total MB mode conversion loss in a WG [25] and can be determined by integrating the exponential, knowing the fit parameters. Similarly, at the output of the 2nd WG the temperature rise is due solely to the baseline heat flux in the WG. From the fit parameters, we find the baseline WG absorption rate is a factor of 7.3 to 8.1 times theory, and the VHOM generation is between 1.2 and 2.3 times the theoretical predictions. The relatively large range in the values is a result of the different MOU alignments.

Determination of the decay length was not possible using the CuCr1Zr or Al WGs, for example, as it becomes of the same order as the length of the WG components (scaling inversely with the surface resistivity i.e. $\sqrt{\sigma_{CuCr1Zr}/\sigma_{SS}}$, where $\sigma$ is the electrical conductivity). Similarly, the presence of HOMs is not evident in the WGs with high heat diffusivity; no significant azimuthal temperature variation was measured as these WGs cannot sustain the high temperature gradients seen in Fig. 5 (e.g front to back as well as axially).

Determination of the decay length (and baseline losses) in materials other than SS depends on scalings based on the relative conductivity (1/resistivity) of the material.

On the other hand because the VHOMs are eliminated by the SS WGs, it may be possible to directly measure the baseline heating in other WG materials placed downstream from the SS WGs, effectively "far" from the MB (source of VHOMs).

## 5 WGs far from MBs

COMSOL™ [26] thermal simulations were carried out for three WGs placed after the SS WGs, where the internal heat flux is due solely to the enhanced baseline heating. The room temperature of 20°C has been subtracted from the simulation temperature presented in Fig. 7. They are compared to measured temperature-difference profiles for 3 shots at different powers. A constant heat flux (along the axis) is applied in each simulation to match the measurements of the first WG (in CuCr1Zr, given in the figure legend as hfluxCCZ). The simulated loss rate is equivalent to 0.052%/m for each shot. The second WG is in Al and the third in (thinner-walled) CuCr1Zr, followed by a short, thick walled CuCr1Zr waveguide and the RFL.20. (The hfluxCCZ is scaled to simulate higher losses in Al, as described below.)

The heating appears to be significantly over-estimated in the simulation for the Al WG (centered at 6m in Fig. 7). We note that by chance the simulated temperature (dark blue line) for the first shot matches the measured temperature (magenta line) for the second shot. This indicates that the simulation overestimates the temperature in Al by a factor $\frac{302kW}{211kW} = 1.43$ relative to measurements.

The theoretical ratio of losses ($\sqrt{\sigma_{CuCr1Zr}/\sigma_{Al}} = 1.30$) and the inverse ratio of the thermal masses ($(mc_p)_{CuCr1Zr}/(mc_p)_{Al} = 1.43$), where $m$ is the mass and $c_p$ the heat capacity, have already been taken into account in the simulations (i.e. $\Delta T_{Al} \approx 1.86\Delta T_{CuCr1Zr}$, as seen in Fig. 7). So, it appears that surface resistances of the two materials are nearly the same (i.e. the factor 1.3 is not required for the simulation to match the measured Al temperature). Alternately, the losses in the CuCr1Zr WG could be underestimated.

Subsequent conductivity measurements in the 80-100GHz frequency range in a Fabry-Perot resonance cavity configuration using samples of the 3 WG materials, as well as a reference oxygen-free high-conductivity Cu sample, confirm the higher than expected surface resistance (aternatively, lower than expected conductivity) of the CuCr1Zr.

Using the measured conductivity, the expected $LP_{01}$ loss rate in the corrugated CuCr1Zr WG is 0.0046%/m; so, the simulations indicate a baseline heating enhancement of 11, perhaps even higher nearer the load. This exceeds the enhancement factor of 7.7 in SS WG and suggests an additional $\approx$ 47% contribution from reflected power from the load even at a distance of $\approx$ 4.5m.

Without a knowledge of the absorption rate of the reflected power in the WG, it is difficult to integrate the total reflected power and thereby quantify the load reflection coefficient ($P_{reflected}/P_{incident}$). Low power load reflection measurements, using a vector network analyser with extension heads for 170GHz operation attached to a Gaussian Optics Lens Antenna (GOLA) to couple to $\varnothing$=50mm WG at the input of the load, appropriately time-gated to remove the effects of the GOLA and associated WG, gave 0.010% at 170GHz and 0.1% at 169.68GHz (covering the gyrotron frequency-drift range during pulses of up to 1MW). However, this measurement methodology may not capture well all the HOMs generated by the RFL.20; therefore, these constitute a lower bound for the reflections.

Previous modeling [12] postulated 0.7% reflection. And measurements in a test setup for JT60SA [13] indicate that the influence of an additional preload extends $\approx$17m from the preload location back towards the gyrotron in $\varnothing$60.3mm Al 6061-T6 waveguide. Our CuCr1Zr waveguide has similar losses to this material. From this, if we postulate a decay length of $\approx$ 4m for load reflections, we cannot resolve the difference in the enhancement factors between the stainless steel and the other waveguides; a much shorter decay of 1.4m is required. In that case, we derive the pair (load reflection, baseline loss enhancement): $\approx$ (0.52%, 6.8×) i.e. a load reflectivity of $\approx$ -23dB. However, if this pair of parameters is used in the COMSOL™ model, does it match the measured temperatures along all of the waveguides?

Using this pair of parameters for the load reflection and baseline loss enhancement, COMSOL™ modeling shows a reasonably good fit to the measured temperatures only at the same waveguides used to determine the enhancement (i.e. the SS WGs and the CuCr1Zr waveguide at the left in Fig. 7). The simulated waveguide temperatures nearer the load (at the right in Fig. 7) strongly exceed the measured temperatures making this an unreliable measurement of the load reflection.

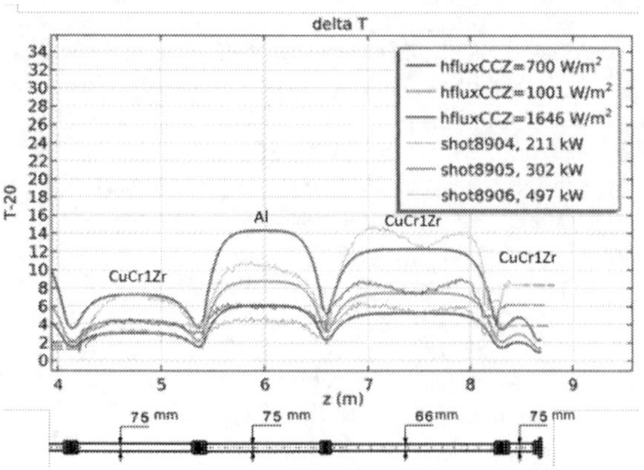


**Figure 7.** COMSOL™ simulated temperature (thick lines) and measured temperature (thin lines) for three different transmitted powers (see legend). Three WGs are simulated at their axial locations (z) relative to the MB: left - CuCr1Zr (outer diameter = 75mm), middle - Al (o.d. = 75mm), right - CuCr1Zr (o.d. = 66mm). The couplings are seen as notches (not as steep as with SS WGs - see Fig. 6). The heat flux in CuCr1Zr is 520W/MW/m. The heat flux in Al is scaled with conductivity as described in the text. A sketch of the waveguides with couplings is shown below the graph.

## 6 Conclusions

We have shown that MB manufacturing is very reproducible allowing MBs to be used as diagnostics when equipped with TCs in the mirror and body.

Stainless steel waveguides are also useful as diagnostics, showing localised heating due to mode beating that is masked in high-conductivity components. The thus-determined mode structure can be modified by altering the beam alignment at the TL input.

Because of its relatively high resistivity, decay lengths of MB-generated VHOMs are measurable in relatively short SS components. Scaling the measured decay length to CuCr1Zr WGs yields 1.7m - longer than most of the available components. It was not possible to measure the decay length in these components reliably.

Two SS WGs together dampen all VHOMs from the MB allowing a good determination of the baseline heating "far" from a MB - both in the SS and, with a different measurement setup, in CuCr1Zr. Similarly the total forward traveling VHOM power from the MB was determined by integration of the best-fit simplified model curves. An enhancement of between 1.3 and 2.3 is found above theoretical expectations (depending on the beam alignment).

Comparison of the measurements from different materials has proved useful. First, a high-frequency measurement of the CuCr1Zr surface resistance, or conductivity, helped explain the lack of significant difference between Al and CuCr1Zr component heating. Second, resolving the difference in baseline enhancement leads to an estimate of the reflected power from the RFL.20 of $\approx$ -23dB; albeit raising a different discrepancy in the measurements (over-estimate of WG temperatures near the RFL.20). Despite the latter, it is clear that even small power reflections can dominate power loss measurements in the short transmission line tested at FALCON, as evoked in experiments elsewhere (e.g. [11] and references therein). Estimates of the reflection coefficient of the ITER Upper Launcher [27] are similar to those discussed here for the load.

These enhancement results were used to calculate the expected losses in the EW of all five EC launchers for ITER, showing that the present design meets the ITER allocation of losses.

## References

[1] S. Alberti, K.A. Avramidis, W. Bin, A. Bertinetti, J. Dubray, D. Fasel, S. Garavaglia, J. Genoud, T. Goodman, J.P. Hogge et al., *High-efficiency, long-pulse operation of MW-level dual-frequency gyrotron, 84/126GHz, for the TCV Tokamak*, in *2019 44th International Conference on Infrared, Millimeter, and Terahertz Waves (IRMMW-THz)* (2019), pp. 1–2

[2] S. Moriyama, T. Kobayashi, A. Isayama, M. Terakado, M. Sawahata, S. Suzuki, K. Yokokura, M. Shimono, K. Hasegawa, S. Hiranai et al., Nuclear Fusion **49**, 085001 (2009)

[3] M. Cengher, J. Lohr, P. Simmerling, Y. Gorelov, A. Torrezan, D. Ponce, J. Doane, D. Su, X. Chen, IEEE Transactions on Plasma Science **48**, 1698 (2020)

[4] D. Wagner, J. Stober, M. Kircher, F. Leuterer, F. Monaco, M. Münich, M. Schubert, H. Zohm, G. Gantenbein, J. Jelonnek et al., *Commissioning of the extended multi-frequency ECRH system at ASDEX upgrade*, in *2017 42nd International Conference on Infrared, Millimeter, and Terahertz Waves (IRMMW-THz)* (2017), pp. 1–2

[5] H. Laqua, A. Leggieri, C. Lechte, D. Moseev, F. Legrand, G. Gantenbein, H. Oosterbeek, H. Braune, I. Chelis, J. Jin et al., *The ECRH-Power Upgrade at the Wendelstein 7-X Stellarator*, in *21st joint workshop on electron cyclotron emission (ECE) and electron cyclotron resonance heating (ECRH)* (2022)

[6] L. Delpech, *A new 3MW ECRH system at 105 GHz for WEST*, in *21st joint workshop on electron cyclotron emission (ECE) and electron cyclotron resonance heating (ECRH)* (2022)

[7] B. Bigot, Nuclear Fusion **62**, 042001 (2022)

[8] M. Tran, P. Agostinetti, G. Aiello, K. Avramidis, B. Baiocchi, M. Barbisan, V. Bobkov, S. Briefi, A. Bruschi, R. Chavan et al., Fusion Engineering and Design **180**, 113159 (2022)

[9] A. Romano, B. Baiocchi, L. Balbinot, W. Bin, A. Bruschi, D. Busi, A. Bussolan, M. De Nardi, F. Fanale, P. Fanelli et al., *The design of the ECRH system of DTT*, in *21st joint workshop on electron cyclotron emission (ECE) and electron cyclotron resonance heating (ECRH)* (2022)

[10] S.J. Freethy, L. Figini, M. Henderson, H. El-Haroun, S. Gibson, K.K. Kirov, A. Köhn-Seeman, I. Konoplev, S. Saarelma, R. Sharma et al., EPJ Web Conf. (2022)

[11] J. Anderson, J. Doane, H. Grunloh, R. O'Neill, R. Ikeda, Y. Oda, K. Takahashi, K. Sakamoto, Nuclear Fusion **57**, 056030 (2017)

[12] T.P. Goodman, M. Cavinato, R. Chavan, A. Más-Sanchez, P.S. Silva, M. Vagnoni, T. Tersztyanszky, S. Carre, IEEE Transactions on Plasma Science **48**, 1537 (2020)

[13] T. Kobayashi, H. Yamazaki, S. Hiranai, M. Sawahata, M. Terakado, K. Ishita, J. Hinata, F. Sato, K. Wada, R. Ikeda et al., Fusion Engineering and Design **175**, 113009 (2022)

[14] R. Callis, J. Doane, H. Grunloh, K. Kajiwara, A. Kasugai, C. Moeller, Y. Oda, R. Olstad, K. Sakamoto, K. Takahashi, *Testing of ITER-class ECH transmission line components at the JAEA radio-frequency test stand*, in *Proc. 23rd IAEA Fusion Energy Conf., Republic of Korea* (2010)

[15] K. Takahashi, K. Kajiwara, Y. Oda, A. Kasugai, N. Kobayashi, K. Sakamoto, J. Doane, R. Olstad, M. Henderson, Review of Scientific Instruments **82**, 063506 (2011)

[16] J.P. Anderson, J.L. Doane, H.L. Grunloh, R.W. Callis, R. Ikeda, Y. Oda, K. Takahashi, K. Sakamoto, Journal of Infrared, Millimeter, and Terahertz Waves **37**, 55 (2016)

[17] M.C. Kaufman, C. Lau, G.R. Hanson, Journal of Infrared, Millimeter, and Terahertz Waves **39**, 456 (2018)

[18] E.A. Nanni, S.K. Jawla, M.A. Shapiro, P.P. Woskov, R.J. Temkin, Journal of Infrared, Millimeter, and Terahertz Waves **33**, 695 (2012)

[19] A. Leggieri, D. Bariou, V. Hermann, F. Legrand, G. Lietaer, C. Lievin, R. Marchesin, P. Thouvenin, F. Albajar, F. Sanchez et al., *Th1509U European 170 Ghz 1 Mw Cw Industrial Gyrotron Upgrade* (IEEE, New York, 2021), IEEE International Vacuum Electronics Conference IVEC

[20] T.P. Goodman, S. Alberti, J. Genoud, H. Torreblanca, F. Albajar, F. Sanchez, A. Leggieri, F. Legrand, C. Liévin, Z. Ioannidis et al., 47th International Conference on Infrared, Millimeter and Terahertz Waves (IRMMW-THz) **Delft, The Netherlands** (2022, 28 August – 2 September)

[21] W. Bin, A. Bruschi, F. Fanale, M. Francesca, F. Lucca, F. Albajar, S. Alberti, G. Carannante, M. Cavinato, I. Chelis et al., Fusion Engineering and Design **146**, 36 (2019)

[22] Anderson, James, Doane, John, Moeller, Charles, Grunloh, Howard, O'Neill, Raymond, Brookman, Michael, Smiley, Matthew, Su, David, EPJ Web Conf. **203**, 04001 (2019)

[23] A. Xydou, T.P. Goodman, R. Chavan, M. Vagnoni, H. Torreblanca, M. Cavinato, Fusion Engineering and Design **170**, 112457 (2021)

[24] M.K. Thumm, W. Kasparek, IEEE Transactions on plasma science **30**, 755 (2002)

[25] J. Doane, J. Anderson, H. Grunloh, W. Wu, Fusion Engineering and Design **93**, 1 (2015)

[26] *Comsol multiphysics® v. 6.0, comsol ab, stockholm, sweden*, `www.comsol.com`.

[27] T.P. Goodman, *Rf stray radiation estimation in upper launcher (yc4ntv v1.0)*, `https://user.iter.org/default.aspx?uid=YC4NTV`

# AUTHOR INDEX

# AUTHOR INDEX

# AUTHOR INDEX